Fundamentals of Modern Unsteady Aerodynamics

Ülgen Gülçat

Fundamentals of Modern Unsteady Aerodynamics

Third Edition

Springer

Ülgen Gülçat
Faculty of Aeronautics and Astronautics
Istanbul Technical University
İstanbul, Turkey

ISBN 978-3-030-60776-0 ISBN 978-3-030-60777-7 (eBook)
https://doi.org/10.1007/978-3-030-60777-7

This Springer imprint is published by the registered company Springer Nature Switzerland AG
The registered company address is: Gewerbestrasse 11, 6330 Cham, Switzerland

Preface to the Third Edition

The second edition of this book was made available on the shelves and also as an e-book about a half a decade ago. Since then, there have been quite a few contributions to the field concerning the unsteady applications of the flapping wing technology. This made it necessary to add to the book a new chapter, Chap. 9, under the title of Unsteady Applications related to the thrust optimization, aerodynamic stability and trim.

The thrust optimization applications cover the constraints on the magnitude of the motion and the aerodynamic efficiency, optimization of the efficiency and non-sinusoidal path optimization for the flapping. Furthermore, topics covered are the optimum thrust at zero free-stream and the optimum power extraction from the flapping motion of a windmill turbine.

The state space representation is implemented for the fast and efficient prediction of the unsteady air loads, which are present in the equations of motion used for flight dynamics. The relevant stability derivatives contain the state variables related to the pitch and the pitch rate in a coupled manner with the linear and the rotational acceleration terms of the motion respectively. The time averaged stability matrix is used for the determination of the trimming for a given flapping motion. In hover, however, trim is possible only if the flapping is anti-symmetric with full unsteady aerodynamics. On the other hand, by implementing the quasi-steady aerodynamics, trim is possible also with symmetric flapping. The validity of the results obtained is discussed in detail in Chap. 9.

Additional material is presented in the Appendix for evaluating the stability derivatives so that no derivation of equations is left incomplete but not overdone in the text.

Needless to say, in the second edition there were a few typographical errors which have been detected and corrected for the third edition.

Dr. Christoph Baumann read the most recent material and took the necessary steps for the third edition, R. P. Chandrasekar and B. Sreenivasan prepared the chapters for processing. Bayram Çelik helped me for the online correction after proof reading. My wife Zeliha, once again, stood by me and help me for downloading and printing the whole book. I would also like to express my gratitude and appreciation to all who made this publication possible.

İstanbul, Turkey
July 2020

Ülgen Gülçat

Preface to the Second Edition

The first edition of this book appeared in the fall of 2010 both as a hard copy and e-book. Since then there has appeared in literature numerous unsteady aerodynamics related material, which deserves to be presented in a graduate textbook. Most of the new material is relevant to Chapter 8: Modern Topics. Here, a calculation method for propulsive force, lift generation and induced drag of a pitching-plunging thin finite wing is provided with a numerical example as an additional material. The unsteady 3-D Boundary Layer solution technique is introduced for prediction of the viscous drag to see if the propulsive force overcomes the drag. In addition, the ground effect on the air vehicles performing near ground is formulated to see how the lift and the propulsive forces are altered for the high and low aspect ratio wings.

The state-space representation of aerodynamics was introduced briefly in the first edition. In the present edition, more detailed discussion of the method is provided via numerical solutions for airfoils and finite wings of various aspect ratios even in the presence of ground.

Additional material, including bio-inspired and biological flows, related to the unsteady flows is also provided at the end of Chapter 9 to emphasize on the present developments and future prospects.

Some more material is added to Appendix so that no derivation of equations is left incomplete but not overdone in the text.

Needless to say, in the first edition there was a few typographical errors which are detected and corrected for the second edition.

Dr. Christoph Baumann read the new material and took the necessary steps for the second edition, and K. M. Govardhana prepared the metadata of the book. Mehmet Tan provided the figure for the cover page. My wife Zeliha, once more, stood by me in all these times with great patience. Finally, I would like to express my gratitude and appreciation to all who made publication of the book possible.

İstanbul
July 2015

Ülgen Gülçat

Preface to the First Edition

The flying animate objects were present in earth's atmosphere about hundreds of million years before the appearance of human kind on earth. Only at the beginning of 20th century, the proper analysis of the lifting force was made to provide the possibility of powered and manned flight. Prior to that, one of the pioneers of mechanics, Sir Isaac Newton had used 'his impact theory' in an attempt to formulate the lifting force created on a body immersed in a free stream. In late 17th century, his theory was a failure due to calculation of insufficient lift generation and made him come to the conclusion that 'flying is a property of heavenly bodies'. In a similar manner, almost after two centuries, William Thomson (Lord Kelvin) whose contributions to thermo and gas dynamics are well known, then proved that 'only objects lighter than air' can fly!

Perhaps it was the adverse influence of these two pioneers of mechanics on Western Europe, where contributions to the discipline of hydrodynamics is unquestionable, that delayed the true analysis of the lift generation. The proper analysis of lifting force, on the other hand, was independently made at the onset of 20th century by the theoretical aerodynamicists Martin Kutta and Nicolai Joukowski of Central and Eastern Europe respectively. At about the same years, the Wright brothers, whose efforts on powered flight were ridiculed by authorities of their time, were able to fly a short distance. Thereafter, in a time interval little more than a century, which is a considerably short span compared to the dawn of civilization, we see not only tens of thousands of aircrafts flying in earth's atmosphere at a given moment but we also witnessed unmanned or manned missions to the moon, missions to almost every planet in our solar system and to deeper space to let the existence of life on earth be known by the other possible intelligent life forms.

The foundation of the century old discipline of aeronautics and astronautics undoubtedly lies in the progress made in aerodynamics. The improvement made on the aerodynamics of wings, based on satisfying the Kutta condition at the trailing edge to give a circulation necessary for lift generation, was so rapid that in less than a quarter century it led to the breaking of the sound barrier and to the discovery of sustainable supersonic flight, which was unprecedented in nature and once thought to be not possible! In many engineering applications involving motion we encounter either forced or velocity induced oscillatory motion at high speeds. If the changes in the excitations are rapid, the response of the system lags considerably. Similarly, the response of the aerodynamic systems cannot be determined using steady aerodynamics for rapidly changing excitations. The unsteady aerodynamics, on the other hand, has sufficient tools to give accurately the phase lag between the rapid motion change and the response of the aerodynamic system. As we observe the performances of perfect aerodynamic structures of nature, we understand the effect of unsteady phenomena to such an extent that lift can be generated with apparent mass even without a free stream. In some cases, when the classical unsteady aerodynamics does not suffice, we go beyond the conventional concepts, with observing nature again, to utilize the extra lift created by the suction force of strong vorticies shed from the sharp leading edge of low aspect ratio wings at high angles of attack. We implement this fact in designing highly maneuverable aircrafts at high angles of attack and low free stream velocities. If we go to angles of attack higher than this, we observe aerodynamically induced but undesirable unsteady phenomena called wing rock. In addition, quite recently the progress made in unsteady aerodynamics integrated with electronics enable us to design and operate Micro Air Vehicles (MAVs) based on flapping wing technology having radio controlled devices.

This book, which gives the progress made in unsteady aerodynamics in about less than a century, is written to be used as a graduate textbook in Aerospace Engineering. Another important aim of this work is to provide the project engineers with the foundations as well as the knowledge needed about the most recent developments involving unsteady aerodynamics. This need emerges from the fact that the design and the analysis tools used by the research engineers are treated as black boxes providing results with inadequate information about the theory as well as practice. In addition, the models of complex aerodynamic flows and their solution methodologies are provided with examples, and enhanced with problems and questions asked at the end of each chapter. Unlike this full text, the recent developments made in unsteady aerodynamics together with the fundamentals have not appeared as a textbook except in some chapters of books on aeroelasticity or helicopter dynamics!

The classical parts of this book are mainly based on 'not so terribly advanced' lecture notes of Alvin G. Pierce and basics of vortex aerodynamics knowledge provided by James C. Wu while I was a PhD student at Georgia Tech. What was then difficult to conceive and visualize because of the involvement of special functions, now, thanks to the software allowing symbolic operations and versatile numerical techniques, is quite simple to solve and analyze even on our PCs. Although the problems become more challenging and demanding by time, however, the development of novel technologies and methods render them possible to solve provided that the fundamentals are well taught and understood by well informed users. The modern subjects covered in the book are based on the lecture notes of 'Unsteady Aerodynamics' courses offered by me for the past several years at Istanbul Technical University.

The first five chapters of the book are on the classical topics whereas the rest covers the modern topics, and the outlook and the possible future developments finalize the book. The examples provided at each chapter are helpful in terms of application of relevant material, and the problems at the end of each chapter are useful for the reader towards understanding of the subject matter and its future usage. The main idea to be delivered in each chapter is given as a verbal summary at chapters' end together with the most up to date references. There are ten Appendixes appearing to supplement the formulae driven without distracting the uniformity of the text.

I had the opportunity of reusing and borrowing some material from the publications of Joseph Katz, AIAA, NATO-AGARD/RTO and Annual Review of Fluid Mechanics with their kind copyright permissions. Dr. Christoph Baumann read the text and made the necessary arrangements for its publication by Springer. Zeliha Gülçat and Canan Danışman provided me with their kind help in editing the entire text. N. Thiyagarajan prepared the metadata of the book. Aydın Mısırlıoğlu and Fırat Edis helped me in transferring the graphs into word documents. I did the typing of the book, and obtained most of the graphs and plots despite the 'carpal tunnel syndrome' caused by the intensive usage of the mouse. Furthermore, heavy concentration on subject matter and continuous work hours spent on the text showed itself as developing 'shingles'! My wife Zeliha stood by me in all these difficult times with great patience. I would like to extend my gratitude, once more, to all who contributed to the realization of this book.

Datça and İstanbul
August 2010

Ülgen Gülçat

Contents

Chapter 1
Introduction

Flights in Earth's atmosphere existed long before the presence of mankind. 300 million years ago it was performed by insects with wings, 60 million years ago by birds and 50 million years ago by bats as flying mammals, (Hitching 1982). Man, on the other hand, being the most recently emerged species among the living things first realized the concept of flight by depicting the flying animals in his creative works related to mythology or real life, (Gibbs-Smith 1954). Needles to say, as a discipline, the science of Aerodynamics provides the most systematic and fundamental approach to the concept of flight.

The Aerodynamics discipline which determines the basic conditions of flying made great progress during the past hundred years, which is slightly longer than the average life span of a modern man (Anderson 2001). The reason of this progress is mainly the existence of wide range of aerospace applications in military and civilian industries. In the civilian aerospace industries, the demand for development of fast, quiet and more economical passenger planes with long ranges, and in the military the need for fast and agile fighter planes made this progress possible. The space race, on the other hand, had an accelerating effect on the progress during the last fifty years.

Naturally, the faster the planes get the more complicated the related aerodynamics become. As a result of this fast cruising, the lifting surfaces like wings and the tail planes start to oscillate with higher frequencies to cause in turn a phase lag between the motion and the aerodynamic response. In order to predict this phase lag, the concept of unsteady aerodynamics and its underlying principles were introduced. In addition, at higher speeds the compressibility of the air plays an important role, which in turn caused the emergence of a new branch of aerodynamics called compressible aerodynamics. At cruising speeds higher than the speed of sound, completely different aerodynamic behavior of lifting surfaces is observed. All these aerodynamical phenomena were first analyzed with mathematical models, and then observed experimentally in wind tunnels before they were tested on prototypes undergoing real flight conditions. Nature, needless to say, inspired many aerodynamicists as well. In recent years, the leading edge vortex formation which

Ü. Gülçat, *Fundamentals of Modern Unsteady Aerodynamics*,
https://doi.org/10.1007/978-3-030-60777-7_1

gives extra lift for highly swept wings at high angles of attack has been studied extensively. During the last decade, the man made flight has no longer been based on a fixed wing. The flapping wing aerodynamics which utilizes the unsteady aerodynamic concepts is used in designing and building micro air vehicles to serve mankind in various fields.

First, let us introduce various pertinent definitions in order to establish a firm convention in studying the topics of unsteady aerodynamics in general.

1.1 Definitions

Aerodynamics: It is the branch of science which studies the forces and moments necessary to have a controlled and sustainable flight in air. These forces are named the *lift* in the direction normal to the flight and the *drag* or the *propulsive* force in the direction of the flight. In addition, it studies the effect of the velocity fields induced by the motion during flight. On the other hand, the study of the forces created by the motion of an arbitrarily shaped body in any fluid is the concern of the Fluid Mechanics in general. It is necessary to make this distinction at this stage.

Aerodynamic Coefficients: These are the non-dimensional values of pressure, force and moment which affect the flying object. In non-dimensionalization, the free stream density ρ and the free stream velocity U are used as characteristic values. One half of the dynamic pressure, $\frac{1}{2}\rho U^2$ is utilized in obtaining pressure coefficient, c_p. As the characteristic length, half of the chord length and as the characteristic area the wing surface area are considered. Hence, the product of dynamic pressure with the half chord is used to obtain the sectional lift coefficient c_l, the drag coefficient c_d, and the moment coefficient c_m, wherein the square of the half chord is used. For the finite wing, however, the coefficient of lift reads as C_L, the drag C_D and the moment coefficient C_M.

Center of pressure: x_{cp}, The location at which the resultant aerodynamic moment is zero. If we consider the profile (the wing section) as a free body, this point can be assumed as the center of gravity for the pressure distribution along the surface of the profile.

Aerodynamic center: x_{ac}, This is the point where the aerodynamic moment acting on the wing is independent of the angle of attack. The aerodynamic center is essential for the stability purposes. For a finite wing it is the line connecting the aerodynamic centers of each section along the span.

Steady Aerodynamics: If the flow field around a flying body does not change with respect to time, the aerodynamic forces and moments acting on the body remain the same all the time. This type of aerodynamics is called steady aerodynamics.

Unsteady Aerodynamics: If the motion of the profile or the wing in a free stream changes by time, so do the acting aerodynamic coefficients. When the changes in the motion are fast enough, the aerodynamic response of the body will have a phase lag. For faster changes in the motion, the inertia of the displaced air will contribute as the apparent mass term. If the apparent mass term is negligible, this type of analysis is called the quasi-unsteady aerodynamics.

Compressible Aerodynamics: When the free stream speeds become high enough, the compressibility of the air starts to change the aerodynamic characteristics of the profile. After exceeding the speed of sound, the compressibility effects changes the pressure distribution so drastically that the center pressure for a thin airfoil moves from quarter chord to midchord.

Vortex Aerodynamics: A vortex immersed in a free stream experiences a force proportional to density, vortex strength and the free stream speed. If the airfoil or the wing in a free stream is modeled with a continuous vortex sheet, the total aerodynamic force acting can be evaluated as the integral effect of the vortex sheet. In rotary aerodynamics, the returning effect of the wake vorticity on the neighboring blade can also be modeled with vortex aerodynamics. At high angles of attack, at the sharp leading edge of highly swept wings the leading edge vortex generation causes such suction that it generates extra lift. Further angle of attack increase causes asymmetric generation of leading edge vortices which in turn causes wing rock. The sign of the leading edge vortices of unswept oscillating wings, on the other hand, determines whether power or propulsive force generation, depending on the frequency and the center of the pitch. For these reasons, the vortex aerodynamics is essential for analyzing, especially the unsteady aerodynamic phenomenon.

1.2 Generation of Lift

The very basic theory of aerodynamics lies in the Kutta-Joukowski theorem. This theorem states that for an airfoil with round leading and sharp trailing edge immersed in a uniform stream with an effective angle of attack, there exists a lifting force proportional to the density of air ρ, free stream velocity U and the circulation Γ generated by the bound vortex. Hence, the sectional lifting force l is equal to

$$l = \rho U \Gamma \tag{1.1}$$

Figure 1.1 depicts the pertinent quantities involved in generation of lift.

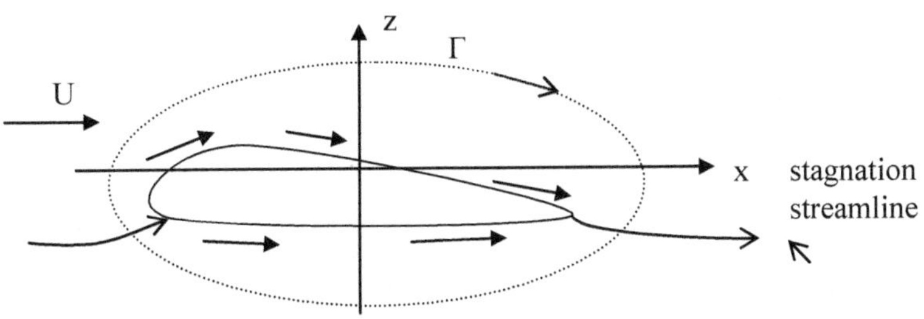

Fig. 1.1 An airfoil immersed in a free stream generating lift

The strength of the bound vortex is given by the circulation around the airfoil, $\Gamma = \oint \mathbf{V}.\mathbf{ds}$.

If the effective angle of attack is α, and the chord length of the airfoil is $c = 2b$, with the Joukowski transformation the magnitude of the circulation is found as Γ = 2 π α b U. Substituting the value of Γ into Eq. 1.1 gives the sectional lift force as

$$l = 2\,\rho\,\pi\,\alpha\,b\,U^2 \tag{1.2}$$

Using the definition of sectional lift coefficient for the steady flow we obtain,

$$c_l = \frac{l}{\rho\,U^2 b} = 2\,\pi\,\alpha \tag{1.3}$$

The very same result can be obtained by integrating the relation between the vortex sheet strength γ_a and the lifting surface pressure coefficient c_{pa} along the chord as follows.

$$c_{pa}(x) = c_{pl} - c_{pu} = 2\,\gamma_a(x)/U$$

The lifting pressure coefficient for an airfoil with angle of attack reads as

$$c_{pa}(x) = 2\alpha\sqrt{\frac{b-x}{b+x}}, \quad -b \le x \le b \tag{1.4}$$

Equation 1.4 is singular at the leading edge, $x = -b$, as depicted in Fig. 1.2. Integrating Eq. 1.4 along the chord and non-dimensionalizing the integral with b gives Eq. 1.3. The singularity appearing in Eq. 1.4 is an integrable singularity which, therefore, gives a finite lift coefficient. 1.4 In Fig. 1.2, the comparison of the theoretical and experimental values of lifting pressure coefficients for a thin airfoil are given. This comparison indicates that around the leading edge the experimental values suddenly drop to a finite value. For this reason, the experimental value of the lift coefficient is always slightly lower than the theoretical value predicted with a mathematical model. The derivation of Eq. 1.4 with the aid of a distributed vortex sheet will be given in detail in later chapters.

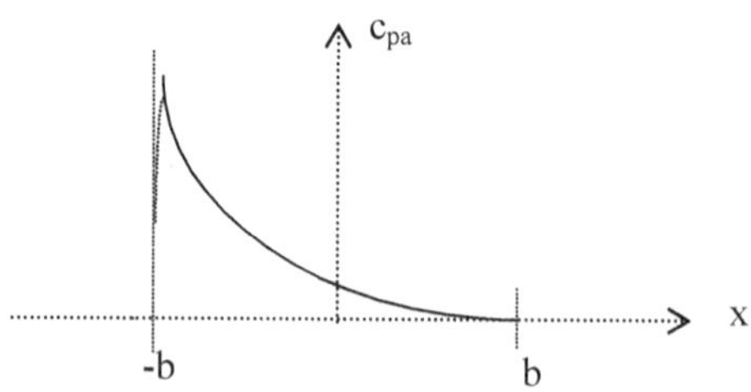

Fig. 1.2 Lifting surface pressure coefficients c_{pa}: ___ theoretical, — experimental

For steady aerodynamic cases, the center of pressure for symmetric thin airfoils can be found by the ratio the first moment of Eq. 1.1 with the lifting force coefficient, Eq. 1.3. The center of pressure and the aerodynamic centers are at the quarter chord of the symmetrical airfoils.

Abbot and Von Deonhoff (1959) give the geometrical and aerodynamic properties of so many conventional airfoils even utilized in the present time.

1.3 Unsteady Lifting Force Coefficient

During rapidly changing unsteady motion of an airfoil the aerodynamic response is no longer the timewise slightly changing steady phenomenon.

For example, let us consider a thin airfoil with a chord length of 2b undergoing a vertical simple harmonic motion in a free stream of U with zero angle of attack. If the amplitude of the vertical motion is $\overline{h}$ and the angular frequency is ω then the profile location at any time t reads as

$$z_a(t) = \overline{h}e^{i\omega t} \tag{1.5}$$

If we implement the pure steady aerodynamics approach, because of Eq. 1.3 the sectional lift coefficient will read as zero. Now, we write the time dependent sectional lift coefficient in terms of the reduced frequency $k = \omega b/U$ and the non-dimensional amplitude $\overline{h}^* = \overline{h}/b$.

$$c_l(t) = [-2\,i\,kC(k)\overline{h}^* + k^2\overline{h}^*]\pi\, e^{i\omega t} \tag{1.6}$$

Let us now analyze each term in Eq. 1.6 in terms of the relevant aerodynamics.

(i) *Unsteady Aerodynamics*: If we consider all the terms in Eq. 1.6 then the analysis is based on unsteady aerodynamics. *C*(*k*) in the first term of the expression is a complex function and called the Theodorsen function which is the measure of the phase lag between the motion and aerodynamic response. The second term, on the other hand, is the acceleration term based on the inertia of the air parcel displaced during the motion. It is called the apparent mass term and is significant for the reduced frequency values larger than unity.

(ii) *Quasi Unsteady Aerodynamics*: If we neglect the apparent mass term in Eq. 1.6 the aerodynamic analysis is then called quasi unsteady aerodynamics. Accordingly, the sectional lift coefficient reads as

$$c_l(t) = [-2\pi\,i\,kC(k)\overline{h}^*]e^{i\omega t} \tag{1.7}$$

Since the magnitude of the Theodorsen function is less than unity for the values of k larger than 0, quasi unsteady lift coefficient is always reduced.

The Theodorsen function is given in terms of the Haenkel functions. An approximate expression for small values of k is: $C(k) \cong 1 - \pi k/2 + ik(\ln(k/2) + .5772),\ 0.01 \leq k \leq 0.1$.

(iii) *Quasi Steady Aerodynamics*: If we take $C(k) = 1$, then the analysis becomes a quasi steady aerodynamics to give

$$c_l(t) = [-2\,\pi\, i\, k\, \bar{h}^*]e^{i\omega t} \tag{1.8}$$

In this case, there exists a 90° phase difference between the motion and the aerodynamic response.

(iv) *Steady Aerodynamics*: Since the angle of attack is zero, we get zero lift!

So far, we have seen the unsteady aerodynamics caused by simple harmonic airfoil motion. When the unsteady motion is arbitrary, there are new functions involved to represent the aerodynamic response of the airfoil to unit excitations. These functions are the integral effect of the Theodorsen function represented by infinitely many frequencies. For example, the Wagner function gives the response to a unit angle of attack change and the Küssner function, on the other hand, provides the aerodynamic response to a unit sharp gust.

1.4 Steady Aerodynamics of Thin Wings

The finite wing aerodynamics, for special wing geometries, can yield analytical expressions for the aerodynamic coefficients in terms of the sectional properties of the wing. A special case is the elliptical span wise loading of the wing which is proportional to $\sqrt{l^2 - y^2}$, where y is the span wise coordinate and l is the half span. For the wings with large span, using the Prandtl's lifting line theory the wing's lift coefficient C_L becomes equal to the constant sectional lift coefficient c_l. Hence,

$$C_L = c_l \tag{1.9}$$

Another interesting aspect of the finite wing theory is the effect of the tip vortices on the overall performance of the wing. The tip vortices induce a vertical velocity which in turn induces additional drag on the wing. Hence, the total drag coefficient of the wing reads

$$C_D = C_{D_o} + \frac{C_L^2}{\pi AR} \tag{1.10}$$

Here the aspect ratio is $AR = l^2/S$, and S is the wing area. For the symmetric and untwisted wings to have elliptical loading the plan form geometry also should be elliptical as shown in Fig. 1.3.

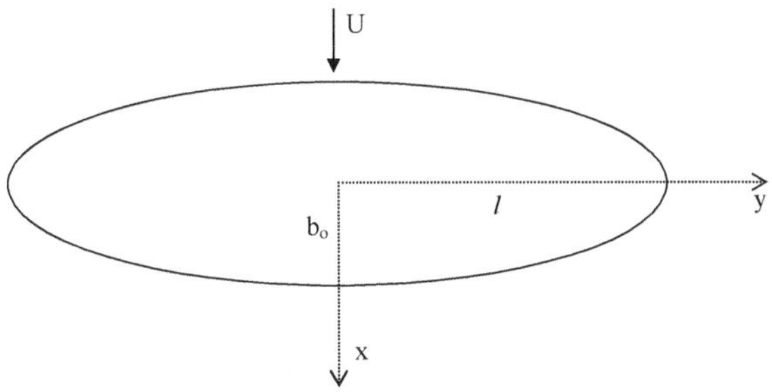

Fig. 1.3 Elliptical plan form

For the case of non-elliptical wings, we use the Glauert's Fourier series expansion of the span wise variation of the circulation given by the lifting line theory. The integration of the numerically obtained span wise distribution of the circulation gives us the total lift coefficient.

If the aspect ratio of a wing is not so large and the sweep angle is larger than 15°, then we use the Weissenger's L-Method to evaluate the lift coefficient of the wing.

For slender delta wings and for very low aspect ratio slender wings, analytical expressions for the lift and drag coefficients are also available. The lift coefficient for a delta wing without a camber in spanwise direction is

$$C_L = \frac{1}{2}\pi AR\,\alpha \tag{1.11}$$

The induced drag coefficient for delta wings having elliptical load distribution along their span is given as

$$C_{D_i} = C_L\alpha/2 \tag{1.12}$$

The lift and drag coefficients for slender delta wings are almost unaffected from the cross flow. Therefore, even at high speeds the cross flow behaves incompressible and the expressions given by Eqs. 1.11–1.12 are valid even for the supersonic ranges. In the 4th chapter, the Weissenger's L-Method and the derivation of Eq. 1.11–1.12 will be seen in a detail.

1.5 Unsteady Aerodynamics of Slender Wings

It is also customary to start the unsteady aerodynamic analysis of wings with simple harmonic motion and obtain analytical expressions for the amplitude of the aerodynamic coefficients of the large aspect ratio wings which have elliptical span wise load distribution. In addition, Reissner's approach for the large aspect ratio rectangular wings numerically provides us with the aerodynamic characteristics. As a more general approach, the doublet lattice method handles wide range of aspect ratio wings with large sweeps and with span wise deflection in compressible subsonic flows. In later chapters, the necessary derivations and representative examples of these methods will be provided.

1.6 Compressible Steady Aerodynamics

It is a well known fact that at high speeds comparable with the speed of sound the effect of compressibility starts to play an important role on the aerodynamic characteristics of airfoil. At subsonic speeds, there exists a similarity between the compressible and incompressible external flows based on the Mach number $M = U/a_\infty$, $a_\infty =$ free stream speed of sound. This similarity enables us to express the compressible pressure coefficient in terms of the incompressible pressure coefficient as follows

$$c_p = \frac{c_{p_o}}{\sqrt{1 - M^2}} \tag{1.13}$$

Here,

$$c_{p_o} = \frac{p_o - p_\infty}{\frac{1}{2}\rho_\infty U^2}$$

is the surface pressure coefficient for the incompressible flow about a wing which is kept with a fixed thickness and span but stretched along the flow direction, x, with the following rule

$$x_0 = \frac{x}{\sqrt{1 - M^2}}, \; y_0 = y, \; z_0 = z \tag{1.14}$$

as shown in Fig. 1.4. The Prandtl-Glauert transformation for the wings is summarized by Eq. 1.14 and Eq. 1.13 is used to obtain the corresponding surface pressure coefficient. By this transformation, once we know the incompressible pressure coefficient at a point x, y, z, Eq. 1.13 gives the pressure coefficient for the known free stream Mach number at the stretched coordinates x_0, y_0, z_0. As seen from Fig. 1.4, it is not practical to build a new plan form for each Mach number. Therefore, we need to find more practical approach in utilizing Prandtl-Glauert transformation.

For this purpose, assuming that the free stream density does not change for the both flows, we integrate Eq. 1.13 in chord direction to obtain the same sectional lift coefficient for the incompressible and compressible flow. While doing so, if we keep the chord length same, i.e., divide x_0 with $(1-M^2)^{1/2}$, then the compressible sectional lift coefficient c_l and moment coefficient c_m become expressible in terms of the incompressible c_{lo} and c_{mo} as follows.

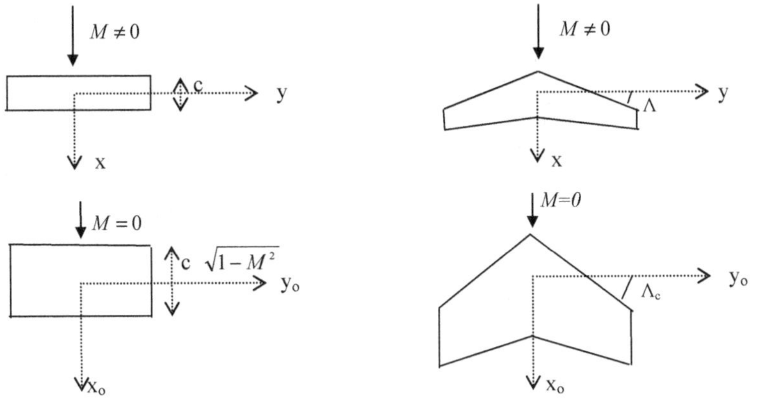

Fig. 1.4 Prandtl-Glauert transformation, before $M = 0$, and after $M \neq 0$

$$c_l = \frac{c_{l_o}}{\sqrt{1 - M^2}} \qquad c_m = \frac{c_{m_o}}{\sqrt{1 - M^2}} \tag{1.15a, b}$$

The result obtained with Eq. 1.15a, b is applicable only for the wings with large aspect ratios and as the aspect ratio gets smaller the formulae given by 1.15a, b fails to give correct results. For two dimensional flows Eq. 1.15a, b gives good results before approaching critical Mach numbers. The critical Mach number is the free stream Mach number at which local flow on the airflow first reaches the speed of sound. Equation 1.15a, b are known as the Prandtl-Glauert compressibility correction and they give the compressible aerodynamic coefficients in terms of the Mach number of the flow and the incompressible aerodynamic coefficients. The drag coefficient, on the other hand, remains the same until the critical Mach number is reached.

The total lift coefficient for the finite thin wings with the sectional lift slope a_o, and aspect ratio AR reads as

$$C_L = a_o \frac{AR}{\sqrt{1 - M^2}AR + 2} \alpha \tag{1.16}$$

Formula 1.16 is applicable until the critical Mach number is reached at the surface of the wing.

In case of finite wings, there is a way to increase the critical Mach number by giving sweep at the leading edge. If the leading edge sweep angle is Λ, then the sectional lift coefficient at angle of attack which is measured with respect to the free stream direction, reads as

$$c_l = \frac{a_o \cos\Lambda}{\sqrt{1 - M^2\cos^2\Lambda}}\alpha \tag{1.17}$$

The effect of Mach number and the sweep angle combined reduces the sectional lift coefficient as compared to the wings having no sweep. Now, if we consider the aspect ratio of the finite wing, the Diederich formula becomes applicable for the total lift coefficient for considerably wide range of aspect ratios,

$$C_L = \frac{a_a \cos\Lambda_e}{\sqrt{1-M^2}} \frac{AR\sqrt{1-M^2}}{AR\sqrt{1-M^2}\sqrt{1+(\frac{a_a\cos\Lambda_e}{\pi AR\sqrt{1-M^2}})^2}+\frac{a_a\cos\Lambda_e}{\pi}}\alpha \tag{1.18}$$

Here, the effective sweep angle Λ_e is given by

$$\cos\Lambda_e = \frac{\sqrt{1-M^2}}{\sqrt{1-M^2\cos^2\Lambda}}\cos\Lambda.$$

For the case of supersonic external flows, we encounter a new type of aerodynamic phenomenon wherein the Mach cones whose axes are parallel to the free stream send the disturbance only in downstream. The lifting pressure coefficient for a thin airfoil, in terms of the mean camber line $z = z_a(x)$, reads as

$$c_{p_a} = -\frac{4}{\sqrt{M^2-1}}\frac{dz_a}{dx} \tag{1.19}$$

Figure 1.5 gives the lifting pressure coefficient distribution for a flat plate at angle of attack α.

In order to obtain the sectional lift for the flat plate airfoil we need to integrate Eq. 1.19 along the chord

$$c_l = \frac{4\alpha}{\sqrt{M^2-1}} \tag{1.20}$$

The sectional moment coefficient with respect to a point whose coordinate is a on the chord reads

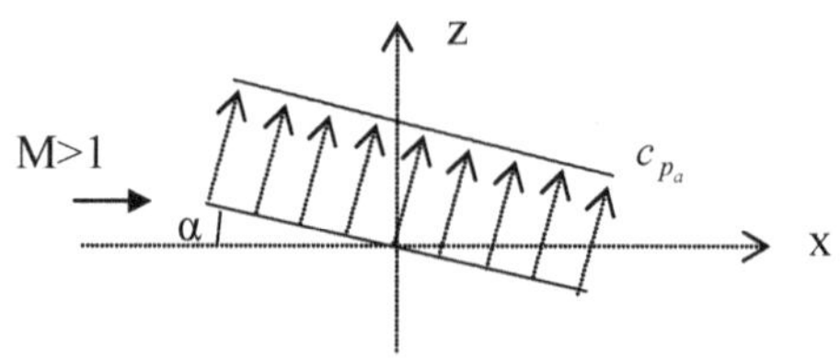

Fig. 1.5 Supersonic lifting pressure distributions along the flat plate

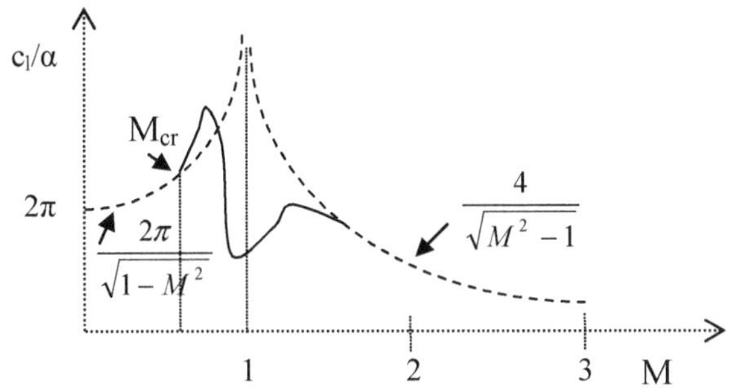

Fig. 1.6 The change of the sectional lift coefficient with the Mach number. (The transonic flow region is shown with dark lines, adapted from Küchemann (1978))

$$c_m = \frac{2a\alpha}{\sqrt{M^2 - 1}} = \frac{a}{2} c_l \tag{1.21}$$

Using 1.20 and 1.21, the center of pressure is found at the half chord point as opposed to the quarter chord point for the case of subsonic flows. The effect of compressibility on the sectional lift coefficient is shown in Fig. 1.6 with the necessary modification near M = 1 area.

An important characteristic of the supersonic flow is its wavy character. The reason for this is the hyperbolic character of the model equations at the supersonic speeds. The emergence of the disturbances with wavy character from the wing surface requires certain energy. This energy appears as wave drag around the airfoil. The sectional wave drag coefficient can be evaluated in terms of the equations for the mean camber line and the thickness distribution along the chord as follows.

$$c_{d_w} = \frac{4}{\sqrt{M^2 - 1}} \int_{-1}^{1} \left[\left(\frac{dz_a^*}{dx^*} \right)^2 + \left(\frac{dz_t^*}{dx^*} \right)^2 \right] dx^* \tag{1.22}$$

According to Eq. 1.22 the sectional drag coefficient is always positive and this is in agreement with the physics of the problem.

1.7 Compressible Unsteady Aerodynamics

The evaluation methods for the sectional as well as the total lift and moment coefficients for unsteady subsonic and supersonic flows will be given in Chap. 5. It is, however, possible to obtain approximate expressions for the amplitude of the sectional lift coefficients at high reduced frequencies and at transonic regimes where M approaches to unity as limiting value. For steady flow on the other hand, the analytical expression is not readily available since the equations are nonlinear. However, local linearization process is applied to obtain approximate values for the aerodynamic coefficients.

Now, we can give the expression for the amplitude of the sectional lift coefficient for a simple harmonically pitching thin airfoil in transonic flow,

$$\overline{c}_l \approx 4(1+ik)\overline{\alpha}, \quad k > 1 \tag{1.23}$$

Here, $\overline{\alpha}$ is the amplitude of the angle of attack. Let us consider the same airfoil in a vertical motion with amplitude of $\overline{h}$.

$$\overline{c}_l \approx 8ik\overline{h}/b, \quad k > 1 \tag{1.24}$$

All these formulae are available from (Bisplinghoff et al. 1996).

Aerodynamic response to the arbitrary motion of a thin airfoil in transonic flow will be studied in Chap. 5 with aid of relevant unit response function in different Mach numbers.

1.8 Slender Body Aerodynamics

Munk-Jones airship theory is a good old useful tool for analyzing the aerodynamic behavior of slender bodies at small angles of attack even at supersonic speeds. The cross flow of a slender wing at a small angle of attack is approximately incompressible. Therefore, according to the Newton's second law of motion, during the vertical motion of a slender body, the vertical momentum change of the air parcel with constant density displaced by the body motion is equal to the differential force acting on the body. Using this relation, we can decide on the aerodynamic stability of the slender body if we examine the sign of the aerodynamic moment about the center of gravity of the body. Expressing the change of the vertical force L, as a lifting force in terms of the cross sectional are S and the equation of the axis $z = z_a(x)$ of the body we obtain the following relation

$$\frac{dL}{dx} = -\rho U^2 \frac{d}{dx}(S\frac{dz_a}{dx}) \tag{1.25}$$

In Fig. 1.7, shown are the vertical forces affecting the slender body whose axis is at an angle of attack α with the free stream direction. Note that the vertical forces are non zero only at the nose and at the tail area because of the cross sectional area increase in those regions. Since there is no area change along the middle portion of the body, there is no vertical force generated at that portion of the body.

As we see in Fig. 1.7, the change of the moment with angle of attack taken around the center of gravity determines the stability of the body. The net moment of the forces acting at the nose and at the tail of the body counteracts with each other to give the sign of the total moment change with α. The area increase at the tail section contributes to the stability as opposed to the apparent area increase at the nose region.

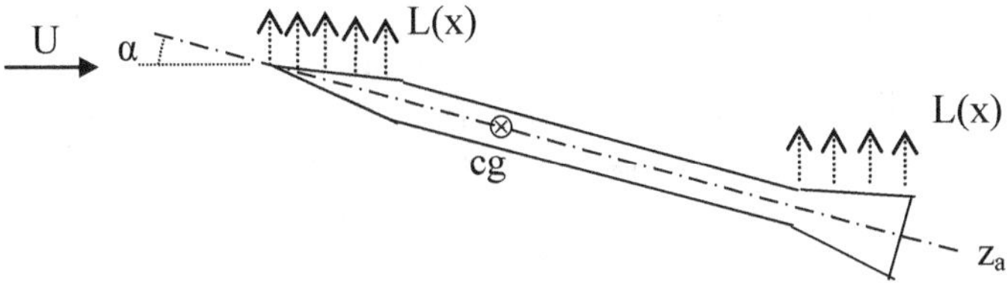

Fig. 1.7 Vertical forces acting on the slender body at angle of attack α

1.9 Hypersonic Aerodynamics

According to Newtonian impact theory, which fails to explain the classical lift generation, the pressure exerted by the air particles impinging on a surface is equal to the time rate of change of momentum vertical to the wall. Using this principle we can find the pressure exerted by the air particles on the wall which is inclined with free stream with angle θ_w. Since the velocity, as shown in Fig. 1.8, normal to the wall is U_n the time rate of change of momentum becomes $p = \rho\, U_n^2$.

If we write $U_n = U \sin \theta_w$, the surface pressure coefficient reads as

$$c_p = \frac{p - p_\infty}{\frac{1}{2}\rho_\infty U^2} = 2\sin^2\theta_w - \frac{2}{\gamma M^2} \tag{1.26}$$

The free stream Mach number M is always high for hypersonic flows. Therefore, its square $M^2 >> 1$ is always true. If the wall inclination under consideration is sufficiently large i.e. θ_w is greater than 35°–40°, the second term in Eq. 1.26 becomes negligible compared to the first term. This allows us to obtain a simple expression for the surface pressure at hypersonic speeds as follows

$$c_p \cong 2\sin^2\theta_w \tag{1.27}$$

Now, we can find the lift and the drag force coefficients for hypersonic aerodynamics based on the impact theory. According to Fig. 1.8 the sectional lift coefficient reads as

$$c_L = 2\sin^2\theta_w \cos\theta_w, \tag{1.28}$$

and the sectional drag coefficient becomes

$$c_D = 2\sin^3\theta_w \tag{1.29}$$

Fig. 1.8 Velocity components for the impact theory

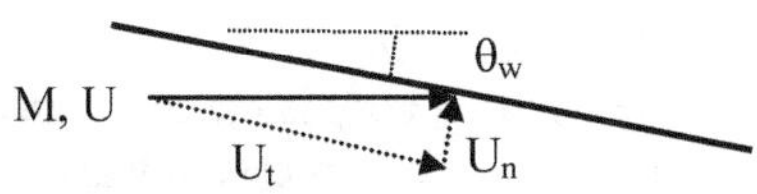

Starting with Newton until the beginning of twentieth century, the lifting force was unsuccessfully explained by the impact theory. Because of $\sin^2$ term in Eq. 1.28 there was never sufficient lift force to be generated in small angles of attack. For this reason, even though Eq. 1.28 has been known since Newton's time, it is only valid at hypersonic speeds and at high angles of attack. The drag coefficient expressed with Eq. 1.29, gives reasonable values at high angles of attack but gives small values at low angles of attack. We have to keep in mind that these formulae are obtained with perfect gas assumption.

The real gas effects at upper levels of atmosphere at hypersonic speeds play a very important role in physics of the external flows. At high speeds, the heat generated because of high skin friction excites the nitrogen and oxygen molecules of air to release their vibrational energy which increases the internal energy. This internal energy increase makes the air no longer a calorically perfect gas and therefore, the specific heat ratio of the air becomes a function of temperature. At higher speeds, when the temperature of air rises to the level of disassociation of nitrogen and oxygen molecules into their atoms, new species become present in the mixture of air. Even at higher speeds and temperatures, the nitrogen and oxygen atoms react with the other species to create new species in the air. Another real gas effect is the diffusion of species into each other. The rate of this diffusion becomes the measure of the catalyticity of the wall. At the catalytic walls, since the chemical reactions take place with infinite speeds the chemical equilibrium is established immediately. Because of this reason, the heat transfer at the catalytic walls is much higher compared to that of non-catalytic walls.

For a hypersonically cruising aerospace vehicle, there exists a heating problem if it is slender, and low lift/drag ratio problem if it has a blunt body. The solution to this dilemma lies in the concept of 'wave rider'. The wave rider concept is based on a delta shaped wing which is immersed in a weak conic shock of relevant to the cruising Mach number. Necessary details will be given in following chapters.

1.10 The Piston Theory

The piston theory is an approximate theory which works for thin wings at high speeds and at small angles of attack. If the characteristic thickness ratio of a body is τ and $M\tau$ is the hypersonic similarity parameter then for $M\tau >> 1$ the Newtonian impact theory works well. For the values of $M\tau < 1$ the piston theory becomes applicable. Since τ is small for thin bodies, at high Mach numbers the shock generated at the leading edge is a highly inclined weak shock. This makes the flow region between the surface and the inclined shock a thin boundary layer attached to the surface. If the surface pressure at the boundary layer is p and the vertical velocity on the surface is w_a, then the flow can be modeled as the wedge flow as shown in Fig. 1.9.

The piston theory is based on an analogy with a piston moving at a velocity w in a tube to create compression wave. The ratio of compression pressure created in the

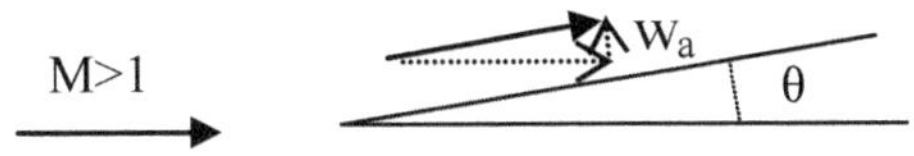

Fig. 1.9 Flow over a wedge for the piston theory

tube to the pressure before passing of the piston reads as Lieppmann and Roshko (1963); Hayes and Probstien (1966),

$$\frac{p}{p_\infty} = [1 + \frac{\gamma - 1}{2}\frac{w}{a_\infty}]^{\frac{2\gamma}{\gamma-1}} \tag{1.30}$$

Here, a_∞ is the speed of sound for the gas at rest. If we linearize Eq. 1.30 by expanding into the series and retain the first two terms, the pressure ratio reads as

$$\frac{p}{p_\infty} \cong 1 + \gamma\frac{w_a}{a_\infty} \tag{1.31}$$

wherein, w_a is the time dependent vertical velocity which satisfies the following condition: $w_a \ll a_\infty$. The expression for the vertical velocity in terms of the body motion and the free stream velocity is given by

$$w_a = \frac{\partial z_a}{\partial t} + U\frac{\partial z_a}{\partial x} \tag{1.32}$$

Equation 1.31 is valid only for the hypersonic similarity values in, $0 < M\tau < 0.15$, and as long as the body remains at small angles of attack during the motion while the vertical velocity changes according to Eq. 1.32. For higher values of the hypersonic similarity parameter, the higher order approximations will be provided in the relevant chapter.

1.11 Modern Topics

Hitherto, we have given the summary of so called classical and conventional aerodynamics. Starting from 1970's, somewhat unconventional analyses based on numerical methods and high tech experimental techniques have been introduced in the literature to study the effect of leading edge separation on the very high lifting wings or on unsteady studies for generating propulsion or power extraction. Under the title of modern topics we will be studying (i) vortex lift, (ii) different sorts of wing rock, and (iii) flapping wing aerodynamics.

(i) *Vortex lift*: The additional lift generated by the sharp leading edge separation of highly swept wings at high angles of attack is called the vortex lift. This additional lift is calculated with the leading edge suction analogy and introduced by Polhamus, (Polhamus 1971). This theory which is also validated by experiments is named Polhamus theory for the low aspect ratio delta wings.

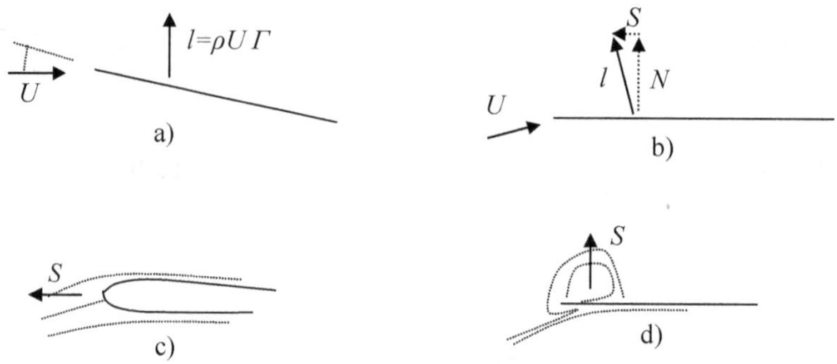

Fig. 1.10 Leading edge suction: **a** lift, **b** and **c** suction *S*, attached flow, **d** suction *S*, detached flow

Now, let us analyze the generation of vortex lift with the aid of Fig. 1.10. According to the potential theory, the sectional lifting force was given in terms of the product of the density, free stream speed and bound circulation as in Eq. 1.1. We can resolve the lifting force into its chord wise component S and the normal component *N*. Here, *S* is the suction force generated by the leading edge portion of the upper surface of the airfoil. Accordingly, if the angle of attack is α then the suction force $S = \rho U\Gamma \sin\alpha$. Now, let us denote the effective circulation and the effective span of the delta wing, shown in Fig. 1.11, Γ and h respectively. Here, we define the effective span as the length when multiplied with the average sectional lift that gives the total lifting force of the wing. This way, the total suction force of the wing becomes as simple as S.h. Because of wing being finite, there is an induced drag force which opposes the leading edge suction force of the wing.

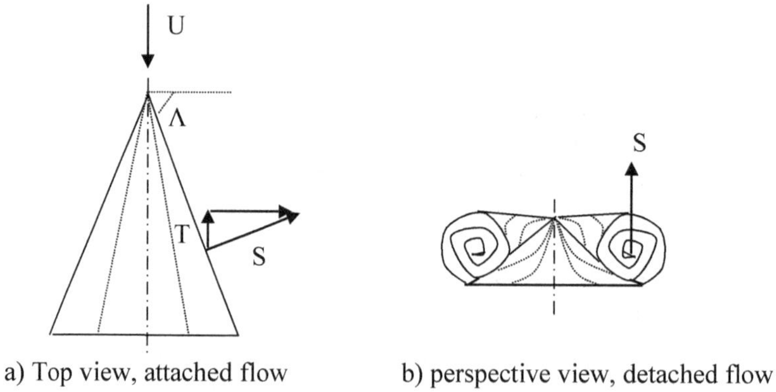

Fig. 1.11 Delta Wing and the suction force, **a** attached, **b** detached flow

Accordingly, the thrust force T in terms of the leading edge suction and the down wash w_i reads $T = \rho\Gamma h\,(U \sin\alpha - w_i)$. Let us define a non dimensional coefficient K_p emerging from potential considerations in terms of the area A of the wing,

$$K_p = 2\Gamma\, h/(A\, U\sin\alpha)$$

The total thrust coefficient can be expressed as

$$C_T = (1 - \frac{w_i}{U \sin\alpha}) K_p \sin^2\alpha$$

The potential lift coefficient now can be expressed in terms of K_p and the angle of attack α as

$$C_{L,p} = C_{N,p}\, cos\, \alpha = K_p \sin\alpha \cos^2\alpha$$

According to Fig. 1.11, the relation between the suction S and the thrust T reads as $S = T/\cos\Lambda$. Hence the vortex lift coefficient $C_{L,v}$ after the leading edge separation becomes

$$C_{L,v} = C_{N,v} \cos\alpha = (1 - \frac{w_i}{U \sin\alpha}) K_p \sin^2\alpha\, \frac{\cos\alpha}{\cos\Lambda}$$

Potential and the vortex lift added together gives the total lift coefficient as

$$C_L = K_p \sin\alpha \cos^2\alpha + K_v \sin^2\alpha \cos\alpha \tag{1.33}$$

Here,

$$K_v = (1 - \frac{w_i}{U \sin\alpha}) K_p / \cos\Lambda.$$

In Eq. 1.33, at the low angles of attack the potential contribution and at high angles of attack the vortex lift term becomes effective. For the low aspect ratio wings at angles of attack less than 10°, the total lift coefficient given by Eq. 1.11 is proportional to the angle of attack. Similarly, Eq. 1.33 also gives the lift coefficient proportional with the angle of attack at low angles of attack. For the case of low aspect ratio delta wings as shown in Fig. 1.11 if the angle of attack is further increased, the symmetry between the two vortices becomes spoiled. As a result of this asymmetry, the suction forces at the left and at right sides of the wing become unequal to create a moment with respect to the wing axes. This none zero moment in turn causes wing to rock along its axes.

(ii) *Wing-Rock*: The symmetry of the leading edge vortices for the low aspect ratio wings is sustained until a critical angle of attack. The further increase of angle of attack beyond the critical value for a certain wing or further reduction of the aspect ratio causes the symmetry to be spoiled. This in turn results in an almost periodic motion with respect to wing axis and this self induced motion is called wing-rock. The wing-rock was first observed during the stability experiments of delta wings performed in wind tunnels and then was validated with numerical investigations. During 1980s the vortex lattice method was extensively used to predict the wing-rock parameters for a single degree of freedom in rolling motion only. After those years however, two more degrees of freedom, displacements in vertical and span wise directions, are added to the studies based on Euler solvers. The Navier-Stokes solvers are expected to give the effect of viscosity on the wing-rock. The basics of wing-rock however, are given with the experimental data. Accordingly, the onset of wing-rock starts for the wings whose sweep angle is more than 74°, (Ericsson). For the wings having less then 74° sweep angle, instead of asymmetric vortex roll up, the vortex burst occurs at the left and right sides of the wing. In Fig. 1.12, the enveloping curve for the stable region, wing-rock and the vortex burst are given as functions of the aspect ratio and the angle of attack. The leading edge vortex burst causes a sudden suction loss at one side of the wing which causes a dynamic instability called roll divergence, (Ericksson 1984). After the onset of roll divergence, the wing starts to turn continuously around its own axis.

Let us now give the regions for the wing-rock, vortex burst and the 2-D separation in terms of the aspect ratio and the angle of attack by means of Fig. 1.12.

The information summarized in Fig. 1.12 also includes the conventional aerodynamics region for fixed wings having large aspect ratios.

The effect of roll angle and its rate on the generation of roll moment will be given in detail in later chapters.

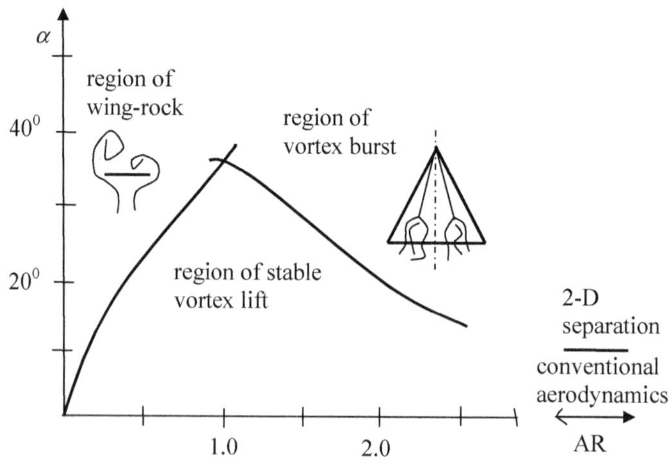

Fig. 1.12 The enveloping curve for the wing-rock

(iii) *Flapping wing theory*, (ornithopter aerodynamics): The flight of birds and their wing flapping to obtain propulsive and lifting forces have been of interest to many aerodynamicists as well as the natural scientists called ornithologists. After long and exhausting years of research and development only recently the prototypes of micro air vehicles are being flown for a short duration of experimental flights, (Mueller and DeLaurier 2003). In this context, a simple model of a flight tested ornithopter prototype was given by its designer and producer, (DeLaurier 1993).

The overall propulsive efficiency of flapping finite wing aerodynamics, which is only in vertical motion, was first given in 1940s with the theoretical work of Kucheman and von Holst as follows

$$\eta = \frac{1}{1 + 2/AR} \tag{1.34}$$

Although their approach was based on quasi steady aerodynamics, according to Eq. 1.34 the efficiency was increasing with increase in aspect ratio. As we have stated before, the quasi steady aerodynamics is valid for the low values of the reduced frequency. This is only possible at considerably high free stream speeds. Because of speed limitations and geometry, the reduced frequency values must be greater than 0.3, which makes the unsteady aerodynamic treatment necessary. If the unsteady aerodynamics is utilized, with the leading edge suction the propulsion efficiency becomes inversely proportional with the reduced frequency. For the vertically flapping thin airfoil the efficiency value is 90% for $k = 0.07$ and becomes 50% as k approaches infinity, (Garrick 1936). Using the Garrick's model for pitching and heaving-plunging airfoil, with certain phase lag between two degrees of freedom, it is possible to evaluate the lifting and the propulsive forces by means of strip theory. In addition, if we impose the span wise geometry and the elastic behavior of the wing to include the bending and torsional deflections, necessary power and the flapping moments are calculated for a sustainable flight, (DeLaurier and Harris 1993). While making these calculations, the dynamic stall and the leading edge separation effects are also considered. The progress made and the challenges faced in determining the propulsive forces obtained via wing flapping, including the strong leading edge separation studies, are summarized in an extensive work of Platzer et al. (2008) Exactly opposite usage of wing flapping in a pitch-plunge mode is for the purpose of power extraction through efficient wind milling. The relevant conditions of power extraction via pitch-plunge oscillations are discussed in a detail by Kinsey and Dumas (2008). More detailed information on proper applications of wing flapping will be given in the following chapters.

In recent years, the ground effect is studied extensively to determine changes in aerodynamic peformances of vehicles operating near ground. These changes are due to the distance to the ground and the angle of attack, acting in combination, as shown in the following equation for the lift of a thin airfoil in steady case:

$$C_l = 2\pi\alpha(1 - \alpha/h_g + 0.25/h_g^2) \tag{1.35}$$

For the unsteady case, the wake and its image plays an important role as shown in the following figure (Fig. 1.13).

The details of the derivations for unsteady cases concerning the effect of the ground on aerodynamic forces for airfoils and wings of various aspect ratios will be given at the end of Chap. 8.

Summary: *A brief but general review of Aerodynamics, starting from the basic definitions including the steady and unsteady aerodynamics notions are introduced. The famous Kutta-Joukowski theorem and as its consequence, the sectional lift coefficient is presented. After introducing the reduced frequency concept, four different types of aerodynamics; (i) unsteady, (ii) quasi-unsteady, (iii) quasi-steady, and (iv) steady aerodynamics are defined utilizing the Theodorsen function for a plunging flat plate. Then steady lift coefficient and the induced drag coefficient for a finite elliptical wing are given. Compressible steady and unsteady flows past two and three dimensional lifting surface are given from subsonic to supersonic flow range. The flow past slender bodies is briefly introduced to predict the stability derivatives of the missile like configurations. Hypersonic flows past blunt bodies are examined via Newtonian impact theory and piston theory is introduced for the hypersonic flows past thin surfaces. For the basis of modern subjects, the leading edge suction analogy is presented to analyze vortex lift generated by the leading edge separation for the low aspect ratio delta wings at high angle of attacks. The lateral stability considerations of vortical flows at higher angle of attacks lead us to observe the wing-rock phenomenon because of asymmetric vortex shedding from delta wings. Further angle of attack increase causes vortex bursting where the lift is no longer sustained. Finally, flapping wing aerodynamics is presented for the ornithopter technology of recent years applied to design and manufacture small aerial vehicles which are recently called Micro Air Vehicles, MAVs.*

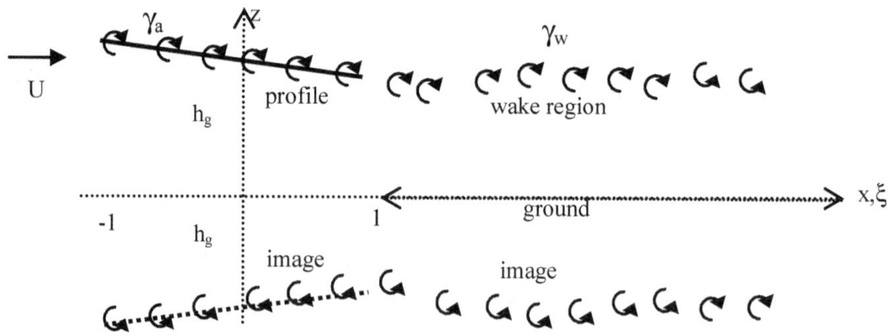

Fig. 1.13 Bound and image vorticies and their wake

1.12 Questions and Problems

1.1 Find the sectional lift coefficient for a thin symmetric airfoil with integrating the lifting pressure coefficient.

1.2 Find the sectional moment coefficient of a thin symmetric airfoil with respect to the mid chord. Then find (i) the center of pressure and (ii) the aerodynamic center of the airfoil considered.

1.3 Using the approximate expression of the Theodorsen function for the vertical motion of an airfoil given by $z_a(t) = \mathrm{h}\cos(\mathrm{ks})$ where $s = Ut/b$, find the sectional lift coefficient change and plot it for $k = 0.1$ and for s, with (i) Unsteady aerodynamics, (ii) Quasi unsteady aerodynamics, and (iii) Quasi steady aerodynamics.

1.4 The exact expression for the Theodorsen is $C(k) = \mathrm{H}_1^2(k)/[\mathrm{H}_1^2(k) + i\mathrm{H}_o^2(k)]$. Plot the real and imaginary parts of the Theodorsen function with respect to the reduced frequency for $0.01 < k < 5$.

1.5 The graph of the lift versus drag coefficient is called the drag polar. Plot a drag polar for a thin wing for incompressible flow.

1.6 Define the critical Mach number for subsonic flows. Describe how it is determined for an airfoil.

1.7 Plot the lift line slope change of a thin wing with respect to the aspect ratio.

1.8 Plot the lift line of a swept wing with a low aspect ratio using Diederich formula with respect to sweep angle for AR = 2,3,4.

1.9 Find the wave drag of an 8% thick biconvex airfoil at free stream Mach number of M = 2.

1.10 For a thin airfoil pitching simple harmonically about its leading edge, plot the amplitude and phase curves with respect to the reduced frequency at transonic regime.

1.11 Compare the amplitude of a sectional lift coefficient of a thin airfoil in vertical oscillation in transonic regime with the same airfoil oscillating in incompressible flow in terms of the reduced frequency.

1.12 By definition, if the change of the moment about the center of gravity of a slender body with respect to angle of attack is negative then the body is statically stable, Fig. 1.7. Comment on the position of the center of gravity and the tail shape as regards to the static stability of the body.

1.13 Compare the hypersonic surface pressure expression with the incompressible potential flow surface pressure of a flow past a circular cylinder. Comment on the validity of both surface pressures.

1.14 Find the surface pressure for the frontal region of the capsule during its reentry. Assume the shape of the frontal region as a half circle and comment on the region of validity of your result.

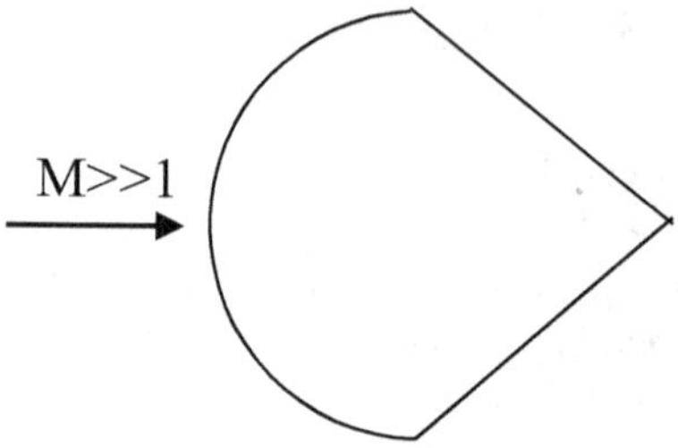

1.15 The Newtonian impact theory is valid at high angles of attack. The wall inclination for a blunt body gradually decreases along the free stream direction. For such cases, when this angle is less than 35° the Prandtl-Meyer expansion is applicable. Solve Problem 1.14 using the Newtonian impact theory together with the Prandtl-Meyer expansion to obtain the surface pressure for the half circle.

1.16 Find the amplitude of the surface pressure coefficient for a flat plate simple harmonically oscillating in hypersonic flow with amplitude $\bar{h}$. Define an interval for the hypersonic similarity parameter wherein validity of your answer is assured.

1.17 For the attached flows over slender delta wings, show that at low angles of attack Eqs. 1.11 and 1.33 are identical.

1.18 For a delta wing with a sharp leading edge separation plot the non dimensional potential K_p and vortex lift coefficient K_v changes with respect to the aspect ratio AR.

1.19 Explain why we need to resort to unsteady aerodynamic concepts for ornithopter studies.

References

Abbott, I.H., Von Doenhoff, A.E.: Theory of Wing Sections. Dover Publications Inc. New York (1959)

Anderson, J.D., Jr.: Fundamentals of Aerodynamics, 3rd edn. Mc-Graw Hill, Boston (2001)

Bisplinghoff, R.L., Ashley, H., Halfman, R.L.: Aeroelasticity. Dover Publications Inc., New York (1996)

DeLaurier, J.D, Harris, J.M.: A study of mechanical wing flapping wing flight. Aeronaut. J. (1993)

DeLauerier, J.D.: An aerodynamic model for flapping-wing flight. Aeronaut. J. April (1993)

Ericksson, L.E.: The fluid mechanics of slender wing rock. J. Aircraft **21**, 322–328 (1984)

Garrick, L.E.:Propulsion of a flapping and oscillating airfoil, NACA-TR 567 (1936)

Gibbs-Smith, C.H.: A History of Flight. Frederic A. Praeger Pub, New York (1954)

Hayes, W.D., Probstein, R.F.: Hypersonic Flow Theory, Inviscid Flows, Vol. 1, 2nd Ed., Academic Press, New York (1966)

Hitching, F.: The Neck of Giraffe. Pan Books, London (1982)

Kinsey, T, Dumas, G.: Parametric study of an oscillating airfoil in a power-extraction regime, AIAA J. **46**(6), June (2008)

Küchemann, D.: Aerodynamic Design of Aircraft. Pergamon Press, Oxford (1978)
Lieppmann, H.W., Roshko, A.: Elements of Gasdynamics. Wiley, New York (1963)
Mueller, T.J., DeLaurier, J.D.: Aerodynamics of small vehicles, Ann. Rev. Fluid Mech. (2003)
Platzer, M.F., Jones, K.D., Young, J., Lai, J.S.: Flapping-wing aerodynamics: progress and challenges, AIAA J. **46**(9) September (2008)
Polhamus, E.C.: Predictions of vortex-lift characteristics by a leading-edge suction analogy. J. Aircraft **8**, 193–199 (1971)

Chapter 2
Fundamental Equations

The mathematical models, which simulate the physics involved, are the essential tools for the theoretical analysis of aerodynamical flows. These mathematical models are usually based on the equations which are nothing but the fundamental conservation laws of mechanics. The conservation equations are usually satisfied locally as differential equations; therefore, their unique solution requires initial and boundary conditions which are described with the farfield conditions and the time dependent motion of the body. Let us follow the historical development of the aerodynamics, and start our analysis with potential flow theory. The potential theory will help us to determine the aerodynamic lifting force which is in the direction normal to the flight and necessary to balance the weight of the body in flight. Since the viscous forces are neglected in potential theory, the drag force which is in the direction of flight cannot be calculated. On the other hand, the potential theory can determine the lift induced drag for three dimensional flows past finite wings. Now, in order to perform our aerodynamical analysis let us introduce further definitions and the simplification of the equations for first, (A) The Potential Theory with its assumptions and limitations, and then for the (B) Real Gas Flow which covers all sorts of viscous effects and the effect of composition changes in the gas because of high altitude flows with high speeds.

A. Potential Flow.

2.1 Equation of Motion

Let us write the velocity vector **q** in Cartesian coordinates as **q** = u **i** + v **j** + w **k**. Here, u, v and w denotes the velocity components in x, y, z directions, and **i, j, k** shows the corresponding unit vectors. At this stage it is useful to define the following vector operators.

Ü. Gülçat, *Fundamentals of Modern Unsteady Aerodynamics*,
https://doi.org/10.1007/978-3-030-60777-7_2

The divergence of the velocity vector is given by

$$\text{div}\,\mathbf{q} = \nabla \cdot \mathbf{q} = \frac{\partial u}{\partial x} + \frac{\partial v}{\partial y} + \frac{\partial w}{\partial z}$$

and the curl

$$\text{curl}\,\mathbf{q} = \nabla \times \mathbf{q} = \begin{vmatrix} \mathbf{i} & \mathbf{j} & \mathbf{k} \\ \frac{\partial}{\partial x} & \frac{\partial}{\partial y} & \frac{\partial}{\partial z} \\ u & v & w \end{vmatrix}$$

The gradient of any function, on the other hand, reads as

$$\text{grad f} = \nabla \text{f} = \frac{\partial f}{\partial x}\mathbf{i} + \frac{\partial f}{\partial y}\mathbf{j} + \frac{\partial f}{\partial z}\mathbf{k}$$

The material or the total derivative as an operator is shown with

$$\frac{D}{Dt} = \frac{\partial}{\partial t} + u\frac{\partial}{\partial x} + v\frac{\partial}{\partial y} + w\frac{\partial}{\partial z}$$

Here, t denotes the time. Now, we can give the equations associated with the laws of classical mechanics.

$$\text{Equation of Continuity:} \frac{D\rho}{Dt} + \rho \nabla \cdot \mathbf{q} = 0 \tag{2.1}$$

$$\text{Momentum Equation:} \frac{D\mathbf{q}}{Dt} + \frac{1}{\rho}\nabla \text{p} = 0 \tag{2.2}$$

$$\text{Energy Equation:} \frac{D}{Dt}\left(\frac{a^2}{\gamma - 1} + \frac{q^2}{2}\right) = \frac{1}{\rho}\frac{\partial p}{\partial t} \tag{2.3}$$

$$\text{Equation of State :} \quad \text{p} = \rho \text{R T} \tag{2.4}$$

Here, the pressure is denoted with p, density with ρ, temperature with T, speed of sound with a, specific heat ratio with γ and the gas constant with R.

In addition, the air is assumed to be a perfect gas and the body and frictional forces are neglected. It is also assumed that no chemical reaction takes place during the motion. The energy equation is given in BAH (Bisplinghoff et al. 1996).

Let us now see the useful results of Kelvin's theorem under the assumptions made above, (Batchelor 1979). The following line integral on a closed path defines the

Circulation: $\Gamma = \oint q.ds$.

The Kelvin's theorem: $\frac{D\Gamma}{Dt} = -\oint \frac{dp}{\rho}$.

For incompressible flow or a barotropic flow where $p = p(\rho)$ the right hand side of Kelvin's theorem vanishes to yield

$$\frac{D\Gamma}{Dt} = 0.$$

This tells us that the circulation under these conditions remains the same with time. Now, let us analyze the flow with constant free stream which is the most referred flow case in aerodynamics. Since the free stream is constant then its circulation $\Gamma = 0$. The stokes theorem states that

$$\oint q.\,ds = \oiint \nabla \mathbf{x} \mathbf{q} \cdot \mathbf{dA} = 0 \tag{2.5}$$

The integrand of the double integral must be zero in order to have Eq. 2.5 equal to zero for arbitrary differential area element. This gives $\nabla \mathbf{x}\ \mathbf{q} = 0$.

$\nabla \mathbf{x}\ \mathbf{q} = 0$, on the other hand, implies that the velocity vector $\mathbf{q}$ can be obtained from the gradient of a scalar potential ϕ, i.e.

$$\mathbf{q} = \nabla\phi \tag{2.6}$$

At this stage if we expand the first term of the momentum equation into its local and convective derivative terms, and express the convective terms with its vector equivalent we obtain

$$\frac{\partial \mathbf{q}}{\partial t} + (\mathbf{q}\cdot\nabla)\mathbf{q} = -\frac{1}{\rho}\nabla \mathrm{p} \quad \text{and} \quad (\mathbf{q}\cdot\nabla)\mathbf{q} = \nabla\frac{q^2}{2} - \mathbf{q}\,\mathbf{x}(\nabla\,\mathbf{x}\,\mathbf{q}).$$

From Eq. 2.5 we obtained $\nabla \mathbf{x}\ \mathbf{q} = \mathbf{0}$. Utilizing this fact the momentum equation reads as

$$\frac{\partial \mathbf{q}}{\partial t} + \nabla\frac{q^2}{2} + \frac{1}{\rho}\nabla \mathrm{p} = 0 \tag{2.7}$$

Now, we can use the scalar potential ϕ in the momentum equation in terms of Eq. 2.6.

For a baratropic flow we have the 3rd term of Eq. 2.7 as

$$\frac{1}{\rho}\nabla \mathrm{p} = \nabla\int\frac{dp}{\rho}.$$

Then collecting all the terms of Eq. 2.7 together

$$\nabla \cdot \left(\frac{\partial \phi}{\partial t} + \frac{q^2}{2} + \int \frac{dp}{\rho} \right) = 0$$

we see that the scalar term under gradient operator is in general only depends on time, i.e,

$$\frac{\partial \phi}{\partial t} + \frac{q^2}{2} + \int \frac{dp}{\rho} = \mathrm{F}(t) \tag{2.8}$$

According to Eq. 2.8, F(t) is arbitrarily chosen, and if we set it to be zero we obtain the classical Kelvin's equation

$$\frac{\partial \phi}{\partial t} + \frac{q^2}{2} + \int \frac{dp}{\rho} = 0 \tag{2.9}$$

Let us try to write the continuity equation, Eq. 2.1, in terms of ϕ only,

$$\frac{\partial \rho}{\partial t} + (\mathbf{q} \cdot \nabla)\, \rho + \rho \nabla \cdot \mathbf{q} = 0 \tag{2.10}$$

The gradient of the velocity vector now reads as

$$\nabla . \mathbf{q} = \nabla^2 \phi.$$

Dividing Eq. 2.10 by ρ we obtain

$$\frac{1}{\rho} \frac{\partial \rho}{\partial t} + \left(\frac{\mathbf{q}}{\rho} \cdot \nabla \right) \rho + \nabla^2 \phi = 0 \tag{2.11}$$

Note that Eq. 2.11 becomes the Laplace equation for incompressible flow

$$\nabla^2 \phi = 0 \tag{2.12}$$

We know that Laplace equation by itself is independent of time. The time dependent boundary conditions make us seek the time dependent solutions of Eq. 2.12.

Now, we can obtain the simplified version of Eq. 2.11 for the compressible flows. Let us rearrange Kelvin's equation, Eq. 2.9 in following form

$$\frac{\partial}{\partial t} \int \frac{dp}{\rho} = - \frac{\partial}{\partial t} \left(\frac{\partial \phi}{\partial t} + \frac{q^2}{2} \right)$$

and the integral on the left hand side can be differentiated to give

$$\frac{\partial}{\partial t}\int\frac{dp}{\rho}=\frac{1}{\rho}\frac{\partial p}{\partial \rho}\frac{\partial \rho}{\partial t} \tag{2.13}$$

In Eq. 2.13, the speed of sound is related to the pressure and density changes: $\frac{\partial p}{\partial \rho}=a^2$.

Hence, we obtain the following for the first term of the Eq. 2.11

$$\frac{1}{\rho}\frac{\partial \rho}{\partial t}=-\frac{1}{a^2}\frac{\partial}{\partial t}\left(\frac{\partial \phi}{\partial t}+\frac{q^2}{2}\right) \tag{2.14}$$

Now, let us write Eq. 2.7 in terms of ϕ and the pressure gradient. Furthermore, expressing the pressure gradient in terms of the density gradient and the local speed of sound we obtain

$$\frac{1}{\rho}\nabla \mathrm{p}=\frac{a^2}{\rho}\nabla \rho=-\nabla\left(\frac{\partial \phi}{\partial t}+\frac{q^2}{2}\right)$$

and with the aid of 2.14 and the multiplying term $\mathbf{q}/a^2$, the final form of Eq. 2.11 reads as

$$\nabla^2\phi-\frac{1}{a^2}\left(\frac{\partial^2\phi}{\partial t^2}+\frac{\partial q^2}{\partial t}+\mathbf{q}\cdot\nabla\frac{q^2}{2}\right)=0 \tag{2.15}$$

In Eq. 2.15, we express the velocity vector in terms of the velocity potential. This way, the scalar non linear equation has the scalar function as the only unknown except the speed of sound. The equation itself models many kinds of aerodynamic problems. We need to impose, however, the boundary conditions in order to model a specific problem.

2.2 Boundary Conditions

Equation 2.15 as a fundamental equation is solved with the proper boundary conditions. In general the external flow problems will be studied. Therefore, we need to impose the boundary conditions accordingly as follows.

(i) At infinity, all disturbances must die out and only free stream conditions prevail.
(ii) The time dependent boundary conditions at the body surface must be given as the time dependent motion of the body.

The equation of a surface for a 3-D moving body in Cartesian coordinate system is given as follows

$$B(x, y, z, t) = 0 \tag{2.16}$$

Let us take the material derivative of this surface in the flow field $\mathbf{q} = u\,\mathbf{i} + v\,\mathbf{j} + w\,\mathbf{k}$.

$$\frac{DB}{Dt} = \frac{\partial B}{\partial t} + u\frac{\partial B}{\partial x} + v\frac{\partial B}{\partial y} + w\frac{\partial B}{\partial z} = 0 \tag{2.17}$$

For the steady flow it simplifies to

$$u\frac{\partial B}{\partial x} + v\frac{\partial B}{\partial y} + w\frac{\partial B}{\partial z} = 0$$

The external flows studied here require to find the pressure distribution at the lower and upper surfaces of the body immersed in a free stream. For this purpose, we need to know the upper and lower surface equations of a body in a free stream in x direction. If we show the direction normal to the flow with z, then the single valued surface equation, with the aid of Eq. 2.16, reads as

$$B(x, y, z, t) = z - z_a(x, y, t) = 0 \tag{2.18}$$

Now, we can take the material derivative of Eq. 2.18 with the aid of Eq. 2.17

$$w = \frac{\partial z_a}{\partial t} + u\frac{\partial z_a}{\partial x} + v\frac{\partial z_a}{\partial y} \tag{2.19}$$

Note that, $\frac{\partial B}{\partial z} = 1$ is used for the convective term in z direction. Here, the explicit expression of vertical velocity component w is named 'downwash' in aerodynamics. This downwash at the near wake is the indicative of the lifting force on the body. The direction of the force and the downwash are the same but their senses are opposite. Accordingly, for the downward downwash the force is then upward. In other words, downward velocity component at the wake region creates a clockwise circulation which in turn generates the lifting force together with the free stream.

Equations 2.15 and 2.19 are not linear. In order to solve those equations together, linearization is necessary. Once the equations are linearized we can also employ the superpositioning technique for solving them.

2.3 Linearization

Let us begin the linearization process with the boundary conditions. The small perturbations approach will be used here. Accordingly, let U be the free stream speed in positive x direction, Fig. 2.1.

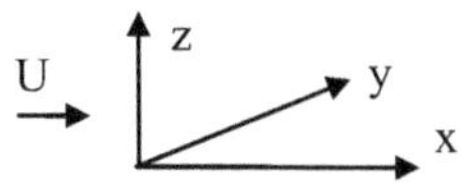

Fig. 2.1 Coordinate system and the free stream U

Let u' be the perturbation velocity component in x direction which makes the total velocity component in x direction: $u = U + u'$. In addition, defining function ϕ' as the perturbation potential gives us the relation between the two potentials as follows: $\phi = \phi' + U\,x$. As a result, we can write the relation between the perturbation potential and the velocity components in following form

$$\frac{\partial \phi'}{\partial x} = u', \quad \frac{\partial \phi'}{\partial y} = v \quad \text{and} \quad \frac{\partial \phi'}{\partial z} = w.$$

The small perturbation method is based on the assumption that the perturbation speeds are quite small compared to the freestream speed, i.e. u', v, $w \ll U$. In addition, because of thin wing theory the slopes of the body surface are small therefore we can write

$$\frac{\partial z_a}{\partial x} \ll 1 \quad \text{and} \quad \frac{\partial z_a}{\partial y} \ll 1$$

Then the boundary condition 2.19 become

$$w = \frac{\partial z_a}{\partial t} + U\,\frac{\partial z_a}{\partial x} + u'\,\frac{\partial z_a}{\partial x} + v\frac{\partial z_a}{\partial y} \quad \text{where } u'\frac{\partial z_a}{\partial x}, v\frac{\partial z_a}{\partial y} << U\frac{\partial z_a}{\partial x}$$

which gives the approximate expression for the boundary condition

$$w = \frac{\partial z_a}{\partial t} + U\,\frac{\partial z_a}{\partial x} \tag{2.20}$$

Equation 2.20 is valid at angles of attack less than 12° for thin airfoils whose thickness ratio is less than 12%. For the upper and lower surfaces, the linearized downwash expression will be denoted as follows.

$$\text{Upper surface } (u): \quad w = \frac{\partial z_u}{\partial t} + U\frac{\partial z_u}{\partial x}; \quad z = 0^+$$

$$\text{Lower surface } (l): \quad w = \frac{\partial z_l}{\partial t} + U\frac{\partial z_l}{\partial x}; \quad z = 0^-.$$

Now, let us obtain an expression for the linearized surface pressure coefficient. For this purpose we are going to utilize the linearized version of Eq. 2.8. The second term of the equation is linearized as follows

$$\frac{q^2}{2} \cong \frac{U^2}{2} + 2U\,u'$$

For the right hand side of Eq. 2.8 if we arbitrarily choose F(t) = U^2/2 then the term with the integral reads as

$$\int \frac{dp}{\rho} = -\frac{\partial \phi}{\partial t} - 2Uu'.$$

The relation between the velocity potential and the perturbation potential gives: $\frac{\partial \phi}{\partial t} = \frac{\partial \phi'}{\partial t}$. If we now evaluate the integral from the free stream pressure value p_∞ to any value p and omit the small perturbations in pressure and in density we obtain

$$\int_{p_\infty}^{p} \frac{dp}{\rho} \cong \frac{p - p_\infty}{\rho_\infty} = -\left(\frac{\partial \phi'}{\partial t} + \mathrm{U}\frac{\partial \phi'}{\partial x}\right)$$

Using the definition of pressure coefficient

$$C_p = \frac{p - p_\infty}{\frac{1}{2}\rho_\infty U^2} = -\frac{2}{U^2}\left(\frac{\partial \phi'}{\partial t} + \mathrm{U}\frac{\partial \phi'}{\partial x}\right) \tag{2.21}$$

Here, the pressure coefficient is expressed in terms of the perturbation potential only.

Example: Let the equation of the surface of a body immersed in a free stream U be

$$z_{u,l} = \pm a\sqrt{\frac{x}{l}}\,(0 \le x \le l)$$

If this body pitches about its nose simple harmonically with a small amplitude, find the downwash at the upper and the lower surfaces of the body in terms of a, l and the amplitude and the frequency of the oscillatory motion.

Answer: Let $\alpha = \bar{\alpha} \sin \omega t$ ($\bar{\alpha}$: small amplitude and ω: angular frequency) be the pitching motion, let x, z be the stationary coordinate and x', z' be the moving coordinate system attached to the body. The relation between the fixed and the moving coordinate system is given by Fig. 2.2 in terms of α.

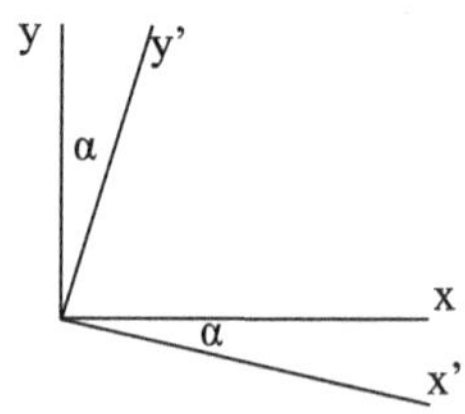

Fig. 2.2 α pitch angle and the coordinate systems

The coordinate transformation gives

$$x' = x\cos\alpha - z\sin\alpha$$
$$z' = x\sin\alpha + z\cos\alpha$$

In body fixed coordinates the surface equations $z'_{u,l} = \pm a\sqrt{\frac{x'}{l}}\ \ (0 \le x \le l)$

In terms of the stationary coordinate system

B(x,z,t) = $z' - z'_{u,l}(x') = x\sin\alpha + z\cos\alpha \mp a\left(\frac{x\cos\alpha - z\sin\alpha}{l}\right)^{1/2}$ for small α $\sin\alpha$ $\cong \alpha$ *and* $\cos\alpha \cong 1$. Then B(x,z,t) = $x\alpha + z \mp a\left(\frac{x - z\alpha}{l}\right)^{1/2}$.

Equation 2.17 gives

$$w_{u,l} = -\{x\,\dot{\alpha} \pm \frac{a\,\dot{\alpha}\,z}{2l}\left(\frac{x - z\alpha}{l}\right)^{-1/2} + U[\alpha \mp \frac{a}{2l}\left(\frac{x - z\alpha}{l}\right)^{-1/2}]\}/[1 \pm \frac{a\,\alpha}{2l}\left(\frac{x - z\alpha}{l}\right)^{-1/2}]$$

Here $\dot{\alpha} = \omega\bar{\alpha}\cos\omega$ t. Now, let us express the downwash for t = 0.

$w_{u,l} = -[\bar{\alpha}\omega\,x \pm a\bar{\alpha}\omega\frac{z}{l}\left(\frac{x}{l}\right)^{-1/2} \mp \frac{Ua}{2l}\left(\frac{x}{l}\right)^{-1/2}]$. If we divide both sides with U and divide x and z with l the non dimensional form of the downwash expression becomes

$$\frac{w_{u,l}}{U} = -[\bar{\alpha}\,l\omega\frac{x}{Ul} \pm \frac{a}{Ul}\bar{\alpha}\,l\omega\frac{z}{l}\left(\frac{x}{l}\right)^{-1/2} \mp \frac{a}{2l}\left(\frac{x}{l}\right)^{-1/2}].$$

If we write the reduced frequency: $k = \frac{\omega l}{U}$, and the nondimensional coordinates $a^* = \frac{a}{l}$: $x^* = \frac{x}{l}$ *ve* $z^* = \frac{z}{l}$, new form of the downwash becomes

$$\frac{w_{u,l}}{U} = -[\bar{\alpha}\,k\,x^* \pm a^*\,\bar{\alpha}\,k\,z^*\,(x^*)^{-1/2} \mp \frac{a^*}{2}(x^*)^{-1/2}].$$

In the last expression, the first two terms are time dependent and the last term is the term due to the steady flow.

Now, we can linearize Eq. 2.15 for the scalar potential with small perturbation approach. The nonlinear terms are the second and third terms in parentheses. The velocity vector in the second term is $\mathbf{q} = \mathrm{Ui} + \nabla\phi' = \mathrm{Ui} + u'\mathbf{i} + v\mathbf{j} + w\mathbf{k}$

$$\frac{\partial q^2}{\partial t} = 2\mathbf{q}.\frac{\partial \mathbf{q}}{\partial t} = 2(U\mathbf{i} + \nabla\phi')\cdot\frac{\partial}{\partial t}(U\mathbf{i} + \nabla\phi')$$

If we include the time dependent derivative under the gradient operator we obtain

$$2\mathbf{q}\cdot\frac{\partial \mathbf{q}}{\partial t} = 2(U\mathbf{i}+u'\mathbf{i}+v\mathbf{j}+w\mathbf{k})\cdot\left(\frac{\partial^2\phi'}{\partial t\partial x}\mathbf{i}+\frac{\partial^2\phi'}{\partial t\partial y}\mathbf{j}+\frac{\partial^2\phi'}{\partial t\partial z}\mathbf{k}\right)$$
$$= 2(\mathrm{U}+u')\frac{\partial u'}{\partial t}+2v\frac{\partial v}{\partial t}+2w\frac{\partial w}{\partial t}$$

Ignoring the second order perturbation terms, the approximate but linear form of the time derivative of the velocity reads

$$\frac{\partial q^2}{\partial t} \cong 2U\frac{\partial u'}{\partial t} = 2U\frac{\partial^2\phi'}{\partial t\partial x} \tag{2.22}$$

Now, let us linearize the third term in parentheses

$$\mathbf{q}\cdot\nabla\frac{q^2}{2} = (U\mathbf{i}+\nabla\phi')\cdot\nabla\left(\frac{U^2}{2}+U\mathbf{i}\cdot\nabla\phi'+\nabla\frac{\phi'^2}{2}\right)$$
$$= (U+u')\left(U\frac{\partial u'}{\partial x}+u'\frac{\partial u'}{\partial x}+v\frac{\partial v}{\partial x}+w\frac{\partial w}{\partial x}\right)+v\left(U\frac{\partial u'}{\partial y}+u'\frac{\partial u'}{\partial y}+v\frac{\partial v}{\partial y}+w\frac{\partial w}{\partial y}\right)+$$
$$+w(U\frac{\partial u'}{\partial z}+u'\frac{\partial u'}{\partial z}+v\frac{\partial v}{\partial z}+w\frac{\partial w}{\partial z})$$

Neglecting the second and third order terms, the approximate convective term reads

$$\mathbf{q}\cdot\nabla\frac{q^2}{2} \cong U^2\frac{\partial u'}{\partial x} = U^2\frac{\partial^2\phi'}{\partial x^2} \tag{2.23}$$

Remembering $\frac{\partial^2\phi}{\partial t^2} = \frac{\partial^2\phi'}{\partial t^2}$ with the aid of Eq. 2.22 and 2.23 Eq. 2.15 becomes d

$$\nabla^2\phi' - \frac{1}{a^2}\left(\frac{\partial^2\phi'}{\partial t^2}+2U\frac{\partial^2\phi'}{\partial t\partial x}+U^2\frac{\partial^2\phi'}{\partial x^2}\right) = 0$$

If we write second term in the form of an operator square we obtain

$$\nabla^2\phi' - \frac{1}{a^2}\left(\frac{\partial}{\partial t}+U\frac{\partial}{\partial x}\right)^2\phi' = 0 \tag{2.24a}$$

In Eqs. 2.15 and 2.24a, one of the non linear quantities is the square of the local speed of sound a^2, which will be linearized next, to give us totally linear potential.

Let us start the linearization with the energy equation, Eq. 2.3 given in (Liepmann and Roshko 1963). The energy equation:

$$\frac{D}{Dt}\left(\frac{a^2}{\gamma-1}+\frac{q^2}{2}\right)=\frac{1}{\rho}\frac{\partial p}{\partial t}$$

Writing the material derivative at the left hand side of the equation in its approximate form reads

$$\frac{D}{Dt}\,(\frac{a^2}{\gamma-1}+\frac{q^2}{2})=(\frac{\partial}{\partial t}+U\frac{\partial}{\partial x})\,(\frac{a^2}{\gamma-1}+\frac{q^2}{2})$$

If we take the time derivative of the Kelvin's equation, Eq. 2.9, for the integral term we get

$$\frac{\partial}{\partial t}\int\frac{dp}{\rho}=\frac{\partial}{\partial t}\int f(p)\,dp=\frac{\partial F(p)}{\partial t}=\frac{dF(p)}{dp}\frac{\partial p}{\partial t}=\frac{1}{\rho}\frac{\partial p}{\partial t}=-\frac{\partial^2\phi}{\partial t^2}-\frac{1}{2}\frac{\partial q^2}{\partial t}$$

With the last line the energy equation reads

$$\left(\frac{\partial}{\partial t}+U\frac{\partial}{\partial x}\right)\left(\frac{a^2}{\gamma-1}+\frac{q^2}{2}\right)=-\frac{\partial^2\phi}{\partial t^2}-\frac{1}{2}\frac{\partial q^2}{\partial t}$$

Rearranging the equation gives

$$\left(\frac{\partial}{\partial t}+U\frac{\partial}{\partial x}\right)\left(\frac{a^2}{\gamma-1}\right)+\frac{\partial^2\phi}{\partial t^2}=-\frac{\partial q^2}{\partial t}-\frac{U}{2}\frac{\partial q^2}{\partial x}$$

If we take the derivative of the right hand side of the last equation we obtain

$$\begin{aligned}-2\,\mathbf{q}.\tfrac{\partial \mathbf{q}}{\partial t}-U\,\mathbf{q}.\tfrac{\partial \mathbf{q}}{\partial x}&=-2\,(\,\mathrm{U}\,+u')\tfrac{\partial u'}{\partial t}-2\,v\tfrac{\partial v}{\partial t}-2w\tfrac{\partial w}{\partial t}\\&\quad-U(U+u')\tfrac{\partial u'}{\partial x}-U\,v\tfrac{\partial v}{\partial x}-U\,w\tfrac{\partial w}{\partial x}\\&\cong-2U\tfrac{\partial u'}{\partial t}-U^2\tfrac{\partial u'}{\partial x}\end{aligned}$$

Now, the energy equation reads as

$$(\frac{\partial}{\partial t}+U\frac{\partial}{\partial x})\,(\frac{a^2}{\gamma-1})=-(\frac{\partial}{\partial t}+U\frac{\partial}{\partial x})^2\phi' \tag{2.24b}$$

Let us denote the perturbation of the local speed of sound as $a=a_\infty+a'$, and multiply the energy equation with $(\gamma-1)/a_\infty^2$

$$\begin{aligned}-\tfrac{\gamma-1}{a_\infty^2}(\tfrac{\partial}{\partial t}+U\tfrac{\partial}{\partial x})^2\phi'&=(\tfrac{\partial}{\partial t}+U\tfrac{\partial}{\partial x})\,(1+\tfrac{a'}{a_\infty})^2=(\tfrac{\partial}{\partial t}+U\tfrac{\partial}{\partial x})\,[2\tfrac{a'}{a_\infty}+(\tfrac{a'}{a_\infty})^2]\\&\cong(\tfrac{\partial}{\partial t}+U\tfrac{\partial}{\partial x})\,[2\tfrac{a'}{a_\infty}]\end{aligned}$$

Here, $(\frac{a'}{a_\infty})^2 << 1$ is assumed. The linearization process has then given

$$\frac{a'}{a_\infty} = -\frac{\gamma - 1}{2\,a_\infty^2}(\frac{\partial}{\partial\,t} + U\frac{\partial}{\partial x})\phi'$$

The presence of speed of sound at the denominator of the right hand side of the last line implies that the perturbation speed of sound is very small compared to the free stream speed of sound. Therefore, it can be neglected near the free stream speed of sound to give approximate value of the local speed of sound as the free stream speed of sound. Hence, the final form of the linearized potential flow equation reads as

$$\nabla^2\phi' - \frac{1}{a_\infty^2}\left(\frac{\partial}{\partial t} + U\frac{\partial}{\partial x}\right)^2\phi' = 0 \tag{2.24}$$

2.4 Acceleration Potential

Another useful potential function which is used in aerodynamics is the acceleration potential. If we recall the momentum equation for barotropic flows:

$$\frac{D\mathbf{q}}{\mathrm{Dt}} = -\nabla\int\frac{dp}{\rho}$$

As seen in the left hand side of the equation, the material derivative of the velocity vector is obtained from the gradient of a function of pressure and density only. Hence, we can define the acceleration potential as follows

$$\frac{D\mathbf{q}}{\mathrm{D}t} = \nabla\,\psi.$$

As a result of last line the momentum equation reads as,

$$\nabla\,\psi + \nabla\int\frac{dp}{\rho} = 0$$

The integral form of the last equation becomes

$$\psi = -\int\frac{dp}{\rho} + \mathrm{F}(t)$$

The pressure term integrated at the right hand side of the equation from free stream to the point under consideration gives,

$$\psi = \frac{p_\infty - p}{\rho}$$

Because of the direct relation between the pressure and the acceleration potential, this potential is also called the pressure integral. Let us rewrite the Kelvin's equation in gradient form

$$\nabla\left[\frac{\partial\phi}{\partial t} + \frac{q^2}{2} + \int\frac{dp}{\rho}\right] = 0.$$

We can now find the relation between the velocity potential an the acceleration potential as follows

$$\nabla\left[\frac{\partial\phi}{\partial t} + \frac{q^2}{2} - \psi\right] = 0.$$

The integral of the last equation

$$\frac{\partial\phi}{\partial t} + \frac{q^2}{2} - \psi = \mathrm{F}(t)$$

Once again if we choose $\mathrm{F}(t) = \mathrm{U}^2/2$ we can satisfy the flow conditions at infinity. Hence, the acceleration potential becomes,

$$\psi = \frac{\partial\phi}{\partial t} + \frac{q^2}{2} - \frac{U^2}{2}$$

With small perturbation approach, the linear form of the last line reads

$$\psi = \left(\frac{\partial}{\partial t} + U\frac{\partial}{\partial x}\right)\phi' \tag{2.25}$$

If the linear operator

$$\left(\frac{\partial}{\partial t} + U\frac{\partial}{\partial x}\right)$$

operates on Eq. 2.24 to give

$$\left(\frac{\partial}{\partial t} + U\frac{\partial}{\partial x}\right)\left[\nabla^2\phi' - \frac{1}{a_\infty^2}\left(\frac{\partial}{\partial t} + U\frac{\partial}{\partial x}\right)^2\phi'\right] = 0,$$

Interchanging the operators and utilizing Eq. 2.25 gives us the final form of the equation for the acceleration potential

$$\left[\nabla^2\psi - \frac{1}{a_\infty^2}\left(\frac{\partial}{\partial t} + U\frac{\partial}{\partial x}\right)^2 \Psi\right] = 0 \tag{2.26}$$

2.5 Moving Coordinate System

The linearized equations which are obtained previously enable us to analyze aerodynamical problems more conveniently. Let us now elaborate on the coordinate systems which will further simplify the equations. The type of external flows we study usually considers a constant free stream velocity U at the far field. The reference frame used for this type analysis is a body fixed coordinate system which moves in the negative x direction with velocity U. Another type of analysis is based on the moving reference system which moves with the free stream. With this type analysis, the form of the equations looks simpler to handle. Let us write Eq. 2.24 in the moving coordinate system which moves with the free stream. Let x, y, z be the body fixed coordinate system and, $x^{'}$, $y^{'}$, $z^{'}$ be the flow fixed coordinate system. As seen from Fig. 2.3, since the free stream velocity is U, after the time interval t the flow fixed coordinate system translates in x direction by an amount Ut.

The relation between the two coordinate system reads as

$$x' = x - U\,t, \quad y' = y, \quad z' = z, \quad t' = t.$$

The derivative with respect to t' becomes

$$\frac{\partial}{\partial t'} = \frac{\partial}{\partial t} + \frac{\partial}{\partial x'}\frac{\partial x'}{\partial t} = \frac{\partial}{\partial t} + \frac{\partial}{\partial x'}(-U)$$

Here, $\frac{\partial x'}{\partial t} = -U.$

The partial derivatives with respect to body fixed coordinates in terms of the flow fixed coordinates then become:

$$\frac{\partial}{\partial t} + U\frac{\partial}{\partial x} = \frac{\partial}{\partial t'} \quad \frac{\partial}{\partial x} = \frac{\partial}{\partial x'} \quad \frac{\partial}{\partial y} = \frac{\partial}{\partial y'} \quad \frac{\partial}{\partial z} = \frac{\partial}{\partial z'}$$

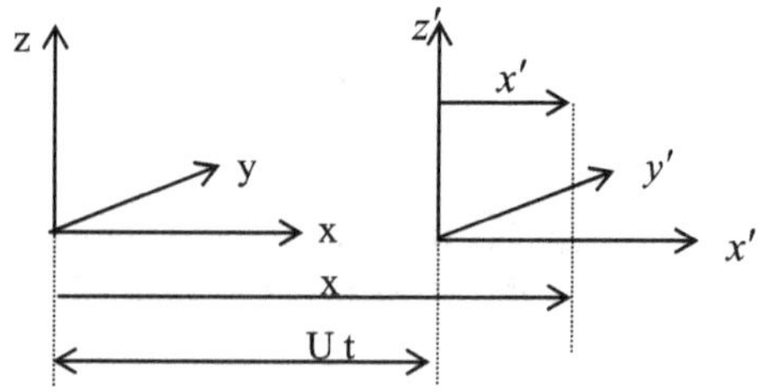

Fig. 2.3 Body fixed x, y, z and the flow fixed x', y', z' coordinate systems

Equation 2.24 in the flow fixed coordinate system reads as

$$\nabla^2\phi' - \frac{1}{a_\infty^2}\frac{\partial^2\phi'}{\partial t'^2} = 0$$

The last equation is in the form of the classical wave equation whose solutions are well known in mathematical physics. The boundary conditions and the pressure coefficient expressions, Eq. 2.20 and 2.21, become:

Boundary condition: $w = \frac{\partial z_a}{\partial t'}$.

Pressure coefficient: $C_p = -\frac{2}{U^2}\frac{\partial \phi'}{\partial t'}$.

Summary:

Hitherto, we have given the linearized form of the potential equations which are applicable to various problems of classical aerodynamics. In order for these equations to be valid in our modeling, the following assumptions must be true:

1. The air is considered as a perfect gas.
2. Mass, momentum and the energy conservations are used.
3. Body forces, viscous forces and the chemical reactions are ignored.
4. The flowfield is assumed to be either incompressible or barotropic.
5. The slopes of the body surfaces and all the flowfield perturbations are assumed to be small.
6. The time rate of change of the flow parameters are assumed to be small.

In addition, the linearized form of the compressible flow is only valid for subsonic and supersonic flows. The nonlinear approaches for the transonic and the hypersonic flows will be seen separately in relevant chapters.

B. Real Gas Flow.

The real gas flow equations are free of all the restrictions given above. Therefore, they are first introduced in their weak form, integral form, in terms of the system and control volume approaches.

2.6 System and Control Volume Approaches

Let **V**(x,y,z,t) be the velocity vector field given in a stationary space coordinate system x, y, z and time coordinate t. Shown in Fig. 2.4 is the closed system composed of air coalescing with a control volume at time t. The control volume remains the same at time $t + \Delta t$ the system, however, as the collection of same particles, moves and deforms with the flow as shown in Fig. 2.4.

Let N be the total thermodynamical property in our system. Because of the flow field, there will be a change with time in the property N as DN/Dt. Let η be the specific and local value of property N, which is distributed throughout the control volume. The total value of this property can be represented as an integral as

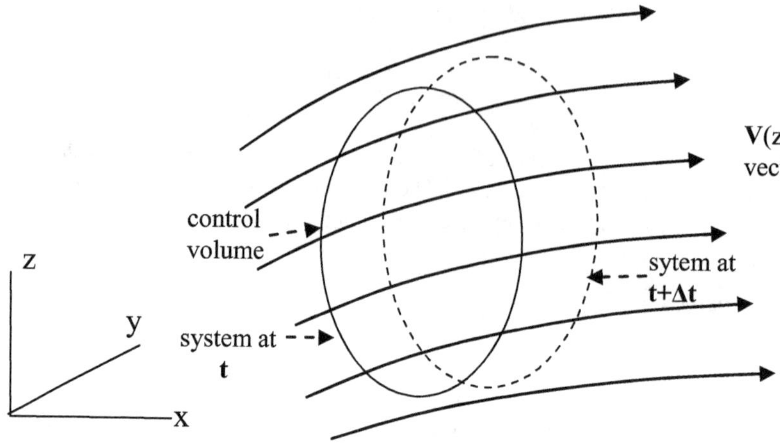

Fig. 2.4 The velocity vector field **V**(x,y,z,t), the system and the control volume

follows: N = $\iiint \eta\rho\, d\forall$. Here, $d\forall$ shows the infinitesimal volume element in the control volume. Now, we can relate the time rate of change of property η in the control volume in terms of its flux through the control surface as the control volume coincides with the system as Δt approaches zero. Under this condition, the flux of η from the control surface will be $\oiint \eta\rho(\vec{V}.d\vec{A})$, (Fox and McDonald 1992). If we consider the limiting case as the system coinciding with the control volume, the total derivative of the property N in the system can be related to the control volume as follows

$$\frac{DN}{Dt} = \frac{\partial}{\partial t}\iiint \eta\rho\, d\forall + \oiint \eta\rho(\vec{V}.d\vec{A}) \tag{2.27}$$

where $\mathbf{V} = \vec{V}$. Now, we can apply the conservation laws of mechanics to Eq. 2.27 and obtain the strong forms of the governing equations.

2.7 Global Continuity and the Continuity of the Species

Continuity equation: If M is the total mass in the system then N = M and for the system.

DN/Dt = DM/Dt = 0. In addition, since η = M/N = 1 Eq. 2.27 reads

$$0 = \frac{\partial}{\partial t}\iiint \rho\, d\forall + \oiint \rho(\vec{V}.d\vec{A}). \tag{2.28}$$

Using the divergence theorem, the second term at the right hand side of Eq. 2.28 reads as (Hildebrand 1976),

$$\iiint \nabla.(\rho\vec{V})\,d\forall = \oiint \rho(\vec{V}.d\vec{A}) \tag{2.29}$$

The new form of Eq. 2.28 becomes

$$\frac{\partial}{\partial t}\iiint \rho\,d\forall + \iiint \vec{\nabla}.(\rho\vec{V})\,d\forall = \iiint (\frac{\partial \rho}{\partial t} + \vec{\nabla}.(\rho\vec{V}))d\forall = 0 \tag{2.30}$$

In Eq. 2.30, the control volume does not change with time therefore, the time derivative can be taken into inside of the first term without causing any alteration. Since the volume element d $\forall$ is arbitrary and different from 0, to satisfy Eq. 2.30 the integrand must be zero to give the differential form, strong form, of the continuity equation.

$$\frac{\partial \rho}{\partial t} + \vec{\nabla}.(\rho\vec{V}) = 0 \tag{2.31}$$

At high temperatures when the real gas effects take place, the air starts to disassociate and chemical reactions create new species. Because of this, we may need to write continuity of the species for each specie separately. If we consider specie i whose density is ρ_i and its production rate is $\dot{w}_i$ in a control volume, then we have to have a source term at the left hand side of Eq. 2.27.

$$\iiint \dot{w}_i\,d\forall = \frac{\partial}{\partial t}\iiint \rho_i\,d\forall + \oiint \rho_i(\vec{V}_i.d\vec{A}) \tag{2.32}$$

Here, the velocity $\mathbf{V}_i$ is the mass velocity of specie i. The differential form of Eq. 2.32 reads as

$$\frac{\partial \rho_i}{\partial t} + \vec{\nabla}.(\rho_i\vec{V}_i) = \dot{w}_i \tag{2.33}$$

Defining the mass fraction or the concentration of a specie with $c_i = \rho_i/\rho$, the total density then becomes $\rho = \Sigma\, c_i\, \rho_i$. The mass velocity $\mathbf{V}_i$ of a specie in a mixture is related with the global velocity as follows: $\mathbf{V} = \Sigma\, c_i\, \mathbf{V}_i$. A mass velocity of a specie is found with adding its diffusion velocity $\mathbf{U}_i$ to the global velocity $\mathbf{V}$ i.e., $\mathbf{V}_i = \mathbf{V} + \mathbf{U}_i$. According to the Ficks law of diffusion, the diffusion speed of a specie is proportional with its concentration. If we denote the proportionality constant with D_{mi} the diffusion velocity of i reads

$$\vec{U}_i = -c_i D_{mi} \vec{\nabla} c_i \tag{2.34}$$

If we combine Eq. 2.34 with 2.31 and use it in 2.31, we obtain the continuity of the species in terms of their concentrations as follows, (Anderson 1989),

$$\rho \frac{D c_i}{D t} = \vec{\nabla}.(\rho D_{mi} \vec{\nabla} c_i) + \dot{w}_i \tag{2.35}$$

2.8 Momentum Equation

The Newton's second law of motion, based on the conservation of momentum, is applicable only on the systems. According to this law, the forces acting on the system cause a change in their momentum. For a system which is not under the influence of any non-inertial force, let $\mathbf{F}_S$ be the surface force acting at time t. This surface force changes the $\mathbf{N} = \mathbf{MV}$ momentum of the system. Here, if we let the momentum be independent of mass, then we find for the relevant property $\eta = \mathbf{N}/\mathrm{M} = \mathbf{V}$. We can now write the balance between the surface forces and the corresponding moment changes at the system which coincides with the control volume at time t.

$$\vec{F}_S = \frac{\partial}{\partial t} \iiint \rho \vec{V} \, d\forall + \oiint \rho \vec{V} (\vec{V}.d\vec{A}) \tag{2.36}$$

The forces at the surface of the system can be considered as the integral effect of the stress tensor $\boldsymbol{\tau}$ over the entire surface of the control volume: $\vec{F}_S = \oiint \vec{\tau}.d\vec{A}$. If we use this on the left hand side of Eq. 2.36 and change the surface integrals to volume integrals with the aid of divergence theorem we obtain

$$\iiint \vec{\nabla}.(\vec{\vec{\tau}}) \, d\forall = \frac{\partial}{\partial t} \iiint \rho \vec{V} \, d\forall + \iiint \vec{\nabla}.(\rho \vec{V}\vec{V}) \, d\forall \tag{2.37}$$

Here, the double arrow and the velocity vector multiplied by itself indicate the tensor quantities. Equation 2.37 can also be expressed in differential form to give the local expression of the momentum equation as

$$\frac{\partial \rho \vec{V}}{\partial t} + \vec{\nabla}.(\rho \vec{V}\vec{V} - \vec{\vec{\tau}}) = 0 \tag{2.38}$$

In Eq. 2.38, the stress tensor includes in itself the pressure, velocity gradient and for the turbulent flows the Reynolds stresses and reads like

$$\vec{\vec{\tau}} = (-p + \lambda \vec{\nabla}.\vec{V})\vec{\vec{I}} + \mu\, sim\vec{V} - <\rho \vec{v}'\vec{v}'> \tag{2.39}$$

Here, **I** is the unit tensor and sim**V** is the symmetric part of the gradient of the velocity vector. According to Stoke's hypothesis, the coefficient $\lambda = -2/3\ \mu$, wherein the average viscosity of the species is denoted by μ. Eq. 2.38 is valid only for the inertial reference frame. If we include the inertial forces, we consider a control volume in a local non-inertial coordinate system xyz accelerating in a fixed reference frame XYZ. Let the non-inertial coordinate system xyz move with a linear acceleration **R**" and rotate with angular speed **Ω** and the angular acceleration **Ω'** in the fixed coordinate system XYZ as shown in Fig. 2.5.

Let the control volume in Fig. 2.5 be attached to the non-inertial frame of reference xyz. The infinitesimal mass element $\rho\, d\forall$ considered in the control volume in the fixed reference frame XYZ has the acceleration $\mathbf{a}_{XYZ}$. At this stage, the relation between the acceleration $\mathbf{a}_{xyz}$ in the non-inertial frame and the acceleration $\mathbf{a}_{XYZ}$ in the inertial frame in terms of linear acceleration: **R**", Coriolis **force**: 2 **Ω xV**$_{xyz}$, centripetal force: **Ω x(Ω xr)** and **Ω**' **x r** reads as given in (Shames 1969)

$$\mathbf{a}_{XYZ} = \mathbf{a}_{xyz} + \mathbf{R}'' + 2\mathbf{\Omega x V}_{xyz} + \mathbf{\Omega x}(\mathbf{\Omega x r}) + \mathbf{\Omega' x r} \tag{2.40}$$

Here, $\mathbf{V}_{xyz}$ is the velocity vector in xyz and **r** is the position of the infinitesimal mass $\rho\, d\forall$ in xyz coordinate system. If we write the Newton's second law of motion in the fixed reference frame for the infinitesimal mass at time *t* using Eq. 2.40 we obtain

$$\mathrm{d}\mathbf{F} = \rho d\forall \mathbf{a}xyz = \rho d\forall \left[\mathbf{a}_{xyz} + \mathbf{R}'' + 2\mathbf{\Omega x V}_{xyz} + \mathbf{\Omega x}(\boldsymbol{\omega} \mathbf{x r}) + \mathbf{\Omega' x r}\right] \tag{2.41}$$

Equation 2.41 can be written for the acceleration in the non-inertial reference frame in terms of the inertial forces

$$\rho d\forall\ \mathbf{a}_{xyz} = \mathrm{d}\mathbf{F} - \rho d\nabla \left[\mathbf{R}'' + 2\mathbf{\Omega x V}_{xyz} + \mathbf{\Omega x}(\mathbf{\Omega x r}) + \mathbf{\Omega' x r}\right] \tag{2.41a}$$

We know that $\mathbf{F} = \int \mathrm{d}\mathbf{F}$. As the new form of the momentum equation expressed in the non-inertial reference frame xyz we obtain

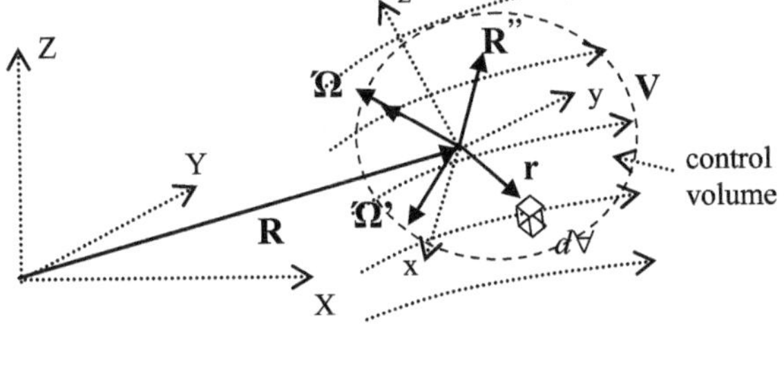

Fig. 2.5 The non-inertial coordinate system xyz in the inertial system XYZ

$$\vec{F}_S - \iiint \left[\vec{R}'' + 2\vec{\Omega} x \vec{V} + \vec{\Omega} x (\vec{\Omega} x \vec{r}) + \vec{\Omega}' x r\right] \rho \, d\forall = \frac{\partial}{\partial t} \iiint \rho \vec{V} \, d\forall + \oiint \rho \vec{V} \, (\vec{V}.dA) \tag{2.42}$$

If we consider the surface forces expressed in terms of stress tensor the differential form of the momentum equation becomes

$$\frac{\partial \rho \vec{V}}{\partial t} + \vec{\nabla}.(\rho \vec{V}\vec{V} - \vec{\tau}) = -\rho\left[(\vec{R}'' + 2\vec{\Omega} x \vec{V} + \vec{\Omega} x (\vec{\Omega} x \vec{r}) + \vec{\Omega}' x r\right] \tag{2.43}$$

Equation 2.43, can be used, in general, for studying the pitching and heaving-plunging airfoils and finite wings in roll and viscous analysis for drag prediction of fuselages.

2.9 Energy Equation

The conservation of energy can be formulated with applying the first law of thermodynamics on systems. The system here is in the flow field and receives the heat rate of $\dot{Q}$. If the work done by the system to the surroundings is $\dot{W}$ then the change of total energy in the system becomes

$$\frac{DE}{Dt} = \dot{Q} - \dot{W} \tag{2.44}$$

At a given time t, let the system under consideration coincide with the control volume we choose. If we let E_i denote the internal energy and $E_k = ½MV^2$ the kinetic energy of the total mass in the system, then as the mass independent transferable quantities the specific internal energy becomes $e = E_i/M$ and the specific kinetic energy reads as $E_k/M = ½V^2$. Which means the total specific energy in the control volume is $\eta = e + ½ V^2$. Now, we can relate the energy changes of the system and the control volume using Eq. 2.44 in Eq. 2.27 to obtain the integral form of the energy equation

$$\dot{Q} - \dot{W} = \frac{\partial}{\partial t} \iiint (e + V^2/2)\rho \, d\forall + \oiint (e + V^2/2)\rho \, \vec{V}.d\vec{A} \tag{2.45}$$

During the flow if we do not provide heat from outside, the system will heat the surroundings by the flux of internal heat from the control surface as follows $\dot{Q} = -\oiint \vec{q}.d\vec{A}$. On the other hand, the work of the stress tensor throughout the whole control surface will become $\dot{W} = -\oiint (\vec{V}.\vec{\tau}).d\vec{A}$. Now, if we substitute the integral forms of the heat flux to the surroundings and the work done by the system on the surrounding, Eq. 2.45 becomes

$$-\oiint \vec{q}.d\vec{A} + \oiint (\vec{V}.\vec{\tau}).d\vec{A} = \frac{\partial}{\partial t}\iiint (e + V^2/2)\rho\, d\forall + \oiint (e + V^2/2)\rho\, \vec{V}.d\vec{A} \tag{2.46}$$

In Eq. 2.46 we have three surface integral terms. If all three area integrals are changed to volume integrals using the divergence theorem, and all the all volume integrals are collected together over the same control volume, we can write the differential form of the energy equation as follows

$$\frac{\partial(\rho\,\varepsilon)}{\partial t} + \vec{\nabla}.(\rho\,\varepsilon\vec{V} - \vec{V}.\vec{\tau} + \vec{q}) = 0 \tag{2.47}$$

Here, $\varepsilon = e + \frac{1}{2}V^2$ denotes the specific total energy and Eq. 2.39 defines the stress tensor. The heat flux from a unit surface area reads as

$$\vec{q} = -k\vec{\nabla}T + \sum_i \rho_i \vec{U}_i h_i + \vec{q}_R + <\varepsilon' \vec{v}'> \tag{2.48}$$

Wherein, k denotes the heat conduction coefficient, the second term indicates the heat of diffusion, the third term represents radiative heat flux and the last term shows the turbulence heating. In summary, the global continuity is given by Eq. 2.31, continuity of species by 2.35, global momentum by 2.38 and the energy Equation by 2.47. Let us express these equations in Cartesian coordinates in conservative forms.

2.10 Equation of Motion in General Coordinates

Continuum equations of motion written in vector form are suitable for implementing the numerical solution of aerodynamical problems. In these equations the unknown vector **U** the flux vectors **F**, **G** and **H,** and the right hand side vector **R** are written as follows

$$\vec{U} = \begin{pmatrix} \rho \\ \rho\, u \\ \rho\, v \\ \rho\, w \\ \rho\, \varepsilon \\ \rho\, c_i \end{pmatrix} . \vec{F} = \begin{pmatrix} \rho\, u \\ \rho\, uu + \tau_{xx} \\ \rho\, uv + \tau_{yx} \\ \rho\, uw + \tau_{zx} \\ \rho\, u\varepsilon + q_x + u\tau_{xx} + v\tau_{xy} + w\tau_{xz} \\ \rho\, u\, c_i + D_{mi} \dfrac{\partial c_i}{\partial x} \end{pmatrix},$$

$$\vec{G} = \begin{pmatrix} \rho\, v \\ \rho\, uv + \tau_{xy} \\ \rho\, vv + \tau_{yy} \\ \rho\, vw + \tau_{zy} \\ \rho\, v\varepsilon + q_y + u\tau_{yx} + v\tau_{yy} + w\tau_{yz} \\ \rho\, v\, c_i + D_{mi} \dfrac{\partial c_i}{\partial y} \end{pmatrix},$$

$$\vec{H} = \begin{pmatrix} \rho\, w \\ \rho\, uw + \tau_{xz} \\ \rho\, vw + \tau_{yz} \\ \rho\, ww + \tau_{zz} \\ \rho\, w\varepsilon + q_z + u\tau_{zx} + v\tau_{zy} + w\tau_{zz} \\ \rho\, v\, c_i + D_{mi} \dfrac{\partial c_i}{\partial z} \end{pmatrix}, \vec{R} = \begin{pmatrix} 0 \\ 0 \\ 0 \\ 0 \\ 0 \\ \dot{w}_i \end{pmatrix},$$

Here, τ_{xx}, τ_{xy}, …, τ_{zz} are the components of the stress tensor and q_x, q_y and q_z are the components of the heat flux vector. Now, we can write the equation of motion in compact form as follows

$$\frac{\partial \vec{U}}{\partial t} + \frac{\partial \vec{F}}{\partial x} + \frac{\partial \vec{G}}{\partial y} + \frac{\partial \vec{H}}{\partial z} = \vec{R} \tag{2.49}$$

In many aerospace applications the Cartesian coordinates are not adequate to represent the surface equations of the body on which the boundary conditions are imposed. For this reason we have to write the equation of motion in body fitted coordinates which are generally referred as the generalized coordinates. Let the transformation from Cartesian coordinates xyz to the generalized coordinates $\xi\eta\varsigma$ be given as

$$\mathrm{x} = \mathrm{x}(\xi, \eta, \zeta), \quad \mathrm{y} = \mathrm{y}(\xi, \eta, \zeta), \quad \mathrm{z} = \mathrm{z}(\xi, \eta, \zeta)$$

With this information in hand, Eq. 2.49 is written in generalized coordinates in terms of the product of flux vectors with the metrics of transformation as follows, (Anderson et al. 1984).

$$\frac{\partial \vec{U}}{\partial t} + (\frac{\partial \xi}{\partial x}\ \frac{\partial \xi}{\partial y}\ \frac{\partial \xi}{\partial z}) \begin{pmatrix} \frac{\partial \vec{F}}{\partial \xi} \\ \frac{\partial \vec{G}}{\partial \xi} \\ \frac{\partial \vec{H}}{\partial \xi} \end{pmatrix} + (\frac{\partial \eta}{\partial x}\ \frac{\partial \eta}{\partial y}\ \frac{\partial \eta}{\partial z}) \begin{pmatrix} \frac{\partial \vec{F}}{\partial \eta} \\ \frac{\partial \vec{G}}{\partial \eta} \\ \frac{\partial \vec{H}}{\partial \eta} \end{pmatrix} + (\frac{\partial \varsigma}{\partial x}\ \frac{\partial \varsigma}{\partial y}\ \frac{\partial \varsigma}{\partial z}) \begin{pmatrix} \frac{\partial \vec{F}}{\partial \varsigma} \\ \frac{\partial \vec{G}}{\partial \varsigma} \\ \frac{\partial \vec{H}}{\partial \varsigma} \end{pmatrix} = \vec{R} \tag{2.50}$$

Shown in Fig. 2.6a,b are two different external flow regions: (a) wing upper surface and the boundaries of its computational domain, and (b) half a fuselage and the computational domain transformed from xyz, Cartesian coordinates to $\xi\eta\zeta$, generalized coordinate system. Both flow domains, after the transformation in $\xi\eta\zeta$ coordinate system, are mapped into the cube denoted by 12,345,678 for which the discretization of the computational domain becomes straight forward.

In Fig. 2.6, the $\xi\eta$ surfaces of physical domain transforms into the square denoted with 1234, wherein, ζ coordinate of the physical domain is inclined with the body surface, i.e. it is not necessarily normal to the surface. After knowing one to one correspondence of the discrete points of both domains, we can numerically calculate the derivative terms for ξ_x, ξ_y, …, ζ_z to be used for solving Eq. 2.50 in the discretized cube 12,345,678. There are quite a few numbers of literature published about the mesh generation and coordinate transformation techniques, however, two separate works by Anderson and Hoffman can be recommended for beginners and the intermediate level users, (Anderson et al. 1984) and (Hoffman 1992).

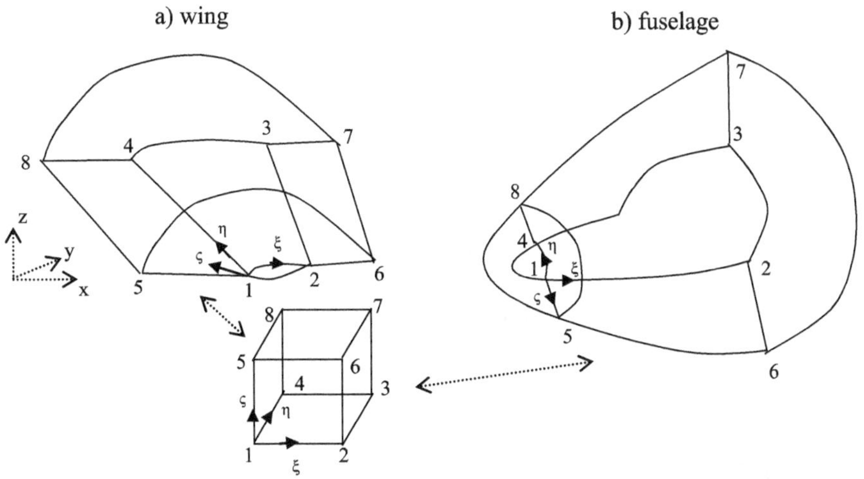

Fig. 2.6 The coordinate transformation **a** the wing, **b** the fuselage: ξ-η: surface coordinates, ς: the coordinate which is inclined with the surface

2.11 Navier-Stokes Equations

In its most general form, including the chemical reactions at high temperatures, Eq. 2.49 was introduced as the set of equations for external flows. Global continuity equation and the conservation of momentum equations deal with the average values of flow parameters, therefore they are of mechanical nature, whereas the energy equation deals with the effect of heating as well the enthalpy increase caused by the diffusion of species. If we do not consider the chemical reactions, then there will not be diffusion terms present and the related specie conservation terms disappear. Therefore, Eq. 2.49 reduces to the well known Navier-Stokes Equations, (Schlichting 1968). Since the Navier-Stokes equations can model all laminar and turbulent flows, they have a wide range of their implementation in aerodynamical applications. For the case of turbulent flows, we have to include the effective viscosity μ_T into the constitutive relations to model the Reynolds stresses. Now, we can re-write the constitutive relation 2.39 and the heat flux term 2.48 with the turbulent Prandtl number P_{rT} as follows

$$\begin{aligned} \vec{\tau} &= (-p+\lambda\vec{\nabla}.\vec{V})\vec{I}+(\mu+\mu_T)\,sim\vec{V}, \\ \vec{q} &= -(k+c_p\mu_T/P_{tT})\vec{\nabla}T+\sum_i \rho_i\vec{U}_i h_i+\vec{q}_R \end{aligned} \tag{2.51a, b}$$

Let us separate the molecular viscosity and the heat transfer terms to rearrange Eq. 2.49 for chemically non-reacting flows to give the new right hand side vectors

$$\vec{S}_1 = \begin{pmatrix} 0 \\ 2\mu(\frac{\partial u}{\partial x}) - \frac{2}{3}\mu\vec{\nabla}.\vec{V} \\ \mu(\frac{\partial u}{\partial y}+\frac{\partial v}{\partial x}) \\ \mu(\frac{\partial u}{\partial z}+\frac{\partial w}{\partial x}) \\ k\frac{\partial T}{\partial x}+u\tau'_{xx}+v\tau'_{xy}+w\tau'_{xz} \end{pmatrix}, \vec{S}_2 = \begin{pmatrix} 0 \\ \mu(\frac{\partial u}{\partial y}+\frac{\partial v}{\partial x}) \\ 2\mu(\frac{\partial v}{\partial y}) - \frac{2}{3}\mu\vec{\nabla}.\vec{V} \\ \mu(\frac{\partial v}{\partial z}+\frac{\partial w}{\partial y}) \\ k\frac{\partial T}{\partial x}+u\tau'_{xy}+v\tau'_{yy}+w\tau'_{yz} \end{pmatrix}, \vec{S}_1$$

$$= \begin{pmatrix} 0 \\ \mu(\frac{\partial u}{\partial z}+\frac{\partial w}{\partial x}) \\ \mu(\frac{\partial v}{\partial z}+\frac{\partial w}{\partial y}) \\ 2\mu(\frac{\partial w}{\partial z}) - \frac{2}{3}\mu\vec{\nabla}.\vec{V} \\ k\frac{\partial T}{\partial x}+u\tau'_{xz}+v\tau'_{yz}+w\tau'_{zz} \end{pmatrix}$$

and to obtain the in final form of the equations

$$\frac{\partial \vec{U}}{\partial t} + \frac{\partial \vec{F}_1}{\partial x} + \frac{\partial \vec{G}_1}{\partial y} + \frac{\partial \vec{H}_1}{\partial z} = \frac{\partial \vec{S}_1}{\partial x} + \frac{\partial \vec{S}_2}{\partial y} + \frac{\partial \vec{S}_3}{\partial z} \tag{2.52}$$

Here, $\overline{\overline{\tau}}' = \overline{\overline{\tau}} - p\vec{I}$ is the pressure free stress tensor, $\mathbf{F}_1$, $\mathbf{G}_1$ and $\mathbf{H}_1$ are the flux terms which are free of viscous effects. That is if we let the right hand side of Eq. 2.52 be zero, we obtain the Euler equations which are already given by Eqs. 2.1–2.3.

The non-dimensional form of the Navier-Stokes equations are usually more convenient to apply to problems of aerodynamics. For this purpose, we use characteristic parameters of the flow. The free stream values for the density, speed, pressure, viscosity, conductivity and the temperature which are $\rho_\infty, V_\infty, p_\infty, \mu_\infty, k_\infty \, and \, T_\infty$ respectively. The corresponding non dimensional quantities become

$$\hat{\rho} = \rho/\rho_\infty \qquad \hat{p} = p/p_\infty \qquad \hat{e} = e/V_\infty^2 \qquad \hat{\mu} = \mu/\mu_\infty \qquad \hat{k} = k/k_\infty$$

$$\hat{T} = T/T_\infty \qquad \hat{t} = t V_\infty/c \qquad \hat{x} = x/c \qquad \hat{y} = y/c \qquad \hat{z} = z/c$$

The non dimensional form of the Navier Stokes equations reads as

$$\frac{\partial \hat{U}}{\partial t} + \frac{\partial \hat{F}_1}{\partial x} + \frac{\partial \hat{G}_1}{\partial y} + \frac{\partial \hat{H}_1}{\partial z} = \hat{S}_1 + \hat{S}_2 + \hat{S}_3 \tag{2.53}$$

The non dimensional quantities in Eq. 2.53

$$\hat{U} = \begin{pmatrix} \hat{\rho} \\ \hat{\rho}\,\hat{u} \\ \hat{\rho}\,\hat{v} \\ \hat{\rho}\,\hat{w} \\ \hat{\rho}\,\hat{\varepsilon} \end{pmatrix}, \quad \hat{F}_1 = \begin{pmatrix} \hat{\rho}\,\hat{u} \\ \hat{\rho}\,\hat{u}\hat{u} + \hat{p} \\ \hat{\rho}\,\hat{u}\hat{v} \\ \hat{\rho}\,\hat{u}\,\overleftarrow{w} \\ (\hat{\rho}\,\hat{\varepsilon} + \hat{p})\,\overleftarrow{u} \end{pmatrix}, \quad \hat{G}_1 = \begin{pmatrix} \hat{\rho}\,\hat{v} \\ \hat{\rho}\,\hat{u}\hat{v} \\ \hat{\rho}\,\hat{v}\hat{v} + \hat{p} \\ \hat{\rho}\,\hat{v}\,\overleftarrow{w} \\ (\hat{\rho}\,\hat{\varepsilon} + \hat{p})\hat{v} \end{pmatrix},$$

$$\hat{H}_1 = \begin{pmatrix} \hat{\rho}\,\hat{w} \\ \hat{\rho}\,\hat{u}\hat{w} \\ \hat{\rho}\,\hat{v}\hat{w} \\ \hat{\rho}\,\hat{w}\,\overleftarrow{w} + \hat{p} \\ (\hat{\rho}\,\hat{\varepsilon} + \hat{p})\hat{w} \end{pmatrix}$$

Here, the total non dimensional specific energy is $\hat{\varepsilon} = \hat{e} + (\hat{u}^2 + \hat{v}^2 + \hat{w}^2)/2$. The viscous terms on the other hand becomes

$$.\hat{S}_1 = \begin{pmatrix} 0 \\ \hat{\tau}_{xx} \\ \hat{\tau}_{xy} \\ \hat{\tau}_{xz} \\ \hat{u}\hat{\tau}_{xx} + \hat{v}\hat{\tau}_{xy} + \hat{w}\hat{\tau}_{xz} - \hat{q}_x \end{pmatrix}, \quad \hat{S}_2 = \begin{pmatrix} 0 \\ \hat{\tau}_{xz} \\ \hat{\tau}_{yy} \\ \hat{\tau}_{yz} \\ \hat{u}\hat{\tau}_{xy} + \hat{v}\hat{\tau}_{yy} + \hat{w}\hat{\tau}_{yz} - \hat{q}_y \end{pmatrix}, \quad \hat{S}_3 = \begin{pmatrix} 0 \\ \hat{\tau}_{xz} \\ \hat{\tau}_{yz} \\ \hat{\tau}_{zz} \\ \hat{u}\hat{\tau}_{xz} + \hat{v}\hat{\tau}_{yz} + \hat{w}\hat{\tau}_{xz} - \hat{q}_z \end{pmatrix}$$

The open form of these viscous terms in terms of velocity components reads

$$\hat{\tau}_{xx} = \frac{\hat{\mu}}{R_e}\left[2\,(\frac{\partial \hat{u}}{\partial \hat{x}}) - \frac{2}{3}(\frac{\partial \hat{u}}{\partial \hat{x}} + \frac{\partial \hat{v}}{\partial \hat{y}} + \frac{\partial \hat{w}}{\partial z})\right], \quad \hat{\tau}_{xy} = \frac{\hat{\mu}}{R_e}(\frac{\partial u}{\partial y} + \frac{\partial v}{\partial x})$$

$$\hat{\tau}_{yy} = \frac{\hat{\mu}}{R_e}\left[2\,(\frac{\partial \hat{v}}{\partial \hat{y}}) - \frac{2}{3}(\frac{\partial \hat{u}}{\partial \hat{x}} + \frac{\partial \hat{v}}{\partial \hat{y}} + \frac{\partial \hat{w}}{\partial z})\right], \quad \hat{\tau}_{xz} = \frac{\hat{\mu}}{R_e}(\frac{\partial \hat{u}}{\partial \hat{z}} + \frac{\partial \hat{w}}{\partial \hat{x}})$$

$$\hat{\tau}_{zz} = \frac{\hat{\mu}}{R_e}\left[2\,(\frac{\partial \hat{w}}{\partial \hat{z}}) - \frac{2}{3}(\frac{\partial \hat{u}}{\partial \hat{x}} + \frac{\partial \hat{v}}{\partial \hat{y}} + \frac{\partial \hat{w}}{\partial z})\right], \quad \hat{\tau}_{yz} = \frac{\hat{\mu}}{R_e}(\frac{\partial \hat{v}}{\partial \hat{z}} + \frac{\partial \hat{w}}{\partial \hat{y}})$$

Heat conduction terms become

$$\hat{q}_x = -\frac{\hat{\mu}}{(\gamma-1)M_\infty^2 R_e P_r}\frac{\partial \hat{T}}{\partial \hat{x}}, \quad \hat{q}_y = -\frac{\hat{\mu}}{(\gamma-1)M_\infty^2 R_e P_r}\frac{\partial \hat{T}}{\partial \hat{y}},$$
$$\hat{q}_z = -\frac{\hat{\mu}}{(\gamma-1)M_\infty^2 R_e P_r}\frac{\partial \hat{T}}{\partial \hat{z}}$$

The non dimensional similarity parameters appearing in the equations are well known Reynolds, Mach and Prandtl numbers which are defined with their physical meanings attached as follows.

Reynolds number: $R_e = \rho_\infty V_\infty c/\mu_\infty$, (inertia forces/viscous forces).

Mach number: $M_\infty = V_\infty/a_\infty$, (kinetic energy of the flow/internal energy).

Prandtl number: $P_r = c_{p\infty}\mu_\infty/k_\infty$, (energy dissipation/heat conduction).

From the perfect gas assumption: $\hat{p} = (\gamma-1)\hat{\rho}\,\hat{e}$ and $\hat{T} = \gamma M_\infty^2 \hat{p}/\hat{\rho}$ relations among the non dimensional parameters are obtained.

In most of the aerodynamics applications there is high free stream speed involved. For the classical applications usually unseparated flows are considered. Regardless of flow being attached or separated, for the flows with high free stream speeds we can apply some approximations to Eq. 2.53 to obtain simpler solutions. Let us now, see this approximations and conditions for their applicability.

2.12 Thin Shear Layer Navier-Stokes Equations

In the open form of Navier-Stokes Eqs. (2.53) , we observe the existence of second derivatives for the velocity and the temperature. This implies that the Navier-Stokes equations are second order partial differential equations. When the freestream speed is high, the Reynolds number is high. This makes the gradients of the flow parameters to be high normal to the surface as compared to the gradients parallel to the surface. Therefore, we can neglect the effect of the viscous terms which are parallel to the flow surface and simplify Eqs. 2.53. Let us now, perform some order of magnitude analysis for the simplification process on a simple wing surface immersed in a high free stream speed given in Fig. 2.7.

Since we consider the air flowing over the wing as a real gas, the boundary conditions on the surface will be (i) no slip condition and (ii) the wall temperature specification. According to Fig. 2.7, the wing surface is almost parallel to xy plane where the molecular diffusion parallel to the xy plane is negligible compared to the diffusion taking place normal to the surface. This is because of high free stream speed transporting the properties in the parallel direction much faster than the molecular diffusion. On the other hand, because of no slip condition, the gradients which are normal to the surface are much higher than the gradients parallel to the surface. The order of magnitude analysis performed on the terms of Eq. 2.53 gives

$$\frac{1}{R_e}\frac{\partial \hat{\mu}}{\partial \hat{z}}(\frac{\partial}{\partial \hat{z}}) \gg \frac{1}{R_e}\frac{\partial \hat{\mu}}{\partial \hat{x}}(\frac{\partial}{\partial \hat{x}}), \ldots, \frac{1}{R_e}\frac{\partial \hat{\mu}}{\partial \hat{y}}(\frac{\partial}{\partial \hat{y}}).$$

The approximate form of the equations result in modeling an external real gas flow which takes place in a thin shear layer around the wing surface. Therefore, the first approximate form of Eq. 2.53 is called 'Thin Shear Layer Navier-Stokes Equations' which are to be introduced next

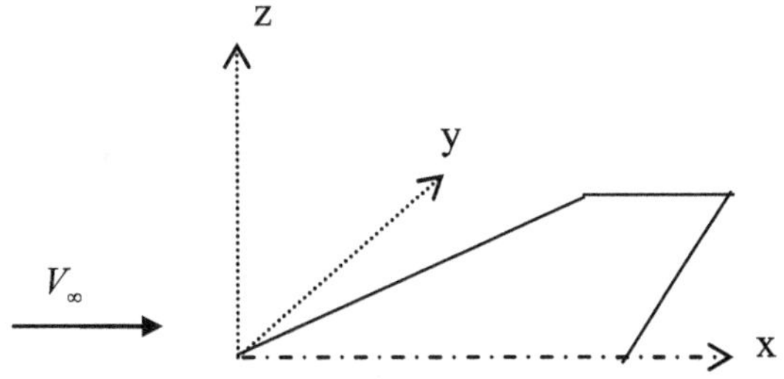

Fig. 2.7 Thin wing in a high freestream speed

$$\frac{\partial}{\partial \hat{t}}\begin{pmatrix}\hat{\rho}\\ \hat{\rho}\hat{u}\\ \hat{\rho}\hat{v}\\ \hat{\rho}\hat{w}\\ \hat{\rho}\hat{\varepsilon}\end{pmatrix}+\frac{\partial}{\partial \hat{x}}\begin{pmatrix}\hat{\rho}\hat{u}\\ \hat{\rho}\hat{u}\hat{u}+\hat{p}\\ \hat{\rho}\hat{u}\hat{v}\\ \hat{\rho}\hat{u}\hat{w}\\ (\hat{\rho}\hat{\varepsilon}+\hat{p})\hat{u}\end{pmatrix}+\frac{\partial}{\partial \hat{y}}\begin{pmatrix}\hat{\rho}\hat{v}\\ \hat{\rho}\hat{u}\hat{v}\\ \hat{\rho}\hat{v}\hat{v}+\hat{p}\\ \hat{\rho}\hat{v}\hat{w}\\ (\hat{\rho}\hat{\varepsilon}+\hat{p})\hat{v}\end{pmatrix}+\frac{\partial}{\partial \hat{z}}\begin{pmatrix}\hat{\rho}\hat{w}\\ \hat{\rho}\hat{u}\hat{w}\\ \hat{\rho}\hat{v}\hat{w}\\ \hat{\rho}\hat{w}\hat{w}+\hat{p}\\ (\hat{\rho}\hat{\varepsilon}+\hat{p})\hat{w}\end{pmatrix}$$
$$=\frac{1}{R_e}\frac{\partial}{\partial \hat{z}}\begin{pmatrix}0\\ \hat{\mu}\frac{\partial \hat{u}}{\partial \hat{z}}\\ \hat{\mu}\frac{\partial \hat{v}}{\partial \hat{z}}\\ \frac{4}{3}\hat{\mu}\frac{\partial \hat{w}}{\partial z}\\ \hat{\mu}(\ddot{u}\frac{\partial \hat{u}}{\partial \hat{z}}+\hat{v}\frac{\partial \hat{v}}{\partial \hat{y}}+\frac{4}{3}\frac{\partial \hat{w}}{\partial \hat{z}})+\frac{\hat{\mu}}{(\gamma-1)M_\infty^2 P_r}\frac{\partial \hat{T}}{\partial \hat{z}}\end{pmatrix} \tag{2.54}$$

Equations 2.54 are written in Cartesian coordinates without considering the wing thickness effect. If we consider the thickness effect and high angles of attack, Eq. 2.54 can be written in $\xi\eta\zeta$ coordinates where only the viscous terms in ζ coordinate, which is normal to the wing surface are retained. With these assumptions and with the additional assumption that the general coordinate system changes with time, the transformation of coordinates from Cartesian to generalized reads

$$\xi=\xi(x,y,z,t),\quad \eta=\eta(x,y,z,t),\quad \zeta=\zeta(x,y,z,t),\quad \tau=t \tag{2.55}$$

Using 2.55, we can write the open form of the non-dimensional Thin Shear Layer Navier-Stokes equations in generalized coordinates where ς is the direction normal to the wing surface

$$\frac{\partial}{\partial \tau}\frac{1}{J}\begin{pmatrix}\hat{\rho}\\ \hat{\rho}\hat{u}\\ \hat{\rho}\hat{v}\\ \hat{\rho}\hat{w}\\ \hat{\rho}\hat{\varepsilon}\end{pmatrix}+\frac{\partial}{\partial \xi}\frac{1}{J}\begin{pmatrix}\hat{\rho}U\\ \hat{\rho}\hat{u}U+\xi_x\hat{p}\\ \hat{\rho}\hat{v}U+\xi_y\hat{p}\\ \hat{\rho}\hat{w}U+\xi_z\hat{p}\\ (\hat{\rho}\hat{\varepsilon}+\hat{p})U-\xi_t\hat{p}\end{pmatrix}+\frac{\partial}{\partial \eta}\frac{1}{J}\begin{pmatrix}\hat{\rho}V\\ \hat{\rho}\hat{u}V+\eta_x\hat{p}\\ \hat{\rho}\hat{v}V+\eta_y\hat{p}\\ \hat{\rho}\hat{w}V+\eta_z\hat{p}\\ (\hat{\rho}\hat{\varepsilon}+\hat{p})V+\xi_t\hat{p}\end{pmatrix}$$
$$+\frac{\partial}{\partial \varsigma}\frac{1}{J}\begin{pmatrix}\hat{\rho}W\\ \hat{\rho}\hat{u}W+\varsigma_x\hat{p}\\ \hat{\rho}\hat{v}W+\varsigma_y\hat{p}\\ \hat{\rho}\hat{w}W+\varsigma_z\hat{p}\\ (\hat{\rho}\hat{\varepsilon}+\hat{p})W-\varsigma_t\hat{p}\end{pmatrix}=\frac{1}{R_e}\frac{\partial S}{\partial \varsigma} \tag{2.56}$$

Here, $J=\frac{\partial(\xi,\eta,\varsigma,\tau)}{\partial(x,y,z,t)}$ is the Jakobian determinant of the transformation, U, V and W are the contravariant velocity components which are normal to the curvilinear surfaces given with constant ξ, η and ς coordinates respectively. They read

$$U=\xi_t+\xi_x\hat{u}+\xi_y\hat{v}+\xi_z\hat{w},\ V=\eta_t+\eta_x\hat{u}+\eta_y\hat{v}+\eta_z\hat{w},\ W=\varsigma_t+\varsigma_x\hat{u}+\varsigma_y\hat{v}+\varsigma_z\hat{w} \tag{2.57}$$

The viscous terms at the right hand side of Eq. 2.56 become

$$\hat{S} = \begin{pmatrix} 0 \\ \hat{\mu}(\varsigma_x^2+\varsigma_y^2+\varsigma_z^2)\hat{u}_\varsigma + \frac{\hat{\mu}}{3}(\varsigma_x\hat{u}_\varsigma+\varsigma_y\hat{v}_\varsigma+\varsigma_z\hat{w}_\varsigma)\eta_x \\ \hat{\mu}(\varsigma_x^2+\varsigma_y^2+\varsigma_z^2)\hat{v}_\varsigma + \frac{\hat{\mu}}{3}(\varsigma_x\hat{u}_\varsigma+\varsigma_y\hat{v}_\varsigma+\varsigma_z\hat{w}_\varsigma)\eta_y \\ \hat{\mu}(\varsigma_x^2+\varsigma_y^2+\varsigma_z^2)\hat{w}_\varsigma + \frac{\hat{\mu}}{3}(\varsigma_x\hat{u}_\varsigma+\varsigma_y\hat{v}_\varsigma+\varsigma_z\hat{w}_\varsigma)\eta_z \\ \hat{\mu}(\varsigma_x^2+\varsigma_y^2+\varsigma_z^2)\left[\frac{1}{2}(\hat{u}^2+\hat{v}^2+\hat{w}^2)_\varsigma + \frac{1}{(\gamma-1)M_\infty^2 P_r}\hat{T}_\varsigma\right] + \frac{\hat{\mu}}{3}(\varsigma_x\hat{u}+\varsigma_y\hat{v}+\varsigma_z\hat{w})(\varsigma_x\hat{u}_\varsigma+\varsigma_y\hat{v}_\varsigma+\varsigma_z\hat{w}_\varsigma) \end{pmatrix}$$

The convective terms in Eq. 2.56 contain the Jacobian determinant in the denominator. This form of the equations are called 'strong conservative forms' and their derivations are provided in Appendix 1.

2.13 Parabolized Navier-Stokes Equations

In numerous aerospace applications we encounter the steady flow cases for which the time dependent terms of the equations are discarded. The thin shear layer equations written for steady flows without time dependent terms are called 'Parabolized Navier-Stokes Equations', (Anderson). According to this definition, from Eq. 2.54 we write the parabolized Navier-Stokes equations in Cartesian coordinates as follows

$$\frac{\partial}{\partial \hat{x}}\begin{pmatrix} \hat{\rho}\hat{u} \\ \hat{\rho}\hat{u}\hat{u}+\hat{p} \\ \hat{\rho}\hat{u}\hat{v} \\ \hat{\rho}\hat{u}\hat{w} \\ (\hat{\rho}\hat{\varepsilon}+\hat{p})+\overleftarrow{u} \end{pmatrix} + \frac{\partial}{\partial \hat{y}}\begin{pmatrix} \hat{\rho}\hat{v} \\ \hat{\rho}\hat{u}\hat{v} \\ \hat{\rho}\hat{v}\hat{v}+\hat{p} \\ \hat{\rho}\hat{v}\hat{w} \\ (\hat{\rho}\hat{\varepsilon}+\hat{p})\hat{v} \end{pmatrix} + \frac{\partial}{\partial \hat{z}}\begin{pmatrix} \hat{\rho}\hat{w} \\ \hat{\rho}\hat{u}\hat{w} \\ \hat{\rho}\hat{v}\hat{w} \\ \hat{\rho}\hat{w}\hat{w}+\hat{p} \\ (\hat{\rho}\hat{\varepsilon}+\hat{p})\hat{w} \end{pmatrix}$$

$$= \frac{1}{R_e}\frac{\partial}{\partial \hat{z}}\begin{pmatrix} 0 \\ \hat{\mu}\dfrac{\partial \hat{u}}{\partial \hat{z}} \\ \hat{\mu}\dfrac{\partial \hat{v}}{\partial \hat{z}} \\ \dfrac{4}{3}\hat{\mu}\dfrac{\partial \hat{w}}{\partial z} \\ \hat{\mu}(\ddot{u}\dfrac{\partial \hat{u}}{\partial \hat{z}} + \hat{v}\dfrac{\partial \hat{v}}{\partial \hat{y}} + \dfrac{4}{3}\dfrac{\partial \hat{w}}{\partial \hat{z}}) + \dfrac{\hat{\mu}}{(\gamma-1)M_\infty^2 P_r}\dfrac{\partial \hat{T}}{\partial \hat{z}} \end{pmatrix}$$

In curvilinear coordinates, we neglect the $\partial(\,)/\partial t$ terms as well as the time dependency of ξ, η and ζ coordinates. Thus, we obtain the parabolized Navier–Stokes equations in curvilinear coordinates. In addition if we can, somehow, impose the pressure from the outside of shear layer then we obtain the well known boundary layer equations.

2.14 Boundary Layer Equations

In the attached or slightly detached external flow cases, we can obtain the surface pressure distribution using the methods described in Sect. 2-A and further simplify set of Eqs. 2.49 and 2.54. In these simplifications we again resort to the order of magnitude analysis. Assuming again that the viscous effects are only in the vicinity of the surface of the body, we can consider the gradients and the diffusion normal to the surface we obtain

$$\text{Continuity: } \frac{\partial \rho}{\partial t} + \frac{\partial \rho u}{\partial x} + \frac{\partial \rho w}{\partial z} = 0 \tag{2.58}$$

Continuity of

$$\text{The species: } \rho\frac{\partial c_i}{\partial t} + \rho u\frac{\partial c_i}{\partial x} + \rho w\frac{\partial c_i}{\partial z} = \frac{\partial}{\partial z}(\rho D_{12}\frac{\partial c_i}{\partial z}) + \dot{w}_i \tag{2.59}$$

$$\text{x} - \text{momentum: } \rho\frac{\partial u}{\partial t} + \rho u\frac{\partial u}{\partial x} + \rho w\frac{\partial u}{\partial z} = -\frac{\partial p}{\partial x} + \frac{\partial}{\partial z}(\mu\frac{\partial u}{\partial z}) \tag{2.60}$$

$$\text{z} - \text{momentum : } \frac{\partial p}{\partial z} = 0 \tag{2.61}$$

Energy:

$$\rho\frac{\partial h}{\partial t} + \rho u\frac{\partial h}{\partial x} + \rho w\frac{\partial h}{\partial z} = \frac{\partial p}{\partial t} + u\frac{\partial p}{\partial x} + \mu(\frac{\partial u}{\partial z})^2 + \frac{\partial}{\partial z}(k\frac{\partial T}{\partial z}) + \frac{\partial}{\partial z}(\rho D_{12}\sum_i h_i\frac{\partial c_i}{\partial z}) \tag{2.62}$$

Here, x is the direction parallel to the surface, z is the normal direction and h_i in Eq. 2.62 is the enthalpy of species i.

The real gas effect in an external flow can be measured with the change caused in the stagnation enthalpy. If we neglect the effect of vertical velocity component, the stagnation enthalpy of the boundary layer flow reads: $h_o = h + u^2/2$. The normal gradient of the stagnation enthalpy at a point then reads

$$\frac{\partial h_0}{\partial z} = \frac{\partial h}{\partial z} + u\frac{\partial u}{\partial z}$$

Hence the new form of the energy equation becomes

$$\rho\frac{\partial h_o}{\partial t} + \rho u\frac{\partial h_o}{\partial x} + \rho w\frac{\partial h_o}{\partial z} = \frac{\partial p}{\partial t} + \mu(\frac{\partial u}{\partial z})^2 + \frac{\partial}{\partial z}(k\frac{\partial T}{\partial z}) + \frac{\partial}{\partial z}(\rho D_{12}\sum_i h_i\frac{\partial c_i}{\partial z} \tag{2.63}$$

During the non dimensionalization process of the boundary layer equations, we introduce the Lewis number to represent the magnitude of diffusion in terms of heat conduction as a non dimensional number: $L_e = \rho\ D_{12}\ c_p/k$. The non dimensional form of Eq. 2.63 reads as

$$\rho\frac{\partial h_o}{\partial t} + \rho u\frac{\partial h_o}{\partial x} + \rho w\frac{\partial h_o}{\partial z} = \frac{\partial p}{\partial t} + \frac{\partial}{\partial z}\left[(1 - \frac{1}{P_r})\mu u\frac{\partial u}{\partial z} + \frac{\mu}{P_r}\frac{\partial h_0}{\partial z} + (1 - \frac{1}{L_e})\rho D_{12}\sum_i h_i\frac{\partial c_i}{\partial z}\right] \tag{2.64}$$

In Eq. 2.64 the local value 1 for the Lewis number makes the contribution of diffusion vanish and as the Lewis number gets higher the diffusion gets stronger. The c_p value in the Lewis number is obtained from the average c_{pi} values of the species involved in the boundary layer under the frozen flow assumption.

2.15 Incompressible Flow Navier-Stokes Equations

In a wide region of aerodynamical applications low subsonic speeds are encountered. Since the free stream Mach number for these types of are very low, the flow is assumed incompressible. The continuity equation for the incompressible flow becomes

$$\vec{\nabla}.\vec{V} = 0 \tag{2.65}$$

Equation 2.65 implies that the flow is divergenless which in turn simplifies the constitutive relations, Eq. 2.51a, b. In addition, because of low speeds the temperature changes in the flow field will also be low which makes the viscosity remain constant. Since the viscosity is constant, the momentum equation is simplified also to take the following form

$$\rho\frac{D\vec{V}}{Dt} = -\vec{\nabla}p + \mu\nabla^2\vec{V} \tag{2.66}$$

In case of turbulent flows, we use the effective viscosity: $\mu_e = \mu + \mu_T$ in Eq. 2.66 which undergoes an averaging process after Reynolds decomposition which makes the final form of the equations to be called 'Reynolds Averaged Navier-Stokes Equations'.

Another convenient form of incompressible Navier-Stokes equations is written in terms of a new variable called vorticity. The vorticity vector is derived from the velocity vector as

$$\vec{\omega} = \vec{\nabla} x \vec{V} \tag{2.67}$$

The vorticity transport equation obtained from two dimensional version of Eq. 2.66 reads as

$$\frac{\partial \omega}{\partial t} + (\vec{V}.\vec{\nabla})\,\omega = \nabla^2 \omega \tag{2.68}$$

Here, ω as the third component of the vorticity appears as a scalar quantity in Eq. 2.68, which does not have any pressure term involved. The integral form of Eqs. 2.65 and 2.67 reads as, (Wu and Gulcat 1981),

$$\vec{V}(\vec{r},t) = -\frac{1}{2\pi}\int_R \frac{\vec{\omega}_o x(\vec{r}_o - \vec{r})}{|\vec{r}_o - \vec{r}|^2} dR_0 + \frac{1}{2\pi}\int_B \frac{(\vec{V}_0.\vec{n}_0)(\vec{r}_o - \vec{r}) - (\vec{V}_0 x \vec{n}_0) x (\vec{r}_o - \vec{r})}{|\vec{r}_o - \vec{r}|^2} dB_0 \tag{2.69}$$

Here, R shows the region for vortical flow, B the boundaries, **r** and $\mathbf{r_o}$ the position vectors and $\mathbf{n_o}$ the unit vector pointing outwards to the boundaries. The boundary B contains the airfoil surface and the far field boundary. While solving Eq. 2.68, we only consider the vertical region confined around the airfoil. Same is done for the evaluation of the velocity field via Eq. 2.69. The integro-differential formulation presented here, therefore, enables us to work with small computational domains. Another use of Eq. 2.69 comes into picture while determining the surface vortex sheet strength through the no-slip boundary condition.

2.16 Aerodynamic Forces and Moments

The aim in performing the real gas flow analysis over bodies is to determine the aerodynamic forces, moments and the heat loads acting. For this purpose the computed pressure and stress fields are integrated over whole surface of the body. The surface stresses are obtained from the velocity gradients calculated at the surface. Let us now write down the x, y and z components of the infinitesimal surface force d**F** acting on the infinitesimal area dA of the surface

$$
\begin{aligned}
dF_x &= (n_x\tau_{xx} + n_y\tau_{xy} + n_z\tau_{xz})\, dA \\
dF_y &= (n_x\tau_{yx} + n_y\tau_{yy} + n_z\tau_{yz})\, dA \\
dF_z &= (n_x\tau_{zx} + n_y\tau_{zy} + n_z\tau_{zz})\, dA
\end{aligned} \tag{2.70}
$$

Here, n_x, n_y and n_z are the direction cosines of the vector normal to the infinitesimal surface dA Let us now express the area dA in curvilinear coordinates. We can express the integral relations which give the total force components in xyz in terms of the differential area given in curvilinear coordinates ξη as shown in Fig. 2.8.

As seen in Fig. 2.8 the differential area dA can be computed in terms of the product of two infinitesimal vectors given as the changes of the position vector **r** = *x***i** + *y***j** + *z***k** in directions of ξ and η coordinates as dA = |(d**r**/dξ) x (d**r**/dη)| dξ dη. The vector product of these two vectors also give the direction of the unit normal **n** of dA. In explicit form we find

$$
\begin{aligned}
dA &= \left\| \begin{vmatrix} \vec{i} & \vec{j} & \vec{k} \\ x_\xi & y_\xi & z_\xi \\ x_\eta & y_\eta & z_\eta \end{vmatrix} \right\| d\xi\, d\eta \\
&= \sqrt{(y_\xi z_\eta - z_\xi y_\eta)^2 + (x_\xi z_\eta - z_\xi x_\eta)^2 + (y_\xi y_\eta - x_\xi y_\eta)^2}\, d\xi\, d\eta
\end{aligned} \tag{2.71}
$$

Here, the term under the square root is named reduced Jacobian I. The unit normal vector in open form becomes

$$
\vec{n} = \left[(y_\xi z_\eta - z_\xi y_\eta)\vec{i} - (x_\xi z_\eta - z_\xi x_\eta)\vec{j} + (y_\xi y_\eta - x_\xi y_\eta)\vec{k} \right] \tag{2.72}
$$

We can write the components of the stress tensor in terms of the velocity gradients expressed in curvilinear coordinates as follows for example for τ_{xy}

$$
\tau_{xy} = \mu(\frac{\partial u}{\partial y} + \frac{\partial v}{\partial x}) = \mu(\frac{\partial u}{\partial \xi}\xi_y + \frac{\partial u}{\partial \eta}\eta_y + \frac{\partial u}{\partial \varsigma}\varsigma_y + \frac{\partial v}{\partial \xi}\xi_x + \frac{\partial v}{\partial \eta}\eta_x + \frac{\partial v}{\partial \varsigma}\varsigma_x) \tag{2.73}
$$

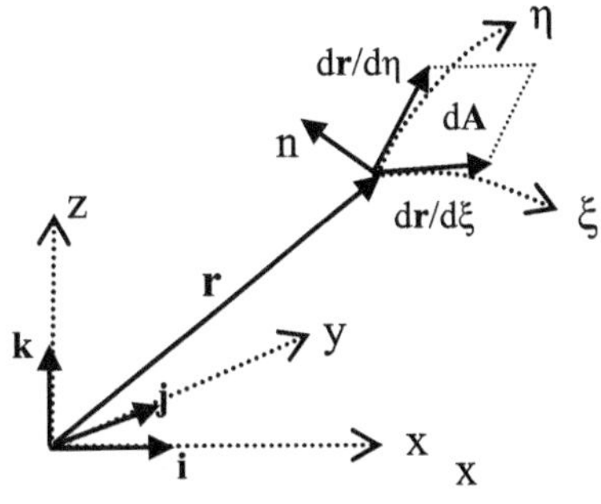

Fig. 2.8 Expressing dA in Curvilinear coordinates ξη

If we consider Eqs. 2.71–2.73 to form the differential force elements and integrate them numerically over the differential area, we obtain the total force components as follows

$$
\begin{aligned}
F_x &= \int_A dF_x = \int_A (n_x\tau_{xx} + n_y\tau_{xy} + n_z\tau_{xz}) I \, d\xi \, d\eta \\
F_y &= \int_A dF_y = \int_A (n_x\tau_{yx} + n_y\tau_{yy} + n_z\tau_{yz}) I \, d\xi \, d\eta \\
F_z &= \int_A dF_z = \int_A (n_x\tau_{zx} + n_y\tau_{zy} + n_z\tau_{zz}) I \, d\xi \, d\eta
\end{aligned} \tag{2.74}
$$

Computations of the moments with respect to a point can be performed similarly with considering the moment arm of the point to the differential area dA.

In case of two dimensional incompressible external flows if we know the vorticity field $\boldsymbol{\omega}$, first the surface vortex sheet strength $\gamma = \int_0^\delta \omega \, dy$ is determined. Afterwards, we can compute the aerodynamic force acting on an airfoil as follows, (Wu)

$$
\vec{F} = -\rho \frac{d}{dt} \int_{B_s} \vec{r} \, x \, (\gamma - \vec{V} x \vec{n}_s) \, dB_s - \rho \frac{d}{dt} \int_W \vec{r} \, x \, \vec{\omega} \, dR \tag{2.75}
$$

Here, $\mathbf{n_s}$ is the unit normal to the airfoil surface and $\mathbf{Vxn_s}$ is the velocity tangent to the surface. For a pitching and plunging airfoil, the value of the tangential velocity is computed at every discrete point on the surface and used in Eq. 2.75.

2.17 Turbulence Modeling

At high free stream speeds external flows are likely to go through a transition from laminar to turbulence on the airfoil surface close to the leading edge. Depending on the value of the Reynolds number most of the flow on the airfoil becomes turbulent. The Reynolds decomposition technique applied to the Navier-Stokes equations results in new unknowns of the flow field called Reynolds stresses. These new unknowns introduce more unknowns than the existing equations which is called the closure problem of turbulence. In order to close the problem, the Reynolds stresses are empirically modeled in terms of the velocity gradients. All these models aim at finding the suitable value of turbulence viscosity μ_T applicable for different flow cases. The empirical turbulence models are in general based on the wind tunnel tests and some numerical verification. The simplest models of turbulence are the algebraic models. More complex models are based on differential equations. Although so many models have been introduced, there has not been a satisfactory model developed to reflect the main characteristics of a turbulent flow. Now, we present the well known Baldwin-Lomax model which is used for the numerical solution of attached or separated, incompressible or compressible flows of aerodynamics. This model is a simple algebraic model which assumes the turbulent region to be composed of two different layers. Accordingly the turbulence viscosity reads

$$\mu_T = \begin{cases} (\mu_T)_i\,,\ for\ z \geq z_c \\ (\mu_T)_o\,,\ for\ z < z_c \end{cases}. \tag{2.76}$$

Here, z is the normal distance to the surface, z_c is the shortest distance where inner and outer viscosity values are equal. The inner viscosity value in terms of the mixing length *l* and the vorticity ω reads as

$$(\mu_T)_i = \rho\, l^2 |\omega|\, R_e \quad \text{and} \quad l = \kappa z\,[1 - \exp(-z^+/A^+)] \tag{2.77a, b}$$

Here, κ = 0.41 is the von Karman constant, $A^+ = 26$ damping coefficient and $z^+ = z\sqrt{|\omega|\, R_e}$. The outer viscosity, on the other hand

$$(\mu_T)_o = K\, C_{cp} F_w F_{kl}(z), \quad F_w = z_{maks} F_{maks} \tag{2.78a, b}$$

Here, $K = 0.0168$ is the Clauser constant and $C_{cp} = 1.6$. F_{maks} maximum of F(z) where z_{maks} is the z value at which F_{maks} is found. For this purpose,

$$F(z) = z\,|\omega|\,[1 - \exp(-z^+/A^+)] \quad \text{and} \quad F_{kl}(z) = \left[1 + 5.5\,(\frac{z\,C_{kl}}{z_{maks}})^6\right]^{-1} \tag{2.79a, b}$$

Here, $C_{kl} = 0.3$ (Baldwin and Lomax 1978).

The research on turbulence models are of interest to many branches of fluid mechanics. The Baldwin-Lomax model is implemented for the aerodynamic applications of attached or separated flows considered here. More complex models based on the differential equation solutions are utilized even in commercial softwares of CFD together with the necessary documentations. Detailed information, scientific basis and their application areas for different turbulent models are provided by Wilcox (1998).

2.18 Initial and Boundary Conditions

The study of aerodynamical problems with real gas effects requires solution of a system of partial differential equations which are first order in time and second order in space coordinates. In order to solve Eq. 2.49 to determine the flow field, all dependent variables must be prescribed at time $t = 0$, and for all times t at the boundaries of the computational domain. All the prescribed values must be in accordance with the physics of the problem. As the initial conditions for the unknown values of **U** we prescribe the undisturbed flow conditions, i.e., $u = 1$, $v = w = 0$ which represents the impulsive start of the flow. Under these conditions the initial values for the unknown vector in generalized coordinates become

$$\vec{U}(t = 0,\ \xi, \eta, \varsigma) = \begin{pmatrix} \rho^0 \\ \rho^0 \\ 0 \\ 0 \\ \varepsilon^0 \\ c_i^0 \end{pmatrix} \tag{2.80}$$

Here, ρ^0 is the initial value for the density, ε^0 is the initial value for the energy and c_i^0 is the initial value of the *i*th specie.

As for the boundary conditions: (i) the unknowns at the surface, and (ii) farfield boundary conditions must be provided.

Accordingly:

(i) As the no slip condition at the surface: $\mathbf{U}(t,\xi,\eta,\varsigma = 0) = \mathbf{0}$ is prescribed. (In Fig. 2.6, $\varsigma = 0$ prescribes the surface). In reactive flows the catalicity of the surface determines the value of the concentration gradients,

(ii) At the farfield: for $\varsigma = \varsigma_{\text{maks}}$ $\mathbf{U}(t, \xi, \eta, \zeta = \zeta_{\text{maks}}) = \vec{U}_\infty$ is prescribed, and the flux condition at $\xi = \xi_{\text{maks}}$ is $\frac{\partial \vec{U}}{\partial \xi} = \left(\vec{0}\right)$,

(iii) If there is a symmetry condition as shown in Fig. 2.6b, we prescribe the flux normal to the symmetry as $\frac{\partial \vec{U}}{\partial \eta} = \left(\vec{0}\right)$.

Summary: *In order to analyze the problems of aerodynamics we have constructed mathematical models of the physical phenomena. For this purpose, conservation equations are used for the development of the potential theory. The equation which is satisfied by the velocity potential is developed using the state, continuity, momentum and the energy equations with assuming the flow of a perfect fluid without any viscous and body forces and without any chemical reactions. The potential equation will be solved with proper boundary conditions. The boundary conditions for an external flow are applied at the infinity and at the body surface. We assumed that at the infinity all disturbances die out, only the free stream conditions prevail and on the body surface the boundary conditions are given according to the time dependent motion of the body surface. The material derivative of the surface equation is utilized to get the downwash which is the vertical velocity component solely responsible for the generation of the lift from the body. Both the potential equation and the boundary conditions are nonlinear in nature. It is very difficult to solve this nonlinear equation except for a few cases where the series* *solutions are obtained. Therefore, the equations are linearized through the concept of small disturbances. A linear equation satisfied by the perturbation potential and a linear form of the downwash expression are obtained. It is now possible to solve this linear equation with a linear boundary condition for variety of external flow cases using the superposition techniques. Another useful potential function is the acceleration potential derived from the definition of the acceleration vector. It is related to the velocity potential and gives a convenient expression for the surface pressure coefficient. Finally, the concept of body fixed and the flow fixed coordinate systems are introduced. The coordinate transformation from body fixed to flow fixed coordinate systems render the potential equation into the form of classical wave equation which has variety of known solutions.*

Real gas effects are also considered in deriving the new form of equations for the viscous, thermally conducting, chemically reacting and diffusing flows which are of interest to aerodynamics. In derivation, the Reynolds transport theorem which relates the system to the control volume for the conservation laws is utilized. Then, the differential form of these conservation laws are derived from the original integral representations. The conservation of species is added to the system of equations for the case of reacting flows which occur at very high speeds. The equations are cast in conservative form. In order to impose the surface boundary conditions properly the equations are expressed in generalized curvilinear coordinates. The Navier-Stokes equations and their simplified versions like thin shear layer N-S and boundary layer equations are derived through order of magnitude analysis. Incompressible Navier- Stokes equations are presented for the flows having low free stream velocities. The formulas for the evaluation of aerodynamic forces and moments involving curved surfaces are provided. For high-speed aerodynamic flows the turbulence is inevitably present. Therefore, an algebraic turbulence modeling which is suitable for the separated and unseparated flows is given. Finally, the initial and the boundary conditions for the aerodynamic flow are presented.

2.19 Questions and Problems

2.1 In a barotropic flow show that $\frac{1}{\rho}\nabla p = \nabla \int \frac{dp}{\rho}$.

2.2 Eq. 2.15 is written in terms of the velocity potential. Express the same equation with partial derivatives of velocity potential.

2.3 An oblate ellipsoid is undergoing vertical simple harmonic motion with amplitude $\overline{a}$. Express the equation of upper and lower surfaces of the airfoil.

2.4 The ellipsoid given in Problem 2.3 is also undergoing a pulsative major axis change with the same period but with phase difference ϕ. Express the equation of surfaces.

2.5 Comment on the steady or unsteady lift generation by referring to the downwash expression given by 2.19.

2.6 The equation of a paraboloid of length l and whose axis is in line with x axis is given as $c(x/l) = (y^2 + z^2)/a^2$, $0 \le x \le l$ and $0 \le y, z \le a$. Obtain the downwash expression at the surface. If a slender paraboloid undergoes SHM about its nose in a vertical y–z plane, find the unsteady downwash expression at the surface.

2.7 A lighter than air prolate ellipsoid moves in air with constant speed U. If this air vehicle oscillates simple harmonically about its center with a small amplitude A in a vertical plane then find the time dependent surface downwash expression at the (i) shoulders, and) (ii) at the front end rear ends.

2.8 We do not need to define perturbation potential for the acceleration potential. Why?

2.9 From the non linear relation between the velocity and the acceleration potential, obtain the linear relation given by Eq. 2.25.

2.10 Obtain the surface pressure and downwash expressions in terms of accelerationpotential.

2.11 Derive the Reynolds Transport theorem, 2.27, which interlaces the system and control volume approaches.

2.12 Obtain Equation 2.35 which gives the continuity of the species

2.13 Express the conservation of momentum in open form in Cartesian coordinates.

2.14 Obtain the expression given by 2.50 by means of the transformation from Cartesian to generalized coordinates.

2.15 For a tapered wing with half-span of 4 units let x be the chordwise and y be the spanwise directions. The equation for the leading edge is given as: $x = 0.15y$, $0 < y < 4$ and the trailing edge: $x = -0.025y + 4$, $0 < y < 4$. Using the two dimensional numerical transformation with $0 < \xi < 1 \times 0 < \eta < 1$ for 11×11 equally spaced discrete points transform the wing surface from x–y coordinates to ξ-η generalized coordinates. Find the metrics of transformation and Jacobian determinant at each discrete location.

2.16 In generalized coordinates, obtain the Navier-Stokes equations for the thin shear layercase in terms of the contravariant velocity components.

2.17 Express the components of stress tensor in generalized coordinates in terms ofvelocity gradients.

References

Anderson, D.A., Tannehill, J.C., Pletcher, R.H.: Computational Fluid Mechanics and Heat Transfer. Hemisphere, New York (1984)

Anderson, J.D.: Hypersonic and High Temperature Gas Dynamics. McGraw-Hill, New York (1989)

(BAH) Bisplinghoff, R.L., Ashley, H., Halfman, R.L.: Aeroelasticity. Dover Publications Inc., New York (1996)

Baldwin, B.S., Lomax, H.: Thin Layer Approximations and Algebraic Model for Separated Flows. AIAA Paper 78–0257, January (1978)

Batchelor, G.K.: An Introduction to Fluid Dynamics. Cambridge University Press, London (1979)

Fox, R.F., McDonald, A.T.: Introduction to Fluid Mechanics, 4th ed. John Wiley, New York (1992)

Hildebrand, F.B.: Advanced Calculus for Applications. Prentice-Hall Inc., Engelwood Cliffs, New Jersey (1976)

Hoffman, J.D.: Numerical Methods for Engineers and Scientists. McGraw-Hill, New York (1992)

Liepmann, H.W., Roshko, A.: Elements of Gas Dynamics. Wiley, New York (1963)

Schlichting, H.: Boundary-Layer Theory, 6th edn. McGraw-Hill, New York (1968)

Shames, I.H.: Engineering Mechanics: Statics and Dynamics. Prentice-Hall, New Delhi (1969)

Wilcox, D.C.: Turbulence Modelling for CFD. DCW Industries, California (1998)

Wu, J.C., Gulcat, U.: Separate treatment of attached and detached flow regions in general viscous flows. AIAA J. **19**, 20–27, No1 (1981)

Chapter 3
Incompressible Flow About an Airfoil

The physical characteristics of external flow past a thin airfoil at a small angle of attack enables us to build a simple mathematical model of the flow. We assume here our profile starts to move impulsively from the rest and reaches the constant speed of U in zero time. If the viscous forces exist, their resistance to the impulsive motion will be so large that the required force to move the airfoil will also be incredibly large. However, if we neglect the viscous effects at the beginning, the assumption of impulsive start of a motion will be more meaningful. Under this assumption, we can model the external flow in connection with the creation of lift via the bound vortex formation around the airfoil in a free stream and explain the whole phenomenon by means of Kelvin's theorem which was introduced in Chap. 2.

3.1 Impulsive Motion

When the airfoil starts its translational motion impulsively, as observed from the body fixed coordinates, the air suddenly starts rushing towards it with the speed U and creates a velocity field $\mathbf{V} = \mathbf{V}(x, z)$ parallel to the surface of the airfoil as shown in Fig. 3.1. The fluid particles move along the streamlines of the flow field. The characteristic streamline of the flow is the stagnation streamline which comes at the front stagnation point and branches into two on the surface and leaves the surface of the airfoil at the rear stagnation point. The fluid particles which move on the stagnation streamlines have naturally zero velocities at the stagnation points. There are two stagnation points for this external flow.

The fluid particles on the frontal stagnation streamline first decelerate towards the frontal stagnation point and after passing the branch point they accelerate over the upper and lower surfaces until they reach their maximum velocity. The particles moving on the upper surface move faster in a narrow passage because of the thickness of the airfoil and they slow down to zero velocity until they reach the rear

Ü. Gülçat, *Fundamentals of Modern Unsteady Aerodynamics*,
https://doi.org/10.1007/978-3-030-60777-7_3

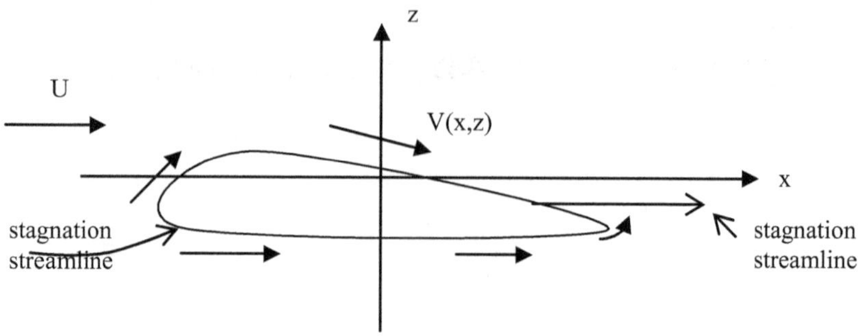

Fig. 3.1 Impulsively started airfoil

stagnation. At the lower surface, however, the accelerating flow particles move towards the trailing edge and almost circle around it and reach the rear stagnation point which for the time being is at the upper surface. The flow picture looks very unsymmetrical and the location of the stagnation points are different from the leading and trailing edges.

The velocity vector which is parallel to the surface and nonzero except at the stagnation points will be used as the edge velocity of the boundary layer which is introduced by Prandtl for analyzing the viscous effects. In the boundary layer, however, the velocity values will go to zero at the surface because of the no slip condition as shown in Fig. 3.2.

At the onset of impulsive motion very high velocity gradients take the shape shown in Fig. 3.2 in a short duration and the outside of the thin boundary layer a very large flow field remains potential. In the boundary layer the viscous effects are likely to generate a part of a circulation which contributes to the overall circulation used in generation of lift. Now, we can use our model in a boundary layer of thickness δ with calculating the infinitesimal circulation $d\Gamma$ over a rectangular boundary whose length is ds as shown in Fig. 3.3. At the left face of the rectangle, the vertical velocity is v and with increment ds its value at the right face is $v + dv$, and $V(x, y)$ is the edge velocity. The infinitesimal circulation in clockwise direction becomes.

$d\Gamma = 0 \cdot ds + v \cdot \delta + V(x, z) \cdot ds - (v + dv) \cdot \delta$. If we neglect the second order term $dv \cdot \delta$ we find: $d\Gamma = V(x, z) \cdot ds$. Accordingly, the rate of increase of circulation reads as

Fig. 3.2 Velocity profile in a boundary layer

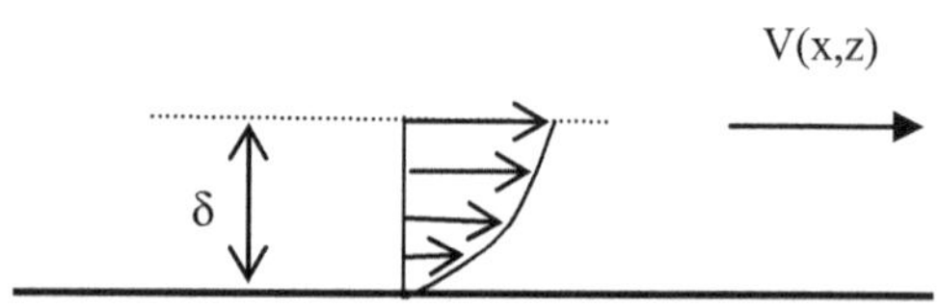

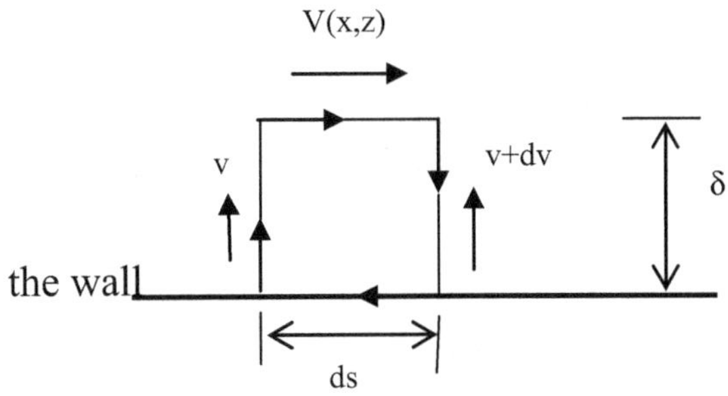

Fig. 3.3 Local circulation dΓ

$$\frac{d\Gamma}{ds} = V(x,z) = \gamma.$$

The boundary layer at the surface can be modeled as a vortex sheet with strength γ while the outside of boundary layer is the potential flow region. This modeling represents the physics of the external flow. Let us use Kelvin's theorem to find the total time variation of the circulation value in the flow field of the impulsively started airfoil According to Kelvin's theorem, the total circulation remains constant throughout the motion. Since the motion starts from rest, the total circulation at the beginning is zero and remains zero to give

$$\Gamma = \oint \gamma . ds = 0$$

The closed integral here is evaluated around the airfoil on an arbitrary closed loop. For the sake of convenience, the closed loop can be chosen as the airfoil surface. As stated before, right after the start accelerating air particles at the bottom surface turn around the trailing edge with a very high velocity. The sharper the trailing edge, the more the speed of turning. Therefore, there is a limit to the turning speed after which the increase is not physically possible because of the pressure drop around the trailing edge. For the physically possible case, after the onset of motion the counter clockwise rotating vortex sheet of the bottom surface tries to turn around the sharp trailing edge but separates from the surface and gets carried downstream while the clockwise rotating vortex sheet of upper surface moves toward the trailing edge. The lifting force which was zero initially starts growing. In Fig. 3.4a, shown is the $t = 0^+$ time depiction of the flow field with the upper and lower vortex sheets mentioned above. A short while after the start, the upper surface vortex sheet moves at the sharp trailing edge, and pushes the lower surface vortex sheet down to wake until the rear stagnation point reaches the trailing edge. After this time, the flow becomes stable on the airfoil with the constant bound circulation Γ_a as shown in Fig. 3.4b at time $t > 0$.

As seen in Fig. 3.4b, there are two different circulations in the flowfield. The first one is due to the bound circulation on the airfoil and the second one is due to the wake circulation. We calculate both of the circulations on clockwise paths as shown

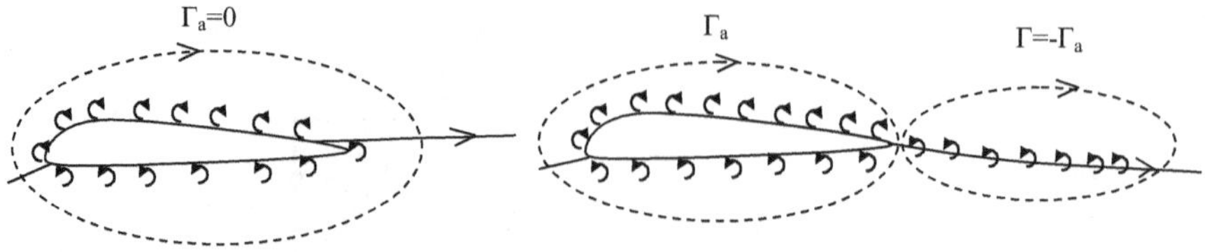

Fig. 3.4 **a** and **b** Surface and wake vortex sheet at $t = 0^+$ and $t > 0$

with dashed lines. The value of bound circulation can be simply found by adding the upper and lower vortex sheet strengths. The wake circulation, on the other hand, consists of only the counter clockwise rotating vortices which then add up to $-\Gamma_a$. According to the Kelvin's theorem, the total circulation must be zero which makes the bound circulation value Γ_a. The picture on the upper surface remains the same, meaning that the bound circulation is present all the time moving with the airfoil to keep the rear stagnation point at the trailing edge.

At the wake, however, the counter clockwise vortices shed into the downstream, get together and form the starting vortex of strength $-\Gamma_a$ and stay at the far wake. Although it retains the same strength for a long time, its effect on the airfoil is negligible according to the Biot–Savart law since it is far away from the airfoil.

As the velocity at the trailing edge becomes zero, the vortex sheet strength of upper and lower surfaces around the trailing edge becomes equal in magnitude and opposite in sign. That means as the steady state is reached, the shed vortices into the wake cancel each other to result in no vortex sheet in near wake. Having zero velocity at the sharp trailing edge is called Kutta condition. It is the Kutta condition which generates a positive circulation and in turn creates the lifting force on the airfoil. It has been observed experimentally that 90% of the lift on the airfoil is generated with 3 chord travel of the airfoil after the impulsive start, (Kuethe and Chow 1998). The early Computational Fluid Dynamics studies with Navier–Stokes solutions had indicated that almost all the lift is generated within the 4 chord of travel of an airfoil after the impulsive start (Gulcat 1981).

Now, we can study the steady flow thin airfoil aerodynamics by considering vortex sheet present at the surface of the profile.

3.2 Steady Flow

Once the Kutta condition is satisfied, the picture of the flow field remains the same, which means the flow is steady. In a steady flow around airfoil as stated before, there is a bound vortex and the starting vortex. Since the starting vortex is located far away from the profile it has practically no effect. The only vortex in effect is the vortex sheets of upper and lower surfaces. If the thickness of the profile is less than

%12, it is assumed that the upper and lower surface vortex sheets are close enough and they add up to a single vortex sheet which is easily modeled as a vortex sheet of strength γ_a. That means, for γ_u showing the upper surface vortex sheet strength and γ_1 showing the lower surface then they add up to

$$\gamma_a(x) = \gamma_u(x) + \gamma_1(x).$$

With this mathematical modeling the Kutta condition and the Laplace's equations, Eq. 2.15, are both satisfied. Figure 3.5 shows the vortex sheet modeling an airfoil with its chord is in line with x axis and has length of $2b$.

According to the Biot–Savart law (Kuethe and Chow 1998), the vortex sheet of strength $\gamma_a(x)$ and length $\mathrm{d}\xi$ induces the differential velocity of $\mathrm{d}V$ at a point (x, z).

$$\mathrm{d}V = \frac{\gamma_a(x)\,\mathrm{d}\xi}{2\pi\, r}$$

The x and z components of dV reads as

$$\mathrm{d}u' = \mathrm{d}V \sin\theta, \quad \mathrm{d}w = -\mathrm{d}V\cos\theta, \quad \sin\theta = z/r \text{ and } \cos\theta = (x-\xi)/r.$$

The induced components, from Fig. 3.5, can be expresses as

$$\mathrm{d}u' = \frac{z\gamma_a(\xi)\mathrm{d}\xi}{2\pi r^2} \quad \text{and} \quad \mathrm{d}w = -\frac{(x-\xi)\gamma_a(\xi)\mathrm{d}\xi}{2\pi r^2}, \quad r^2 = (x-\xi)^2 + z^2$$

Now, we can take the integral over the total length of the vortex sheet

$$u'(x,z) = \frac{1}{2\pi}\int_{-b}^{b} \frac{z\gamma_a(\xi)\mathrm{d}\xi}{(x-\xi)^2 + z^2} \quad \text{and} \quad w(x,z) = -\frac{1}{2\pi}\int_{-b}^{b} \frac{(x-\xi)\gamma_a(\xi)\mathrm{d}\xi}{(x-\xi)^2 + z^2}$$

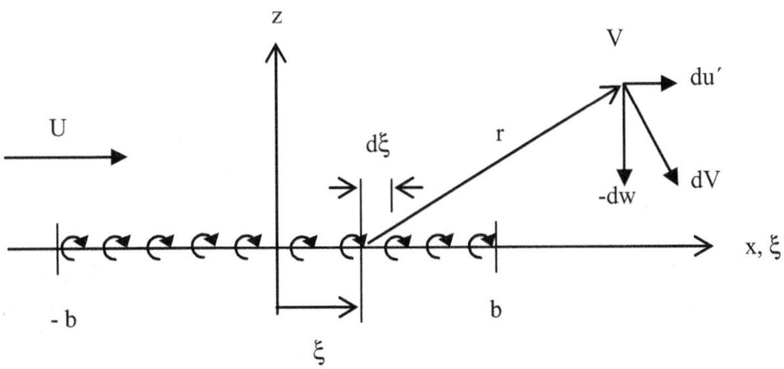

Fig. 3.5 Vortex sheet modeling of the airfoil

At this stage, it is essential to note that there is no contribution to the perturbation velocities from the free stream speed. If we closely examine the sign of z in the integrands of the above integrals we see that u' is antisymmetric and w is symmetric. That is

$$u'(x,0^+) = -u'(x,0^-) \quad \text{and} \quad w(x,0^+) = w(x,0^-)$$

Let us now find the relation between the perturbation speed u' and the vortex sheet strength as follows.

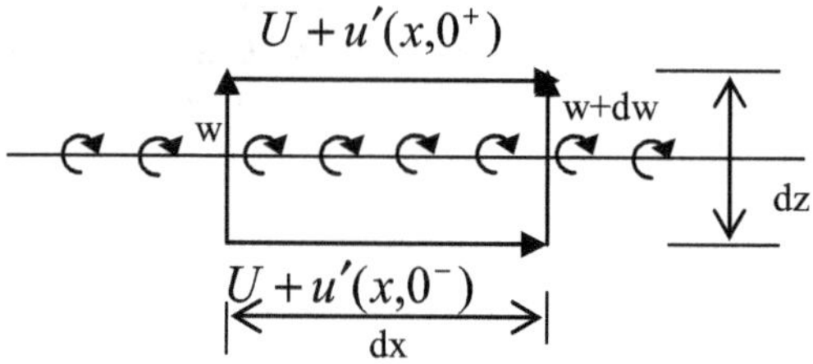

The rectangle shown with the dimensions of $\mathrm{d}x \cdot \mathrm{d}z$ has the circulation given as

$$\begin{aligned}\gamma_a(x)\,\mathrm{d}x &= [U + u'(x,0^+)]dx - (w + \mathrm{d}w)\,\mathrm{dd}z - [U + u'(x,0^-)]\mathrm{d}x + w\,\mathrm{d}z \\ &= [u'(x,0^+) - u'(x,0^-)]\mathrm{d}x - \mathrm{d}w\,\mathrm{d}z\end{aligned}$$

Neglecting the second order terms we get

$$\gamma_\alpha(x) = u'(x,0^+) - u'(x,0^-) = 2\,u'(x,0^+) \tag{3.1}$$

Equation 3.1 tells us that the perturbation speed in x direction is given by the local vortex sheet strength. In addition, the physical meaning of a vortex sheet strength is that it is the discontinuity of the velocity between the upper and lower surfaces.

Let us now find the downwash at the surface, $z = 0$,

$$w(x,0) = -\frac{1}{2\,\pi}\int_{-b}^{b}\frac{\gamma_a(\xi)\,\mathrm{d}\xi}{x-\xi} \tag{3.2}$$

The integral given in Eq. 3.2 has an integrable singularity at $x = \xi$. These type of singular integrals are called the Cauchy type integrals and in Appendix 3 we show how to evaluate this type of integrals at the complex plane.

Equation 3.2 is an integral equation if we consider $\gamma_a(x)$ as unknown function and $w(x)$ as the known downwash. This type of integral equation is called Fredholm type and its inversion is provided in Appendix 2. Accordingly, if we use the non dimensional coordinates $x^* = x/b$ *and* $\xi^* = \xi/b$ and utilize the Eq. 3a, b of Appendix 3 we obtain the inverted form of Eq. 3.2 as

$$\gamma_a(x^*) = \frac{2}{\pi}\sqrt{\frac{1-x^*}{1+x^*}} \int_{-1}^{1} \sqrt{\frac{1+\xi^*}{1-\xi^*}} \frac{w(\xi)}{x-\xi^*} \, \mathrm{d}\xi \tag{3.3}$$

Equation 3.3 satisfies the Kutta condition at the trailing edge because it has a zero value as x^* takes the value of 1. The integrand in the equation is obtained from Eq. 2.20. In case of steady flows the downwash is function of angle of attack, free stream speed U and camber of the airfoil.

After finding an expression for the bound vortex sheet we can now relate it with the lifting pressure coefficient. For the steady flow the pressure coefficient was given by Eq. 2.21 as

$$c_p(x) = -\frac{2}{U}\frac{\partial \varphi'}{\partial x}$$

Let us now find the lifting pressure c_{p_a} as the pressure difference between the lower and upper pressures

$$c_{p_a} = \frac{p_l - p_u}{\frac{1}{2}\rho_\infty U^2} = \frac{2}{U}\left(\frac{\partial \phi_u'}{\partial x} - \frac{\partial \phi_l'}{\partial x}\right) \tag{3.4}$$

The lifting pressure coefficient can be expressed in terms of the upper and lower perturbation speeds. With the aid of Eq. 3.1

$$c_{p_a}(x) = \frac{2\gamma_a(x)}{U} \tag{3.5}$$

According to Eq. 3.5, the lifting pressure coefficient behaves similar to that of the vortex sheet strength. This behavior can be seen with a limiting process applied at the leading and the trailing edges as follows

$$\lim_{x\to -b}[c_{p_a}(x)] = \lim_{\varepsilon\to 0}\frac{1}{\sqrt{\varepsilon}} \quad \text{and} \quad \lim_{x\to b}[c_{p_a}(x)] = \lim_{\varepsilon\to 0}\sqrt{\varepsilon} = 0$$

With these limiting values we see that the singular lifting pressure at the leading edge becomes zero at the trailing edge. Now, the sectional lifting force l can be found using Eq. 3.5 with integration

$$l = \frac{1}{2}\rho_\infty U^2 \int_{-b}^{b} c_{p_a}(x)\,\mathrm{d}x = \rho_\infty U \int_{-b}^{b} \gamma_a(x)\,\mathrm{d}x \quad \text{where} \quad \Gamma_a = \int_{-b}^{b} \gamma_a(x)\,\mathrm{d}x \tag{3.6}$$

gives: $l = \rho_\infty U \, \Gamma_a$ which is known as the Kutta–Joukowski theorem which gives the lifting force acting on a vortex immersed in a free stream speed U and has a strength Γ, (Kuethe and Chow 1998). If we combine Eqs. 3.3 and 3.6 we obtain the bound vortex sheet strength given in terms of the downwash distribution as follows.

$$\Gamma_a = -2\,b \int_{-1}^{1} \sqrt{\frac{1+\xi^*}{1-\xi^*}}\; w(\xi^*)\; \mathrm{d}\xi^* \tag{3.7}$$

Example 3.1 For an airfoil at an angle of attack α find (i) sectional lift coefficient, (ii) moment coefficient and (iii) center of pressure and aerodynamic center.

Solution This the flat plate immersed in a free stream of U with angle of attack α as shown in the following figure.

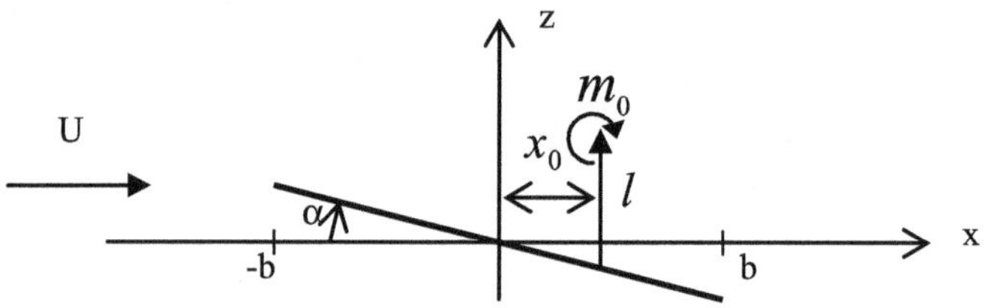

If we use Eq. 2.20 then the downwash w for steady flow reads:

$$w = U\,\frac{\partial z_a}{\partial x} = -U\alpha$$

The Kutta–Joukowski theorem gives the sectional lifting force as

$$l = 2\,b\,\rho_\infty U^2\alpha \int_{-1}^{1} \sqrt{\frac{1+\xi^*}{1-\xi^*}} \mathrm{d}\xi* = 2\,\pi\,b\,\rho_\infty U^2\alpha.$$

(i) The sectional lift coefficient then becomes

$$c_l = \frac{l}{\rho_\infty U^2 b} = 2\,\pi\,\alpha$$

(ii) As a convention we adopt the leading edge up gives + moment. Sectional moment about x_0 reads as:

$$m_0 = -\frac{1}{2}\rho_\infty U^2 \int_{-b}^{b} (x - x_0)\, c_{p_a}(x)\, \mathrm{d}x \tag{3.8}$$

Equation 3.8 for $x_0 = 0$ gives

$$c_{mo} = \pi\,\alpha$$

(iii) The center of pressure x_{cp} in terms of this moment reads as

$$x_{cp} = x_0 - \frac{m_0}{l}$$

in general. Using the result of (ii) for the flat plate

$$x_{cp} = -\frac{b}{2}$$

This result proves that for a symmetric thin airfoil the center of pressure is located at the quarter chord point. Now, let us find the aerodynamic center xac. By definition the aerodynamic center is the point where the sectional moment is independent of the angle of attack α. With Eq. 3.8 and $\partial_{mo}/\partial\alpha = 0$ gives us $x_{ac} = -b/2$. This again proves that the aerodynamic center and the center of pressure are at the same points for a symmetric thin airfoil.

Hitherto, we have given the formulation for the steady flow for which flow conditions remain the same with respect to time. When the flow conditions change slowly with time, we can assume quasi steady flow as it happens for a slow change in the angle of attack so the force and the moment change in phase with the angle of attack. The picture is not the same when the changes are fast because we observe a lag between the motion and the response of the airfoil to the motion. Let us now extend our external flow model for the unsteady treatments which gives us the lag as well as the deviations from the steady flow conditions because the presence of near wake effects.

3.3 Unsteady Flow

Our unsteady analysis of the flow is going to be similar to that of steady flow except now, we are going to assume a vortex sheet strength $\gamma_a = \gamma_a(x, t)$ as the function of two variables x and t. There will also be continuous vortex shedding to the wake from the trailing edge because of having unequal vortex sheet strength from the lower and upper surfaces right at the trailing edge. Since there is a vortex sheet at the near wake there will be a velocity field induced by it as well as its effect on the bound vorticity. Let us now see the effect of the both vortex sheets on the induced downwash with the aid of Fig. 3.6.

Denoting the near wake vortex sheet strength with γ_w, the downwash w at $z = 0$ with the aid of Biot–Savart law

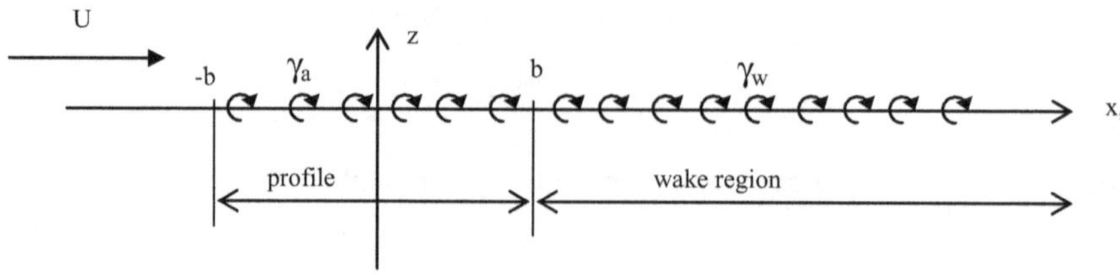

Fig. 3.6 Simple model for the profile and its wake

$$w(x,0,t) = -\frac{1}{2\pi}\int_{-b}^{b}\frac{\gamma_a(\xi,t)\,\mathrm{d}\xi}{x-\xi} - \frac{1}{2\pi}\int_{b}^{\infty}\frac{\gamma_w(\xi,t)\,\mathrm{d}\xi}{x-\xi} \tag{3.9}$$

The first integral at the right hand side of Eq. 3.9 is singular but the second integral is not. We can write the equivalent of the second integral in terms of the bound vortex using the unsteady Kutta condition.

In case of steady flow we have expressed the Kutta condition as the zero velocity at the trailing edge or no vortex sheet at the near wake or no pressure difference at the wake region. In case of unsteady flow, however, there is a non zero velocity at the trailing edge and non zero vortex sheet at the near wake. Therefore, the unsteady Kutta condition is expressed as the zero pressure difference at the wake. Accordingly, the unsteady Kutta condition is more restrictive, and therefore in formulation it reads

$$c_{p_a}(x) = \frac{p_l - p_u}{\frac{1}{2}\rho_\infty U^2} = 0, \quad x \geq b$$

In terms of perturbation potential, using Eq. 2.21 it becomes

$$c_{pa}(x) = \frac{2}{U^2}\frac{\partial}{\partial t}(\phi'_u - \phi'_l) + \frac{2}{U}\left(\frac{\partial \phi'_u}{\partial x} - \frac{\partial \phi_l}{\partial x}\right)$$

Equation 3.1 gives the relation between the perturbation potential and the vortex sheet strength for the steady flow case. Similarly, we can write this relation for the unsteady flows at any time t as follows

$$\gamma_w(x,t) = u'(x,0^+,t) - u'(x,0^-,t) = \left(\frac{\partial \phi'_u}{\partial x} - \frac{\partial \phi'_l}{\partial x}\right)_w, \quad x \geq b$$

The integral relation between the perturbation potential and the perturbation velocities are

$$\phi'_u = \int_{-\infty}^{x} \frac{\partial \phi'_u}{\partial \xi} \mathrm{d}\,\xi = \int_{-\infty}^{x} u'_u\, \mathrm{d}\xi \quad \text{and} \quad \phi'_l = \int_{-\infty}^{x} \frac{\partial \phi'_l}{\partial \xi} \mathrm{d}\xi = \int_{-\infty}^{x} u'_l \mathrm{d}\,\xi.$$

Before the leading edge we do not have any velocity discontinuity between upper and lower surfaces therefore, for $x < -b$ there is not any contribution to the integrals evaluated for $x > b$

$$\begin{aligned}\phi'_u - \phi'_l &= \int_{-b}^{x} (u'_u - u'_l)\mathrm{d}\xi = \int_{-b}^{b} (u'_u - u'_l)\mathrm{d}\xi + \int_{b}^{x} (u'_u - u'_l)\mathrm{d}\xi \\ &= \int_{-b}^{b} \gamma_a(\xi, t)\mathrm{d}\xi + \int_{b}^{x} \gamma_w(\xi, t)\mathrm{d}\xi\end{aligned}$$

If we take the derivatives of the above expression with respect to t and x, the unsteady Kutta condition becomes

$$\frac{\partial}{\partial t}\int_{-b}^{b} \gamma_a(\xi, t)\mathrm{d}\xi + \frac{\partial}{\partial t}\int_{b}^{x} \gamma_w(\xi, t)\mathrm{d}\xi + U\gamma_w(x, t) = 0$$

The first integral at the left hand side is evaluated to the bound vortex $\Gamma_a\,(t)$. Hence, the final form of the unsteady Kutta condition reads

$$\frac{\mathrm{d}\Gamma_a}{\mathrm{d}t} + \frac{\partial}{\partial t}\int_{b}^{x} \gamma_w(\xi, t)\mathrm{d}\xi + U\gamma_w(x, t) = 0 \tag{3.10}$$

Equation 3.10 is an integro-differential equation which relates the bound vortex to the vortex sheet strength of the wake. Our aim here is to eliminate the wake vorticity appearance from the downwash expression so that all the terms in Eq. 3.9 are expressed in terms of the bound vortex sheet strength. If we transform time coordinate to some other coordinate and then differentiate the result with respect to x we can succeed to do so. Let us now take the Laplace transform of Eq. 3.10, remembering the definition and a property of the Laplace transform (Hildebrand 1976),

$$L\{\mathrm{f(t)}\} = \int_{0}^{\infty} e^{-st} f(t)\,\mathrm{d}\,t = \bar{f}\,(s) \lim_{x\to\infty} \quad \text{and} \quad L\left\{\frac{\mathrm{d}f(t)}{\mathrm{d}t}\right\} = s\bar{f}\,(s) - f(0^+).$$

The Laplace transform of Eq. 3.10 then becomes

$$s\overline{\Gamma}_a + \int_b^x s\overline{\gamma}_w(\xi, s)\mathrm{d}\,\xi + U\overline{\gamma}_w(x, s) = 0 \tag{3.11}$$

Here, at $t = 0^+$, Γ_a and $\gamma_w(x)$ are both zero. If we take the derivative of Eq. 3.11 with respect to x, the first term becomes zero and we end up with a first order differential equation in x.

$$S\,\overline{\gamma}_w(x, s) + U\frac{\partial}{\partial x}\,\overline{\gamma}_w(x, s) = 0 \tag{3.12}$$

The solution to this equation becomes

$$\overline{\gamma}_w(x, s) = B(s)\,e^{-\frac{sx}{U}}.$$

In order to determine $B(s)$ we utilize the value of Eq. 3.10 at $x = b$. This gives

$$s\overline{\Gamma}_a + U\overline{\gamma}_w(b, s) = 0 \text{ and } \overline{\gamma}_w(b, s) = B(s)\,e^{-\frac{sb}{U}} \quad \text{combined} \quad B(s) = -\frac{s\overline{\Gamma}_a}{U}\,e^{\frac{sb}{U}}.$$

substituting $B(s)$ gives

$$\begin{aligned} \overline{\gamma}_w(x, s) &= -\frac{s\overline{\Gamma}_a}{U}\; e^{-\frac{s}{U}(x-b)} \text{ or with } x^* = x/b \\ \overline{\gamma}_w(x^*, s) &= -\frac{s\overline{\Gamma}_a}{U}\; e^{-\frac{sb}{U}(x^*-1)} \end{aligned} \tag{3.13}$$

Now, we can express Eq. 3.9 in non dimensional coordinates and in its Laplace transformed form as follows

$$\begin{aligned} \overline{w}(x^*, s) &= -\frac{1}{2\pi}\int_{-1}^{1}\frac{\overline{\gamma}_a(\xi^*, s)\,\mathrm{d}\xi^*}{x^* - \xi^*} - \frac{1}{2\pi}\int_{1}^{\infty}\frac{\overline{\gamma}_w(\xi^*, s)\,\mathrm{d}\xi^*}{x^* - \xi^*} \\ &= -\frac{1}{2\pi}\int_{-1}^{1}\frac{\overline{\gamma}_a(\xi^*, s)\,\mathrm{d}\xi^*}{x^* - \xi^*} + \frac{s\overline{\Gamma}_a}{2\pi U}e^{\frac{sb}{U}}\int_{1}^{\infty}\frac{e^{-\frac{sb}{U}\xi^*}\,\mathrm{d}\xi^*}{x^* - \xi^*} \end{aligned} \tag{3.14}$$

Equation 3.14 can be rearranged to give a Fredholm type but non homogeneous equation as follows

$$\overline{w}(x^*, s) - \frac{s\overline{\Gamma}_a}{2\pi U}e^{\frac{sb}{U}}\int_{1}^{\infty}\frac{e^{-\frac{sb}{U}\xi^*}\,\mathrm{d}\xi^*}{x^* - \xi^*} = -\frac{1}{2\pi}\int_{-1}^{1}\frac{\overline{\gamma}_a(\xi^*, s)\,\mathrm{d}\xi^*}{x^* - \xi^*} \tag{3.15}$$

In Eq. 3.15 the second term at the right hand side of the equation is the non homogeneous term. Inverting the integral as described in Appendix 1 we obtain

$$\overline{\gamma}_a(x^*,s) = \frac{2}{\pi}\sqrt{\frac{1-x^*}{1+x^*}}\int_{-1}^{1}\sqrt{\frac{1+\xi^*}{1-\xi^*}}\frac{\overline{w}(\xi^*,s)\,\mathrm{d}\xi^*}{x^*-\xi^*}$$
$$-\frac{2}{\pi}\sqrt{\frac{1-x^*}{1+x^*}}\int_{-1}^{1}\sqrt{\frac{1+\xi^*}{1-\xi^*}}\frac{s\overline{\Gamma}_a}{2\pi U}e^{\frac{sb}{U}}\int_{1}^{\infty}\frac{e^{-\frac{sb}{U}\lambda}\mathrm{d}\xi^*\mathrm{d}\lambda}{(x^*-\xi^*)(\xi^*-\lambda)}$$

If we interchange the order of integration then we have

$$\overline{\gamma}_a(x^*,s) = \frac{2}{\pi}\sqrt{\frac{1-x^*}{1+x^*}}\int_{-1}^{1}\sqrt{\frac{1+\xi^*}{1-\xi^*}}\frac{\overline{w}(\xi^*,s)\,\mathrm{d}\xi^*}{x^*-\xi^*}$$
$$-\frac{1}{\pi}\sqrt{\frac{1-x^*}{1+x^*}}\frac{s\overline{\Gamma}_a}{\pi U}e^{\frac{sb}{U}}\int_{1}^{\infty}e^{-\frac{sb}{U}\lambda}\int_{-1}^{1}\sqrt{\frac{1+\xi^*}{1-\xi^*}}\frac{\mathrm{d}\xi^*\mathrm{d}\lambda}{(x^*-\xi^*)(\xi^*-\lambda)}$$

Denoting the double integral with I_1, we get

$$I_1 = \int_{1}^{\infty}e^{-\frac{sb}{U}\lambda}\int_{-1}^{1}\sqrt{\frac{1+\xi^*}{1-\xi^*}}\frac{\mathrm{d}\xi^*\mathrm{d}\lambda}{(x^*-\xi^*)(\xi^*-\lambda)}$$

Let us also write the denominator of the integrand as partial fractions

$$I_1 = \int_{1}^{\infty}\frac{e^{-\frac{sb}{U}\lambda}}{x^*-\lambda}\left[\int_{-1}^{1}\sqrt{\frac{1+\xi^*}{1-\xi^*}}\left(\frac{1}{x^*-\xi^*}+\frac{1}{\xi^*-\lambda}\right)d\xi^*\right]\mathrm{d}\lambda$$

and evaluate the inner integrals as follows

$$\int_{-1}^{1}\sqrt{\frac{1+\xi^*}{1-\xi^*}}\frac{d\xi^*}{x^*-\xi^*} = -\pi;\,(-1\le x\le 1)\quad\text{and}$$
$$\int_{-1}^{1}\sqrt{\frac{1+\xi^*}{1-\xi^*}}\frac{\mathrm{d}\xi^*}{\xi^*-\lambda} = \pi\left(1-\sqrt{\frac{\lambda+1}{\lambda-1}}\right);\,(\lambda\ge 1)$$

Adding those two together

$$\mathrm{I}_1 = -\pi \int_1^\infty \sqrt{\frac{\lambda+1}{\lambda-1}} \frac{e^{-\frac{sb}{U}\lambda}}{x^* - \lambda} \mathrm{d}\lambda.$$

Substituting the expression for I_1 in vortex sheet strength formula gives

$$\overline{\gamma}_a(x^*, s) = \frac{2}{\pi}\sqrt{\frac{1-x^*}{1+x^*}} \int_{-1}^{1} \sqrt{\frac{1+\xi^*}{1-\xi^*}} \frac{\overline{w}(\xi^*, s)\,\mathrm{d}\xi^*}{x^* - \xi^*} + \sqrt{\frac{1-x^*}{1+x^*}} \frac{s\overline{\Gamma}_a}{\pi U} e^{\frac{sb}{U}} \int_1^\infty \sqrt{\frac{\lambda+1}{\lambda-1}} \frac{e^{-\frac{sb}{U}\lambda}}{x^* - \lambda} \mathrm{d}\lambda \tag{3.16}$$

In Eq. 3.16, the bound vortex $\overline{\Gamma}_a$ plays the role of a coefficient at the right hand side to determine the bound vortex sheet strength itself. Therefore, if we integrate Eq. 3.16 with respect to full chord we obtain the bound vortex also. In non dimensional coordinates the integral reads as

$$\int_{-1}^{1} \overline{\gamma}_a(x^*, s)\,\mathrm{d}x^* = \frac{\overline{\Gamma}_a}{b} = \frac{2}{\pi} \int_{-1}^{1} \sqrt{\frac{1-x^*}{1+x^*}} \int_{-1}^{1} \sqrt{\frac{1+\xi^*}{1-\xi^*}} \frac{\overline{w}(\xi^*, s)\,\mathrm{d}\xi^*}{x^* - \xi^*} \mathrm{d}x^*$$

$$+ \frac{s\overline{\Gamma}_a}{\pi U} e^{\frac{sb}{U}} \int_{-1}^{1} \sqrt{\frac{1-x^*}{1+x^*}} \int_1^\infty \sqrt{\frac{\lambda+1}{\lambda-1}} \frac{e^{-\frac{sb}{U}\lambda}}{x^* - \lambda} \mathrm{d}\lambda \mathrm{d}x^*$$

If we interchange the order of integrals at the right hand side, we can then perform the integrations with respect to x^* and obtain the following equation $\overline{\Gamma}_a$ in terms of the downwash

$$\overline{\Gamma}_a = -2b \int_{-1}^{1} \sqrt{\frac{1+\xi^*}{1-\xi^*}} \overline{w}(\xi^*, s) d\xi^* - \overline{\Gamma}_a \frac{sb}{U} e^{\frac{sb}{U}} \int_1^\infty \left(\sqrt{\frac{\lambda+1}{\lambda-1}} - 1\right) e^{-\frac{sb}{U}\lambda} \mathrm{d}\lambda \tag{3.17}$$

The second term at the right hand side of Eq. 3.17 can be integrated with respect to λ. The resulting integral is expressible as an Hankel function of second kind in terms of the complex argument ($-isb/U$). A useful relation between the Bessel functions and the Haenkel functions are provided in Appendix 5.

Denoting the integral at the second term of Eq. 3.17 by I_2 with the help from Theodorsen (1935), we obtain

$$\mathrm{I}_2 = \int_1^\infty (\sqrt{\frac{\lambda+1}{\lambda-1}} - 1) e^{-\frac{sb}{U}\lambda} \mathrm{d}\lambda = -\frac{\pi}{2}\left[H_1^{(2)}\left(-i\frac{sb}{U}\right) + iH_0^{(2)}\left(-i\frac{sb}{U}\right)\right] - \frac{e^{-\frac{sb}{U}}}{sb/U} \tag{3.18}$$

Substituting Eq. 3.18 in Eq. 3.17, we get

$$\overline{\Gamma}_a = -2b\int_{-1}^{1}\sqrt{\frac{1+\xi^*}{1-\xi^*}}\overline{w}(\xi^*,s)\mathrm{d}\xi^* + \overline{\Gamma}_a\frac{sb}{U}e^{\frac{sb}{U}}\frac{\pi}{2}[H_1^{(2)} + iH_0^{(2)}] + \overline{\Gamma}_a$$

and write the result for the bound circulation in terms of the downwash

$$\overline{\Gamma}_a\frac{s}{U}e^{\frac{sb}{U}} = \frac{4/\pi}{H_1^{(2)} + iH_0^{(2)}}\int_{-1}^{1}\sqrt{\frac{1+\xi^*}{1-\xi^*}}\overline{w}(\xi^*,s)d\xi^* \tag{3.19}$$

The relation between the downwash w and the time dependent motion of the airfoil was given by Eq. 2.20. We need the Laplace transformed form of Eq. 2.20 to implement in Eq. 3.19, which is

$$\overline{w}(x^*,s) = s\overline{z}_a(x^*,s) + \frac{U}{b}\frac{\partial}{\partial x^*}[\overline{z}_a(x^*,s)] \tag{3.20}$$

At this stage, we can use Eq. 3.20 in Eq. 3.19 and obtain the bound circulation in s domain. After inverting the result to time domain by inverse Laplace transform, we can get the time dependent bound circulation and the lift. For more detailed analysis, the relation between the lifting pressure coefficients and the bound vortex sheet strength we obtain

$$\begin{aligned} c_{pa}(x,t) &= \frac{2}{U^2}\frac{\partial}{\partial t}(\phi'_u - \phi'_l) + \frac{2}{U}\left(\frac{\partial\phi'_u}{\partial x} - \frac{\partial\phi'_l}{\partial x}\right) \\ &= \frac{2}{U^2}\frac{\partial}{\partial t}\int_{-b}^{x}\gamma_a(\xi,t)\mathrm{d}\xi^* + \frac{2}{U}\gamma_a(x,t) \end{aligned} \tag{3.21}$$

We can now take the Laplace transform of Eq. 3.21 which in s domain reads as

$$\overline{c}_{P_a}(x^*,s) = \frac{2}{U}\left[\frac{sb}{U}\int_{-1}^{x^*}\overline{\gamma}_a(\xi^*,s)\,\mathrm{d}\xi^* + \overline{\gamma}_a(x^*,s)\right] \tag{3.22}$$

Substituting Eq. 3.16 in Eq. 3.20 and integrating the fist term on the right hand side we obtain

$$\bar{c}_{p_a}(x^*,s) = \frac{4}{\pi}\sqrt{\frac{1-x^*}{1+x^*}}\int_{-1}^{1}\sqrt{\frac{1+\xi^*}{1-\xi^*}}\frac{\overline{w}(\xi^*,s)\,\mathrm{d}\xi^*}{(x^*-\xi^*)\,U} - \frac{4}{\pi}\frac{sb}{U}\int_{-1}^{1}\Lambda(x^*,\xi^*)\,\frac{w(\xi^*,s)}{U}\mathrm{d}\,\xi^*$$
$$+\frac{4}{\pi}\left[1-\frac{H_1^{(2)}}{H_1^{(2)}+i\,H_0^{(2)}}\right]\sqrt{\frac{1-x^*}{1+x^*}}\int_{-1}^{1}\sqrt{\frac{1+\xi^*}{1-\xi^*}}\frac{\overline{w}(\xi^*,s)\,\mathrm{d}\xi^*}{U} \tag{3.23}$$

Here, $\Lambda\,(x^*,s) = \frac{1}{2}\ln\frac{1-x^*\xi^*+\sqrt{(1-x^{*2})(1-\xi^{*2})}}{1-x^*\xi^*-\sqrt{(1-x^{*2})(1-\xi^{*2})}}$ is given by Bisplinghoff et al. (1996).

The coefficient of the third term contains a new function called the Theodorsen function

$$C\left(-i\frac{sb}{U}\right) = \frac{H_1^{(2)}\left(-i\frac{sb}{U}\right)}{H_1^{(2)}\left(-i\frac{sb}{U}\right)+iH_0^{(2)}\left(-i\frac{sb}{U}\right)} = F\left(-i\frac{sb}{U}\right)+i\,G\left(-i\frac{sb}{U}\right) \tag{3.24}$$

Functions F and G are real although their arguments are imaginary. The Theodorsen function takes the value of unity for s approaching to zero, i.e.

$$\lim_{s\to 0}\left\{C\left(i\frac{sb}{U}\right)\right\} = 1$$

which simplifies the pressure coefficient for $s = 0$ as follows

$$\lim_{s\to 0}\{\bar{c}_p(x^*,s)\} = \frac{4}{\pi}\sqrt{\frac{1-x^*}{1+x^*}}\int_{=1}^{1}\sqrt{\frac{1+\xi^*}{1-\xi^*}}\frac{1}{x^*-\xi^*}\frac{\overline{w}(\xi^*)}{U}\,\mathrm{d}\xi^*$$

This term is called the quasi steady pressure term and it is equivalent to the steady pressure term.

As is well known for steady flow that the zero free stream means zero lift. For unsteady flow however, during the vertical translation of the airfoil we expect to have a lift generation even under zero free stream. We can show this with a limiting process performed on the second term of Eq. 3.23 with multiplying the term with U^2 and letting U go to zero as follows.

$$\lim_{U\to 0}\{U^2\bar{c}_p(x^*,s)\} = -\tfrac{4}{\pi}s\,b\int_{-1}^{1}\Lambda(x^*,\xi^*)\;\overline{w}(\xi^*,s)\,\mathrm{d}\xi^* \quad \text{and} \quad \lim_{U\to 0}\overline{w}(x^*,s) = s\,\bar{z}_a,$$

(from Eq. 3.20).

From the last line we see that the vertical force is proportional with $s^2\,\bar{z}_a$. Since $\bar{z}_a$ is independent of s then inverse Laplace transform of $s^2\,\bar{z}_a$ gives us

$$L^{-}\{s^2\bar{z}_a\} = \frac{\partial^2 z_a}{\partial t^2}$$

The last expression shows that even at zero free stream speed there exists a lifting force which is proportional to the acceleration in vertical translation. This force is an inertial force generated by the motion of the profile and it is called the apparent mass. Since there is no circulation attached to it, it is also called non circulatory term.

The third term at the right hand side of Eq. 3.23 is the circulatory term due to wake vortex sheet. For unsteady flows we do not have to take into consideration all three terms of Eq. 3.23. Depending on the unsteadiness we can ignore some of the terms in our analysis depending on the accuracy we look after. Now, we can discuss which term to neglect under what physical condition. According to a classical classification:

(i) 'Unsteady Aerodynamics': All three terms are included. Motions with about 40 Hertz frequencies are analyzed by this approach.
(ii) 'Quasi Unsteady Aerodynamics': The apparent mass term is neglected. Motions with 5–15 Hz frequencies are analyzed using this approach.
(iii) 'Quasi Steady Aerodynamics': Motions with frequency of 1 Hz or below is analyzed using the circulatory term only.

After making this classification, we can now derive a formula for the lifting pressure coefficient for simple harmonic motions and obtain the relevant aerodynamic coefficients such as sectional lift and moment coefficients.

3.4 Simple Harmonic Motion

In the previous section we have obtained the lifting pressure coefficient in Laplace transformed domain s. In order to express the pressure coefficient in time domain we have to invert Eq. 3.23 either with the Bromwich integral or use some other technique for some type of time dependent motions. One of the special types of motion is a simple harmonic motion of the airfoil for which we can invert Eq. 3.23 directly. Let us now find the lifting pressure coefficient, sectional lift and moment coefficients for an airfoil which undergoes a simple harmonic motion. If we let $\bar{z}_a$ be the amplitude and ω be the frequency of the motion then the equation of the motion for the chordline in its exponential form reads as

$$z_a(x,t) = \bar{z}_a(x)\, e^{i\varpi t} \tag{3.25}$$

According to Eq. 2.20 the downwash expression becomes

$$w(x,t) = \frac{\partial z_a}{\partial t} + U\frac{\partial z_a}{\partial x} = \left(i\omega\,\overline{z}_a + U\frac{\partial \overline{z}_a}{\partial x}\right)e^{i\omega t} = \overline{w}(x)\,e^{i\omega t} \tag{3.26}$$

In Eq. 3.26 the complex downwash amplitude is defined as

$$\overline{w}\,(x) = \left(i\omega\,\overline{z}_a + U\frac{\partial \overline{z}_a}{\partial x}\right)$$

The $\overline{z}_a(x)$ is a real valued function of x in Eq. 3.25, whereas in Eq. 3.26 the amplitude of the downwash, $\overline{w}\,(x)$ expression becomes a complex function. That is when the flow is unsteady there is a phase difference φ between the motion and the its response as a downwash.

This phase difference is somewhat a measure of the unsteadiness and can be represented in the complex plane as shown in Fig. 3.7.

Let us compare the two downwash expressions, the Laplacetransformed one, Eq. 3.20, and the simple harmonic one, Eq. 3.26. The comparison shows that there is a resemblance between the variables (s) and (i ω). On the other hand, the nondimensional parameter (sb/U) of pressure coefficient can be identified with another nondimensional parameter $i(b\,\omega/U) = ik$, where $k = b\,\omega/U$ is the previously defined reduced frequency. We can now give a physical meaning of reduced frequency as 'number of oscillations in radians per half chord travel of the airfoil'. Hence, the reduced frequency is regarded as the nondimensional measure of the unsteadiness.

Instead of the variable (sb/U) of Eq. 3.23, if we use (ik) then for the amplitude of lifting pressure coefficient we obtain

$$\begin{aligned}\overline{c}_{p_a}(x^*,k) &= \frac{4}{\pi}\sqrt{\frac{1-x^*}{1+x^*}}\int_{-1}^{1}\sqrt{\frac{1+\xi^*}{1-\xi^*}}\frac{\overline{w}(\xi^*)\,d\xi^*}{(x^*-\xi^*)\,U} - \frac{4ik}{\pi}\int_{-1}^{1}\Lambda(x^*,\xi^*)\,\frac{\overline{w}(\xi^*)}{U}d\,\xi^* \\ &+ \frac{4}{\pi}[1-C(k)]\sqrt{\frac{1-x^*}{1+x^*}}\int_{-1}^{1}\sqrt{\frac{1+\xi^*}{1-\xi^*}}\frac{\overline{w}(\xi^*)\,d\xi^*}{U}\end{aligned} \tag{3.27}$$

The time dependent form of it reads as

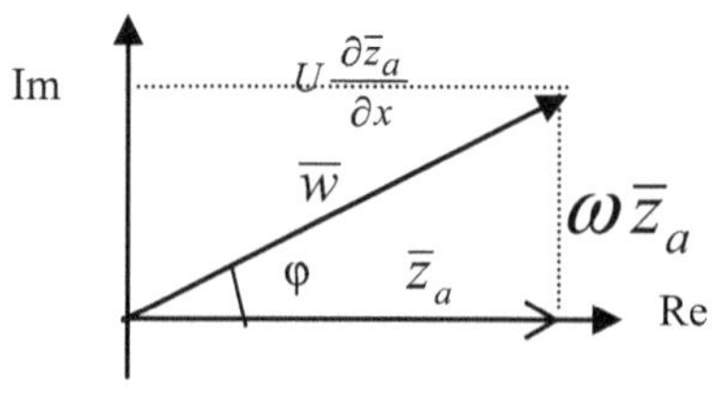

Fig. 3.7 Phase difference φ between the motion and the downwash

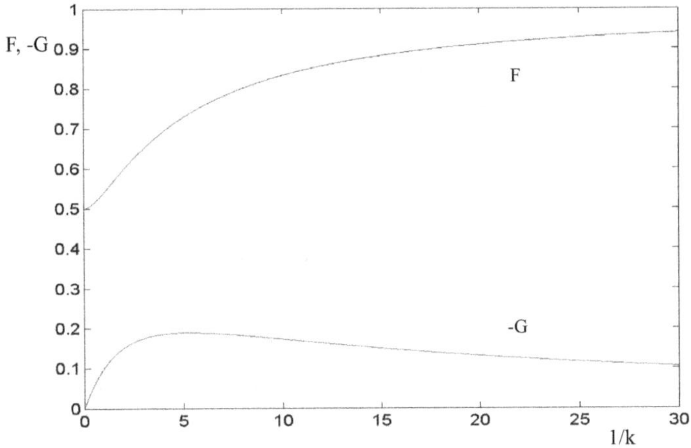

Fig. 3.8 *F*, the real and *G* the imaginary parts of the Theodorsen function

$$c_{pa}(x^*, k, t) = \overline{c}_{p_a}(x^*, k)\, e^{i\varpi t} \tag{3.27a}$$

The Theodorsen function, $C(k) = F(k) + i\, G(k)$, in the last term of Eq. 3.27 is the complex function of the real valued reduced frequency k. In Fig. 3.8, shown is the graph of the real and the imaginary parts of the Theodorsen function in terms of $1/k$.

Equations 3.25–3.27a are expressed in their exponential terms for their time dependency. This means because of their different amplitudes there is a phase difference between the motion, the downwash and the corresponding lifting pressure coefficient.

The sectional lift and moment coefficients of a profile now can be found by integrating the lifting pressure coefficient along the chordline, i.e. the lift coefficient becomes

$$c_l = \frac{l}{\rho_\infty U^2 b} = \int_{-b}^{b} \frac{p_l - p_u}{\rho_\infty U^2 b} \mathrm{d}x = \frac{1}{2} \int_{-1}^{1} c_{pa}\, \mathrm{d}x^* \tag{3.28}$$

and the moment coefficient with respect to mid chord reads as

$$c_m = \frac{m}{\rho_\infty U^2 b^2} = -\int_{-b}^{b} \frac{p_l - p_u}{\rho_\infty U^2 b^2} x\, \mathrm{d}x = -\frac{1}{2} \int_{-1}^{1} c_{pa}\, x^* \mathrm{d}x^* \tag{3.29}$$

In Eqs. 3.28 and 3.29, the positive lift is defined as upwards and the positive moment is defined as the leading edge up. Accordingly, the simple harmonic change of the aerodynamic coefficients read as

$$c_l = \overline{c}_l\, e^{i\omega t}, \quad \left(\overline{c}_l = \frac{1}{2}\int\limits_{-1}^{1} \overline{c}_{p_a}\, \mathrm{d}x^*\right), \quad \text{and} \quad c_m = \overline{c}_m\, e^{i\omega t}$$

$$\left(\overline{c}_m = -\frac{1}{2}\int\limits_{-1}^{1} \overline{c}_{p_a}\, x^*\, \mathrm{d}x^*\right).$$

After performing the integrals, the coefficients in terms of the amplitude of the downwash become

$$\begin{aligned}
\overline{c}_l(k) &= -2C(k)\int\limits_{-1}^{1}\sqrt{\frac{1+\xi^*}{1-\xi^*}}\frac{\overline{w}(\xi^*)\,\mathrm{d}\xi^*}{U} - 2ik\int\limits_{-1}^{1}\sqrt{1-\xi^{*2}}\frac{\overline{w}(\xi^*)\,\mathrm{d}\xi^*}{U}\\
\overline{c}_m(k) &= -[1+C(k)]\int\limits_{-1}^{1}\sqrt{\frac{1+\xi^*}{1-\xi^*}}\frac{\overline{w}(\xi^*)\,\mathrm{d}\xi^*}{U} + 2\int\limits_{-1}^{1}\sqrt{\frac{1+\xi^*}{1-\xi^*}}\,\xi^*\frac{\overline{w}(\xi^*)\,\mathrm{d}\xi^*}{U}\\
&\quad + ik\int\limits_{-1}^{1}\sqrt{1-\xi^{*2}}\,\xi^*\frac{\overline{w}(\xi^*)\,\mathrm{d}\xi^*}{U}
\end{aligned} \tag{3.30a, b}$$

The integrals of Eq. 3.30 with (*ik*) as the coefficients are the noncirculatory terms which are the apparent mass terms. The expressions of the aerodynamic coefficients can give us the quasi steady forms if we take the limits while the reduced frequencies go to zero. The limiting process yields

$$\overline{c}_l^{qs} = \lim_{k\to 0}\{\overline{c}_l(k)\} = -2\int\limits_{-1}^{1}\sqrt{\frac{1+\xi^*}{1-\xi^*}}\frac{\overline{w}(\xi^*)\,\mathrm{d}\xi^*}{U}$$

and

$$\overline{c}_m^{qs} = \lim_{k\to 0}\{\overline{c}_m(k)\} = -2\int\limits_{-1}^{1}\sqrt{\frac{1+\xi^*}{1-\xi^*}}\frac{\overline{w}(\xi^*)\,\mathrm{d}\xi^*}{U} + 2\int\limits_{-1}^{1}\sqrt{\frac{1+\xi^*}{1-\xi^*}}\,\xi^*\frac{\overline{w}(\xi^*)\,\mathrm{d}\xi^*}{U} \tag{3.31a, b}$$

We can express the unsteady forms of the coefficients in terms of the quasi steady coefficients as

$$\begin{aligned}
\overline{c}_l &= \overline{c}_l^{qs}\, C(k) - 2ik\int\limits_{-1}^{1}\sqrt{1-\xi^2}\frac{\overline{w}(\xi^*)\,\mathrm{d}\xi^*}{U}\\
\overline{c}_m &= \overline{c}_m^{qs} + \frac{C(k)-1}{2}\,\overline{c}_l^{qs} + ik\int\limits_{-1}^{1}\sqrt{1-\xi^2}\,\xi^*\frac{\overline{w}(\xi^*)\,\mathrm{d}\xi^*}{U}
\end{aligned} \tag{3.32a, b}$$

The aerodynamic coefficients given by Eq. 3.32a, b give us the relation between the quasi steady and the quasi unsteady coefficients in terms of the Theodorsen function as well as the contributions coming from the apparent mass terms. If we consider only the circulatory terms, the ratio of the quasi unsteady lift to the quasi steady lift is given by the Theodorsen function which measures also the phase difference between the two coefficients as the effect of the circulatory wake term. Another significance, attributable to the Theodorsen function is as follows. If we know the quasi steady coefficients from the experiments or through some other means we can obtain the corresponding quasi unsteady coefficient by multiplying the former by the value of Theodorsen function at desired reduced frequency.

Let us now give some examples ranging from simple to more complex flow cases.

Example 3.2 Vertical oscillation of a flat plate in a free stream.

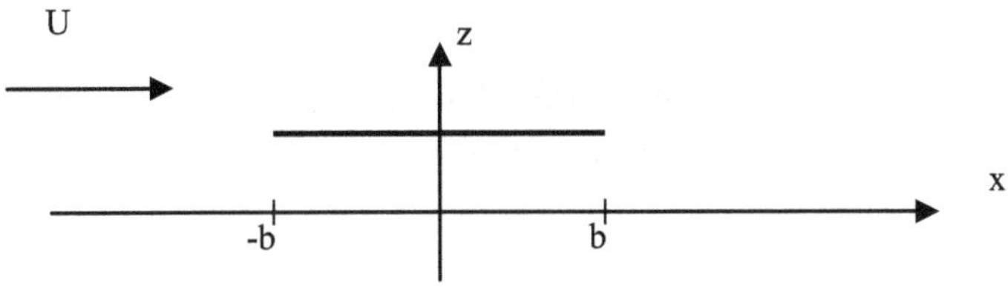

The profile motion is in z direction with amplitude $\overline{z}_a$, therefore the motion for the equation reads as $z_a(x,t) = \overline{z}_a\, e^{i\,\omega t}$. The corresponding downwash becomes

$$w\,(x,t) = i\,\omega\,\overline{z}_a\, e^{i\,\omega t} = \overline{w}\, e^{i\,\omega t}, \quad (\overline{w} = i\,\omega\,\overline{z}_a).$$

As easily seen, the amplitude of downwash differs from the motion with coefficient $i\omega$, which shows that the phase difference between them is 90°.

Substituting the downwash expression in Eq. 3.31a, b we have

$$\overline{c}_l^{qs} = -2\pi \frac{i\,\omega\,\overline{z}_a}{U} = -2\,\pi\,i\,k\,\overline{z}_a^* \quad \text{and} \quad \overline{c}_m^{qs} = -\pi \frac{i\,\omega\,\overline{z}_a}{U} = -\pi\,i\,k\,\overline{z}_a^*.$$

Writing the unsteady aerodynamic coefficients from Eq. 3.32a, b we obtain

$$\overline{c}_l = -2\pi ik\overline{z}_a^* C(k) + \pi k^2 \overline{z}_a^* \quad \text{and} \quad \overline{c}_m = -\pi ik\overline{z}_a^* C(k).$$

From aerodynamic coefficients we observe that the apparent mass contributes to the sectional lift coefficient but not the moment coefficient.

Let us now analyze the response of a thin airfoil to pitch oscillations about its midchord.

Example 3.3 Flat plate pitching about its midchord.

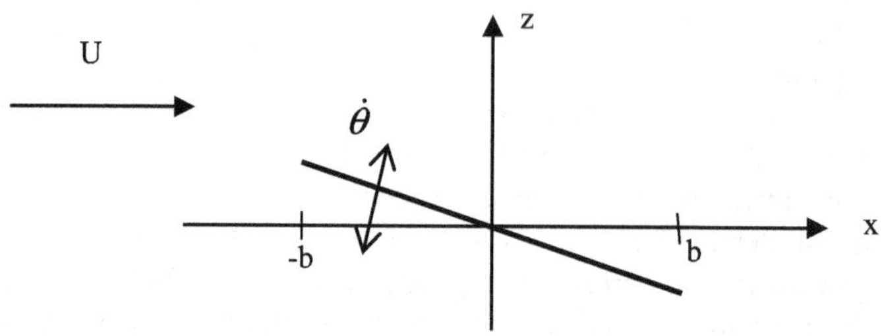

As seen from the picture, the chordline equation of a pitching airfoil reads as $\overline{z}_a(x,t) = -\theta x = -x\overline{\theta}\,e^{i\omega t}$, and the corresponding downwash

$$w\,(x,t) = -\dot{\theta}x - U\theta = (-i\,\omega\,b\,x^* - U)\,\overline{\theta}\,e^{i\omega t}.$$

Considering the steady term $U\theta$ also, Eq. 3.31a, b gives $\overline{c}_l^{qs} = \pi\,i\,k\,\overline{\theta} + 2\pi\overline{\theta}$ and $\overline{c}_m^{qs} = \pi\overline{\theta}$. For the unsteady motion $\overline{c}_l = (\pi\,i\,k\,\overline{\theta} + 2\,\pi\,\overline{\theta})C(k) + \pi\,i\,k^2\overline{\theta}$ and $\overline{c}_m = \pi\,i\,k\,\overline{\theta}\,\frac{C(k)-1}{2} + \pi\,\overline{\theta}\,C(k) + \frac{\pi}{8}k^2$.

The last terms in both amplitudes indicate the effect of apparent mass terms.

We have so far seen the single degree of freedom problems. As a more complex problem we are going to study a two degrees of freedom problem where the airfoil translates vertically and rotates around a fixed point.

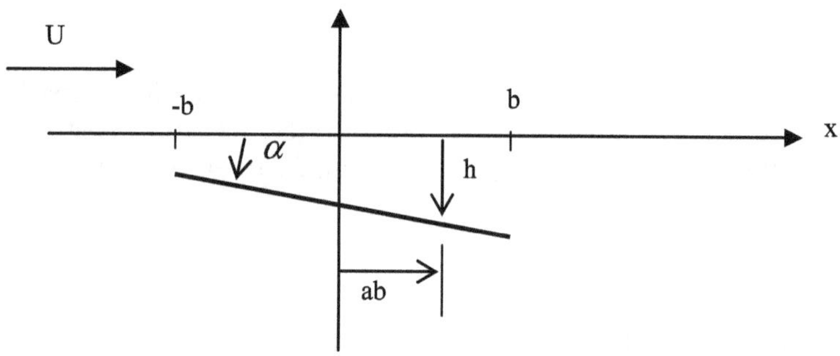

Let the vertical translation in z be $h = \overline{h}\,e^{i\omega t}$, and the rotation about the point ab (where a is a nondimensional number) be $\alpha = \overline{\alpha}\,e^{i\omega t}$ as shown in the Figure. The equation of the profile reads as

$$z_a(x,t) = ab\,\alpha - h - \alpha\,x, \text{ and the downwash}$$

$w\,(x,t) = \{i\,\omega\,[\overline{\alpha}(ab - x) - \overline{h}] - U\overline{\alpha}\}\,e^{i\omega t}$. If we use the downwash expression in Eq. 3.32a, b we obtain the amplitude for the unsteady coefficients we obtain

$$\overline{c}_l = \pi k^2 \left\{ -\left[1 - \frac{2i}{k} C(k)\right]\overline{h}^* - \left[\frac{1}{2} - \frac{i}{k}(1 + 2C(k)) - \frac{2}{k^2} C(k)\right]\overline{\alpha} + \left(\frac{1}{2} + a\right)\left[1 - \frac{2i}{k} C(k)\right]\overline{\alpha} \right\}$$

$$\overline{c}_m = \pi C(k)[(1 - ika)\overline{\alpha} + ik\overline{h}^*] - ik\pi\left[\frac{1}{2} - \frac{1}{2} C(k) + \frac{1}{8} ik\right]\overline{\alpha}. \tag{3.33a, b}$$

The moment coefficient here is computed with respect to mid chord. The moment coefficient about any point α using the coefficients from Eq. 3.33a, b becomes

$$\overline{c}_{m_a} = \overline{c}_m + \overline{c}_l a.$$

Example 3.4 Find the sectional lift coefficient change for an airfoil pitching about its quarter chord with the angle of attack $\alpha = 10°\ \sin\omega t$, and the reduced frequency $k = 0.1$.

Solution Let us consider the terms of Eq. 3.33 which depends on angle of attack only. For the simple harmonic motion for $k = 0.1$ the sectional lift coefficient reads as

$$\overline{c}_l = \pi k^2 \left[-\frac{1}{2} + \frac{i}{k}(1 + 2C(k)) + \frac{2}{k^2} C(k)\right]\overline{\alpha} = 0.92832 - 0.0428i$$

Here, the angle of attack changes with a sinus term. Therefore, we have to write the relation between the sinus term and the exponential form of the angle of attack.

Let us expand the exponential form with Euler's formula as follows

$$\overline{\alpha} e^{i\omega t} = \overline{\alpha} \cos \omega t + i\overline{\alpha} \sin \omega t$$

As seen from the expanded form, the contribution to the lift coefficient will be from the second term which is imaginary and contains sinus term. Therefore, the contribution will come from the second term of Eq. 3.33 which is also imaginary. The general expression of the lift coefficient becomes

$$\begin{aligned}\overline{c}_l e^{i\omega t} &= (\overline{c}_{l_R} + i\overline{c}_{l_I})(\cos\omega t + i \sin \omega t) \\ &= \overline{c}_{l_R} \cos\omega t - \overline{c}_{l_I} \sin \omega t + i(\overline{c}_{l_R} \sin \omega t + \overline{c}_{l_I} \cos\omega t)\end{aligned}$$

Hence, the imaginary part which we are interested, is

$$(\overline{c}_{l_R} \sin \omega t + \overline{c}_{l_I} \cos\omega t).$$

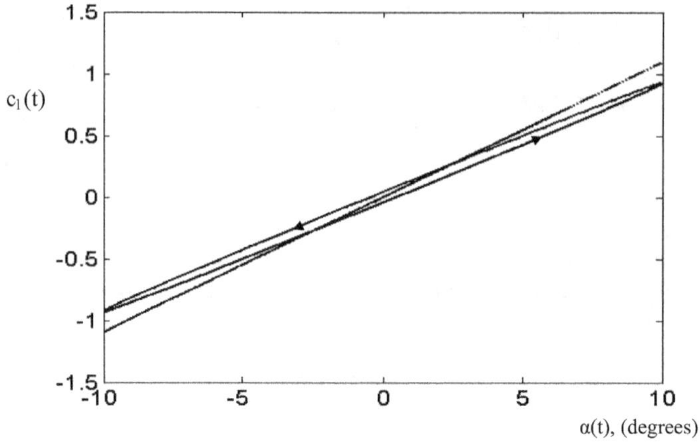

Fig. 3.9 Unsteady sectional lift coefficient change

If we form the linear combination with real and imaginary parts of the sectional Lift coefficient then we obtain

$$c_l = \overline{c}_{l_R} \sin \omega t + \overline{c}_{l_I} \cos \omega t = 0.92832 \sin \omega t - 0.0428 \cos \omega t$$

Figure 3.9 shows the change in the sectional lift coefficient with respect to the angle of attack change.

In Fig. 3.9, the straight line, plotted for the sake of comparison, shows the quasi steady sectional lift change. The comparison with the unsteady coefficient shows that there is a lift loss around the $\pm 10°$ angles of attack. The Theodorsen function is the measure of this lift loss. For unsteady lift curve, on the other hand, there is a hysteresis. This means as the angle of attack increases, the increase in the lift occurs with a lag and at the maximum angle of attack maximum lift has not been achieved yet. As the angle of attack decreases the lift has a higher value than the lift of the same angle which is reached during the angle of attack increase.

Here, The Theodorsen function was utilized for the analysis of unsteady flows about plunging-heaving thin airfoil. The comparison between the theoretical and the experimental studies are given by Leishman (2000) for The NACA 0012 airfoil at low Mach and high Reynolds numbers for the reduced frequency range of $0.07 < k \leq 0.4$, where the lift coefficients are in good agreement. The disagreement for the moment coefficients on the other hand, can be remedied by slightly moving the aerodynamic center in front of the quarter chord. In addition, Leishman gives the experimental results for an airfoil pitching about its quarter chord for the reduced frequency range of $0.05 \leq k \leq 0.6$. The experimental and the theoretical values at low Mach numbers and not so large reduced frequencies agree well.

3.5 Loewy's Problem: Returning Wake Problem

The theory of Theodorsen is developed for an airfoil whose wake extends to undisturbed farfield. On the other hand, more complex motions of an airfoil can be studied by the aid of the Theodorsen function. A representative example for that is the study of a helicopter blade or a blade of a propeller. Loewy and Jones separately studied this problem with the parameters N being the number of blades and h being the distance between the blade and the returning wake. Now, let us give the related formulas for the modified version of the Theodorsen function for a single blade and the multi-blade rotors.

(i) Single blade: The modified Theodorsen function is given in terms of Ω being the rotational speed of the blade in radians per second and h:

$$C'\left(k,\frac{\omega}{\Omega},h\right)=\frac{H_1^{(2)}(k)+J_1(k)\,W}{H_1^{(2)}(k)+i\,H_0^{(2)}(k)+2\,[J_1(k)+i\,J_0(k)]\,W}\quad\text{and}$$
$$\text{where } W\left(\frac{kh}{b},\frac{\omega}{\Omega}\right)=\left(e^{kh/b}e^{i2\pi\,\omega/\Omega)}-1\right)^{-1}. \qquad (3.34\text{a, b})$$

Here, in Eq. 3.34a if we let h go to infinity, we recover the Theodorsen function as expected. In addition, if the ratio given by ω/Ω is an integer, which means the oscillation frequency of the profile is multiples of rotational speed of the blade then the vortices shed are in phase according to Eq. 3.34b.

(ii) N-blades: For this case W as function is altered with number of blades N and $\Delta\psi$ as follows

$$W\left(\frac{kh}{b},\frac{\omega}{\Omega},\Delta\psi,N\right)=\left(e^{kh/b}e^{i2\pi\,\omega/N\,\Omega)}e^{(\Delta\psi\omega/\Omega)}-1\right)^{-1}$$

If we take $\Delta\psi=0$ and study the phase difference only for the distance between the blades the form of W becomes

$$W\left(\frac{kh}{b},\frac{\omega}{\Omega},\Delta\psi,N\right)=\left(e^{kh/b}e^{i2\pi\,\omega/N\,\Omega)}-1\right)^{-1}.$$

Loewy's approach applied to a single blade rotor causes the unsteady lift to increase or decrease depending on the reduced frequency. In Fig. 3.10 given is the change in the amplitude of the Loewy function with *h*/*b* and *k*.

So far we have examined the response of a simple harmonically oscillating airfoil in a free stream or in a returning wake. Now, we can study the unsteady aerodynamic response of an airfoil to its arbitrary motion or to an arbitrary external excitation.

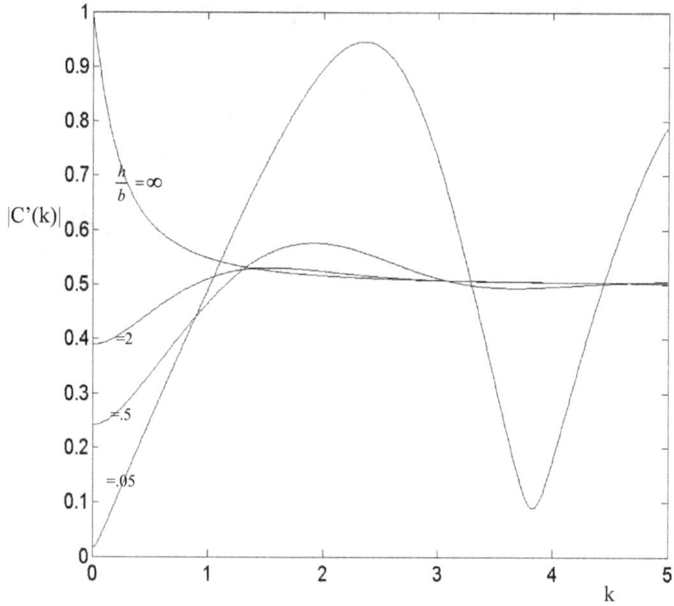

Fig. 3.10 Change in the amplitude of Loewy function with (*h*/*b*) and *k*

3.6 Arbitrary Motion

There will be two different arbitrary motions to be studied. First, we will see the unsteady aerodynamic force and moment created by the arbitrary motion of the airfoil. Afterwards, the response of an airfoil to a sharp edged gust will be studied.

Arbitrary Motion and Wagner Function

The response of the linear system to a unit step function is defined as the indicial admittance function, $A(t)$, see Appendix 6. The response of the same system to the arbitrary excitation is given by the Duhamel integral as $x(t)$

$$x(t) = A(0)f(t) + \int_0^t f(\tau)A'(t-\tau)\Delta\tau.$$

Let us find the indicial admittance, $A(t)$, as the unsteady aerodynamic response of the system for the arbitrary motion of the airfoil. As a two degrees of freedom problem let the airfoil pitch about its midchord while undergoing vertical translation. As is given in the previous section the equation for the chordline for $a = 0$ reads as

$$z_a(x,t) = -h - \alpha x, \quad \text{and} \quad w(x,t) = \frac{\partial z_a}{\partial t} + U\frac{\partial z_a}{\partial x} = -(\dot{h} + \dot{\alpha} x + U\alpha).$$

This downwash expression w can be used in Eq. 3.32a, b, for a simple harmonic motion with regarding the time derivative of the downwash as the apparent mass terms. This gives

$$\begin{aligned} c_l &= c_l^{qs}\, C(k) - \frac{2b}{U^2}\frac{\partial}{\partial t}\int_{-1}^{1}\sqrt{1-\xi^{*2}}w(\xi^*)\,\mathrm{d}\xi^* \\ c_m &= c_m^{qs} + \frac{C(k)-1}{2}\,c_l^{qs} + \frac{b}{U^2}\frac{\partial}{\partial t}\int_{-1}^{1}\sqrt{1-\xi^{*2}}\,\xi^* w(\xi^*)\,\mathrm{d}\xi^* \end{aligned} \qquad (3.35\text{a, b})$$

and the quasi steady terms from Eq. 3.31a, b

$$c_l^{qs} = \frac{2\pi}{U}(\dot{h} + \dot{\alpha} b/2 + U\alpha) \quad \text{and} \quad c_m^{qs} = \frac{\pi}{U}(\dot{h} + U\alpha).$$

In Eq. 3.35a, b regarding noncirculatory terms which are the time derivatives of the vertical translation h and the angle of attack α, the coefficients become

$$\begin{aligned} c_l &= \frac{2\pi}{U}\,C(k)\,(\dot{h} + \dot{\alpha} b/2 + U\alpha) + \frac{\pi b}{U^2}\,(\ddot{h} + U\dot{\alpha}) \\ c_m &= \frac{\pi}{U}C(k)\,(\dot{h} + \dot{\alpha} b/2 + U\alpha) - \frac{\pi b}{2U^2}\,(b\ddot{\alpha}/4 + U\dot{\alpha}) \end{aligned} \qquad (3.36\text{a, b})$$

The first terms of both coefficients given by Eq. 3.36a, b depend on the Theodorsen function and they are valid for simple harmonic motions only. The second terms, on the other hand, are independent of the type of motion and they are just time derivatives of the vertical translation and the rotation. If we closely examine the expression in the parenthesis of the first term, $(\dot{h} + \dot{\alpha} b/2 + U\alpha)$, we observe that this is nothing but the expression for the negative of the downwash at the three quarter chord, i.e.,

$$w(b/2\,,t) = -(\dot{h} + \dot{\alpha} b/2 + U\alpha).$$

We have seen in Eq. 3.36a, b that the circulatory terms of the aerodynamic coefficients are the function of the reduced frequency. Here, the downwash at the three quarter chord point is sufficient to find the sectional coefficients. When there is an arbitrary motion, the downwash will change arbitrarily. Since the problem is linear, we can write the Fourier components of the arbitrary downwash and superimpose the contribution of the each component on integral form to the sectional coefficients. For this purpose let us define the Fourier integral in the frequency domain

$$w\left(\frac{b}{2},t\right)=\frac{1}{2\pi}\int_{-\infty}^{\infty} f(\omega)\,e^{i\omega t}\mathrm{d}\,\omega \tag{3.37}$$

Here, $f(\omega)$ is the Fourier transform of the downwash and covers its full frequency spectrum. The inverse Fourier transform in terms of the downwash value at the three quarter chord becomes

$$f(\omega)=\int_{-\infty}^{\infty} w\left(\frac{b}{2},t\right)e^{-i\omega t}\mathrm{d}\,t \tag{3.38}$$

The circulatory lift at a given frequency ω can be defined as the Fourier component of the total circulatory lift. This component, on the other hand, can be written for a unit amplitude of the downwash as follows

$$w\left(\frac{b}{2},t\right)=e^{i\omega t}$$

The corresponding Fourier component for the circulatory lift at time s becomes

$$\Delta\,c_l^c\,(\omega,t)=-\frac{2\,\pi}{U}\,C(k)\,e^{i\omega t}.$$

This component can be put into the Fourier integral

$$c_l^c\,(t)=\frac{1}{2\,\pi}\int_{-\infty}^{\infty}\Delta\,c_l^c\,(\omega,t)\,e^{i\omega t}\mathrm{d}\,\omega=-\frac{1}{U}\int_{-\infty}^{\infty} f(\omega)\,C(k)\,e^{i\omega t}\mathrm{d}\,\omega.$$

If we employ the same procedure for the moment coefficient, the total coefficients read as

$$\begin{aligned}c_l(t)&=\frac{\pi\,b}{U^2}(\ddot{h}+U\dot{\alpha})-\frac{1}{U}\int_{-\infty}^{\infty} f(\omega)\,C(k)\,e^{i\omega t}\mathrm{d}\,\omega\\ c_m(t)&=-\frac{\pi\,b}{2\,U^2}(b\,\ddot{\alpha}/4+U\dot{\alpha})-\frac{1}{2\,U}\int_{-\infty}^{\infty} f(\omega)\,C(k)\,e^{i\omega t}\mathrm{d}\,\omega\end{aligned} \tag{3.39a, b}$$

Equation 3.39a, b are applicable for the arbitrary downwash and covers piecewise continuous functions with finite Fourier transform.

Since the aerodynamic system we consider is linear, the step function representation of the downwash and the superposition technique will be applied for the aerodynamic effect of the unit change in one of the followings.

(a) for $\alpha = 0$, change in $\dot{h}$,
(b) for $\dot{h} = 0$, change in α for $U =$ constant,
(c) for $\dot{h} = 0$, change in U for $\alpha =$ constant.

Now, let us consider case (b) when U is constant the angle of attack changes from zero to a finite value α_o. The downwash becomes.

$w\left(\frac{b}{2}, t\right) = -\alpha_o\, U\, 1(t)$. The Fourier transform of this reads as

$$f(\omega) = -\alpha_o\, U \int_{-\infty}^{\infty} 1\,(t)\, e^{-i\omega t} \mathrm{d}\, t = -\frac{\alpha_o\, U}{i\,\omega}$$

Substituting this function into the circulatory lift expression we obtain

$$c_l^c\,(t) = \alpha_o \int_{-\infty}^{\infty} \frac{C(k)}{i\omega}\; e^{i\omega t} \mathrm{d}\,\omega.$$

If we use the reduced time $s = Ut/b$ instead of t we get

$$c_l^c\,(s) = \alpha_o \int_{-\infty}^{\infty} \frac{C(k)}{ik}\; e^{i\,ks} \mathrm{d}\,k.$$

From this integral we define a new function

$$\varphi(s) = \frac{1}{2\,\pi} \int_{-\infty}^{\infty} \frac{C(k)}{ik}\; e^{i\,ks} \mathrm{d}\,k$$

as the Wagner function $\varphi(s)$, the circulatory lift coefficient becomes

$$c_l^c(s) = 2\pi\,\alpha_0 \varphi\,(s). \tag{3.40}$$

The Wagner function is a time dependent function whose limit for t going to infinity approaches unity so that according to Eq. 3.40 the lift coefficient goes to $2\pi\,\alpha_0$.

Let us reduce the Wagner function into a numerically integrable form. If we write the complex exponential with sin and cosine terms, take the Fourier transform of the unit step function and consider the symmetry and antisymmetry involved in

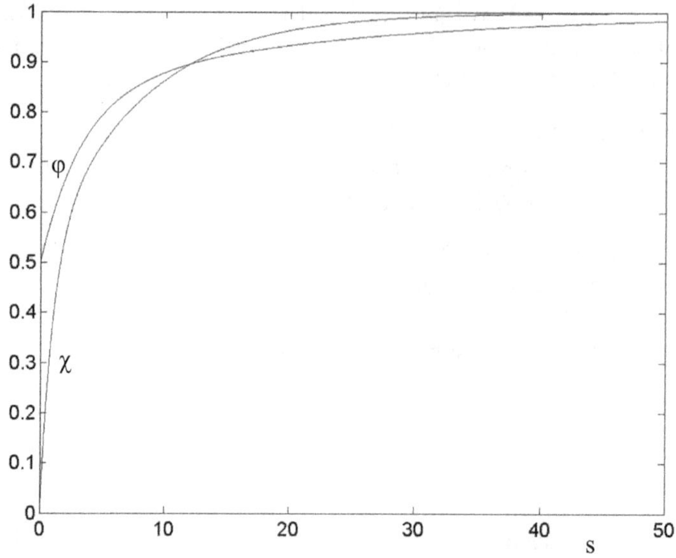

Fig. 3.11 Wagner, φ and Küssner, χ functions

the integrands, we obtain the Wagner function in terms of the real and imaginary parts of the Theodorsen function as follows

$$\varphi(s) = \frac{2}{\pi}\int_0^{\infty} \frac{F(k)}{k} \sin(k\,s)\,\mathrm{d}k = 1 + \frac{2}{\pi}\int_0^{\infty} \frac{G(k)}{k} \cos(k\,s)\,\mathrm{d}k \tag{3.41}$$

For practical uses an approximate form of the Wagner function is given in BAH as

$$\varphi\,(s) \cong 1 - 0.165\,e^{-0.0455s} - 0.335\,e^{-0.3s}. \tag{3.42}$$

The graph of the Wagner function, based on the Jones approach and given by Eq. 3.42 is plotted in Fig. 3.11. The function at zero time takes the value of 0.5 and reaches unity at infinity. This means, after the sudden angle of attack change it takes a long time to reach the steady state value given by Eq. 3.40.

Knowing the expression for the Wagner function, we can give the unsteady aerodynamic coefficients for the arbitrary motion as functions of the reduced time in the form of Duhamel integrals.

$$c_l(s) = \frac{\pi b}{U^2}(\ddot{h} + U\dot{\alpha}) - \frac{2\pi}{U}[w(b/2, s)/2 + \int_0^s w(b/2, \sigma)\,\varphi'(s-\sigma)\,d\sigma]$$

$$c_m(s) = -\frac{\pi b}{2U^2}(b\ddot{\alpha}/4 + U\dot{\alpha}) - \frac{\pi}{U}[w(b/2, s)/2 + \int_0^s w(b/2, \sigma)\,\varphi'(s-\sigma)\,d\sigma] \tag{3.43a, b}$$

We have previously seen that the Wagner function is 0.5 at $t = 0$. This means, the immediate lifting response of an airfoil to a sudden angle of attack change is half the lift value attained steadily. These responses are seen explicitly in the circulatory terms of Eq. 3.43a, b.

Another example for the arbitrary motion of the profile is the response to a sharp edged gust which will be studied next.

Gust Problem, Küssner function

The unsteady aerodynamic response of an airfoil to an arbitrary gust is going to be studied here. An airfoil under the gust load undergoes a motion which consists of arbitrary rotation about any arbitrary point and arbitrary heaving. Therefore, its behavior can not be modeled with the downwash at the three quarter chord point. The downwash changes with respect to time and position on the airfoil as the gust impinges on. Hence, we need a new independent variable to express the downwash on the surface. This new variable depends on the free stream speed with which the gust moves on the surface of the airfoil. For this reason the downwash at the surface becomes:

$$w_a(x, t) = w_a(x - Ut).$$

The gust velocity impinging on the airfoil surface is due to the motion of the air. The downwash on the other hand has an opposite sign to that of gust. If the gust profile is given as w_g then the downwash reads as

$$w_a(x - Ut) = -w_g(x - Ut).$$

In Fig. 3.12 the downwash distribution caused by impinging gust on airfoil surface.

As we did before, let us find the response of the airfoil to unit excitation impinging on to the leading edge as a gust at $t = 0$. If the constant gust speed is w_o, time dependent gust function reads as

$$w_g(x - Ut) = \begin{cases} 0, & Ut < x + b \\ w_o, & Ut \geq x + b \end{cases}$$

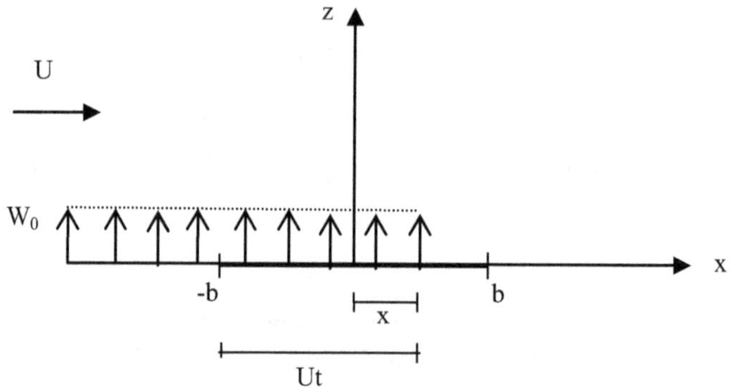

Fig. 3.12 Downwash caused by the gust

This gust function can be rearranged to take the form of the unit function given in Appendix 6 as follows

$$w_g(x - U\,t) = \begin{cases} 0, & U\,t - x - b < 0 \\ w_o, & U\,t - x - b \geq 0 \end{cases}$$

At this stage, it is useful to obtain the Fourier transform of the unit step function

$$1\,(t) = \frac{1}{2\,\pi} \int_{-\infty}^{\infty} \frac{1}{i\omega}\, e^{i\,\omega\,t} \mathrm{d}\,\omega$$

With this transform in mind the Fourier transform of the constant gust reads as

$$w_g(x, t) = \frac{w_o}{2\,\pi} \int_{-\infty}^{\infty} \frac{1}{i\omega}\, e^{i\,\omega(U\,t - x - b)/U} \mathrm{d}\,\omega.$$

In terms of reduced frequency and the reduced time the integral becomes

$$w_g(x^*, s) = \frac{w_o}{2\,\pi} \int_{-\infty}^{\infty} \frac{1}{i\,k}\, e^{ik(s - x^* - 1)}\, \mathrm{d}\,k\,.$$

If we go back and write down Eq. 3.27 for the lifting pressure for the time dependent downwash

$$c_{p_a}(x^*,t) = \frac{4}{\pi}\sqrt{\frac{1-x^*}{1+x^*}}\int_{-1}^{1}\sqrt{\frac{1+\xi^*}{1-\xi^*}}\frac{w(\xi^*,t)\,d\xi^*}{(x^*-\xi^*)\,U} - \frac{4ik}{\pi}\int_{-1}^{1}\Lambda(x^*,\xi^*)\frac{w(\xi^*,t)}{U}d\,\xi^*$$
$$+\frac{4}{\pi}[1-C(k)]\sqrt{\frac{1-x^*}{1+x^*}}\int_{-1}^{1}\sqrt{\frac{1+\xi^*}{1-\xi^*}}\frac{w(\xi^*,t)\,d\xi^*}{U}$$

Here, $w_a(\xi^*,t) = \overline{w}_a(\xi^*)\,e^{i\omega t}$.

If we assume that the gust is simple harmonic in time, the unsteady aerodynamic coefficients can be found in terms of the reduced frequency and time by integrating the lifting pressure coefficient as follows, (BAH)

$$c_l(k,s) = 2\pi\frac{w_o}{U}\{C(k)[J_o(k) - iJ_1(k)] + iJ_1(k)\}e^{iks} \quad \text{and} \quad c_m(k,s) = \frac{1}{2}c_l(k,s)$$

If the gust is not simple harmonic, we have to consider all the harmonics of the gust and integrate the expressions for the aerodynamic coefficients in the frequency domain. The integral representation gives us

$$c_l(s) = \frac{w_o}{U}\int_{-\infty}^{\infty}\frac{1}{ik}\{C(k)[J_o(k) - iJ_1(k)] + iJ_1(k)\}\,e^{ik(s-1)}\,dk$$

and

$$c_m(s) = \frac{1}{2}c_l(s).$$

Now, let us relate the lift coefficient to a new function called the Küssner function as follows

$$c_l(s) = 2\pi\frac{w_o}{U}\chi(s).$$

Here, the Küssner function is the indicial admittance for a sharp edged gust. The Küssner function in terms of the reduced time reads as

$$\chi(s) = \frac{2}{\pi i}\int_{-\infty}^{\infty}\frac{1}{k}\{C(k)[J_o(k) - iJ_1(k)] + iJ_1(k)\}e^{ik(s-1)}\,dk.$$

Let us write the coefficient in the curly bracket of the above integral with its real and imaginary parts as $Fg(k) + i\,Gg(k)$, and write the exponential multiplier with its sin and cosine components, the unilateral integral then reads as a real function of s

$$\chi(s) = \frac{2}{\pi} \int_0^\infty \frac{[F_g(k) - G_g(k)] \ \sin(ks) \ \sin k}{k} \mathrm{d}k. \tag{3.44}$$

The approximate and convenient form of Eq. 3.44 function becomes

$$\chi(s) \cong 1 - 0.5\, e^{-0.13s} - 0.5\, e^{-s}. \tag{3.45}$$

The Küssner function now can be interpreted as the indicial admittance of a sharp edged gust and can be implemented in the Duhamel integral to obtain the responses as the unsteady aerodynamic coefficients expressed in reduced time s,

$$c_l(s) = \frac{2\pi}{U} \int_0^s w_g(\sigma)\ \chi'(s - \sigma)\,\mathrm{d}\sigma] \tag{3.46a, b}$$

and

$$c_m(s) = \frac{1}{2} c_l(s).$$

In Fig. 3.11, also shown is the graph of Küssner function which changes more rapidly in time as compared to Wagner function.

There are two other gust problems which are going to be considered here. These are: (i) sinusoidal gust and, (ii) moving gust problems.

(i) Sinusoidal gust, Sears function: Here, the gust acting on the profile is assumed to change sinusoidally with respect to time and space. The gust intensity with amplitude wo and frequency $\omega_g$has the functional form

$$w_g(x, t) = w_o\, e^{i2\pi\frac{U}{\lambda}\left(t - \frac{x}{U}\right)}$$

Here, λ is the wave length of the gust. For the sake of convenience, we choose the form the gust such a way that at the midchord it starts with a zero effect, i.e.,

$$w_g(t) = w_o\, e^{ik_g t}$$

If we let $kg = 2\pi U/\lambda$ to be the frequency of the gust, the lift coefficient in terms of the Theodorsen and Bessel functions

$$c_l(k_g, t) = 2\pi \frac{w_o}{U} \{C(k)\,[J_o(k) - iJ_1(k)] + iJ_1(k)\}\, e^{ik_g t}$$

A new function, the Sears function, can be defined as

$$S(k_g) = C(k_g)\,[J_o(k_g) - iJ_1(k_g)] + i\,J_1(k_g) = F_s + iG_s$$

whose graph is shown in Fig. 3.13. The corresponding lift coefficient then reads

$$c_l(k_g, t) = 2\,\pi\,\frac{w_o}{U} S(k_g)\,e^{\,i k_g t}.$$

(ii) Moving gust problem, Miles functions. Here, we consider the effect of a gust moving with speed of U_g against or in the direction of free stream speed U. The resulting indicial admittance is the Miles function which is given in terms of a ratio

$$\lambda = \frac{U}{U + U_g}.$$

This function has a significance in rotor aerodynamics. There is a sufficient amount of information about this function and its implementations in Leishman. The parameter λ takes the value between 0 and 1. When the gust speed is zero λ becomes unity and the Miles function becomes Küssner function. On the other hand, for very large gust speeds λ approaches zero and Miles function behaves like Wagner function.

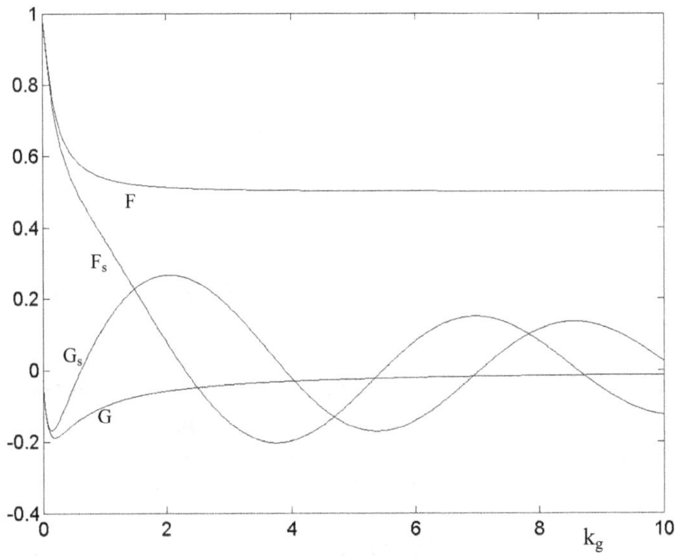

Fig. 3.13 Sears, $S = Fs + i\ Gs$ and Theodorsen functions, real and imaginary parts

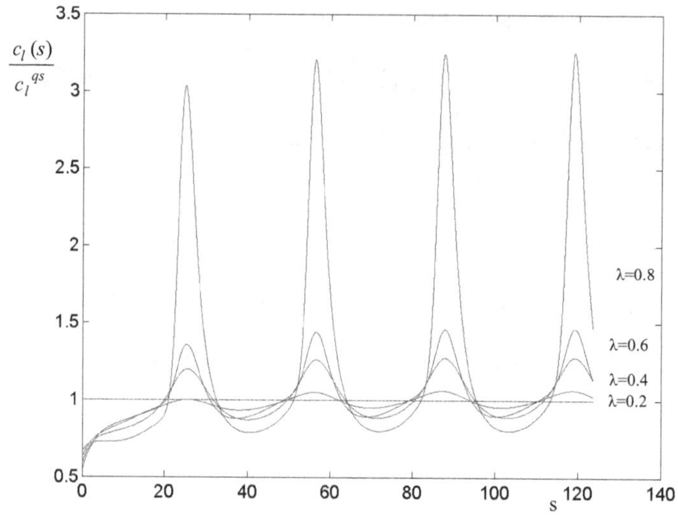

Fig. 3.14 Effect of the time varying free stream on the lift coefficient

We have given, in summary, some analytical expressions involving the Wagner and the Küssner functions. Let us now look at another application for which a 'time varying free stream problem' is considered. This problem can be used to model the unsteady aerodynamics for the forward flight of a single helicopter blade.

Example 3.5 A rotating blade in a forward flight is modeled at its section with a sinusoidally varying free stream speed under constant angle of attack. Obtain the unsteady variation of the sectional lift coefficient in terms of the quasi steady lift coefficient and plot its variation by time for different intensities of the changing sinus term.

Solution We can write the sinusoidally varying free stream speed at a section with $U(t) = Uo(1 + \lambda \sin \omega_t)$. The formulae 3.43a, b for the arbitrary motion can be used to obtain the sectional coefficient as follows

$$c_l(s) = \frac{\pi b}{U^2}\dot{U}\alpha_o + \frac{2\pi}{U}\left[U(s)\,\alpha_o/2 + \int_0^s U(\sigma)\alpha_o\varphi'(s-\sigma)\,d\sigma\right]$$

For $k = 0.2$ and $\lambda = 0.2, 0.4, 0.6, 0.8$ values of the ratios of the unsteady sectional lift coefficient to quasi steady coefficient are plotted with respect to time. The plots are obtained for four period of free stream starting at zero time. The intensity of the change in the free stream causes peaks at the lift coefficient. In each plot, there is a transition period after the onset of the motion.

As observed from the graphs the difference between the minimum and maximum of these curves increase with increasing λ.

Summary: Two dimensional incompressible flows past thin lifting surfaces are studied. The lifting surface is modeled with a vortex sheet. The Biot–Savart law is used to establish the relation between the vortex sheet strength and the downwash w(x, t). The strength of the vortex sheet is calculated via *the inversion of a singular integral equation given in terms of the downwash at the surface as a boundary condition. First, for an airfoil steady state solution after the impulsive start in a uniform stream is considered. Carleman's formula is used to invert the integral relation which explicitly gives the vortex sheet strength satisfying the well known Kutta condition. The lifting pressure distribution and the sectional lift and the moment coefficients are found with integration of lifting pressure along the chord. Furthermore, the concept of center of pressure and aerodynamic center are introduced. The locations of both centers on the chord are then calculated. Unsteady flow case is studied with distribution of a vortex sheet in the wake as well. The wake vorticity and the bound vorticity are tied together with the unsteady Kutta condition which is nothing but imposing zero lifting pressure at the wake region. The Laplace transform technique is employed to establish the relation between the total bound circulation and the downwash. Then the transformed form of the bound vortex sheet strength and the downwash relation is used to obtain the expression for the lifting surface pressure in the Laplace domain. Since the inverse Laplace transform of the pressure expression is quite complex, the inversion is performed for the simple harmonic pressure variation. The lifting surface pressure has three terms each signifying different aspects of aerodynamic phenomena. The first term is the quasi steady term which is identical with the steady pressure term, the second term accounts for the contribution of the wake vorticity and, finally the third term is the 'apparent mass' term which is responsible for the non circulatory lifting pressure without the presence of free stream. Depending on the type of aerodynamics we use, the relevant terms are kept in the expression given for the pressure. Accordingly, (i) if all three terms are retained then the approach is called 'unsteady aerodynamics' and it is used for the problems having oscillations of order of 40 Hz, (ii) for the case of 'quasi unsteady aerodynamics' we neglect the apparent mass term where the approach is applicable for the 5–15 Hz. range, and finally, (iii) 'quasi steady aerodynamics' requires retaining the quasi steady term only where the approach is good for the frequencies of 1 Hz. or less. As a special type of an unsteady flow, Simple Harmonic Motion of an airfoil in pitch and/or in vertical translation is considered. The Theodorsen function here is indicative of the effect of the wake vorticity on the circulatory term in the lift expression. The effect of the wake vorticity on the profile shows itself as reduction of the magnitude of the lift coefficient, and the lagging of the response of the airfoil with the angle of attack change in time. In order to demonstrate this several basic examples of SHM are provided, wherein the hysteresis under the lift vs angle of attack curve is accounted for. In addition, returning wake problem of Loewy is studied with the help of a new function which depends on the wake spacing, reduced frequency and the rotational frequency of the blade, replace the Theodorsen function. Arbitrary unsteady motions are analyzed with the concept of the 'indicial admittance' function applicable to linear systems. First, the Wagner function as the indicial admittance to the arbitrary angle of attack change of an airfoil in a free stream is considered. Then, the Küssner function of the sharp edged gust impingement is obtained as the indicial admittance. In analyzing the response of an airfoil to an arbitrary motion the integral Fourier Transform Technique is utilized. The sinusoidal gust problem is studied by establishing the Sears function as the indicial admittance. For the analysis of the moving gust problem the concept of the Miles function is introduced. The Miles function is generally utilized in rotor aerodynamics. However, when the moving gust velocity becomes zero the Miles function transforms itself into the Küssner function. Finally, as an application of Wagner function an airfoil immersed in a sinusoidally varying free stream velocity is considered. This problem can be utilized in obtaining the estimate of the total lift coefficient of a blade in a forward flight from two dimensional considerations only.*

3.7 Questions and Problems

3.1 Equation 3.3 gives the relation between the downwash and the vortex sheet strength in x–z coordinates for a positive free stream running from left to write. Obtain a similar expression for a free stream running from right to left (Make sure to satisfy the Kutta condition).

3.2 An airfoil is given by a parabolic camber line, i.e., $z_a = -(a/b^2)\, x^2$. Find: (i) sectional lift coefficient, (ii) center of pressure, and (iii) aerodynamic center, at zero angle of attack.

3.3 Find the phase difference between the displacement and the downwash for a flat plate oscillating simple harmonically in a free stream at a zero angle of attack.

3.4 Comment on the physical meaning of the Theodorsen function.

3.5 Find the sectional lift and moment coefficients taken about the midchord for the airfoil given in Problem 3.2 undergoing a simple harmonic motion $h = \overline{h}e^{i\omega t}$.

3.6 Find the sectional moment coefficient taken about the midchord of the Example 3.4. Plot the change with respect to angle of attack.

3.7 Find the lift and moment coefficients about the quarter chord for NACA 0012 profile which is pitching about its quarter chord with $\alpha(t) = 3° + 10° \sin\omega t$. (Compare your results with that of Katz and Plotkin (2001), p 503, given for $k = 0.1$).

3.8 For the returning wake problem, interpret the phase angles for, ω/Ω values being equal to an integer, integer plus a quarter and integer plus one half. Take $h/b = 3$. Find the amplitude variations of the function for the same changes in ω/Ω.

3.9 Use the data of Problem 3.8 to find the phase differences of the Loewy function for a double bladed rotor where only the distance between the blades are counted. Make the same computations for amplitude variations.

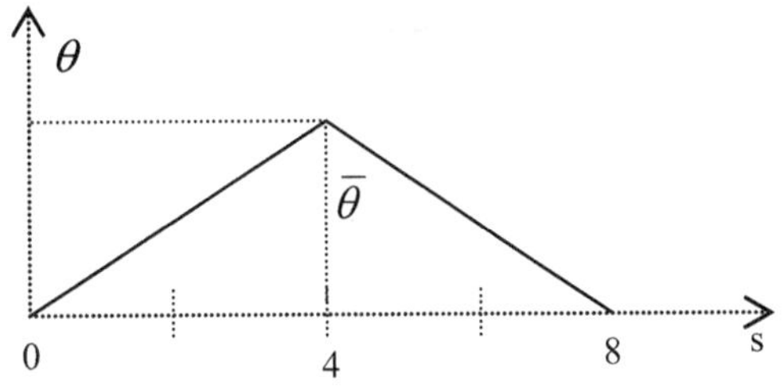

Fig. 3.15 $\theta(s)$ variation

3.10 Obtain the time variation of the sectional lift coefficient for an airfoil which is pitching about its leading edge as shown in Fig. 3.15 using (i) unsteady aerodynamics, (ii) quasi unsteady aerodynamics, and (iii) quasi steady aerodynamics. Plot the lift coefficient vs time curve for all three cases.

3.11 Find the lift and the moment coefficients about the midchord of an airfoil which undergoes a sudden vertical translation under zero angle of attack. Use the Wagner function.

3.12 Find the lift and the moment coefficient at the midchord of an airfoil undergoing sudden velocity change. Use the Wagner function (Derive the lift formula used for Example 3.5).

3.13 If the gust intensity with time varies as given in Fig. 3.16, obtain the lift and the moment coefficient changes about the midchord and their plots with respect to time.

3.14 Express the phase angle of the Sears function as the function of the reduced frequency and compare it with the phase of the Theodorsen function on a graph.

3.15 Consider the simple harmonically varying free stream problem for the reduced values $k = 0.2$, 0.4, 0.6 and 0.8, find the lift coefficient under constant angle of attack. Take the amplitude change as $\lambda = 0.4$ and obtain the graph for the ratio of the unsteady lift coefficient to quasi steady lift coefficient for each reduced frequency values. Comment on the peaks of the lift curves.

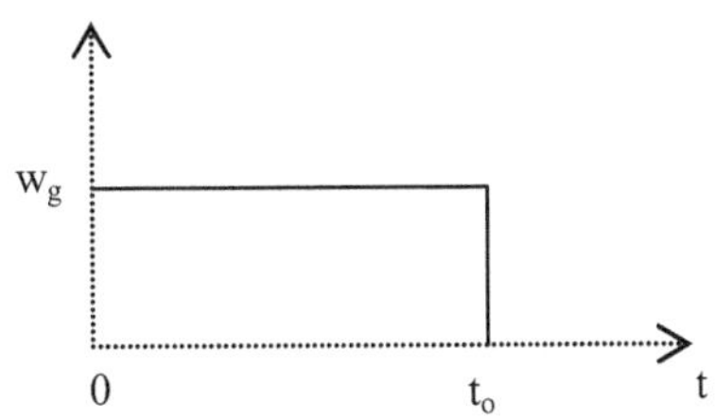

Fig. 3.16 Gust intensity

3.16 For a simple harmonically varying free stream problem obtain the expression for sectional moment coefficient about the midchord in terms of the reduced frequency for the amplitude. Plot the graph for the moment coefficient using the data of Problem 3.15.

3.17 Assume a blade with radius R is rotating with angular speed Ω at a constant forward flight speed U. Show that the problem can be modeled as a variable free stream: $Us = U \sin(\omega t) + \Omega R$. (i) What will be the values of λ and ω in terms of Ω and R? (ii) Assuming that effective span of this blade starts at the 10% span from the root find an expression for the lift coefficient using the strip theory, (iii) comment on the validity of your answer in terms of three dimensionality and the existence of the tip vortices.

References

Bisplinghoff, R.L., Ashley, H., Halfman, R.L.: Aeroelasticity. Dover Publications Inc., New York (1996)

Gulcat, Ü.: Separate numerical treatment of attached and detached flow regions in general viscous flows. Ph.D. Dissertation, Georgia Institute of Technology, School of Aerospace Engineering, Atlanta (1981)

Hildebrand, F.B.: Advanced Calculus for Applications. Prentice-Hall Inc., Engelwood Cliffs (1976)

Katz, J., Plotkin, A.: Low Speed Aerodynamics, 2nd edn. Cambridge University Press (2001)

Kuethe, A.M., Chow, C.-Y.: Foundations of Aerodynamics, 5th edn. Wiley, New York (1998)

Leishman, G.J.: Principles of Helicopter Aerodynamics. Cambridge University Press (2000)

Theodorsen, T.: General Theory of Aerodynamic Instability and Mechanism of Flutter, T.R. No 496, NACA 1935

Chapter 4
Incompressible Flow About Thin Wings

Thin wing theory is an efficient tool for the study of the spanwise variation of aerodynamic characteristics which has effect on the total lift and moment coefficient of a finite wing. This variation is considerably slow except at the tip region of the high aspect ratio wings. For low aspect ratio or delta wings, on the other hand, the aerodynamic characteristics vary rapidly in their short span. Another characteristic of the finite wing theory is the downwash generation because of the tip vortices, which in turn induces drag. The magnitude of the induced drag is proportional with lift and inversely proportional with the aspect ratio. The physical model we use for the three dimensional aerodynamic analysis is based on the two dimensional vortex sheet spread over the wing surface and its wake. In this model, imposing the boundary conditions on the wing the spanwise and the chordwise components of the vortex sheet strength are expressed in terms of the downwash as an integral equation. The remaining task now is the inversion of this integral equation with different assumptions relevant to the flow conditions. Let us now build our model for different wing shapes to find the aerodynamic coefficients.

4.1 Physical Model

Let the unsteady components of the vortex sheet strength on the wing surface immersed in a free stream with angle of attack be given by $\gamma(x,y,t)$ in spanwise direction and be given by $\delta(x,y,t)$ in chordwise direction, respectively. In Fig. 4.1, shown are the wing surface in the free stream and the relevant geometry for the point (x,y,z) at which the vortex sheet induces the downwash under consideration.

According to the Biot-Savart law the infinitesimal vortex with intensity of Γds located at a point (ξ,η) induces a differential velocity dV at a point (x,y,z) as follows

Ü. Gülçat, *Fundamentals of Modern Unsteady Aerodynamics*,
https://doi.org/10.1007/978-3-030-60777-7_4

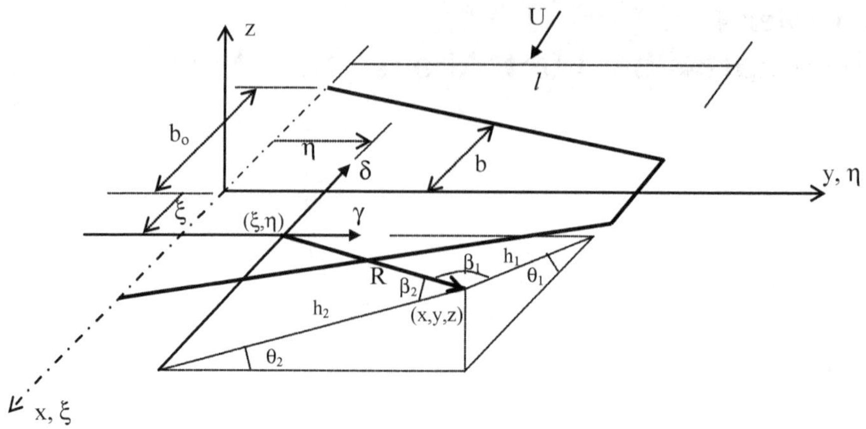

Fig. 4.1 Wing geometry and the position vector R from (ξ,η) to (x,y,z)

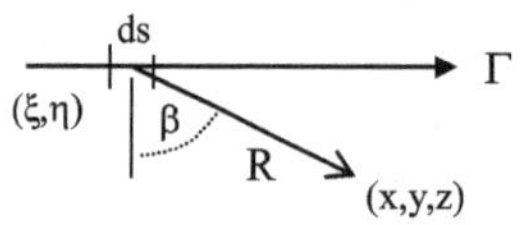

$$dV = \frac{\Gamma}{4\pi}\frac{\cos\beta}{R^2}ds$$

The relations between the distances and the angles become

$$R = \sqrt{(x-\xi)^2 + (y-\eta)^2 + z^2}, \quad h_1 = R\cos\beta_1, \quad h_2 = R\cos\beta_2.$$

Viewing the x–z plane from y axis, we can find the differential velocity dV_1 induced by the spanwise vortex sheet component γ as follows.

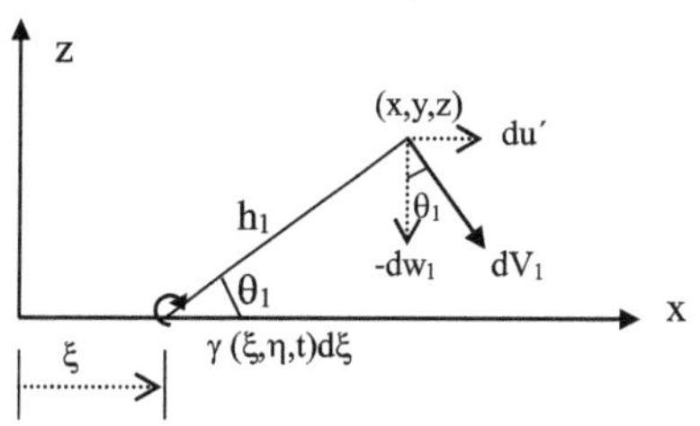

$$dV_1 = \frac{\gamma d\xi}{4\pi}\frac{\cos\beta_1}{R^2}d\eta$$

$$dw_1 = -dV_1\cos\theta_1 = -\frac{1}{4\pi}\frac{\gamma(x-\xi)d\xi d\eta}{R^3}$$

$$du' = dV_1\sin\theta_1 = \frac{1}{4\pi}\frac{\gamma z d\xi d\eta}{R^3}$$

Similarly, looking at y–z plane from x axis, dV_2 component induced by y δ can be written as.

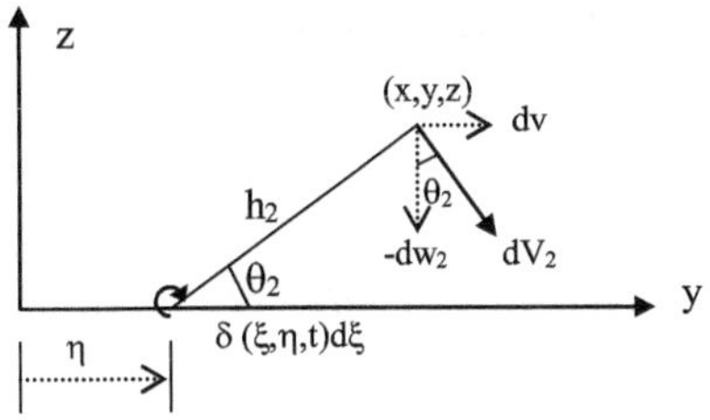

$$dV_2 = \frac{\delta d\eta}{4\pi}\frac{\cos\beta_2}{R^2}d\xi$$

$$dw_2 = -dV_2\cos\theta_2 = -\frac{1}{4\pi}\frac{\delta(y-\eta)d\xi d\eta}{R^3}$$

$$dv = dV_2\sin\theta_2 = \frac{1}{4\pi}\frac{\delta z d\xi d\eta}{R^3}$$

The induced velocities given above are in differential form of the perturbation velocities. If we want to find the effect of whole x–y plane we have to find the integral effect to obtain the total induced velocity components at a point (x,y,z) as follows

$$u'(x,y,z,t) = \frac{1}{4\pi}\iint \frac{\gamma(\xi,\eta,t)z d\xi d\eta}{[(x-\xi)^2+(y-\eta)^2+z^2]^{3/2}},$$

$$v = \frac{1}{4\pi}\iint \frac{\delta(\xi,\eta,t)z d\xi d\eta}{[(x-\xi)^2+(y-\eta)^2+z^2]^{3/2}}$$

$$w = -\frac{1}{4\pi}\iint \frac{\gamma(\xi,\eta,t)(x-\xi)+\delta(\xi,\eta,t)(y-\eta)}{[(x-\xi)^2+(y-\eta)^2+z^2]^{3/2}}d\xi d\eta \quad (4.1\text{a, b, c})$$

The components of the induced velocities at the lower and upper surfaces of the thin wing have the following relations for z = $0^{\pm}$

$$u'(x,y,0^+,t) = -u'(x,y,0^-,t) \quad \text{and} \quad v(x,y,0^+,t) = -v(x,y,0^-,t).$$

Now, we can write the relation vortex sheet strength components and the perturbation speeds.

$$\gamma(x,y,t) = u'(x,y,0^+,t) - u'(x,y,0^-,t) = \frac{\partial\phi'_u}{\partial x} - \frac{\partial\phi'_l}{\partial x} \quad \text{and,}$$

$$\delta(x,y,t) = v(x,y,0^+,t) - v(x,y,0^-,t) = \frac{\partial\phi'_u}{\partial y} - \frac{\partial\phi'_l}{\partial y}.$$

In the last two lines, if we take the derivative of the equation with respect to y, and the second equation with respect to x, they become equal, i.e.,

$$\frac{\partial\gamma}{\partial y} = \frac{\partial\delta}{\partial x} \quad (4.2)$$

Due to the presence of the wing in a free stream, there are three distinct flow regions: i) the wing surface R_a, ii) wake region R_w, and iii) rest of the area in x–y plane.

In addition, we can define the lifting pressure as the pressure difference between the lower and upper surfaces of the wing as follows,

$$\Delta p = p_l - p_u$$

The Kelvin's equation can be written for the pressure differences of the wing surface and the wake region.

$$\begin{aligned} \frac{\Delta p_a}{\rho} &= \frac{\partial}{\partial t}\int_{x_l}^{x} \gamma_a(\xi,\eta,t)d\xi + U\gamma_a(x,y,t) \\ \frac{\Delta p_w}{\rho} &= \frac{\partial}{\partial t}\int_{x_l}^{x_t} \gamma_a(\xi,\eta,t)d\xi + \frac{\partial}{\partial t}\int_{x_t}^{x} \gamma_w(\xi,\eta,t)d\xi + U\gamma_w(x,y,t) \end{aligned} \tag{4.3a, b}$$

At this stage, we can consider the downwash as the velocity induced separately by the vortex sheet of the surface and the wake region.

$$\begin{aligned} w_a(x,y,t) = &-\frac{1}{4\pi}\iint_{R_a} \frac{\gamma_a(\xi,\eta,t)(x-\xi)+\delta_a(\xi,\eta,t)(y-\eta)}{[(x-\xi)^2+(y-\eta)^2]^{3/2}}d\xi d\eta \\ &-\frac{1}{4\pi}\iint_{R_w} \frac{\gamma_w(\xi,\eta,t)(x-\xi)+\delta_w(\xi,\eta,t)(y-\eta)}{[(x-\xi)^2+(y-\eta)^2]^{3/2}}d\xi d\eta \end{aligned} \tag{4.4}$$

If the equation of the wing surface is given as $z = z_a(x,y,t)$, the downwash on the surface becomes,

$$w_a(x,y,t) = \frac{\partial z_a}{\partial t} + U\frac{\partial z_a}{\partial x}; \ (\mathrm{x,y}) \subset \mathrm{R_a} \tag{4.5}$$

In the wake region, since there is no lifting pressure the unsteady Kutta condition becomes.

$$\Delta p_w(x,y,t) = 0; \quad (\mathrm{x,y}) \subset \mathrm{R_w} \tag{4.6}$$

In the rest of the x–y plane, there is no vortex sheet in our model.

The remaining task here is to find the lifting pressure in terms of the surface vortex sheet strength given as 4.3a. The surface vortex sheet and the wake vortex sheet strengths are related to each other via unsteady Kutta condition, 4.3b and 4.6, This relation is used to eliminate the wake vortex from Eq. 4.4. If we now use 4.5 to express the downwash in terms of the equation of surface, we obtain the integral relation giving the surface vortex sheet strength in terms of motion of the wing. The resulting integral equation contains double integral, and quite naturally it is not analytically invertible! Depending on the geometry of a planform we can make simplifying assumptions to this integral equation and find approximate solutions. It is convenient to start inverting the equation for steady flow.

4.2 Steady Flow

Under steady flow conditions the terms involving time derivative vanish in Eq. 4.3b and, since the pressure difference at the wake is zero, the spanwise vortex sheet strength at the wake also vanishes, i.e., $\gamma_w = 0$. This results in the continuity of the vortices.

$$\frac{\partial \gamma_a}{\partial y} = \frac{\partial \delta_a}{\partial x} \quad \text{and,} \qquad \frac{\partial \delta_w}{\partial x} = 0 \tag{4.7a, b}$$

Equation 4.7b dictates that δ_w is only the function of y. At the trailing edge the Kutta condition imposes the following restriction on the chordwise component of the vortex sheet

$$\delta_w(x, y) = \delta_w(x_t, y) = \delta_a(x_t, y)$$

which means its value is constant along x at a constant spanwise station. If we integrate Eq. 4.7a with respect to x and differentiate the result with respect to x, the Leibnitz rule gives the following for the chordwise component of the surface vortex sheet strength

$$\delta_a(x_t, y) = \int_{x_l}^{x_t} \frac{\partial \gamma_a}{\partial y} dx + 0 = \frac{d}{dy} \int_{x_l}^{x_t} \gamma_a dx + \gamma_a(x_l, y)\frac{dx_l}{dy} - \gamma_a(x_t, y)\frac{dx_t}{dy}$$

The last two terms of the last expression vanish because of the character of the vortex sheet. Only contribution comes from the first term which is the derivative of the bound circulation to give.

$$\delta_w(y) = \delta_a(x_t, y) = \frac{d\Gamma}{dy} \tag{4.8}$$

Equation 4.8 tells us that the wake vorticity has a component only in stream wise direction and its strength varies with the bound circulation. The downwash expression then reads as.

$$\begin{aligned} w_a(x, y) = &-\frac{1}{4\pi} \iint_{R_a} \frac{\gamma_a(\xi, \eta)(x - \xi) + \delta_a(\xi, \eta)(y - \eta)}{[(x - \xi)^2 + (y - \eta)^2]^{3/2}} d\xi d\eta \\ &- \frac{1}{4\pi} \iint_{R_w} \frac{\delta_w(\xi, \eta)(y - \eta)}{[(x - \xi)^2 + (y - \eta)^2]^{3/2}} d\xi d\eta \end{aligned} \tag{4.9}$$

Now, we can evaluate the integrals given by 4.9 for a rectangle with a span of 2 *l* and chord of *2b*. We can rewrite the integrals using the constant integral limits based on *b* and *l*, which gives

$$w(x,y) = -\frac{1}{4\pi}\int_{-b}^{b}\int_{-l}^{l}\frac{\gamma_a(\xi,\eta)(x-\xi)}{[(x-\xi)^2+(y-\eta)^2]^{3/2}}d\xi d\eta - \frac{1}{4\pi}\int_{-b}^{b}\int_{-l}^{l}\frac{\delta_a(\xi,\eta)(y-\eta)}{[(x-\xi)^2+(y-\eta)^2]^{3/2}}d\xi d\eta$$
$$-\frac{1}{4\pi}\int_{-l}^{l}\delta_a(\xi,\eta)(y-\eta)\int_{b}^{\infty}\frac{d\xi}{[(x-\xi)^2+(y-\eta)^2]^{3/2}}d\eta$$

If we denote all three integrals with I_1, I_2 and I_3 respectively, the downwash becomes

$$w(x,y) = -\frac{1}{4\pi}(I_1+I_2+I_3).$$

The first integral I_1 can be integrated by parts with respect to η to give

$$I_1 = \int_{-b}^{b}\int_{-l}^{l}\frac{\partial\gamma_a}{\partial\eta}\frac{(y-\eta)}{(x-\zeta)[(x-\xi)^2+(y-\eta)^2]^{1/2}}d\xi d\eta.$$

Similarly, I_2 is integrated by parts with respect to ξ.

$$I_2 = \int_{-l}^{l}\left.\frac{-\delta_a(\xi,\eta)(x-\xi)}{(y-\eta)[(x-\xi)^2+(y-\eta)^2]^{1/2}}\right|_{\xi=-b}^{\xi=b}d\eta + \int_{-b}^{b}\int_{-l}^{l}\frac{\partial\delta_a}{\partial\xi}\frac{(x-\xi)}{(y-\eta)[(x-\xi)^2+(y-\eta)^2]^{1/2}}d\xi d\eta$$

The first term of the right hand side is evaluated at the lower limit right before the leading edge to have zero value. The upper limit value on the other hand cancels with the lower limit value of I_3.

The inner integral of I_3 can be taken directly with respect to ξ to have

$$I_3 = \int_{-l}^{l}\left.\frac{-\delta_a(b,\eta)(x-\xi)}{(y-\eta)[(x-\xi)^2+(y-\eta)^2]^{1/2}}\right|_{\xi=b}^{\xi=\infty}d\eta$$

As seen clearly the lower limit of I_3 cancels the upper limit of I_2. The upper limit value of the integrand gives

$$\left.\frac{-1}{(y-\eta)[1+((y-\eta)/(x-\xi))^2]^{1/2}}\right|^{\xi=\infty} = \frac{-\sqrt{1}}{y-\eta}.$$

At this stage we choose $\sqrt{1} = -1$ to obtain the correct sign for the induced downwash. The summation of all three integrals gives.

$$w(x,y) = -\frac{1}{4\pi}\int_{-b}^{b}\int_{-l}^{l}\frac{\partial\gamma_a}{\partial\eta}\frac{[(x-\xi)^2+(y-\eta)^2]^{1/2}}{(x-\xi)(y-\eta)}d\xi d\eta - \frac{1}{4\pi}\int_{-l}^{l}\frac{d\Gamma}{d\eta}\frac{d\eta}{y-\eta} \quad (4.10)$$

In terms of boundary conditions 4.10 reads as.

$$U\frac{\partial z_a}{\partial x} = -\frac{1}{4\pi}\int_{-b}^{b}\int_{-l}^{l}\frac{\partial\gamma_a}{\partial\eta}\frac{[(x-\xi)^2+(y-\eta)^2]^{1/2}}{(x-\xi)(y-\eta)}d\xi d\eta - \frac{1}{4\pi}\int_{-l}^{l}\frac{d\Gamma}{d\eta}\frac{d\eta}{y-\eta} \quad (4.11a)$$

We can employ the same integral for a wing with a straight line trailing edge by variable leading edge to have.

$$U\frac{\partial z_a}{\partial x} = -\frac{1}{4\pi}\int_{x_l(y)}^{x_t}\int_{-l}^{l}\frac{\partial\gamma_a}{\partial\eta}\frac{[(x-\xi)^2+(y-\eta)^2]^{1/2}}{(x-\xi)(y-\eta)}d\xi d\eta - \frac{1}{4\pi}\int_{-l}^{l}\frac{d\Gamma}{d\eta}\frac{d\eta}{y-\eta} \quad (4.11b)$$

which can be used for the swept wings.

Lifting Line Theory:

The Prandtl's Lifting Line Theory is valid only for the high aspect ratio wings. For high aspect ratio wings, x-ξ value can be neglected compared to y-η in first term of the right hand side of Eq. 4.11b. While making this assumption here, we presume that as x approaches ξ and y approaches η, the vortex sheet strength is not too large. Now, if we use the fact that $(y-\eta)^2$ is much larger than $(x-\xi)^2$ we can simplify the double integral in 4.11b as follows.

$$\int_{x_l(y)}^{x_t}\int_{-l}^{l}\frac{\partial\gamma_a}{\partial\eta}\frac{[(x-\xi)^2+(y-\eta)^2]^{1/2}}{(x-\xi)(y-\eta)}d\xi d\eta = \int_{-b}^{b}\frac{1}{x-\xi}\int_{-l}^{l}\frac{\partial\gamma_a}{\partial\eta}\frac{|y-\eta|}{(y-\eta)}d\xi d\eta$$
$$= 2\int_{-b}^{b}\frac{\gamma_a(\xi,y)}{x-\xi}d\xi \quad (4.12)$$

Substituting 4.12 into 4.11b we obtain.

$$U\frac{\partial z_a}{\partial x} = -\frac{1}{2\pi}\int_{-b}^{b}\frac{\gamma_a(\xi,y)}{x-\xi}d\xi - \frac{1}{4\pi}\int_{-l}^{l}\frac{d\Gamma}{d\eta}\frac{d\eta}{y-\eta} \quad (4.13)$$

In Eq. 4.13, if we neglect the second term at the right hand side we obtain the two dimensional steady state flow relation between the vortex sheet strength and the equation for the profile. The second term, on the other hand, is the contribution of the spanwise circulation change. In order to invert Eq. 4.13 we multiply the equation with $\sqrt{(b+x)/(b-x)}$ and integrate with respect to x we obtain.

$$U\int_{-b}^{b}\sqrt{\frac{b+x}{b-x}}\frac{\partial z_a(x,y)}{\partial x}dx=-\frac{1}{2}\Gamma-\frac{b}{4}\int_{-l}^{l}\frac{d\Gamma}{d\eta}\frac{d\eta}{y-\eta} \tag{4.14}$$

In two dimensional steady flow the sectional lift coefficient obtained before was.

$$c_l=-2\int_{-1}^{1}\sqrt{\frac{1+\xi^*}{1-\xi^*}}\frac{w(\xi^*)}{U}d\xi. \tag{4.15}$$

If we compare the left hand side of 4.14 with the right hand side of 4.15, and consider the spanwise dependence also for any section on the wing we obtain.

$$U\int_{-b}^{b}\sqrt{\frac{b+x}{b-x}}\frac{\partial z_a(x,y)}{\partial x}dx=Ub\int_{-1}^{1}\sqrt{\frac{1+\xi^*}{1-\xi^*}}\frac{w(x^*,y)}{U}dx^*=-\frac{1}{2}Ubc_l(y) \tag{4.16}$$

In small angles of attack the sectional lift coefficient is proportional with the angle of attack. This enables us to define the lift line slope as $a=(\mathrm{y})\partial c_l/\partial\alpha$. The lift coefficient becomes.

$$c_l(y)=\frac{\partial c_l}{\partial\alpha}(y)=a(y)\alpha(y) \tag{4.17}$$

Using 4.16 and 4.17 in Eq. 4.14 we obtain the formula for Prandtl's lifting line theory as follows.

$$\Gamma(y)=Uba(y)\left[\alpha(y)-\frac{1}{2a(y)U}\int_{-l}^{l}\frac{d\Gamma}{d\eta}\frac{d\eta}{y-\eta}\right] \tag{4.18}$$

In Eq. 4.18 the expression given in brackets is a function of y and it is the effective angle of attack. The effective angle of attack is nothing but the difference between the sectional angle of attack α and the angle induced by the downwash which is also induced by the tip vortices of the wing.

An efficient method of solving Eq. 4.18 to find the spanwise circulation is the Glauert's Fourier series method. Let us first transform the spanwise y and η coordinates from *l* to –*l* with

$$y = l\cos\phi \quad \text{and} \quad \eta = l\cos\theta$$

Expansion of the circulation distribution in to sin series only enables us to have the vanishing circulation values at the tips. Having the Fourier coefficient with no dimension suggests the following form for the circulation expression.

$$\Gamma(\phi) = Ua_o b_o \sum_{n=1}^{\infty} A_n \sin n\phi. \tag{4.19}$$

In Eq. 4.19 the coefficient $a_o b_o$ denotes the lift line slope and the half chord values at the root. Using 4.19 and its derivative in 4.18 we obtain.

$$\Gamma(\phi) = Uab\left[\alpha + \frac{a_o b_o}{2al}\sum_{n=1}^{\infty} nA_n \int_0^{\pi} \frac{\cos n\theta d\theta}{\cos\phi - \cos\theta}\right] \tag{4.20}$$

The integral tables give that $\int_0^{\pi} \frac{\cos n\theta d\theta}{\cos\phi - \cos\theta} = -\pi \frac{\sin n\phi}{\sin\phi}$,

Hence, we obtain.

$$\frac{ab}{a_o b_o}\alpha = \sum_{n=1}^{\infty} A_n\left[\sin n\phi + \frac{b\pi n}{2l}\frac{\sin n\phi}{\sin\phi}\right]. \tag{4.21}$$

Equation 4.21 is valid for the whole span from left tip to right tip with A_n being the unknown coefficients once the geometry of the wing is specified. In order to determine these unknown coefficients we have to pick first N terms in the series together with the sufficient number of spanwise stations along the span so that we end up with the number of unknowns being equal to number of equations written for each station. After solving the system of equations for the unknown coefficients, we obtain the circulation value at each station using 4.19. If we examine Eq. 4.19, we observe that for odd values of n, n = 1,3,5,…, the circulation values will be symmetric with respect to wing root and for even n, n = 2,4,6,…, will be antisymmetric. The integration of the circulation along the span with the Kutta-Joukowski theorem gives the total lift and the lift induced drag. For a symmetric but arbitrary wing loading the total lift and the induced drag coefficients in terms of the aspect ratio AR and the wing area S become.

$$C_L = \pi a_0 b_0 l A_1/S, \quad C_{D_i} = C_L^2/(\pi AR)\sum_{n=1}^{\infty} nA_n^2/A_1^2. \tag{4.22a, b}$$

Prandtl's lifting line theory helps us to find the pitching moment distribution along the span of a wing. At a section of a wing, the moment is determined as the summation of the moment acting at the center of pressure $\left(m_{cp}=0\right)$ with the moment at the aerodynamic center (m_{ac}) where the moment is independent of angle of attack. Thus, we place the bound vortex at the quarter chord where the lifting force is acting. To find the moment at the quarter chord, the moment at the aerodynamic center is transferred to the quarter chord.

Shown in Fig. 4.2 is the line of centers of pressure and the line of aerodynamic centers for a swept wing which is symmetric with respect to its root. Let us first find the distance X_{AC} between the aerodynamic center of this wing to the reference line with integrating the sectional characteristics along the span.

$$X_{AC}=\frac{\int_0^l c_l(y)x_{ac}b(y)dy}{\int_0^l c_l(y)b(y)dy} \tag{4.23a}$$

Now, the moment with respect to the aerodynamic center can be found with defining $\Delta x_{ac}(y)=X_{AC}-x_{ac}(y)$ at each section as follows.

$$M_{AC}=\int_0^l (m_{ac}-L'\Delta x_{ac})dy \tag{4.23b}$$

Here, *L'* denotes the sectional lift.

Example 1 A rectangular wing which has an aspect ratio of 7 has a symmetrical profile. Find its lift coefficient in terms of the constant angle of attack α.

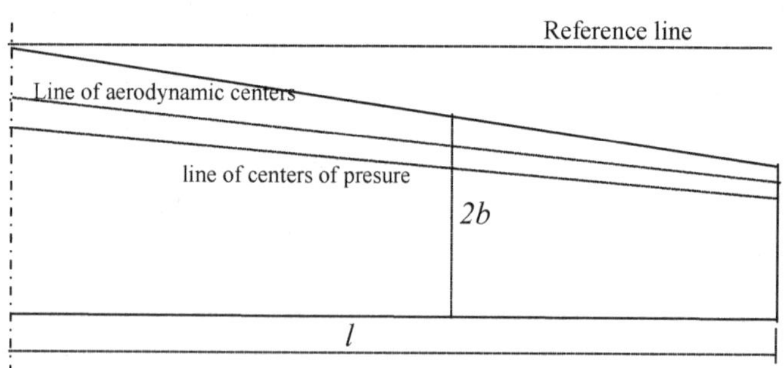

Fig. 4.2 Lines of enter of pressure and aerodynamic centers on a wing

Solution Since the wing is symmetric with respect to its root, we take only the value of odd n. It is sufficient to choose 4 station points with $\phi_i = \pi/8, \pi/4, 3\pi/8$ and $\pi/2$ to find 4 unknown coefficients A_n, n = 1,2,3,4 with four equations written for each station.

For α being constant at each station Eq. 4.21 gives

$$\alpha_i = \sum_{n=1}^{4} A_n \sin n\phi_i \left[1 + \frac{\pi n}{2AR} \frac{1}{\sin \phi_i}\right], i = 1, \ldots, 4$$

Since the angle of attack is constant the solution of the final equation gives $A_1 = 0.9517\alpha$, $A_3 = 0.1247\alpha$, $A_5 = 0.0262\alpha$, $A_7 = 0.0047\alpha$.

The lift coefficient for the wing then becomes

$$C_L = \pi^2 \frac{A_1}{2} = 4.6977\alpha.$$

For wings with moderate aspect ratios and with sweep or no sweep, the Weissinger's theory, which we are going to study next, works well.

Weissinger's L-Method:

The Prandtl's lifting line theory is not valid for the wings which have forward of backward sweep of more than 15°. For highly swept wings the method proposed by Weissenger is widely used. Weissenger's method, rather than ignoring the terms with (x-ξ), it replaces by half chord, b, to simplify Eq. 4.11a. This approximation is justified physically because it considers the average value of term (x-ξ) rather than neglecting it. Rewriting Eq. 4.11a with this simplification we obtain.

$$U\frac{\partial z_a}{\partial x} = -\frac{1}{4\pi}\int_{-b}^{b}\int_{-l}^{l}\frac{\partial \gamma_a}{\partial \eta}\frac{[b^2 + (y-\eta)^2]^{1/2}}{(x-\xi)(y-\eta)}d\xi d\eta - \frac{1}{4\pi}\int_{-l}^{l}\frac{d\Gamma}{d\eta}\frac{d\eta}{y-\eta} \tag{4.24}$$

Multiplying Eq. 4.24 with $\sqrt{(b+x)/(b-x)}$ and integrating the result, as we did before, with respect to x, we obtain the following for the multiple integral term after integration with respect to ξ.

$$\int_{-b}^{b}\sqrt{\frac{b+x}{b-x}}\int_{-b}^{b}\int_{-l}^{l}\frac{\partial \gamma_a}{\partial \eta}\frac{\left[b^2 + (y-\eta)^2\right]^{1/2}}{(x-\xi)(y-\eta)}d\xi d\eta dx = \pi\int_{-l}^{l}\frac{d\Gamma}{d\eta}\frac{\left[b^2 + (y-\eta)^2\right]^{1/2}}{(y-\eta)}d\eta \tag{4.25}$$

Using the last line for the first term of right hand side of 4.25 and remembering that the lift coefficient is proportional with the angle of attack we obtain.

$$\pi\alpha Ub = \frac{b}{4}\int_{-l}^{l} \frac{d\Gamma}{d\eta}\frac{d\eta}{y-\eta} + \frac{1}{4}\int_{-l}^{l} \frac{d\Gamma}{d\eta}\frac{[b^2+(y-\eta)^2]^{1/2}}{(y-\eta)}d\eta \tag{4.26}$$

Nondimensionalizing the circulation and the coordinates as follow

$$G = \frac{\Gamma}{2Ul}, \quad y^* = y/l, \eta^* = \eta/l \quad and \quad l^*(y) = l/b(y)$$

gives us.

$$\alpha(y) = \frac{1}{2\pi}\int_{-1}^{1} \frac{dG}{d\eta^*}\frac{d\eta^*}{y^*-\eta^*} + \frac{l^*}{2\pi}\int_{-1}^{1} \frac{dG}{d\eta^*}\frac{[1+l^{*^2}(y^*-\eta^*)^2]^{1/2}}{l^*(y^*-\eta^*)}d\eta^*. \tag{4.27}$$

With simple algebra the right hand side of Eq. 4.27 reads as.

$$\alpha(y) = \frac{1}{\pi}\int_{-1}^{1} \frac{dG}{d\eta^*}\frac{d\eta^*}{y^*-\eta^*} + \frac{l^*}{2\pi}\int_{-1}^{1} \frac{dG}{d\eta^*}\frac{[1+l^{*^2}(y^*-\eta^*)^2]^{1/2}-1}{l^*(y^*-\eta^*)}d\eta^*. \tag{4.28}$$

This alteration saves the second term on the right hand side of 4.28 being from singular. Now, we define the Weissenger's L function.

$$L(y^*, \eta^*) = \frac{[1+l^{*^2}(y^*-\eta^*)^2]^{1/2}-1}{l^*(y^*-\eta^*)} \tag{4.29}$$

In this form Eq. 4.29 is valid for the wings with their quarter chord line parallel to y axis.

When the wing has a sweep, Weissinger places the bound vortex at the quarter chord line and applies the boundary conditions at the three quarter chord line. The bound vortex is placed at both sides of the wing with the sweep angle Λ as shown in figure below where the wake vortices are also indicated as straight lines parallel to main stream.

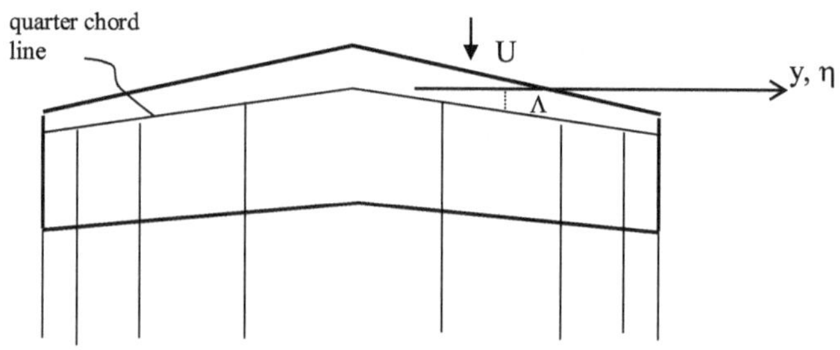

Let us now write down the Weissenger's L(y^*, η^*) function with sweep.

$$\alpha(y^*) = \frac{1}{2\pi}\int_{-1}^{1}\frac{dG}{d\eta^*}\frac{d\eta^*}{y^*-\eta^*} + \frac{l^*}{2\pi}\int_{-1}^{1}\frac{dG}{d\eta^*}L(y^*,\eta^*)d\eta^* \tag{4.30}$$

The L function in 4.30 is more complex compared to the one in 4.29, (BAH). For $y^* \geq 0$ ve $\eta^* \geq 0$ the function becomes.

$$L(y^*,\eta^*) = -\frac{1}{l^*(y^*-\eta^*)} + \frac{\{[1+l^*(y^*-\eta^*)\tan\Lambda]^2 + l^{*^2}(y^*-\eta^*)^2\}^{1/2} - 1}{l^*(y^*-\eta^*)} \tag{4.31}$$

When sweep angle Λ goes to zero, Eq. 4.31 becomes 4.29. If we transform the spanwise coordinates, as we did for the case of lifting line theory with $y^* = \cos\phi$ and $\eta^* = \cos\theta$ and expand the circulation term into Fourier sin series, the necessary aerodynamic coefficients are obtained through solution of 4.30 (BAH).

Low Aspect Ratio Wings:

Prandtl's theory works for high aspect ratio wings and Weissinger's theory works for wings with medium aspect ratios. Jones' theory, on the other hand, is applicable to the wings having low aspect ratio. By studying Jones' theory, we will be covering all ranges of aspect ratios for the thin wings. In low aspect ratio wings we usually study the planforms having curved leading edges as shown in Fig. 4.3.

Since the trailing edge is a straight line, the integral Eq. 4.11b can be inverted. This time we neglect $(y-\eta)^2$ compared to $(x-\xi)^2$ to obtain

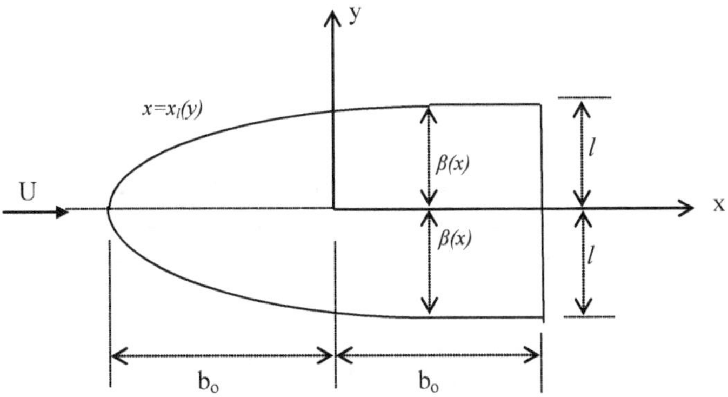

Fig. 4.3 Low aspect ratio wing

$$U\frac{\partial z_a}{\partial x} = -\frac{1}{4\pi}\int_{-l}^{l}\int_{x_l(\eta)}^{b_o}\frac{\partial \gamma_a}{\partial \eta}\frac{|x-\xi|}{(x-\xi)(y-\eta)}d\xi d\eta - \frac{1}{4\pi}\int_{-l}^{l}\frac{d\Gamma}{d\eta}\frac{d\eta}{y-\eta}$$

$$= -\frac{1}{4\pi}\int_{-l}^{l}\frac{1}{y-\eta}\frac{\partial}{\partial \eta}\left[\int_{x_l(\eta)}^{b_o}\gamma_a(\xi,\eta)\frac{|x-\xi|}{(x-\xi)}d\xi\right]d\eta$$

$$-\frac{1}{4\pi}\int_{-l}^{l}\frac{1}{y-\eta}\frac{\partial}{\partial \eta}[\int_{x_l(\eta)}^{b_o}\gamma_a(\xi,\eta)d\xi]d\eta$$

Taking care of the terms with absolute value and breaking the integrals we obtain.

$$U\frac{\partial z_a}{\partial x} = -\frac{1}{4\pi}\int_{-l}^{l}\frac{1}{y-\eta}\frac{\partial}{\partial \eta}[\int_{x_l(\eta)}^{x}\gamma_a(\xi,\eta)d\xi + \int_{x}^{b_o}\gamma_a(\xi,\eta)d\xi$$

$$+\int_{x_l(\eta)}^{x}\gamma_a(\xi,\eta)d\xi - \int_{x}^{b_o}\gamma_a(\xi,\eta)d\xi]d\eta \tag{4.32}$$

$$= -\frac{1}{2\pi}\int_{-l}^{l}\frac{1}{y-\eta}\frac{\partial}{\partial \eta}[\int_{x_l(\eta)}^{x}\gamma_a(\xi,\eta)]d\eta$$

If we write the bound vortex sheet strength in terms of the perturbation potential differences between the upper and lower surface, $\Delta\phi' = \phi_u' - \phi_l'$ we have.

$$\int_{x_l(\eta)}^{x}\gamma_a(\xi,\eta)d\xi = \int_{x_l(\eta)}^{x}(u_u' - u_l')d\xi = \int_{x_l(\eta)}^{x}\frac{\partial}{\partial \xi}(\Delta\phi')d\xi = \Delta\phi'(x,\eta) \tag{4.33}$$

The integral in Eq. 4.33 is taken at a section from leading edge to a point x on the chord. In order to cover the full wing, the spanwise integration must be taken from $\pm l$ to $\pm\beta(x)$ as shown in Fig. 4.3. Equation 4.32 becomes.

$$U\frac{\partial z_a}{\partial x} = -\frac{1}{2\pi}\int_{-\beta(x)}^{\beta(x)}\frac{1}{y-\eta}\frac{\partial}{\partial \eta}\Delta\phi'(x,\eta)d\eta \tag{4.34}$$

Equation 4.34 can be directly inverted. Nondimensionalizing with y* = y/β(x) and η* = η/β(x) Eq. 4.34 then reads as.

$$U\frac{\partial z_a}{\partial x} = -\frac{1}{2\pi}\int_{-1}^{1}\frac{\partial}{\partial \eta}\Delta\phi'(x,\eta)\frac{d\eta^*}{y^* - \eta^*}. \tag{4.35}$$

If we further nondimensionalize the following integral to obtain.

$$\int_{=1}^{1}\frac{\partial \Delta\phi'}{\partial \eta}d\eta^* = \frac{1}{\beta(x)}\int_{=1}^{1}\frac{\partial \Delta\phi'}{\partial \eta^*}d\eta^* = \frac{1}{\beta(x)}\Delta\phi' = 0 \tag{4.36}$$

Using the property as $f(\eta^*) = \frac{\partial}{\partial \eta^*}\Delta\phi'$ having zero integral between –1 and 1 as follows we have

$$g(y^*) = \frac{1}{2\pi}\int_{-1}^{1}\frac{f(\eta^*)}{y^* - \eta^*}d\eta^* \quad \text{and} \quad \text{if} \int_{=1}^{1} f(\eta^*)d\eta^* = 0$$

then.

$$f(y^*) = -\frac{2}{\pi\sqrt{1 - y^{*2}}}\int_{-1}^{1}\frac{g(\eta^*)\sqrt{1 - \eta^{*2}}}{y^* - \eta^*}d\eta^*. \tag{4.37a, b}$$

Taking care of the signs and using $-U\frac{\partial z_a}{\partial x}$ for g in 4.37 we have

$$\frac{\partial \Delta\phi'}{\partial y} = \frac{2U}{\pi\sqrt{1 - y^{*2}}}\int_{-1}^{1}\frac{\partial z_a}{\partial x}\frac{\sqrt{1 - \eta^{*2}}}{y^* - \eta^*}d\eta^*$$

In dimensional form it becomes.

$$\frac{\partial \Delta\phi'}{\partial y} = \frac{2U}{\pi\sqrt{\beta^2(x) - y^2}}\int_{-\beta(x)}^{\beta(x)}\frac{\partial z_a}{\partial x}\frac{\sqrt{\beta^2(x) - \eta^2}}{y - \eta}d\eta. \tag{4.38}$$

The linearized form of Kelvin's equation, 3.5, gives us the relation between the lifting pressure coefficient and the surface vortex sheet as follows.

$$c_{p_a} = \frac{2\gamma_a}{U} = \frac{2}{U}\frac{\partial}{\partial x}\Delta\phi' = \frac{2}{U}\frac{\partial}{\partial x}\int_{-\beta(x)}^{y}\frac{\partial \Delta\phi'(x,\bar{y})}{\partial \bar{y}}d\bar{y} \tag{4.39}$$

Since the integrand of 4.39 is equal to the right hand side of 4.38, for the known wing geometry the lifting pressure coefficient can be found via 4.39.

If we assume that for the low aspect ratio wings the elastic deformations and the camber exist only in the chordwise direction, i.e., $\partial z_a/\partial y = 0$, the integral in 4.39 is easily evaluated. The singular integral given below evaluates to

$$\int_{-\beta(x)}^{\beta(x)} \frac{\sqrt{\beta^2(x)-\eta^2}}{\overline{y}-\eta} d\eta = \beta(x)\int_{-1}^{1} \frac{\sqrt{1-\eta^{*2}}}{\overline{y}^*-\eta^*} d\eta^* = \beta(x)\int_{-1}^{1} \sqrt{\frac{1+\eta^*}{1-\eta^*}} \frac{1-\eta^*}{\overline{y}^*-\eta^*} d\eta^*$$
$$= \pi\overline{y}.$$

If above integral is placed in 4.38 we obtain

$$\int_{-\beta(x)}^{y} \frac{\pi\overline{y}d\overline{y}}{\sqrt{\beta^2(x)-\overline{y}^2}} = -\pi\sqrt{\beta^2(x)-y^2}$$

As a result the lifting pressure coefficient 4.39 reads as.

$$\begin{aligned} c_{p_a}(x,y) &= -4\frac{\partial}{\partial x}\left[\frac{dz_a}{dx}\sqrt{\beta^2(x)-y^2}\right] \\ &= -\frac{d^2z_a}{dx^2}\sqrt{\beta^2(x)-y^2} - \frac{dz_a}{dx}\frac{4\beta(x)}{\sqrt{\beta^2(x)-y^2}}\frac{d\beta}{dx}. \end{aligned} \quad (4.40)$$

Equation 4.40 provides the lifting pressure coefficient explicitly for the low aspect ratio wings. The validity of 4.40 depends on satisfying the Kutta condition at the trailing edge. The first term of 4.40 goes to zero for uncambered wings. The second term on the other hand is zero if the span remains constant at the trailing edge. Satisfying these two conditions makes the Jones' approach applicable, otherwise it will not be applicable. Figure 4.3 has a planform shape which has a constant span at the trailing edge to satisfy the Kutta condition.

Let us find the sectional lift of a low aspect ratio wing by integrating 4.40 along the chord.

$$\begin{aligned} L'(y) &= \frac{1}{2}\rho U^2 \int_{x_l}^{b_o} c_{p_a} dx = -2\rho U^2 \int_{x_l}^{b_o} \frac{\partial}{\partial x}\left[\frac{dz_a}{dx}\sqrt{\beta^2(x)-y^2}\right] dx \\ &= -2\rho U^2\sqrt{l^2-y^2}\left(\frac{dz_a}{dx}\right)_t \end{aligned} \quad (4.41)$$

The end result of 4.41 tells us that a low aspect ratio wing deformable only in chordwise direction is elliptically loaded and this load is proportional with the angle of attack at the trailing edge. The total lift now can be found by integrating 4.41 in spanwise direction.

$$L = \int_{-l}^{l} L'(y)dy = -\pi\rho U^2 l^2 \left(\frac{dz_a}{dx}\right)_t = \pi\rho U^2 l^2 \alpha \tag{4.42}$$

Here, α is the angle of attack for a straight planform wing. If we write the aspect ratio as follows AR = $(2\ l)^2/S$, the lift line slope for the wing becomes.

$$\frac{dc_L}{d\alpha} = \frac{2\pi l^2}{S} = \frac{1}{2}\pi AR. \tag{4.43}$$

Equation 4.43 is used for usually delta wings.

Now, we can also calculate the chordwise variation of lift which is usually done for the delta wings.

$$\begin{aligned} L'(x) &= \frac{1}{2}\rho U^2 \int_{-\beta(x)_l}^{\beta(x)} c_{p_a} dy = -2\rho U^2 \frac{\partial}{\partial x}\left[\frac{dz_a}{dx}\int_{-\beta(x)_l}^{\beta(x)} \sqrt{\beta^2(x) - y^2}\, dy\right] \\ &= -\pi\rho U^2 \frac{d}{dx}\left[\beta^2(x)\frac{dz_a}{dx}\right] \end{aligned} \tag{4.44}$$

Jones' approach gives small downwash values compared to the free stream speed. For low aspect ratio delta wings this means small cross flow velocity even for the high free stream speeds in compressible flows. The cross flow becoming incompressible enables us to apply Eq. 4.43 even for the case of supersonic flows.

As seen from Eq. 4.41, the spanwise load distribution is elliptic which now yields an induced drag for the low aspect ratio wings.

$$C_{Di} = C_L^2/(\pi AR). \tag{4.45}$$

Example 2 For a low aspect ratio delta wing with angle of attack α, plot the chordwise load distribution on the wing.

Solution The equation of leading edge is given by $\beta(x) = (x + b_o)l/(2b_o)$. Equation 4.44 gives the chordwise distribution as follows

$$\begin{aligned} L'(x) &= -\pi\rho U^2 \frac{d}{dx}\left[\beta^2(x)\frac{dz_a}{dx}\right] \\ &= \pi\rho U^2 \alpha \frac{d}{dx}\left[\frac{l^2}{4b_o^2}(x + b_o)^2\right] = \pi\rho U^2 \alpha \frac{l^2}{b_o^2}(x + b_o)/2 \end{aligned}$$

$$\text{and} \quad c_l(x) = \frac{L'}{\frac{1}{2}\rho U^2 2b} = \pi\alpha \frac{l^2}{b_o^2}(x + b_o).$$

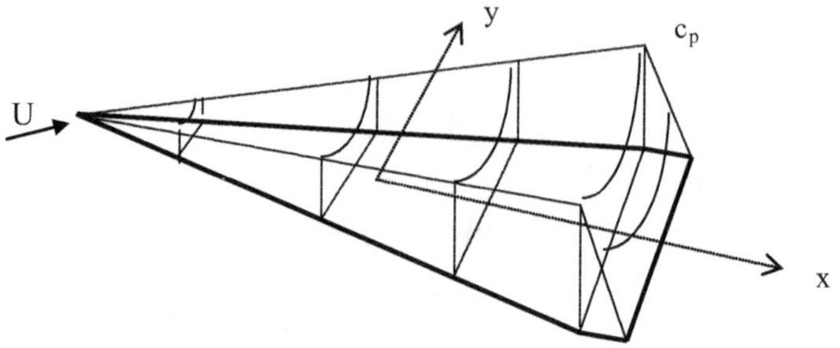

Fig. 4.4 Spanwise load distribution according to Jones' theory

In order to satisfy the Kutta condition the trailing edge ends with a constant span as shown in Fig. 4.4.

4.3 Unsteady Flow

In this section we are going to study, for the sake of completeness of the unsteady aerodynamic theory, the incompressible flow past some special planform undergoing time dependent motions. It has been shown that steady flow past a finite wing created zero spanwise vortex at the wake, $\gamma_w = 0$, and according to 4.7b chordwise vorticity at the wake was constant, i.e.,$\delta_w =$ constant. For two dimensional unsteady flow, the time variation of the effect of wake vorticity on the profile was reflected by Theodorsen function. Now, let us consider the effect of wake vorticity on the finite wing surface for simple harmonic motion. Let R_a denote the wing surface and R_w the wake region for a wing whose surface motion is given by $z_a(x, y, t) = \bar{z}_a(x, y)e^{i\omega t}$. The downwash at the surface reads as

$$w_a(x, y, t) = \bar{w}_a(x, y)e^{i\omega t} = \left[i\omega\bar{z}_a(x, y) + U\frac{\partial \bar{z}_a}{\partial x}\right]e^{i\omega t}$$

With the aid of 4.4, the amplitude of downwash in terms of vortex sheet strength becomes.

$$\begin{aligned}\bar{w}_a(x, y) = &-\frac{1}{4\pi}\iint_{R_a}\frac{\bar{\gamma}_a(\xi,\eta)(x-\xi) + \bar{\delta}_a(\xi,\eta)(y-\eta)}{[(x-\xi)^2 + (y-\eta)^2]^{3/2}}d\xi d\eta \\ &-\frac{1}{4\pi}\iint_{R_w}\frac{\bar{\gamma}_w(\xi,\eta)(x-\xi) + \bar{\delta}_w(\xi,\eta)(y-\eta)}{[(x-\xi)^2 + (y-\eta)^2]^{3/2}}d\xi d\eta\end{aligned} \tag{4.46}$$

As we did before, to obtain the relation between the bound circulation $\bar{\Gamma}_a$ and the vortex sheet strength $\bar{\gamma}_w$, we will, similarly, at a spanwise station η write the following relations in three dimensional case

$$\overline{\gamma}_w(\xi,\eta) = -ik_o \frac{\overline{\Gamma}_a(\eta)}{b_o} e^{i\frac{\omega x_t}{U}} e^{-i\frac{\omega \xi}{U}} \quad \text{with } \overline{\Gamma}_a(\eta) = \int\limits_{x_l(\eta)}^{x_t(\eta)} \overline{\gamma}_a(\xi,\eta) d\xi.$$

Here, the trailing edge is given by $x_t = x_t(\eta)$.

Defining the reduced circulation as $\overline{\Omega}(\eta) = \frac{\overline{\Gamma}_a(\eta)}{b_o} e^{i\frac{\omega x_t}{U}}$, the wake vortex sheet strength reads as $\overline{\gamma}_w(\xi,\eta) = -ik_o\overline{\Omega}(\eta)e^{-i\frac{\omega\xi}{U}}$. The continuity of the vorticity, $\frac{\partial \delta_w}{\partial \xi} = \frac{\partial \gamma}{\partial \eta}$, once integrated with respect to ξ gives,

$$\begin{aligned}
\overline{\delta}_w &= \int\limits_{-\infty}^{\xi} \frac{\partial \overline{\gamma}(\xi',\eta)}{\partial \eta} d\xi' = \frac{\partial}{\partial \eta} \int\limits_{-\infty}^{\xi} \overline{\gamma}(\xi',\eta) d\xi' \\
&= \frac{\partial}{\partial \eta} \int\limits_{x_l(\eta)}^{x_t(\eta)} \overline{\gamma}(\xi',\eta) d\xi' + \frac{\partial}{\partial \eta}[-ik_o\overline{\Omega}(\eta) \int\limits_{x_t(\eta)}^{\xi} e^{-i\frac{\omega\xi'}{U}} d\xi'].
\end{aligned}$$

After performing last two integrals we obtain

$$\begin{aligned}
\overline{\delta}_w &= \frac{\partial}{\partial \eta}[b_o\overline{\Omega}(\eta)e^{i\frac{\omega x_t}{U}}] + \frac{\partial}{\partial \eta}[-ik_o\overline{\Omega}(\eta) \int\limits_{x_t(\eta)}^{\xi} e^{-i\frac{\omega\xi}{U}} d\xi] \\
&= b_o \frac{d}{d\eta}\overline{\Omega}(\eta)e^{-i\frac{\omega\xi}{U}}
\end{aligned}$$

Substituting these into 4.46 gives.

$$\begin{aligned}
\overline{w}_a(x,y) = &-\frac{1}{4\pi} \iint_{R_a} \frac{\overline{\gamma}_a(\xi,\eta)(x-\xi) + \overline{\delta}_a(\xi,\eta)(y-\eta)}{[(x-\xi)^2 + (y-\eta)^2]^{3/2}} d\xi d\eta \\
&-\frac{1}{4\pi} \iint_{R_w} e^{-i\frac{\omega\xi}{U}} \frac{-ik_o\overline{\Omega}(\eta)(x-\xi) + b_o\frac{d\overline{\Omega}(\eta)}{d\eta}(y-\eta)}{[(x-\xi)^2 + (y-\eta)^2]^{3/2}} d\xi d\eta
\end{aligned} \tag{4.47}$$

The first integral of Eq. 4.47, using continuity of vorticity, can be written in terms of $\overline{\gamma}_a$ to obtain the integral equation between the downwash and the unknown bound vortex strength. As we did for the case of steady flow, we make some assumptions to simplify the double integrals. Let us now consider the Reissner's simplifying approach as given in (BAH).

Reissner's Approach:

The following assumptions are going to be made to simplify the integrals.

(i) Similar to the lifting line theory, we assume the wing is loaded as quasi two dimensional at any spanwise station y.

(ii) The chordwise wake vortex is projected forward from the trailing edge to a spanwise line passing through the point where the downwash is to be calculated.

(iii) The spanwise vortex of the wake which deviates from two dimensional behavior can be projected up to a line passing through the calculation point.

Let us see now, the simplifications of the terms of Eq. 4.47 with following assumptions.

Assumption (i) $\mapsto \quad w_a(x,y) = -\frac{1}{2\pi}\int\limits_{x_l(y)}^{x_t(y)} \frac{\overline{\gamma}_a(\xi,y)}{x-\xi}d\xi$

(iii) $\mapsto \quad +\frac{ik_o\overline{\Omega}(y)}{2\pi}\int\limits_{x_t}^{\infty} \frac{e^{i\omega\xi/U}}{x-\xi}d\xi.$

(ii) $\mapsto \quad -\frac{e^{-i\omega x/U}}{4\pi}\int\limits_{-l}^{l} \frac{d\overline{\Omega}}{d\eta}K\left[\frac{\omega}{U}(y-\eta)\right]d\eta.$

The kernel of the integral (ii) reads as

$$K(q) = \frac{k_o}{q} - \frac{ik_o}{q}\int\limits_0^{\infty} e^{-i\lambda}\left[1 + \frac{|q| - \sqrt{\lambda^2+q^2}}{\lambda}\right]d\lambda.$$

The integral in K(q) is named the Cicala function with its argument being $q = \frac{\omega}{U}(y-\eta)$.

Let us define the nondimensional parameters as follows.

$$x^* = \frac{2x - x_t - x_t}{2b}, \quad y^* = \frac{y}{b_o}, \quad k_m = \frac{k_o}{2b_o}(x_t + x_l) \text{ and}$$
$$l^* = \frac{l}{b_o}$$

Here, k_m is the measure of the sweep angle and it is zero for straight mid chord line. In nondimensional coordinates the downwash expression becomes.

$$w_a(x^*,y^*) = -\frac{1}{2\pi}\int\limits_{-1}^{1} \frac{\overline{\gamma}_a(\xi^*,y^*)}{x^*-\xi^*}d\xi^* + \frac{ik_o e^{-ik_m}\overline{\Omega}(y^*)}{2\pi}\int\limits_{1_t}^{\infty} \frac{e^{-ik\xi^*}}{x^*-\xi^*}d\xi^*$$
$$-\frac{e^{-ikx^*}e^{-ik_m}}{4\pi}\int\limits_{-l^*}^{l^*} \frac{d\overline{\Omega}}{d\eta^*}K[k_o(y^*-\eta^*)]d\eta^* \tag{4.48}$$

Here we have $\overline{\Omega}(y^*) = \frac{b}{b_o}e^{i(k+k_m)}\int\limits_{-1}^{1} \overline{\gamma}_a(\xi^*,y^*)d\xi^*.$

The relation between the bound vortex sheet strength and the lifting pressure coefficient was.

$$c_{p_a} = \frac{\Delta \overline{p}_a}{\frac{1}{2}\rho U^2} = -2\overline{\gamma}_a(\xi^*, y^*) - 2ik \int_{-1}^{1} \overline{\gamma}_a(\xi^*, y^*) d\xi \tag{4.49}$$

Let us now invert Eq. 4.48 to be used in 4.49.

$$\begin{aligned}\overline{\gamma}_a(\xi^*, y^*) = \frac{2}{\pi}\sqrt{\frac{1-x^*}{1+x^*}}\{\int_{-1}^{1}\sqrt{\frac{1+\xi^*}{1-\xi^*}}\frac{\overline{w}_a(\xi^*, y^*)}{x^*-\xi^*}d\xi^* + \frac{ik_o e^{-ik_m}\overline{\Omega}(y^*)}{2\pi}\int_{1_t}^{\infty}\sqrt{\frac{\xi^*+1}{\xi^*-1}}\frac{e^{-ik\xi^*}}{x^*-\xi^*}d\xi \\ - \frac{e^{-ikx^*}e^{-ik_m}}{4\pi}\int_{-l^*}^{l^*}\frac{d\overline{\Omega}}{d\eta^*}K[k_o(y^*-\eta^*)]d\eta x\int_{-1_t}^{1}\sqrt{\frac{1+\xi^*}{1-\xi^*}}\frac{e^{-ik\xi^*}}{x^*-\xi^*}d\xi\}\end{aligned} \tag{4.50}$$

The reduced circulation on the other is determined by integrating 4.50 from –1 to 1 as follows.

$$\begin{aligned}&\overline{\Omega}(y^*) + \frac{-J_o(k)+iJ_1(k)}{\pi ik[H_1^{(2)}(k)+iH_o^{(2)}(k)]}\frac{b}{b_o}\int_{-l^*}^{l^*}\frac{d\overline{\Omega}}{d\eta^*}K[k_o(y^*-\eta^*)]d\eta^* \\ &= 4\frac{b}{b_o}e^{ik_m}\frac{\int_{-1}^{1}\sqrt{(1+\xi^*)/(1-\xi^*)}w_a(\xi^*, y^*)d\xi^*}{\pi ik[H_1^{(2)}(k)+iH_o^{(2)}(k)]}\end{aligned} \tag{4.51}$$

Here, we define the coefficient $\mu(k) = \frac{-J_o(k)+iJ_1(k)}{\pi ik[H_1^{(2)}(k)+iH_o^{(2)}(k)]}$ and the right hand side of 4.51 as $\overline{\Omega}^{(2)}(y^*)$.

The lifting pressure expression 4.49, with the aid of 4.50 and 4.51 becomes.

$$\begin{aligned}c_{p_a} = &\frac{4}{\pi}\sqrt{\frac{1-x^*}{1+x^*}}\int_{-1}^{1}\left[\sqrt{\frac{1+\xi^*}{1-\xi^*}}\frac{1}{x^*-\xi^*} + ik\Lambda\right]\overline{w}_a(\xi^*, y^*)d\xi^* \\ &- \frac{4}{\pi}\sqrt{\frac{1-x^*}{1+x^*}}\int_{-1}^{1}\sqrt{\frac{1+\xi^*}{1-\xi^*}}\overline{w}_a(\xi^*, y^*)d\xi\left\{[C(k)-1] - \left[\frac{\overline{\Omega}(y^*)}{\overline{\Omega}^{(2)}(y^*)} - 1\right]\left[C(k) + \frac{iJ_1(k)}{J_o(k)-iJ_1(k)}\right]\right\}\end{aligned} \tag{4.52}$$

Here, Λ is the same as given in 3.23 and C(k) is the Theodorsen function.

In the last term in 4.52 if the coefficient is defined as follows

$$\sigma(y^*) = \left[\frac{\overline{\Omega}(y^*)}{\overline{\Omega}^{(2)}(y^*)} - 1\right]\left[C(k) + \frac{iJ_1(k)}{J_o(k) - iJ_1(k)}\right]$$

We can see the difference between the two dimensional lifting pressure coefficient 3.23. Here, σ is also a function of C(k) and shows us the spanwise variation of the circulation.

The aerodynamic coefficients can be calculated using the Reissner's theory by the following steps.

For simple harmonic motion; i) if only bending is considered: $h(y^*, t) = \overline{h}e^{i\omega t}f_h(y^*)$, ii) if torsion about an axis is considered: $\alpha(y^*, t) = \overline{\alpha}e^{i\omega t}f_\alpha(y^*)$, are employed.

(1) Since the reduced frequency and the wing geometry is known $\mu(k)$ *and* $\overline{\Omega}^{(2)}(y^*)$ are determined to solve 4.51 to find $\overline{\Omega}(y^*)$.
(2) $\overline{\Omega}^{(2)}(y^*)$ and $\overline{\Omega}(y^*)$ are known, σ is determined.
(3) At any station y* the aerodynamic coefficients are found using 2-D theory.
(4) These coefficients are corrected with known values of σ as the 3-D solution, as follows

$$\Delta L_h(y^*, t) = 2\pi\rho U^2 b_o[ik\sigma_h(y^*)]h(y^*, t)/b_o$$

$$\Delta L_\alpha(y^*, t) = 2\pi\rho U^2 b_o[ik(1/2 - a)\sigma_\alpha(y^*)]h\alpha(y^*, t).$$

Summary of the Reissner's Theory:

(i) Compared to a 2-D case, non circulatory term does not change
(ii) At the wing tips non circulatory terms can contribute
(iii) As compared with the experimental values for rectangular wings good agreement is observed for the aspect ratio values down to 2.

During experiments it is difficult to reduce the viscous effects on oscillating wings. However, at high reduced frequencies these effects are expected to be low. In their numerous experimental and computational work, Reissner and Stevens have shown that the finite wing effects can be neglected depending on the reduced frequency and the aspect ratio values. In summary:

(1) For the wings with an aspect ratio around 6 if the reduced frequency is higher than 1, and for the wings with an aspect ratio around 3 if the reduced frequency is higher than 2, 3-D effects can be neglected.
(2) For the wings with an aspect ratio around 6 if the reduced frequency is less than 0.5, and for the wings with an aspect ratio 3 if the reduced frequency is lees than 1, 3-D effects can not be neglected.

Numerical Solution: The aerodynamic coefficients for the wings undergoing simple harmonic oscillations, the integro-differential Eq. 4.51 can be solved to obtain the amplitude of the reduced circulation as we did for the steady case. For this purpose, expanding the reduced circulation into Fourier like series will give us the algebraic system of equations. Before expanding into the series, let us first transform the spanwise coordinates with $y^* = l^* \cos\phi$ ve $\eta^* = l^* \cos\theta$. The series form of the reduced circulation in series can be expressed as follows

$$\overline{\Omega}_j(y^*) = \sum_{n=1}^{N} K_{nj} \frac{\sin n\phi}{n}, \quad j = \alpha, h, \beta$$

Here, α denotes rotation, h vertical displacement and β flap motion. With this notation Eq. 4.51 becomes the following set of equations.

$$\sum_{n=1}^{N} K_{nj}\{\frac{\sin n\phi}{n} + \frac{b}{b_o}\frac{\pi}{l^*}\mu(k)[\frac{\sin n\phi}{n} + \frac{ik_o l*}{\pi}\int_0^{\pi} \frac{\cos\phi - \cos\theta}{|\cos\phi - \cos\theta|} \tag{4.53}$$

$$F(k_o l^* |\cos\phi - \cos\theta|)\cos n\theta d\theta]\} = \overline{\Omega}_j^{(2)}(l^* \cos\phi)$$

Here, $F(q) = \int_0^{\infty} e^{-i\lambda}\left[\frac{1}{q} + \frac{1}{\lambda} - \frac{\sqrt{q^2+\lambda^2}}{q\lambda}\right]d\lambda,\ q \geq 0$, denotes the Cicala function.

In addition, we have for vertical displacement h and for rotation with respect to α.

$\overline{\Omega}_h^{(2)} = \frac{4iC(k)}{kH_1^{(2)}(k)} e^{ik_m} \frac{h(y^*,t)}{b_o e^{i\omega t}}$ and $\overline{\Omega}_\alpha^{(2)} = \frac{4iC(k)}{k_o H_1^{(2)}(k)} e^{ik_m}\left[1 + ik\left(\frac{1}{2} - a\right)\right]\frac{\alpha(y^*,t)}{e^{i\omega t}}$.

Defining function S_n as follows

$$S_n(k_o l^*, \phi) = \frac{\sin n\phi}{n} + \frac{ik_o l*}{\pi}\int_0^{\pi} \frac{\cos\phi - \cos\theta}{|\cos\phi - \cos\theta|}$$

$$F(k_o l^* |\cos\phi - \cos\theta|)\cos n\theta d\theta$$

the algebraic equation becomes.

$$\sum_{n=1}^{N} K_{nj}\{\frac{\sin n\phi}{n} + \frac{b}{b_o}\frac{\pi}{l^*}\mu(k)S_n(k_o l^*, \phi) = \overline{\Omega}_j^{(2)}(l^* \cos\phi). \tag{4.54}$$

As we did for the steady flow case for 4.21, in Eq. 4.54 we change φ between 0 and $\pi/2$ and take n odd for antisymmetric and n even for symmetric loadings. Coefficients K_{nj} can be found by taking j number of stations along the span as follows.

$$\left\langle \left[\frac{\sin n\phi}{n}\right] + \frac{\pi}{l^*}\left[\frac{b}{b_o}\mu(k)\right][S_n]\right\rangle\{K_{nj}\} = \left\{\overline{\Omega}_j^{(2)}\right\} \tag{4.55}$$

For a symmetric loading, taking 2 N-1 number of stations results in

$$\left[\frac{\sin n\phi}{n}\right] = \begin{bmatrix} \sin\phi_1 \frac{\sin 3\phi_1}{3} \dots \frac{\sin(2N-1)\phi_1}{2N-1} \\ \sin\varphi_2 \frac{\sin 3\phi_2}{3} \dots \frac{\sin(2N-1)\phi_2}{2N-1} \\ \cdot \\ \cdot \\ \cdot \\ \sin\phi_N \frac{\sin 3\phi_N}{3} \dots \frac{\sin(2N-1)\phi_N}{2N-1} \end{bmatrix}, [S_n]$$

$$= \begin{bmatrix} S_1(k_o l^*, \phi_1) S_3(k_o l^*, \phi_1) \dots S_{2N-1}(k_o l^*, \phi_1) \\ S_1(k_o l^*, \phi_2) S_3(k_o l^*, \varphi_2) \dots S_{2N-1}(k_o l^*, \phi_2) \\ \cdot \\ \cdot \\ \cdot \\ S_1(k_o l^*, \phi_N) S_1(k_o l^*, \varphi_N) \dots S_{2N-1}(k_o l^*, \phi_N) \end{bmatrix}$$

The entries of the matrix S_n and the right hand side of Eq. 4.55 are complex. Therefore, coefficients K_{nj} are obtained as N complex numbers. These coefficients help us to find the reduced circulation values at each station. From the reduced circulation values we obtain the amplitude of the circulation. Integrating the circulation along the span gives us the amplitude of the total lift. The total lift value being complex gives us the phase difference between the simple harmonic motion of the wing.

For a rectangular planform with a constant chord 2b the reduced frequency along the span remains the same. Therefore, for a given frequency and the mode shape the reduced circulation becomes proportional with the amplitude of the motion. Hence, the right hand side of 4.54 is simplified as follows.

$$\frac{\overline{\Omega}_h^{(2)}}{U} = \frac{4iC(k)}{kH_1^{(2)}(k)}\frac{\overline{h}}{b}f_h(y^*)$$

While computing the coefficients K_{nj} from 4.54 the right hand side of the equation may become real. If we denote the new coefficients with K'_{nh}, we can write.

$K'_{nh} = \frac{K_{nh}}{\frac{4iC(k)}{kH_1^{(2)}(k)}\frac{\overline{h}}{b}}$ to have 4.54 as follows.

$$\sum_{n=1}^{N} K'_{nh}\{\frac{\sin n\phi}{n} + \frac{\pi}{l^*}\mu(k)S_n(kl^*, \phi)\} = f_h(l^* \cos \phi) \qquad (4.56)$$

The Fourier like series expansion of the reduced circulation becomes.

$$\overline{\Omega}_h^{(2)} = \frac{4iUC(k)}{kH_1^{(2)}(k)}\frac{\overline{h}}{b}\sum_{n=1}^{N} K'_{nh}\frac{\sin n\phi}{n} \qquad (4.57)$$

Similarly, knowing the coefficients K'_{nh}, we can calculate the amplitude of circulation at spanwise stations from the reduced circulation values as follows.

$$\frac{\overline{\Omega}_h}{\overline{\Omega}_h^{(2)}} = \frac{1}{f_h(l^* \cos \phi)}\sum_{n=1}^{N} K'_{nh}\frac{\sin n\phi}{n} \qquad (4.58)$$

Example 3 A rectangular wing with an aspect ratio 6 undergoes vertical oscillation with k = 2/3 and amplitude $\overline{h}$. Find the spanwise distribution of lift.

Solution Using the Reissner's tables and the 2-D lift value: $L^{(2)}/\ 2\rho U^2 h$ = -0.425 + 1.19i we find

$$y^* = 0.0, \quad 0.4, \quad 0.8, \quad 1.0$$
$$L/2\rho U^2 h = -0.441 + 1.195i, \quad -0.455 + 1.18i, \quad -0.461 + 1.071i, \quad -0.042 + 0.23i$$

4.4 Arbitrary Motion of a Thin Wing

For elliptically loaded thin wings it is possible to determine the indicial admittance functions like Wagner and Küssner functions for arbitrary motions of wing. Accordingly for the sudden angle of attack change from 0 to α_o we have the Wagner function to give the lift creation.

$$c_L(s) = 2\pi\alpha_o\varphi(s) \qquad (4.59)$$

and, similarly for the effect of the gust with magnitude w_o on the lift change as the Küssner function.

$$c_{Lg}(s) = 2\pi\frac{w_o}{U}\chi(s) \qquad (4.60)$$

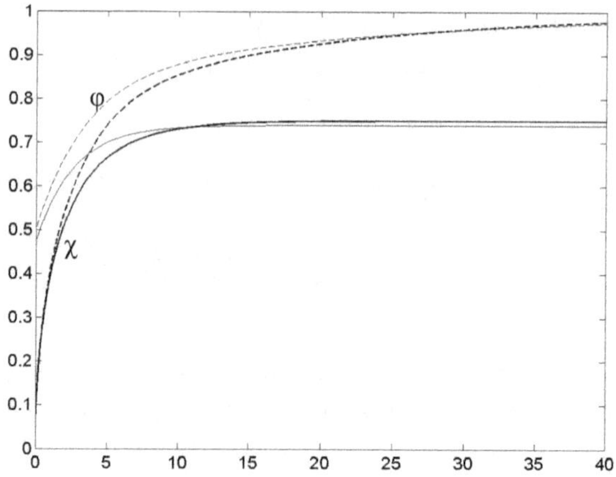

Fig. 4.5 Wagner, φ -and Küssner, χ- functions for a wing with AR = 6

Table 4.1 The Wagner and the Küssner functions variations with respect to aspect ratio

	AR	b_o	b_1	b_2	b_3	β_1	β_2	B_3
$\varphi(s)$	3	0.6	0.17	0	0	0.54	-	-
	6	0.74	0.267	0	0	0.381	-	–
	∞	1.0	0.165	0.335	0	0.0455	0.3	–
$\chi(s)$	3	0.6	0.407	0.136	0	0.558	3.2	–
	6	0.75	0.336	0.204	0.145	0.29	0.725	3.0
	∞	1.0	0.236	0.513	0.171	0.058	0.364	2.42

Here, s is the reduced time based on the root half chord. The Jones approach for the Wagner and Küssner functions are given in exponential form which has coefficients and exponents given in Table 4.1 and their plots are provided for a wing with an aspect ratio of 6 in Fig. 4.5.

$$\left.\begin{matrix}\varphi(s)\\ \chi(s)\end{matrix}\right\} = b_o - b_1 e^{-\beta_1 s} - b_2 e^{-\beta_2 s} - b_3 e^{-\beta_3 s}$$

4.5 Effect of Sweep Angle

The significance of sweep for a wing comes into the picture for compressible flows in achieving high critical Mach numbers. Here, for the sake of completeness we are going to briefly analyze the effect of sweep for incompressible flows.

As we did for the steady flow, let us define the sweep angle Λ as the angle between the quarter chord line of the wing and the line normal to the free stream. It is, on the other hand, possible to find the aerodynamic coefficients via chordwise strip theory for the wings with the constant spanwise twist and downwash distribution. Multiplying Eq. 3.36a,b with cosΛ gives us the aerodynamic coefficients for the swept wings. For this case only, for the nonorthogonal coordinate system having its axis as the free stream direction and the half chord line, we can write downwash expression along the chord as follows.

$$\overline{w}_a(x^*) = -\frac{1}{2\pi\cos\Lambda}\left(\int_{-1}^{1}\frac{\overline{\gamma}_a(\xi^*)d\xi^*}{x^*-\xi^*} - \frac{ik\overline{\Gamma}_a}{b}\int_{1}^{\infty}\frac{e^{-ik\xi^*}d\xi^*}{x^*-\xi^*}\right) \tag{4.61}$$

Inverting Eq. 4.61 and substituting it into the lifting pressure coefficient helps us to find the sectional lift coefficient with chordwise integral of the lifting pressure. At each section, assuming that the strip theory is valid, spanwise integration of the sectional values of lift will give us the total lift, (BAH).

Another approach here is redefining the coordinate system as $\overline{y}$ in spanwise direction and $\overline{x}$ to the normal to spanwise direction. If we now denote the vertical displacement by σ and torsion by τ, we can find the aerodynamic forces as functions of σ and τ, (BAH).

Both of the approaches are not quiet sufficient from the aerodynamical angle. Therefore, in practice a semi-numerical method called ‘doublet lattice’ is used extensively. We will be studying the doublet lattice method in next chapter.

4.6 Low Aspect Ratio Wing

Let us study the unsteady aerodynamic forces for the time dependent motions of low aspect ratio wingss.

For the thin and low aspect ratio wing as shown in Fig. 4.6 with its top and side views, we can make our simplifying assumptions as we did for the steady case to obtain the time dependent downwash expression in terms of the perturbation potential difference as follows.

$$w_a(x,y,t) = -\frac{1}{2\pi}\int_{-\beta(x)}^{\beta(x)}\frac{\partial\Delta\phi'}{\partial\eta}\frac{1}{y-\eta}d\eta \tag{4.62}$$

Employing Eq. 4.37a,b similar to 4.38 we.

Fig. 4.6 Low aspect ratio wing

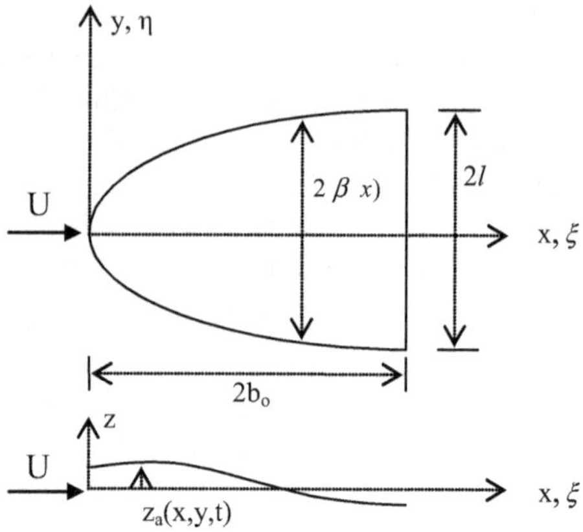

$$\frac{\partial \Delta\phi'(x,y,t)}{\partial y} = \frac{2}{\pi\sqrt{\beta^2(x)-y^2}} \int\limits_{-\beta(x)}^{\beta(x)} \overline{w}_a(x,y,t) \frac{\sqrt{\beta^2(x)-\eta^2}}{y-\eta} d\eta \tag{4.63}$$

The lifting pressure coefficient for the unsteady flow was.

$$c_{p_a} = \frac{2}{U^2}\frac{\partial}{\partial t}\Delta\phi' + \frac{2}{U}\frac{\partial}{\partial x}\Delta\phi' \tag{4.64a}$$

Here, $\Delta\phi'(x,y,t) = \int\limits_{-\beta(x)}^{y} \frac{\partial \Delta\phi'}{\partial \eta} d\eta.$

For the simple harmonic motion the downwash expression in terms of the surface equation

$$\overline{w}_a(x,y) = i\omega\overline{z}_a(x,y) + U\frac{\partial}{\partial x}\overline{z}_a(x,y).$$

The amplitude of the lifting pressure in terms of the perturbation potential reads as.

$$\overline{c}_{p_a} = \frac{2}{U^2} i\omega\Delta\overline{\phi}' + \frac{2}{U}\frac{\partial}{\partial x}\Delta\overline{\phi}' \tag{4.64b}$$

If we allow elastic deformation and the camber only in chordwise direction, the downwash in Eq. 4.63 becomes independent of y; therefore, the integral becomes πy. Accordingly, from 4.63 for the amplitude of the perturbation potential we obtain.

$$\Delta\overline{\phi}'(x,y) = \int\limits_{-\beta(x)}^{y} \frac{2[i\omega\overline{z}_a(x) + U\frac{d}{dx}\overline{z}_a(x)]}{\sqrt{\beta^2(x) - \overline{y}^2}}\overline{y}d\overline{y}$$
$$= -2[i\omega\overline{z}_a(x) + U\frac{d\overline{z}_a(x)}{dx}]\sqrt{\beta^2(x) - y^2} \tag{4.65}$$

Lifting pressure coefficient from 4.64 reads as.

$$\overline{c}_{p_a} = \frac{2}{U^2}i\omega\Delta\overline{\phi}' + \frac{2}{U}\frac{\partial}{\partial x}\Delta\overline{\phi}' = -4\left[-\frac{\omega^2}{U^2}\overline{z}_a\sqrt{\beta^2(x) - y^2} + \frac{d}{dx}\left(\frac{d\overline{z}_a}{dx}\sqrt{\beta^2(x) - y^2}\right)\right]$$
$$- 4\frac{i\omega}{U}[\frac{d\overline{z}_a}{dx}\sqrt{\beta^2(x) - y^2} + \frac{d}{dx}(\overline{z}_a\sqrt{\beta^2(x) - y^2})] \tag{4.66}$$

In Eq. 4.66 if we take frequency as zero, we obtain Eq. 4.40 which was given for the steady case. The second term of the right hand side of 4.66 gives the phase difference between the lifting pressure coefficient and the wing motion.

Now, we can express 4.66 in more convenient form using the reduced frequency, $k = \omega b_o/U$, and the nondimensional coordinates with superscript * written in term of root half chord, b_o, as follows.

$$\overline{c}_{p_a} = -4[-k^2\overline{z}_a^*\sqrt{\beta^2(x) - y^2/b_o} + \frac{d}{dx^*}(\frac{d\overline{z}_a^*}{dx^*}\sqrt{\beta^2(x) - y^2/b_o}]$$
$$- 4ik\left[\frac{d\overline{z}_a^*}{dx^*}\sqrt{\beta^2(x) - y^2/b_o} + \frac{d}{dx^*}(\overline{z}_a^*\sqrt{\beta^2(x) - y^2/b_o})\right] \tag{4.67}$$

The second term of the right hand side of Eq. 4.67 is the apparent mass term. In order to satisfy the Kutta condition this term needs to go to zero at the trailing edge. To remedy this and to be in accord with the experimental findings the lifting pressure coefficient is multiplied with an empirical factor (BAH) given as.

$$F(x) = (1 - x^{*^2})^{1/2}. \tag{4.68}$$

Example 4 The wing given in Example 2 is undergoing a simple harmonic motion with $h = \overline{h}e^{i\omega t}$. Find the lifting pressure on the wing surface in terms of the wing geometry and the reduced frequency.

Solution Using the fixed vertical amplitude and the wing geometry $\beta(x) = (l/b_o)x/2$ in 4.67 we obtain

$$\overline{c}_{p_a} = 4k^2\overline{h}^*\sqrt{\beta^2(x) - y^2/b_o} - ik\overline{h}^*\frac{d}{dx^*}\sqrt{\beta^2(x) - y^2/b_o})$$

$$\bar{c}_{p_a} = 4k^2\bar{h}^* \sqrt{l^{*2}x^{*2}/4 - y^2/b_o^2} - ik\bar{h}^* \frac{d}{dx^*}\sqrt{l^{*2}x^{*2}/4 - y^2/b_o^2}.$$

The empirical relation 4.68 is used as a multiplier to satisfy the Kutta condition

$$\begin{aligned}\bar{c}_{p_a} = 4(1 - x^{*2})^{1/2}[k^2\bar{h}^* \sqrt{l^{*2}x^{*2}/4 - y^2/b_o^2} \\ - ik\bar{h}^* \frac{d}{dx^*}\sqrt{l^{*2}x^{*2}/4 - y^2/b_o^2}]\end{aligned}$$

Summary *Three dimensional incompressible flows past thin finite wings is studied.A Physical model based on the distributed vortex sheet over the lifting thin surface is utilized to obtain the downwash via. Biot-Savart Law. First, steady state solutions are established with the Kutta condition applied at the trailing edge. The inversion of the integral relation between the boundary condition and the vortex sheet strength requires some simplification in the physical model. In that respect, Prandtl's lifting line concept is utilized for the large aspect ratio wings, the Weissinger's L-Method is made use offor the moderate aspect ratio or swept wings and Jone's simplification is employed for the pointed wings with very low aspect ratio. Assuming the Fourier series solution for the spanwise circulation, the first two methods resulted in matrix equation between the known geometric angle of attack and the unknown coefficients of the series. The Jone's theory however gave explicit and integrable analytical expression for the lifting pressure coefficient. All three methods yield agreeable solutions with the experimental data in their range of applications. For unsteady solutions for the wings undergoing simple harmonic motions, Reissner's approach is implemented for the high aspect ratio wings. This resulted in a similar matrix equation for the amplitude of the angle of attack and the reduced circulation as in the case of lifting line. Jone's approximation to Wagner and Küssner function's as the indicial admittance, are provided for the total lift coefficient expressions of the elliptical wings undergoing arbitrary motions. The effect of sweep is also introduced for the wings having considerable sweep angle. Finally, simple harmonic motion of very low aspect ratio wings with pointed nose is studied. An empirical correcting factor to the lifting pressure expression is given to obtain solutions satisfying the Kutta condition which enables us to find results agreeable with the experimental data provided for the very low aspect ratio wings.*

4.7 Questions and Problems

4.1 The vorticity vector, **ω,** is defined from the velocity vector, **q**, as follows **ω** = **∇x q .**
Show that the vorticity vector satisfies the equation of continuity.

4.2 Evaluate integral I_1 of the downwash expression 4.9 with integration by parts.

4.3 Derive the relation between the wing surface slope and the sectional circulation Integral, 4.14, from 4.13.

4.4 Transform the spanwise y, η coordinates into ϕ, θ, and obtain Eq. 4.21 as Glauert did.

4.5 The Aerodynamic Influence Coefficient Matrix, [A], gives the lift coefficient generated at a section with the angle of attack change at another section, i.e., $\{c_l\} = [A]\{\alpha\}$. Using the Glauert's approach, obtain [A] for a symmetrically loaded wing with choosing 2 N-1 spanwise stations.

4.6 Find the Aerodynamic Influence coefficients for the wing given in Fig. 4.7.

4.7 Show that for an elliptically loaded wing, the total lift line slope in terms of the root lift line slope a_o and the aspect ratio AR reads as

$$\frac{dc_l}{d\alpha} = a_o \frac{AR}{AR+2}.$$

4.8 Glauert's, $G(\phi) = \Gamma(\phi)/(2lU) = \sum_{i=1}^{m} 2A_i \sin i\phi$, series can be written with Multhopp's distribution for ϕ as $\phi_n = \frac{n\pi}{m+1}$, $\quad n = 1, 2, \ldots, m.$, to obtain better resolution of the circulation at the wing tips. Show that with the Multhopp's distribution coefficients, A_i, with integrating for varible φ reads as.

(i) $A_i = \frac{2}{m+1} \sum_{n=1}^{m} \frac{G(\phi_n) \sin i\phi_n}{\sin \phi_n} \sin \phi_n.$

(ii) The induced velocity at station, n=j, is

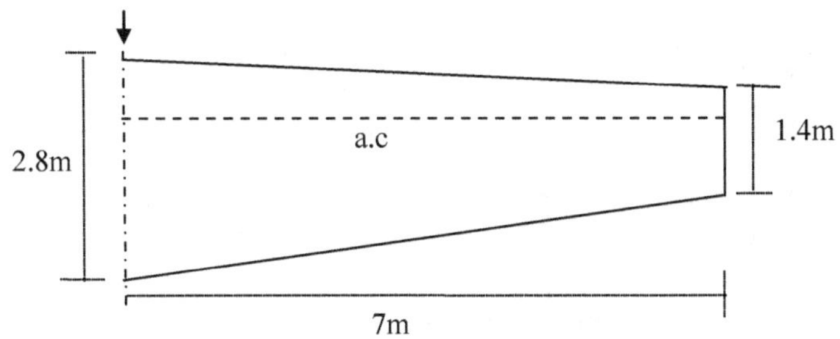

Fig. 4.7 Symmetrical wing geometry

$$\frac{1}{4\pi U}\int_{-l}^{l}\frac{d\Gamma}{d\eta}\frac{d\eta}{y-\eta}=\frac{1}{m+1}\sum_{n=1}^{m}G_n\sum_{i=1}^{m}\frac{i\sin i\phi_n\sin i\phi_j}{\sin\phi_j}$$

(iii) the j th station induces the velocity on to itself bjj Gj and n th station on j induces with, bjn Gn, the total of $b_{jj}G_j - \sum_{\substack{n=1\\ n\neq 1}}^{m} b_{jn}G_n$..

Here, using the definition in (ii) show that $b_{jj}=\frac{m+1}{4\sin\phi_j}$ and $b_{jn}=\frac{1-(-1)^{n-j}}{2(m+1)}\frac{\sin\phi_n}{(\cos\phi_n-\cos\phi_j)^2}$.

(iv) Finally, show that unknowns, Gj , in terms of the angle of attack at j are given by

$$G_j=\frac{(ab)_j}{2l}[\alpha_j-b_{jj}G_j+\sum_{\substack{n=1\\ n\neq j}}^{m}b_{jn}G_n].$$

4.9 Solve Example 1 using the results of Problem 4.8.

4.10 In Weissinger's L-Method the circulation G can be expanded into Fourier series after performing the $y*=\cos\phi_j$ transformation. Using the trapezoidal rule for integration the angle of attack at j th station can be written

$$\alpha_j=\left[2b_{jj}+\frac{l}{b_j}g_{jj}\right]G_j+\sum_{\substack{n=1\\ n\neq j}}^{m}\left[\frac{l}{b_j}-2b_{jn}\right]G_n.$$

Here,

$g_{jn}=\frac{-1}{2(M+1)}\left[\frac{L_{jo}f_{no}+L_{j,M+1}f_{n,M+1}}{2}+\sum_{i=1}^{M}L_{ji}f_{ni}\right]$ and,

$L_{ji}=L(\phi_j,\theta_i)=L(y^*,\eta^*)$ and $f_{ni}=f_n(\phi_i)=\frac{2}{m+1}\sum_{k=1}^{m}k\sin k\phi_n\cos k\phi_i$.

M is a parameter involved in the numerical integration process. Using the formulation given here, solve Problem 4.9 with the unswept L formula and compare your results.

Write the Aerodynamic Influence Coefficient matrix for the Weissinger L-Method in terms of b_{jn} and g_{jn}.

For the low aspect ratio wings, show that the integral given below is

$$\int_{-\beta(x)}^{\beta(x)} \frac{\sqrt{\beta^2(x) - \eta^2}}{\bar{y} - \eta} d\eta = \pi\bar{y}$$

4.13 Find the center of pressure for a thin wing with small aspect ratio.

4.14 If there is a chordwise camber for the wing given in Example 2 find the lift coefficient at zero angle of attack. Does parabolically cambered airfoil satisfy the Kutta condition? How do we have to choose the camber so that the Kutta condition is satisfied?

4.15 If there is a spanwise parabolic camber in Example 2 find the lifting pressure coefficient for the wing.

4.16 Find the lift coefficient for the wing given in Problem 4.14 and analyze the effect of spanwise camber on the coefficient of lift.

4.17 A wing with aspect ratio of 6 is undergoing pitching about its root leading edge as given by Figure P.3.10. Find the time dependent variation of lift coefficient.

4.18 Obtain the time dependent lift coefficient for an elliptic thin wing in a variable free stream. For a wing with an aspect ratio of 6 obtain the unsteady to quasi steady lift ratio for the Example 3.5. Plot the result.

4.19 The wing given in Problem 16 is subject to the gust given by P 3.13. Find and Plot the total lift coefficient variation with respect to time.

4.20 Two high aspect ratio identical rectangular wings ($2\ l\ x\ 2b$) are separated with a distance h. Establish an expression for the downwash in terms of the surface vortex sheet strength of each planform. Propose a method to calculate the total lift coefficient generated by bi-plane.

4.21 The tapered symmetric thin wing geometry is shown in Fig. 4.8. If this wing under goes a simple harmonic motion in vertical direction, find the amplitude of sectional lift coefficients in terms of the amplitude of the motion.

4.22 Find the total lift coefficient for the wing given in Example 4 with i) theoretical approach and, ii) with empirical correction.

4.23 A wing with spanwise simple camber shown in Fig. 4.P23 is undergoing simple harmonic motion in vertical translation. Obtain the total lift coefficient in terms of the amplitude of the motion. Starting from this expression find i) spanwise, and ii) chordwise aerodynamic loadings.

Fig. 4.8 Swept wing

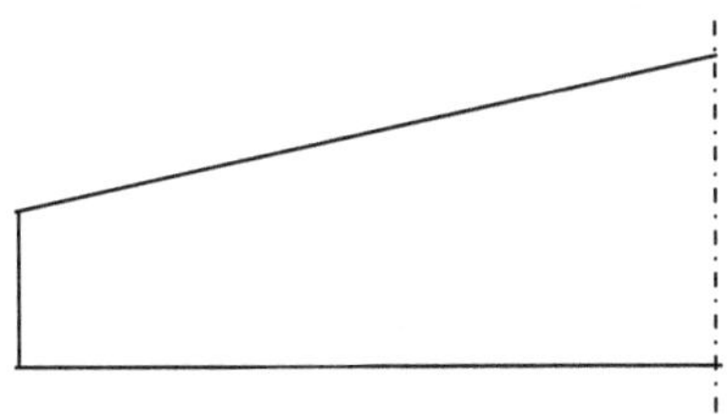

Fig. 4.9 Cambered wing

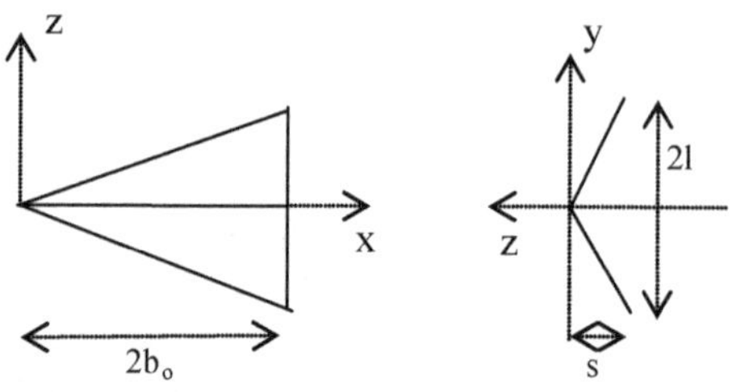

4.24 Obtain the total lift coefficient for Problem 4.21.

4.25 For a delta wing pitching simple harmonically about its nose, with respect to pitch angle find the amplitudes of,

(i) lifting pressure coefficient,
(ii) chordwise and spanwise lift loadings,
(iii) total lift coefficient.

4.26 For a delta wing rolling simple harmonically about its root find the amplitudes of

(i) lifting pressure,
(ii) spanwise and chordwise variation of lift,
(iii) total lift coefficient, with respect to amplitude of roll angle.

4.27 Propose a method to find the amplitude of the total lift coefficient for the wing of Problem 20 plunging simple harmonically. (Fig. 4.9)

References

Bisplinghoff, Raymond L., H. Ashley, R.L. Halfman:*Aeroelasticity*, Dover Publications Inc., New York, 1996

Katz, Joseph, A.Plotkin: *Low Speed Aerodynamics*, 2nd. Ed., Cambridge University Press, 2001.

Kuchemann, D.: The Aerodynamic Design of Aircraft. Pergamon Press, Oxford (1978)

Kuethe, Arnold M, Chow, C-Y,: *Foundations of Aerodynamics*, 5th. Ed. J. Wiley, New York, 1998.

Pierce, A.G.: Advanced Potential Flow I. Lecture Notes, Georgia Institute of Technology, School of Aerospace Engineering, Atlanta (1978)

Reissner, Eric, J.E. Stevens: Effect of Finite Span on the Airloads Distributions for Oscillating Wings II, NACA TN-1195, 1947.

Chapter 5
Subsonic and Supersonic Flows

In a compressible medium like air, the propagation speed of small perturbations is equal to the speed of pressure waves which in turn is equal to the speed of sound, Shapiro (1953). As the velocity of the moving object gets close to the speed of sound in the air, the effect of compressibility can no longer be neglected. In other words, when the flow velocity is in the same order of magnitude with the propagation speed of the perturbations, we have to consider the compressibility effect. The low flow velocity, compared to propagation speed, enables us to neglect all compressibility effects and identify the flow as incompressible. The measure of compressibility in aerodynamics as a parameter is the Mach number which is defined as the ratio of the flow velocity to the local speed of sound. In this chapter we are going to study the compressible flow, ranging from simple to complex, based on the linear potential theory using point sources and sinks with intensities q related to the perturbation potential. Shown in Fig. 5.1, is the point source, with intensity q, having only radial velocity on the spherical surface whose radius is r.

With the aid of Fig. 5.1 and using the definition of the velocity potential, we can obtain the expressions for the velocity potential in terms of the intensity of the point source as follows.

(i) The relation between the velocity potential ϕ and the radial and tangential speeds for the steady incompressible flow:

$$u_\theta = \frac{1}{r}\frac{\partial\phi}{\partial\theta} \quad \text{and} \quad u_r = \frac{\partial\phi}{\partial r} \quad \text{gives} \quad \phi = -\frac{q}{4\pi r}$$

since in Cartesian coordinates $r^2 = x^2 + y^2 + z^2$ then

$$\phi(x, y, z) = -\frac{q}{4\pi\sqrt{x^2 + y^2 + z^2}} \tag{5.1}$$

Ü. Gülçat, *Fundamentals of Modern Unsteady Aerodynamics*,
https://doi.org/10.1007/978-3-030-60777-7_5

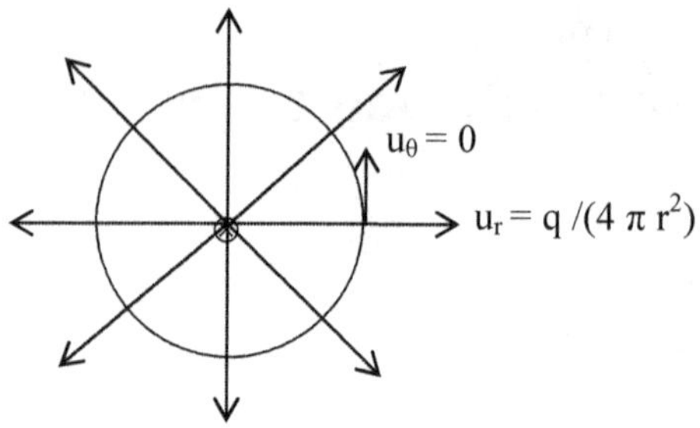

Fig. 5.1 Point source in three dimensions

Here, the intensity of the source, q = q_1 = constant.

(ii) The source expression for incompressible unsteady flow also satisfies the Laplace's equation with time dependent source strength, q = q(t). The time dependent velocity potential then reads as

$$\phi(x, y, z, t) = -\frac{q_2(t)}{4\pi\sqrt{x^2 + y^2 + z^2}}. \tag{5.2}$$

(iii) For compressible steady flow, rewriting Eq. 2.24 without time dependent terms and using $\beta^2 = 1 - M^2$, we obtain

$$\beta^2 \frac{\partial^2 \varphi}{\partial x^2} + \frac{\partial^2 \varphi}{\partial y^2} + \frac{\partial^2 \varphi}{\partial z^2} = 0 \tag{5.3}$$

We can transform Eq. 5.3 into Laplace equation with the following Gallilean transformation performed on the coordinates

$$\bar{x} = x, \quad \bar{y} = \beta y, \quad \bar{z} = \beta z$$

to obtain the following

$$\frac{\partial^2 \phi}{\partial \bar{x}^2} + \frac{\partial^2 \phi}{\partial \bar{y}^2} + \frac{\partial^2 \phi}{\partial \bar{z}^2} = 0. \tag{5.4}$$

Equation 5.4 has the solution in the transformed coordinates, by analogy with Eq. 5.1, as follows

$$\phi(\bar{x}, \bar{y}, \bar{z}) = -\frac{q_3}{4\pi\sqrt{\bar{x}^2 + \bar{y}^2 + \bar{z}^2}}$$

where q_3 is a constant. If we transform back to original coordinates we will have

$$\phi(x, y, z) = -\frac{q_3}{4\pi\sqrt{x^2 + \beta^2(y^2 + z^2)}}. \tag{5.5}$$

(iv) For the compressible unsteady flow we use the full form of Eq. 2.24 as follows

$$\frac{\partial^2 \phi}{\partial x^2} + \frac{\partial^2 \phi}{\partial y^2} + \frac{\partial^2 \phi}{\partial z^2} - \frac{1}{a^2}\left(\frac{\partial}{\partial t} + U\frac{\partial}{\partial x}\right)^2 \phi = 0 \tag{2.24}$$

and perform the coordinate transformation of Sect. 2.5 in moving coordinates we obtain the classical wave equation

$$\frac{\partial^2 \phi}{\partial x'^2} + \frac{\partial^2 \phi}{\partial y'^2} + \frac{\partial^2 \phi}{\partial z'^2} = \frac{1}{a^2}\frac{\partial^2 \phi}{\partial t'^2} \tag{2.26}$$

The well known solution of the classical wave equation in moving coordinates is

$$\phi(x', y', z', t') = -\frac{q_4(t')}{4\pi\sqrt{x'^2 + y'^2 + z'^2}} \tag{5.6}$$

We can go back to original coordinate system in Eq. 5.6 in terms of the free stream speed and the elapsed time t.

Now, let us use the physical models to express the mathematical derivations we have provided in this section.

5.1 Subsonic Flow

When the flow speed is less than the speed of sound which means the Mach number is under unity, the flow is called subsonic. In such a flow with a free stream speed U, a disturbance which was introduced at time τ becomes the spherical front, as shown in Fig. 5.2, after the time duration of Δt at time t.

The disturbance reaches the point r from its origin with $r = a\,\Delta t = a(t - \tau)$ and in terms of the coordinates x, y, z and the times given above we have $a^2(t-\tau)^2 = [x - U(t-\tau)]^2 + y^2 + z^2$. If we solve for the time of introduction of the disturbance, τ, we obtain

$$\tau = t + \frac{1}{a\beta^2}[Mx \pm \sqrt{x^2 + \beta^2(y^2 + z^2)}]. \tag{5.7}$$

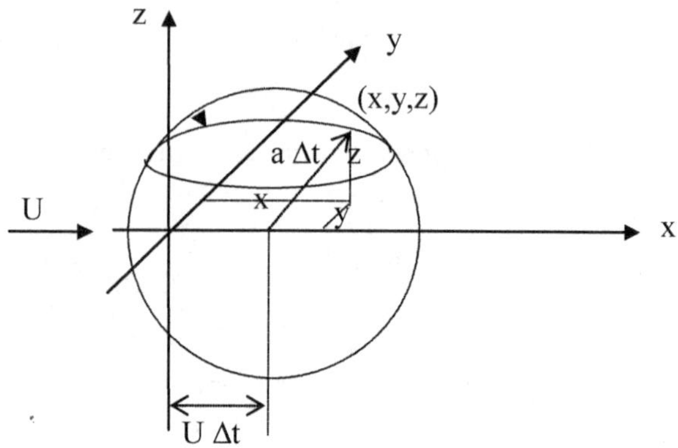

Fig. 5.2 Spherical perturbation front at time t

In Eq. 5.7 we have two different times for τ. For the subsonic flow we have to choose the one which has the smaller value because τ − t must be negative for subsonic flows. This is possible only for the following τ.

$$\tau = t + \frac{1}{a\beta^2}[Mx - \sqrt{x^2 + \beta^2(y^2 + z^2)}] \quad (5.8)$$

Now, we can write the velocity potential for a source generated at time τ and reached the point x, y, z at time t, using Eqs. 5.5 and 5.6.

$$\phi(x, y, z, t) = -\frac{q(\tau)}{4\pi\sqrt{x^2 + \beta^2(y^2 + z^2)}} \quad (5.9)$$

If the intensity of the source varies simple harmonically in time, that is $q(\tau) = \bar{q}e^{i\omega\tau}$, then the potential with Eq. 5.8 reads as

$$\phi(x, y, z, t) = -\frac{\bar{q}e^{i\omega\{t + \frac{1}{a\beta^2}[Mx - \sqrt{x^2 + \beta^2(y^2 + z^2)}]\}}}{4\pi\sqrt{x^2 + \beta^2(y^2 + z^2)}} \quad (5.10)$$

We can also obtain Eq. 5.10 using pure mathematical approach with a Lorentz type of transformation for which time coordinate is no longer absolute and given as follows

$$\bar{x} = x, \quad \bar{y} = \beta y, \quad \bar{z} = \beta z \quad \text{ve } \bar{t} = t + Mx/a\beta^2, \quad \beta = \sqrt{1 - M^2} \quad (5.11)$$

For this transformation, the derivatives in old coordinates in terms of the new ones read as

$$\frac{\partial}{\partial x} = \frac{\partial}{\partial \bar{x}}\frac{\partial \bar{x}}{\partial x} + \frac{\partial}{\partial \bar{y}}\frac{\partial \bar{y}}{\partial x} + \frac{\partial}{\partial \bar{z}}\frac{\partial \bar{z}}{\partial x} + \frac{\partial}{\partial \bar{t}}\frac{\partial \bar{t}}{\partial x} = \frac{\partial}{\partial \bar{x}} + \frac{M}{a\beta^2}\frac{\partial}{\partial \bar{t}}$$

and, $\frac{\partial}{\partial y} = \beta\frac{\partial}{\partial \bar{y}}, \quad \frac{\partial}{\partial z} = \beta\frac{\partial}{\partial \bar{z}}$.

The second derivatives then become:

$$\frac{\partial^2}{\partial x^2} = (\frac{\partial}{\partial \bar{x}} + \frac{M}{a\beta^2}\frac{\partial}{\partial \bar{t}})^2, \quad \frac{\partial^2}{\partial y^2} = \beta^2\frac{\partial^2}{\partial \bar{y}^2}, \quad \frac{\partial^2}{\partial z^2} = \beta^2\frac{\partial^2}{\partial \bar{z}^2}$$

Substituting these derivatives into Eq. 2.24 the equation for potential in transformed coordinates reads as

$$\bar{\nabla}^2\phi = \frac{1}{a^2\beta^4}\frac{\partial^2\phi}{\partial \bar{t}^2} \tag{5.12}$$

In Eq. 5.2 $\bar{\nabla}^2$ is the Laplace operator in the transformed coordinates. Using the method of separation of variables we have

$$\phi(\bar{x}, \bar{y}, \bar{z}, \bar{t}) = g(\bar{x}, \bar{y}, \bar{z})h(\bar{t}) \tag{5.13}$$

Substituting Eq. 5.13 into Eq. 5.12 we obtain

$$\frac{\bar{\nabla}^2 g}{g} = \frac{1}{a^2\beta^4}\frac{h''}{h} = -k^2 \tag{5.14}$$

In Eq. 5.14, k^2 is a positive number and the derivatives, denoted by 'prime', of h is taken with respect to transformed time coordinate. Since the right hand side of Eq. 5.14 is constant, it gives us two separate homogeneous, coupled only with constant k, equations for the functions g and h as follows.

$$\begin{aligned} h'' + a^2\beta^4k^2h &= 0 \\ \bar{\nabla}^2 g + k^2 g &= 0 \end{aligned} \tag{5.15a, b}$$

Equation 5.15a, b, is simple harmonic in time. Therefore, if we take $\omega = a\,\beta^2 k$ the general solution of Eq. 5.15a, b becomes

$$h(\bar{t}) = \bar{h}e^{i\omega\bar{t}} \tag{5.16}$$

Equation 5.15a, b, on the other hand, is the well known Helmholtz equation which has a solution in transformed coordinates as, Korn and Korn (1968),

$$g(\bar{x}, \bar{y}, \bar{z}) = \bar{g}\frac{e^{\pm ik\bar{R}}}{\bar{R}}, \quad \bar{R} = \sqrt{\bar{x}^2 + \bar{y}^2 + \bar{z}^2} \tag{5.17}$$

Combining Eqs. 5.16 and 5.17, the velocity potential in terms of $\bar{\phi} = \bar{g}\bar{h}$ becomes

$$\phi(\bar{x}, \bar{y}, \bar{z}, \bar{t}) = \bar{\phi}\, e^{i\omega(\bar{t} \pm \bar{R}/a\beta^2)}/\bar{R} \tag{5.18}$$

where we have two solutions separated with ±. If we go back to the original (x, y, z, t) coordinates we will have

$$\phi(x, y, z, t) = \bar{\phi} e^{i\omega[t + \frac{1}{a\beta^2}(Mx - R)]}/R, R = \sqrt{x^2 + \beta^2(y^2 + z^2)} \tag{5.19}$$

In Eq. 5.19, for the exponential term we take the one with – sign to have solution in agreement with Eq. 5.10.

In subsonic flows the acceleration potential rather than the velocity potential is preferred for its direct relation with the lifting pressure. Therefore, let us remember the relation between the two, the acceleration and the velocity potentials, as Eq. 2.25

$$\psi = (\frac{\partial}{\partial t} + U\frac{\partial}{\partial t})\phi' \tag{2.25}$$

Utilizing Eq. 2.25 with Eq. 2.21 the acceleration potential in terms the pressure and density of the farfield we obtain

$$\psi = \frac{p_\infty - p}{\rho_\infty} \tag{5.20}$$

As stated before, Eq. 5.20 gives the direct relation between the acceleration potential and the surface pressure which is to be used in determining the aerodynamic coefficients. Recalling Eq. 2.26, reminds us that the acceleration potential also satisfies Eq. 2.24 whose solution for the acceleration potential is

$$\psi(x, y, z, t) = \bar{\psi} e^{i\omega[t + \frac{1}{a\beta^2}(Mx - R)]}/R \tag{5.21}$$

The acceleration potential can directly be related to the surface lifting pressure discontinuity in terms of doublet distribution. We can derive the expression for a potential written in terms of a doublet. Defining a doublet requires a pair of source and a sink which are of equal strength and distance of ε apart from each other as shown in Fig. 5.3.

Now, let us express the potential for a source given by Eq. 5.21 in terms of a function f in the following manner, $\psi = -\bar{\psi} f(x, y, z, t)$. For a sink with the same

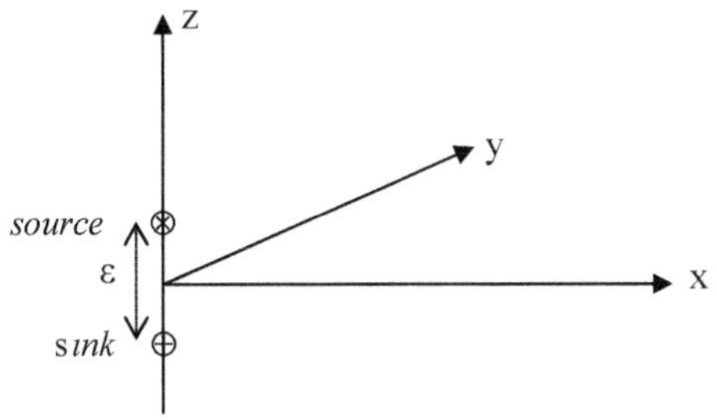

Fig. 5.3 Source and sink pair placed on z axis

strength, the potential becomes $\psi = \bar{\psi} f(x, y, z, t)$. The total effect of these two potentials placed on z axis with a distance ε reads as

$$\psi = \bar{\psi}[f(x, y, z - \varepsilon/2, t) - f(x, y, z + \varepsilon/2, t)] \tag{5.22}$$

If we multiply and divide Eq. 5.22 by ε, and take the double limit of the resulting ratio for the strength going to infinity as ε approaches zero we obtain

$$\psi = \lim_{\substack{\varepsilon \to 0 \\ \bar{\psi} \to \infty}} \left[-\bar{\psi}\varepsilon \frac{f(x, y, z + \varepsilon/2, t) - f(x, y, z - \varepsilon/2, t)}{\varepsilon}\right] \tag{5.23}$$

The limiting process employed on $\bar{\psi}\varepsilon$ results in

$$\lim_{\substack{\varepsilon \to 0 \\ \bar{\psi} \to \infty}} [-\bar{\psi}\varepsilon] = A$$

where, A is a constant having a finite value. The limit on f is nothing but the derivative of f with respect to z, i.e.

$$\psi(x, y, z, t) = A\frac{\partial}{\partial z} f(x, y, z, t) = A\frac{\partial}{\partial z}\{e^{i\omega[t + \frac{1}{a\beta^2}(Mx - R)]}/R\} \tag{5.24}$$

If we take the derivative of the expression in curly parenthesis with respect to z we obtain

$$\psi(x, y, z, t) = -Ae^{i\omega[t + \frac{1}{a\beta^2}(Mx - R)]}/R^2\left(\frac{i\omega}{a} + \frac{\beta^2}{R}\right)z \tag{5.25}$$

Now, we can comment on the physical meaning of acceleration potential given by Eq. 5.25 at the surface where z = 0. At this surface the value of potential is zero except for R = 0 where there is a singularity. Equation 5.20 provided us the relation between the pressure and the acceleration potential. Rearranging Eq. 5.20 to obtain the pressure at a point (x, y, z) for a given time t gives us

$$p(x,y,z,t) = p_\infty - \rho_\infty \psi(x,y,z,t) \tag{5.20}$$

We can express the lifting pressure in terms of the singular doublet strength A given by Eq. 5.24 as $\Delta p = p_l - p_u \propto A$. Dimensional analyses show that A must have the dimensions L^4T^{-2}. Therefore, the strength of the doublet is related to the lifting pressure as follows

$$A = \frac{l^2}{\rho_\infty} \Delta p \tag{5.26}$$

Here, l is the characteristic length to be employed for defining the strength of the acceleration potential as the pressure discontinuity in following form.

$$\psi(x,y,z,t) = \frac{l^2}{\rho_\infty} \Delta p \frac{\partial}{\partial z} \{ e^{i\omega[t + \frac{1}{a\beta^2}(Mx - R)]} / R \} \tag{5.27}$$

We have finally obtained an expression, Eq. 5.27, for our mathematical model for lifting bodies in subsonic flows. Equation 5.27, however, is developed for a doublet placed at the origin. In order to represent lifting surfaces, on the other hand, we need to derive the same expression for the effect of an arbitrary point on the surface.

5.2 Subsonic Flow About a Thin Wing

We are going to use the distributed acceleration potential rather than a single one to model the unknown lifting pressure distribution over the wing surface. For this modeling to work the lifting pressure must go to zero along the trailing edge in order to satisfy the Kutta condition. The lifting pressure as a discontinuity at point (ξ, η) of the surface is related to the acceleration potential at any point (x, y, z) at time t as follows

$$\psi(x,y,z,t) = \frac{\Delta \bar{p}(\xi,\eta)}{\rho_\infty} \frac{\partial}{\partial z} \{ e^{i\omega[t + \frac{1}{a\beta^2}M(x-\xi) - \frac{R}{a\beta^2})]} / R \} d\xi d\eta, \tag{5.28}$$

Here, $R = \sqrt{(x-\xi)^2 + \beta^2[(y-\eta)^2 + z^2]}$.

In Eq. 5.28 the amplitude of the lifting pressure enables us to express the simple harmonic representation in following form.

$$\Delta p(\xi,\eta,t) = \Delta \bar{p}(\xi,\eta) e^{i\omega t} \tag{5.29}$$

We know the relation between the lifting pressure and the acceleration potential. Now, we have to relate the velocity potential to the lifting pressure so that we can impose the boundary conditions to obtain the lifting pressure for a prescribed motion of the wing. Equation 2.25 gives the relation between the two potentials. For a simple harmonic motion 2.25 becomes

$$\bar{\psi} = i\omega\bar{\phi} + U\frac{\partial\bar{\phi}'}{\partial x} \tag{5.30}$$

Equation 5.30 is a first order differential equation for the velocity potential which has an explicit solution in the following form

$$\bar{\phi}' = \frac{1}{U}e^{-i\omega x/U}\int_{-\infty}^{x} \bar{\psi}(\lambda, y, z)e^{i\omega\lambda/U}d\lambda \tag{5.31}$$

Using Eq. 5.28 in Eq. 5.31 gives us the amplitude of the velocity potential in terms of the acceleration potential as follows

$$\bar{\phi}'(x, y, z) = \frac{\Delta\bar{p}(\xi, \eta)}{\rho_\infty U}e^{-i\omega(x-\xi)/U}\frac{\partial}{\partial z}\int_{-\infty}^{x-\xi}\frac{e^{i\omega[\lambda/U+\frac{1}{a\beta^2}M(\lambda-R)]}}{R}d\lambda d\xi d\eta \tag{5.32}$$

Prescribing the simple harmonic equation of motion for the thin wing as $z_a = \bar{z}_a(x, y)e^{i\omega t}$ the boundary condition at the surface reads as

$$\bar{w}(x, y) = (i\omega + U\frac{\partial}{\partial x})\bar{z}_a(x, y) \tag{5.33}$$

Integrating the downwash expression over the whole surface S yields

$$\bar{w}(x, y) = \lim_{z\to 0}(\frac{\partial\bar{\phi}}{\partial z}) = \lim_{z\to 0}[\frac{\partial}{\partial z}\iint_S \bar{\phi}'(x, y, z, \xi, \eta)d\xi d\eta]. \tag{5.34}$$

If we substitute Eq. 5.32 into Eq. 5.34 we obtain the downwash in terms of lifting pressure as follows

$$\bar{w}(x, y) = \frac{1}{\rho_\infty U}\iint_S \Delta\bar{p}(\xi, \eta)\lim_{z\to 0}\{\frac{\partial^2}{\partial z^2}e^{-i\omega(x-\xi)/U}\frac{\partial}{\partial z}\int_{-\infty}^{x-\xi}\frac{e^{i\omega[\lambda/U+\frac{1}{a\beta^2}M(\lambda-R)]}}{R}d\lambda\, d\xi d\eta\}. \tag{5.35}$$

The lifting pressure can be found by solving the integral equation, Eq. 5.35, once the boundary condition Eq. 5.33 is prescribed as the left hand side of Eq. 5.35. In order to simplify Eq. 5.35 let us define new parameters in terms of the old ones as follows.

$$[x' = x - \xi, y' = y - \eta, \omega' = \omega/U\beta^2, r^2 = \beta^2(y'^2 + z^2)]$$

Using the new parameters we obtain

$$\frac{\bar{w}(x,y)}{U} = \frac{1}{\rho_\infty U^2} \iint_S \Delta\bar{p}(\xi,\eta) \lim_{z\to 0}\{\frac{\partial^2}{\partial z^2} e^{-i\omega'\beta^2 x'} \frac{\partial}{\partial z} \int_{-\infty}^{x'} \frac{e^{i\omega'[\lambda - M\sqrt{\lambda^2 + r^2}]}}{R} d\lambda\} d\xi d\eta \tag{5.36}$$

The singular inner integral part of Eq. 5.36 is subject to a limiting process, and it is called the Kernel function. If we denote the Kernel function with K(x′, y′) and the nondimensional pressure discontinuity with $L(\xi,\eta) = \Delta\bar{p}(\xi,\eta)/\rho_\infty U^2$ the downwash expression becomes

$$\frac{\bar{w}(x,y)}{U} = \iint_S L(\xi,\eta) K(x', y') d\xi d\eta \tag{5.37}$$

Direct inversion of Eq. 5.37 is not possible therefore, numerical methods are used for that purpose.

5.3 Subsonic Flow Past an Airfoil

Before studying three dimensional subsonic flow, we are going to start our analysis with two dimensional flows. There are two options to do so. We either rederive the equations for two dimensional flows or we take the integral of the three dimensional acceleration potential in y to make the equations independent of spanwise direction. Here, the latter approach is preferred since we have already obtained the necessary expression to be integrated only. Integrating Eq. 5.36 from $-\infty$ to ∞ in spanwise direction we obtain

$$\psi^{2D} = \int_{-\infty}^{\infty} \psi^{3D} d\eta = A\frac{\partial}{\partial z}\{\frac{i}{4\beta} e^{i\omega[t + \frac{M}{\beta^2}(x-\xi)]} H_o^{(2)}(\frac{\omega M}{U\beta^2}\sqrt{(x-\xi)^2 + \beta^2 z^2})\} \tag{5.38}$$

Let us now write the amplitude of the downwash in terms the lifting pressure using the Kernel function for two dimensional flows, (BAH).

$$\bar{w}(x) = -\frac{\omega}{\rho_\infty U^2} \int_{-b}^{b} \Delta\bar{p}(\xi) K(M, k\frac{x-\xi}{b}) d\xi, \quad -b \le x \le b \tag{5.39a}$$

Here, if we take $k' = k\frac{x-\xi}{b}$ and $u = k\frac{\xi'-\xi}{b\beta}$, the kernel function K, in terms of Mach number and the reduced frequency reads as

$$K(M,k') = \frac{1}{4\beta}\{e^{i\frac{M^2}{\beta^2}k'}[iM\frac{|x-\xi|}{x-\xi}H_1^{(2)}(\frac{M^2}{\beta^2}|k'|) - H_0^{(2)}(\frac{M^2}{\beta^2}|k'|)] + i\beta^2 e^{-i\frac{M^2}{\beta^2}k'}[\frac{2}{\pi\beta}\ln\frac{1+\beta}{M} + \int\limits_0^{k'/\beta} e^{iu}H_0^{(2)}(M|u|)du]\} \tag{5.39b}$$

Equation 5.39a is called the Possio integral equation. Possio tried to solve his equation in terms of the Fourier series, however, his solution technique confronted with the convergence problem. A different approach from Possio to remedy the convergence problem is to employ Fourier like series for the lifting pressure expressed in $\xi = b\cos\theta$ coordinates. A new way of approximating the pressure discontinuity is

$$\Delta\bar{p}(\theta) = A_0\cot\frac{\theta}{2} + \sum_{n=1}^{\infty} A_n\frac{\sin n\theta}{n}, \quad 0 \le \theta \le \pi \tag{5.40}$$

The cotangent term in Eq. 5.40 gives an integrable singularity for the lifting pressure at the leading edge while satisfying the Kutta condition with zero lifting pressure at the trailing edge. The integral of the second term of Kernel function, Eq. 5.39b, can be evaluated numerically. In doing so, keeping the number of control points on the chord equal to the number of terms in Eq. 5.40 enables us to have a number of algebraic equations equal to the number unknowns with complex elements. The right hand sides of the equation, f_i, are the known values of the prescribed airfoil motion to result in following set of linear equations.

$$\sum_{j=0}^{N} K_{ij}(M,\phi,\theta)A_j = f_i(\phi), i = 0,\ldots,N \tag{5.41}$$

Here, $x = b\cos\phi$ denotes the coordinates for the control points.

5.4 Kernel Function Method for Subsonic Flows

The Kernel function method, known also as the pressure mode method, is used for solving the lifting pressure value via the integral equation, 5.37, for the prescribed motion of a thin wing in subsonic flows. The expression for the Kernel function was given in Eq. 5.36 as follows.

$$K(x',y') = \lim_{z\to 0}\{\frac{\partial^2}{\partial z^2}e^{-i\omega'\beta^2 x'}\frac{\partial}{\partial z}\int_{-\infty}^{x'}\frac{e^{i\omega'[\lambda - M\sqrt{\lambda^2+r^2}]}}{R}d\lambda\, d\xi d\eta\}$$

Watkins et al. (1955) gives the open form of the kernel function in terms of the hypergeometric functions; K_1: first order modified Bessel function of the second kind, I_1: first order modified Bessel function of first kind, and L_1: first order modified Struve function as follows.

$$\begin{aligned} K(x',y') = \frac{k^2}{l^2}e^{-kx'}\{ &-\frac{1}{k|y'|}K_1(k|y'|) - \frac{\pi i}{2k|y'|}[I_1(k|y'|) - L_1(k|y'|)] + \frac{iMk|y'|+\beta}{M\beta(ky')^2}e^{-\frac{ik|y'|}{\beta}} \\ &-\int_0^{M/\beta}\sqrt{1+\tau^2}e^{-ik|y'|\tau}d\tau - \frac{Mkx' + \sqrt{(kx')^2+\beta^2(ky')^2}}{(Mky')^2\sqrt{(kx')^2+\beta^2(ky')^2}}e^{\frac{i}{\beta^2}[Mkx' + \sqrt{(kx')^2+\beta^2(ky')^2}]} \\ &+\frac{i}{M(ky')^2}\int_0^{kx'} e^{\frac{i}{\beta^2}[\lambda - M\sqrt{\lambda^2+\beta^2(ky')^2}]}d\lambda\} \end{aligned} \tag{5.42}$$

Here, all the coordinates are nondimensionalized with root half chord, b_o.

In order to solve for the lifting pressure with the kernel function method we need to expand the lifting pressure into the sin series similar to that given for Eq. 5.40. In doing so, we need to transform the chordwise coordinates for an arbitrarily shaped planform as shown in Fig. 5.4.

As seen in Fig. 5.4, at any station along span, we use an angular coordinate θ along the chord. The midchord variation along the span can be expressed in terms of the spanwise coordinate η. Using this approach we can deal with the planforms

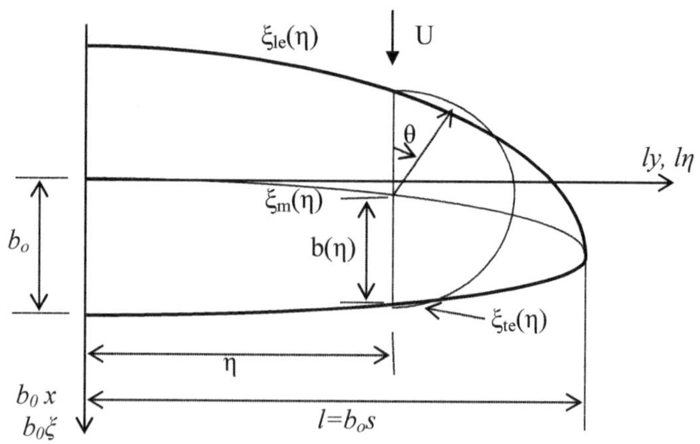

Fig. 5.4 Planform and the variables for transformation

having not only straight but also curved edges. Now, we can write the chordwise ξ coordinate in terms of the angular θ coordinate as shown in Fig. 5.4 as follows

$$\xi = \xi_m - \frac{b}{b_0}\cos\theta, \quad 0 \le \theta \le \pi \quad \text{ve} \quad \xi_m = \frac{\xi_{te} + \xi_{le}}{2}, \quad \frac{b}{b_0} = \frac{\xi_{te} - \xi_{le}}{2}$$

The unknown lifting pressure in Eq. 5.37 can be expressed in terms of the transformed coordinates in a following manner.

$$\Delta\bar{p}(\theta,\eta) = 4\pi\rho U^2 \frac{l}{b_0} L(\theta,\eta) \tag{5.43}$$

Substituting Eq. 5.43 into Eq. 5.37 gives

$$\frac{\bar{w}(x,y)}{U} = \frac{b_0 l}{4\pi\rho U^2} \int_{-1}^{1} \int_{\xi_{le}(\eta)}^{\xi_{te}(\eta)} \Delta\bar{p}(\xi,\eta) K(M,k,x',sy') d\xi d\eta \tag{5.44}$$

Now, we can expand the nondimensional lifting pressure, L(θ, η), into proper series, Ashley et al. (1965) as follows

$$\begin{aligned} L(\theta,\eta) &= \tfrac{b_0}{b}\sqrt{1-\eta^2}[(a_{00} + \eta a_{01} + \eta^2 a_{02} + \ldots)\cot\tfrac{\theta}{2} \\ &+ (a_{10} + \eta a_{11} + \eta^2 a_{12} + \ldots)\sin\theta \\ &\ldots\ldots\ldots\ldots\ldots\ldots\ldots\ldots\ldots\ldots \\ &\ldots\ldots\ldots\ldots\ldots\ldots\ldots\ldots\ldots\ldots \\ &+ \tfrac{4}{2^{2n}}(a_{n0} + \eta a_{n1} + \eta^2 a_{n2} + \ldots)\sin n\theta] \end{aligned} \tag{5.45}$$

Here, coefficients a_{nm} are the unknowns to be determined. Equation 5.45 can be written as the product of two entities as follows

$$L(\theta,\eta) = \frac{b_0}{b}\sum_n l_n(\theta) A_n(\eta) \tag{5.46}$$

In Eq. 5.46, for $n = 0$: $l_0(\theta) = \cot\frac{\theta}{2}$, and $l_n(\theta) = \frac{4}{2^{2n}}\sin n\theta, n \ge 1$,

$$A_n(\eta) = \sqrt{1-\eta^2}(a_{n0} + \eta\, a_{n1} + \ldots + \eta^m a_{nm} + \ldots) = \sqrt{1-\eta^2}\sum_m \eta^m a_{nm} \tag{5.47}$$

If there is a symmetric loading on the wing we take the even powered terms and for the antisymmetric loading the odd powered terms of series Eq. 5.47.

The integral Eq. 5.44 has a kernel with a strong singularity. In order to prevent the difficulties in numerical integration function K is rewritten as $\bar{K}$ with a scaling presented in a following manner

$$\bar{K}[M,k,x',s(y-\eta)] = b_0^2 s^2 (y-\eta)^2 K[M,k,x',s(y-\eta)], s = l/b_0 \tag{5.48}$$

The integral equation with the aid of Eq. 5.48 becomes

$$\frac{\bar{w}(x,y)}{U} = \int_{-1}^{1} \frac{b}{b_0} \frac{d\eta}{(y-\eta)^2} \int_{\xi_{le}(\eta)}^{\xi_{te}(\eta)} \Delta\bar{p}(\xi,\eta)\bar{K}_\theta(M,k,\theta,\eta)\sin\theta d\theta \tag{5.49}$$

Here, in its open form $\bar{K}$ reads as

$$\bar{K}_\theta(M,k,\theta,\eta) = \bar{K}[M,k,(x-\xi_m+\frac{b}{b_0}\cos\theta),s(y-\eta)]$$

The series expressions Eqs. 5.46 and 5.47 for the nondimensional lifting pressure, $L(\theta, \eta)$, can be substituted in integral equation, 5.49, to obtain the following system of linear equations written in double series

$$\frac{\bar{w}(x,y)}{U} = \sum_n \sum_m a_{nm} \int_{-1}^{1} \frac{\eta^m\sqrt{1-\eta^2}}{(y-\eta)^2} d\eta \int_0^\pi l_n(\theta)\bar{K}_\theta(M,k,\theta,\eta)\sin\theta d\theta. \tag{5.50}$$

The right hand side of the Eq. 5.50 is given as the boundary condition. Unknown coefficients, a_{nm}, at the right hand side of the equation can be determined with taking as many control points on the wing as unknowns. The lifting pressure values at these stations are then calculated using Eq. 5.46.

In Fig. 5.5, there are 20 (5 spanwise × 4 chordwise) unequally spaced control points, including the root chord, on a symmetric wing. There are some rules to follow while choosing the control points. These are:

(i) No control points are placed at the wing edges. Sufficient number of points, no less than 20, in chordwise and spanwise directions must be taken.
(ii) In symmetrical loadings even powers of m and antisymmetrical loadings odd powers of m must be considered.
(iii) For high reduced frequencies more number of control points must be taken.
(iv) If there are control surfaces on the wing the control points must be chose accordingly.

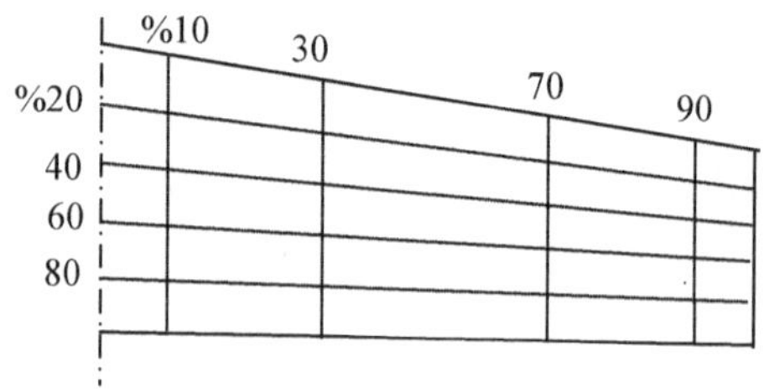

Fig. 5.5 Representative control points on the wing

Hitherto, we have formulated the kernel function method applicable to subsonic flows over thin wings. In next section we are going to study a new method called 'Doublet-Lattice' method applicable to general problems involving more complex wings with control surfaces and tails and tail wings.

5.5 Doublet—Lattice Method

The doublet lattice method is based on a linear theory using a numerical approach to study the subsonic three dimensional flows past complex lifting surfaces (Albano and Rodden 1969). In this method as lifting surfaces, the wing, the horizontal and vertical tails are subdivided into trapezoidal surfaces (the parallel sides of the trapezoids are in line with the free stream direction) to discretize the flow domain. In addition, if there is a tank or a store as an external type body, its surface is also subdivided into trapezoids during discretization. In order to discretize a surface on each panel, a doublet line is placed at quarter chord of the panel and a control point is assigned at three quarter chord point as shown in Fig. 5.6. On the doublet line the unknown but constant value of the doublet strength is considered and at the control point known downwash value is assigned. This way the Kutta condition is satisfied numerically. As seen in Fig. 5.6, the wing surface is divided into n panels where the information about locations of the doublet line and the control points are used to obtain the algebraic equations from Eq. 5.37 using numerical integration.

The downwash, w_i, induced on the control point of the ith panel by the doublet lines of the panels j = 1, 2, 3,..., n, shown in Fig. 5.6, can be expressed as follows

$$\bar{w}_i = \sum_{j=1}^{n} D_{ij} \Delta \bar{p}_j \tag{5.51}$$

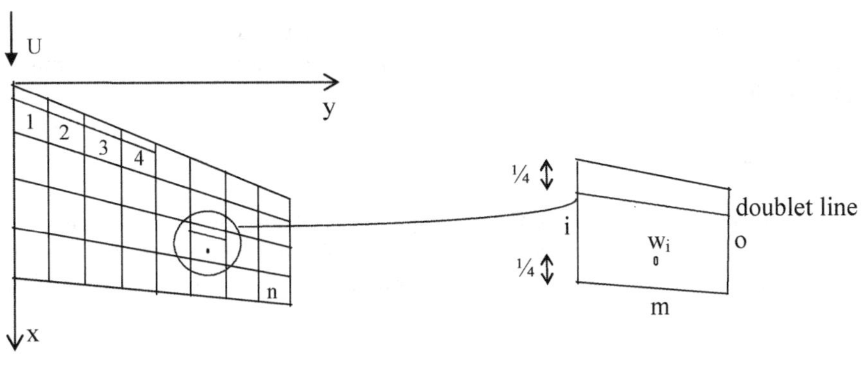

Fig. 5.6 Lattices on wings with their details

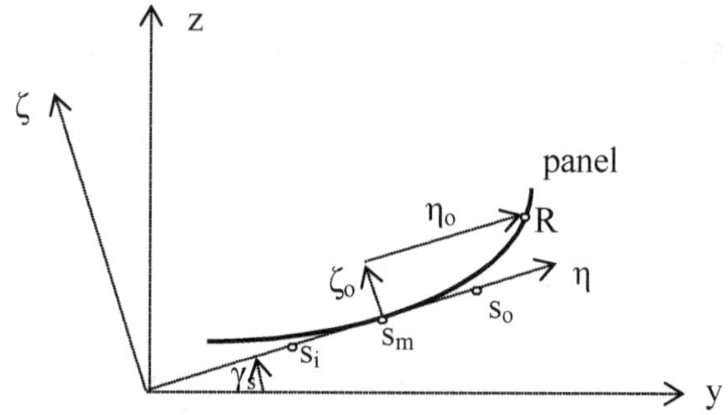

Fig. 5.7 Local coordinate system attached to the panel

Here, D_{ij} is written in terms of the numerical integrals over the Kernel function. In order to perform the numerical integrals for each penal we have to transform the local coordinates, (ξ, η, ζ), into the global coordinates, (x, y, z), as shown in Fig. 5.7.

According to Fig. 5.7 the coordinate transformation in terms of the mid point tangent angle γ_s reads as

$$\eta = y \cos \gamma_s + z \sin \gamma_s$$
$$\varsigma = -y \sin \gamma_s + z \cos \gamma_s$$

Using the geometry of the panel shown in Fig. 5.7, we can obtain the coefficient matrix D_{ij} of Eq. 5.51 as follows. First we determine the coordinates of the control point, $R = (x_R, y_R, z_R)$, then the doublet line coordinates as inner left, s_i, middle: s_m, outer right, s_o, coordinates are considered. If there is a spanwise curvature for the wing the local coordinates from the global ones read as

$$\eta_0 = (y_R - y_{Sm}) \cos \gamma_s + (z_R - z_{Sm}) \sin \gamma_s$$
$$\varsigma_0 = -(y_R - y_{Sm}) \sin \gamma_s + (z_R - z_{Sm}) \cos \gamma_s$$
$$r_1^2 = \eta_0^2 + \varsigma_0^2$$

If the doublet line length is l_j and the angle between the doublet line and the span is λ_j then let us define $e = 1/2\, l_j cos \lambda_j$. The value of the Kernel function in terms of new coordinates becomes

$$\kappa = r_1^2 K \quad \text{and} \quad \kappa_m = \kappa(R, s_m), \quad \kappa_i = \kappa(R, s_i), \quad \kappa_0 = \kappa(R, s_0).$$

If the panel length is small enough the parabolic change of the Kernel function will give us sufficiently accurate approximation. Accordingly, along the doublet line we have the following approximate integral to represent the Kernel integration with respect to dμ

$$I_{ij} = \int_{l_i} K[x_i, s_i, x_j(\mu), s_j(\mu) \cos \lambda_j d\mu \approx \int_{-e}^{e} \frac{A\eta^2 + B\eta + C}{(\eta_0 - \eta)^2 + \varsigma_0^2} d\eta \tag{5.52}$$

Here, A, B and C respectively reads as

$$A = (\kappa_i - 2\kappa_m + \kappa_0)/2e^2, \quad B = (\kappa_0 - \kappa_i)/2e, \quad C = \kappa_m$$

The resulting integral in general becomes

$$I_{ij} = [(\eta_0^2 - \varsigma_0^2)A + \eta_0 B + C]|\varsigma_0|^{-1} \tan^{-}(\frac{2e|\varsigma_0|}{r_1^2 - e^2}) + (\frac{1}{2}B + \eta_0 A) \ln \frac{r_1^2 - 2\eta_0 e + e^2}{r_1^2 + 2\eta_0 e + e^2} + 2eA$$

and, for the wings with straight span

$$\zeta_0 \to 0 : I_{ij} = [\eta_0^2 A + \eta_0 B + C]|\varsigma_0|^{-1} (\frac{1}{\eta_0 - e} - \frac{1}{\eta_0 + e}) + (\frac{1}{2}B + \eta_0 A) \ln(\frac{\eta_0 - e}{\eta_0 + e})^2 + 2eA, \tag{5.53a, b}$$

In the Kernel function for $x' = x - \zeta$ we have two different integrals I_1 and I_2 as follows

$$I_1 = \int_{u_1}^{\infty} \frac{e^{-ik_1 u}}{(1+u^2)^{3/2}} du \quad and$$

$$I_1 = \int_{u_1}^{\infty} \frac{e^{-ik_1 u}}{(1+u^2)^{5/2}} du, u_1 = (MR - x')/\beta^2 r_1, k_1 = \omega r_1/U, R = x'^2 + \beta^2 r_1^2$$

The Kernel function in terms of these integrals reads

$$K = e^{-i\omega x'/U} \{ [I_1 + \frac{e^{-ik_1 u_1}}{(1+u_1^2)^{1/2}}] T_1 - [3I_2 + \frac{ikM^2 r_1^2}{R^2} \frac{e^{-ik_1 u_1}}{(1+u_1^2)^{1/2}} + \frac{Mr_1}{R} (2 + (1+u_1^2) \frac{\beta^2 r_1^2}{R^2} + \frac{Mr_1 u_1}{R}) \frac{e^{-ik_1 u_1}}{(1+u_1^2)^{3/2}}] T_2 \} / r_1^2$$

Here,

$$T_1 = \cos(\gamma_i - \gamma_j) \quad and$$

$$T_2 = (\frac{z - \varsigma}{r_1} \cos \gamma_i - \frac{y - \eta}{r_1} \sin \gamma_i)(\frac{z - \varsigma}{r_1} \cos \gamma_j - \frac{y - \eta}{r_1} \sin \gamma_j)$$

With all these, we have given the necessary information for the 'doublet-lattice' method to be applied numerically. This method is applicable to wing-tail, wing-external store and wing-fuselage interaction problems as well as the wings with curved spans.

5.6 Arbitrary Motion of a Profile in Subsonic Flow

For the case of compressible flow response of an airfoil to the arbitrary motion differs from that of the incompressible flow. Therefore, we have to modify the indicial admittance functions in terms of the Mach number of the flow for the sudden angle of attack change and for the gust impingement problems. Similar to that of incompressible flow we denote the Wagner function, $\varphi(s)$, for the sudden angle of attack change, α_o, and the Küssner function, $\chi(s)$, for the gust effects to give
with the Wagner function

$$c_L(s) = 2\pi\alpha_o\varphi(s). \tag{5.54}$$

For the gust of intensity w_o we write
with the Küssner function

$$c_{Lg}(s) = 2\pi\frac{w_o}{U}\chi(s) \tag{5.55}$$

Here, s denotes the reduced time based on the half chord of the airfoil. For the Wagner and the Küssner functions we have the following general approximation in terms of the exponential functions as follows (BAH).

$$\left.\begin{matrix}\varphi(s)\\ \chi(s)\end{matrix}\right\} = b_o - b_1e^{-\beta_1 s} - b_2e^{-\beta_2 s} - b_3e^{-\beta_3 s}$$

The values for the exponents of each function with respect to the Mach number are given in Table 5.1. Figure 5.8 gives the plots for the Küssner function in terms of three different Mach numbers. The compressibility effect in these plots is evident as the steady state is reached.

Table 5.1 Variations of Wagner an Küssner functions with respect to the Mach number of the flow

	M	b_o	b_1	b_2	b_3	β_1	β_2	β_3
$\varphi(s)$	0	1.0	0.165	0.335	0	0.0455	0.3	–
	0.5	1.155	0.406	0.249	−0.773	0.0753	0.372	1.89
	0.6	1.25	0.452	0.63	−0.893	0.0646	0.481	0.958
	0.7	1.4	0.5096	0.567	−0.5866	0.0536	0.357	0.902
$\chi(s)$	0	1.0	0.5	0.5	0.	0.13	1.0	–
	0.5	1.155	0.45	0.47	0.235	0.0716	0.374	2.165
	0.6	1.25	0.41	0.538	0.302	0.0545	0.257	1.461
	0.7	1.4	0.563	0.645	0.192	0.0542	0.3125	1.474

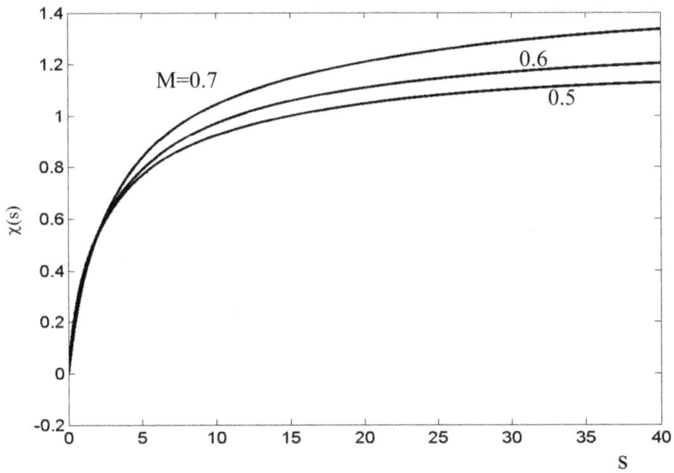

Fig. 5.8 Effect of gust in various Mach numbers

5.7 Supersonic Flow

In a compressible flow if the local speeds of air particles exceed the local speed of sound, the supersonic flow condition emerges. In such cases different types of waves appear in the flow depending on geometry or back pressure change in the flow. The most pronounced among these waves are the shock waves. The simplest shock wave is the normal shock wave which occurs normal to the flow direction of one dimensional flows. The physically possible flow is the one being the supersonic flow before and the subsonic flow after the shock (Shapiro). This way the entropy increases after the shock in accordance with the second law of thermodynamics. The normal shocks are strong shocks therefore the entropy increase is large and this large entropy causes the flow to change its regime from supersonic to subsonic.

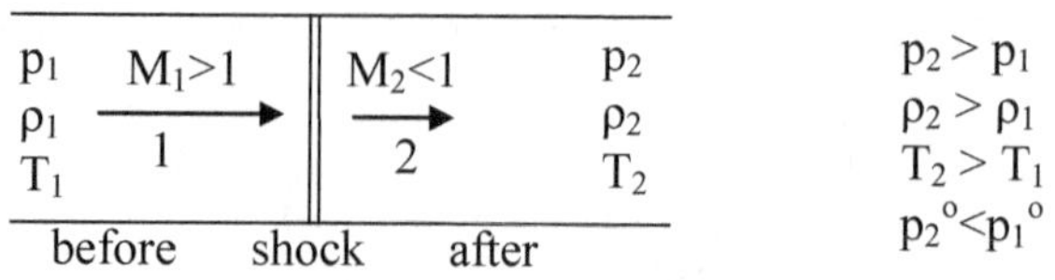

Fig. 5.9 Normal sock, before and after

A simple model of a normal shock is shown in Fig. 5.9. After the normal shock, the pressure temperature and the density of the flow increase whereas the Mach number and the stagnation pressures decrease.

The oblique shock waves, on the other hand, appear to be two dimensional but they can be treated with one dimensional flow analysis. Oblique waves are inclined with the free stream but they are straight like normal shocks. The free stream of the flow deflects and slows down after passing an oblique shock. The amount of deflection in the flow is inversely proportional with the inclination of the shock. The oblique shocks are known as weak shocks because they cause smaller entropy change than that of a normal shock. The reason for the occurrence of the oblique shock is in general the narrowing of the flow area as shown in Fig. 5.10. The flow after the oblique shock goes parallel to the wall which is narrowed down with angle θ. As indicated in Fig. 5.10, after the shock the component of the freestream parallel to the shock remains the same, however, the normal component slows down. Since the shock is weak the entropy change after the shock is small. This fact and the shock being a straight line enable us to assume potential flow before and after the shock.

Another reason for waves to appear in the supersonic flow is the expansion of the flow field. In this case the expansion waves are created in the flow so that the flow passing through this waves is made to turn until it remains parallel to the wall. The expansion process as shown in Fig. 5.11 is an isentropic process.

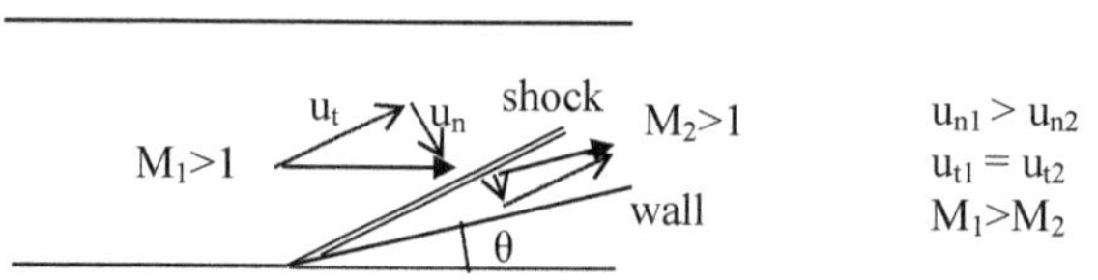

Fig. 5.10 Oblique shock, velocity relations

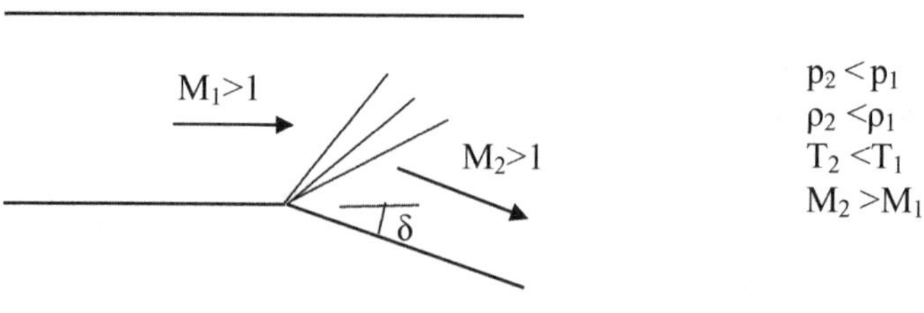

Fig. 5.11 Isentropic expansion waves around a corner

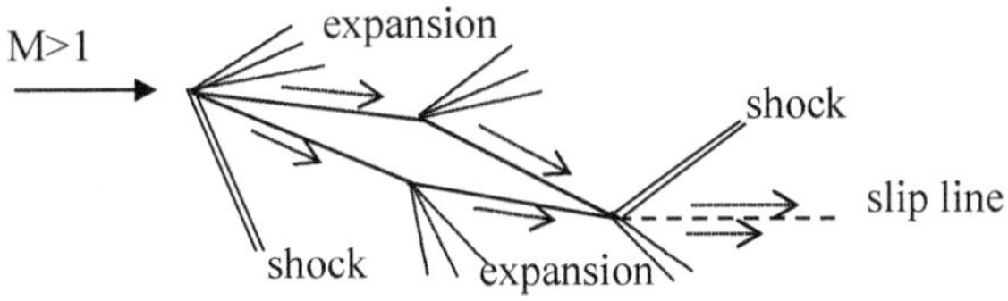

Fig. 5.12 Waves around a thin airfoil in a supersonic flow

After seeing all the wave types likely to occur in supersonic flow, we can analyze the flow about a profile immersed in a supersonic stream. Shown in Fig. 5.12 are the expansion and the shock waves over the upper and lower surfaces of a diamond shaped profile immersed in a supersonic stream with an angle of attack.

According to Fig. 5.12, on the upper surface of the profile the flow expands after the leading edge and continues to expand and speed up at the mid chord until the trailing edge where the oblique shock slows down and changes the direction of the flow. At the lower surface, however, because of compression at the leading edge an oblique shock is generated to slow down the flow and the flow expands after mid chord and at the trailing edge to become parallel with the flow of the upper surface. After the trailing edge, we may have upper and lower surface speeds different from each other so that a slip line is created without any pressure discontinuity across. In supersonic flow the pressure difference at the wake is zero and that is how we satisfy the Kutta condition.

So far we have seen a summary of types of waves which can occur because of presence of a profile in a steady supersonic flow. Let us study furthermore, the time dependent creation of these waves from the lifting surfaces for which we are interested to find unsteady aerodynamic coefficients.

5.8 Unsteady Supersonic Flow

The linearized potential flow equations are exactly the same for the subsonic and supersonic flows. Equation 2.24 in its original open form was obtained as

$$\frac{\partial^2 \phi}{\partial x^2} + \frac{\partial^2 \phi}{\partial y^2} + \frac{\partial^2 \phi}{\partial z^2} - \frac{1}{a^2}\left(\frac{\partial}{\partial t} + U\frac{\partial}{\partial x}\right)^2 \phi = 0 \tag{2.24}$$

This time we can use $\beta^2 = M^2 - 1$ as a parameter and perform a new Lorentz type of transformation (Miles 1959) to obtain the classical wave equation for the perturbation potential. Let us see the transformation based on a complex variables.

$$\bar{x} = x, \quad \bar{y} = i\beta y, \quad \bar{z} = i\beta z \quad \text{ve} \quad \bar{t} = t - Mx/a\beta^2, \quad \beta = \sqrt{M^2 - 1} \tag{5.56}$$

In transformed coordinates the potential equation reads as

$$\bar{\nabla}^2 \phi = \frac{1}{a^2 \beta^4} \frac{\partial^2 \phi}{\partial \bar{t}^2} \tag{5.57}$$

The solution of Eq. 5.57 in original coordinates reads as

$$\phi(x, y, z, t) = \bar{\phi} e^{i\omega[t - \frac{1}{a\beta^2}(Mx \pm R)]} / R, R = \sqrt{x^2 - \beta^2 (y^2 + z^2)} \tag{5.58}$$

The mathematical solution given with Eq. 5.58 is the expression for a source term in a supersonic flow. The ± appearing in exponent of Eq. 5.58 indicates that there are two different solutions for source. In order to identify the source solution which reflects the physics of the problem, we have to study the propagation of a disturbance in a supersonic flow as shown in Fig. 5.13. The speed of a wave front of a perturbation in a supersonic flow placed at the origin is faster than the free stream. Therefore, the wave front is carried down in the direction of the free stream only in the positive x axis. For this reason, the tangent drawn from origin to the front surface at any time t makes a constant acute angle with the x axis as shown in Fig. 5.13. The locus of all these tangent lines is a conical surface called the Mach cone. The cone which is symmetric of this cone with respect to y-z plane is called the upstream Mach cone, and the opposite one is called the downstream Mach cone.

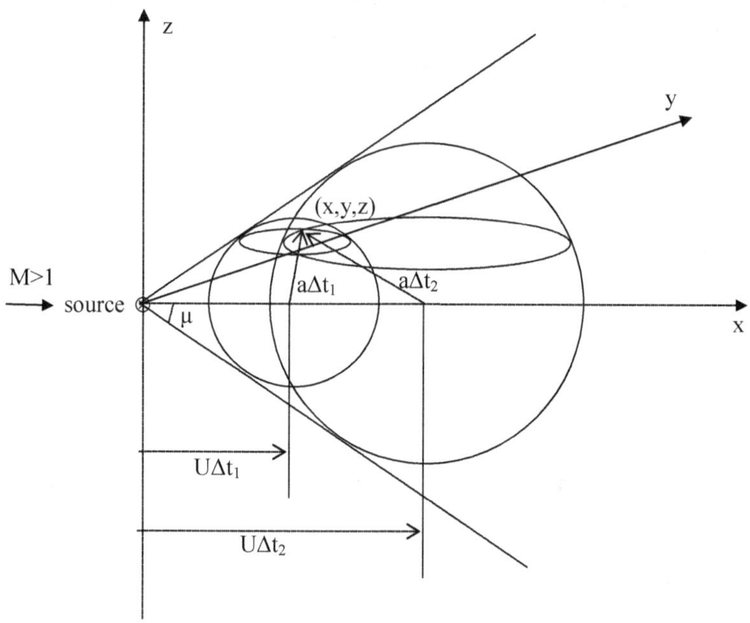

Fig. 5.13 Downstream mach cone and the disturbances reaching to (x, y, z) at two different time

Let Δt denote the time between its introduction to the flow and the present time t. If we call the half angle of the Mach cone as μ = Mach angle, then we have

$$\sin \mu = \frac{a\Delta t}{U\Delta t} \quad \text{ve} \quad \mu = \sin^{-}(\frac{1}{M})$$

Now, we can comment on the physics of the problem via Fig. 5.13. The comment will be based on the kinematic entities such as the free stream speed and the propagation speed of a disturbance placed in the flowfield. Knowing that the disturbance front travel as an expanding spherical surface, we can analyze the conditions under which the spherical front reaches at an arbitrary point (x, y, z) of the flowfield.

As seen in Fig. 5.13, there are two different spherical surfaces which pass through point (x, y, z). Both of these spherical surfaces are the products of the same disturbance created at the origin at time τ and felt at the point (x, y, z) at two different times t_1 and t_2. With these timings the elapsed times are measured as $\Delta t_1 = t_1 - \tau$ and $\Delta t_2 = t_2 - \tau$. The point (x, y, z), first feels the disturbance at time t_1 fro the frontal side, and at later time t_2 the same point is effected by the back side of the disturbance. In other words, at the same instant t, point (x, y, z) feels the effect of two different disturbances created at the origin at two different times τ_1 and τ_2. This time, the durations elapsed between the creation of the disturbances and the present time t reads as $\Delta t_1 = t - \tau_1$ and $\Delta t_2 = t - \tau_2$. In both cases, point (x, y, z) lies in the Mach cone. Any point outside of Mach cone does not feel any disturbance, therefore that region is called the zone of silence. Now, we can obtain the relations between the relevant timings using Fig. 5.13 and the interpretations made above.

$$\begin{aligned} a(t - \tau_1) &= \sqrt{[x - U(t - \tau_1)]^2 + y^2 + z^2} \\ a(t - \tau_2) &= \sqrt{[U(t - \tau_2) - x]^2 + y^2 + z^2} \end{aligned} \tag{5.59a, b}$$

From Eq. 5.59a,b we can solve for τ_1 and τ_2 as follows.

$$\begin{aligned} \tau_1 &= t - \frac{1}{a\beta^2}(Mx - R) \\ \tau_2 &= t - \frac{1}{a\beta^2}(Mx + R) \end{aligned} \tag{5.60a, b}$$

Considering the chronological order of the timings we have $t > \tau_1 > \tau_2$. We have also obtained the solution for the potential of the supersonic flow with expression Eq. 5.58. If we want to obtain a continuous solution in time, we have to include both of the solutions of Eq. 5.60a,b for our perturbation potential as follows

$$\phi(x, y, z, t) = \bar{\phi}(e^{i\omega\tau_1} + e^{i\omega\tau_2})/R. \tag{5.61}$$

The solution we have obtained with Eq. 5.61 is going to be used as continuously distributed sources over the x-y plane to simulate the lifting surfaces undergoing simple harmonic motions. This distributed source has unknown distributed strength over the planform but in time it will oscillate simple harmonically. On the other hand, since the flow is supersonic, any point (x, y, z) in the flowfield is affected only from the points which are in its upstream Mach cone. We can now, write the integral expression accounting for all surface disturbances for the velocity potential as follows

$$\phi(x, y, z, t) = \iint_V A(\xi, \eta)[(e^{i\omega\tau_1} + e^{i\omega\tau_2})/R]d\xi d\eta \tag{5.62}$$

Here, A(ξ, η) is the unknown amplitude of the distributed source. The area over which the integral to be evaluated is a hyperbola defined as the intersection of the x-y plane with the upstream Mach cone of point (x, y, z), i.e. R = 0. This means, while z approaches zero we need to satisfy $(x - \xi)^2 - \beta^2(y - \eta)^2 - z^2 = 0$, which in turn means, it gives an equation of a hyperbola on x-y plane, because intersection of a cone with a plane parallel to its axis is an hyperbola. The hyperbola and the pertinent geometric variables are shown in Fig. 5.14. The lower and upper integration limits for the double integral of Eq. 5.62 shown in Fig. 5.14 are defined as η_1 and η_2 as follows.

$$\eta_{1,2} = y \mp \sqrt{(\frac{x - \xi}{\beta})^2 - z^2} \tag{5.63}$$

The integral limits in the chord direction can be considered as follows. We take the ξ = 0 point as the lower limit, and the apex of the hyperbola as the upper limit.

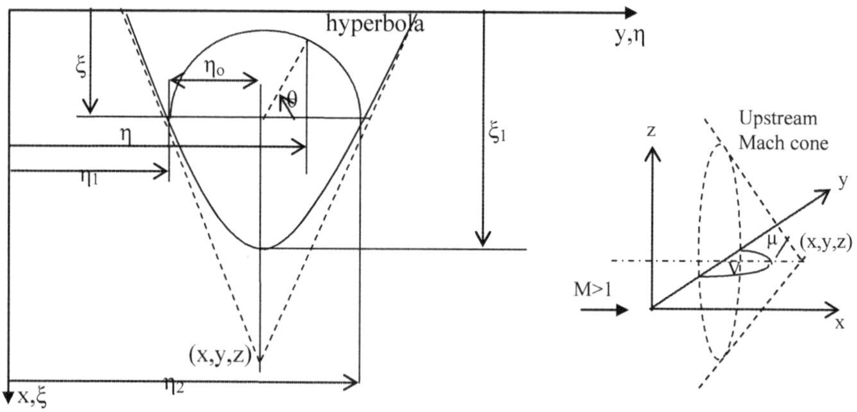

Fig. 5.14 Upstream mach cone, and the integration area V

Equation 5.63 tells us that $\eta_1 = \eta_2$ is the point for the upper limit, where the term under square root must be zero. That gives us

$$\xi_1 = x \pm \beta z \tag{5.64}$$

Here, according to Fig. 5.14 $x > \xi_1$, therefore in Eq. 5.64, +z is used for the lower surface, and –z is used for the upper surface. This information will be useful for evaluation of lifting pressure as lower minus upper surface pressure distributions.

Knowing the upper and lower limits of the integral Eq. 5.62, we can express

$$\phi(x, y, z, t) = \int_0^{\xi_1} \int_{\eta_1}^{\eta_2} A(\xi, \eta)[(e^{i\omega\tau_1} + e^{i\omega\tau_2})/R] d\xi d\eta \tag{5.65}$$

Here, the exponents read as

$$\tau_{1,2} = t - \frac{1}{a\beta^2}[M(x - \xi) \mp R)] \text{ and } R = \sqrt{(x - \xi)^2 - \beta^2 (y - \eta)^2 + \beta^2 z^2}. \tag{5.66a, b}$$

Values for η_1 and η_2 are used to give us $R = \beta\sqrt{(\eta - \eta_1)(\eta_2 - \eta)}$. In spanwise direction, as shown in Fig. 5.14, if we use variable θ we obtain $2\eta = (\eta_2 - \eta_1)\cos\theta + \eta_2 + \eta_1$. With this transformation, and defining $\eta_o = (\eta_2 - \eta_1)/2$ gives us

$$d\eta = -\eta_o \sin\theta d\theta \quad \text{ve} \quad R = \beta\eta_o \sin\theta$$

Here, again according to Fig. 5.14, $\theta = \pi$ at $\eta = \eta_1$, and $\theta = 0$ at $\eta = \eta_2$. If we rewrite integral Eq. 5.65 in terms of new variables, it becomes

$$\phi(x, y, z, t) = \frac{1}{\beta} \int_0^{\xi_1} \int_0^{\pi} A(\xi, y + \eta_o \cos\theta)(e^{i\omega\tau_1} + e^{i\omega\tau_2}) d\xi d\theta, \tau_{1,2} = t - \frac{1}{a\beta^2}[M(x - \xi) \mp \beta\eta_o \sin\theta] \tag{5.67}$$

The velocity potential given by Eq. 5.67 is differentiated with respect to z to obtain the downwash at the wing surface. First, let us employ Eq. 5.67 to study the supersonic flow past a profile.

5.9 Supersonic Flow About a Profile

In a supersonic flow any point in the flowfield is affected by the points lying inside of its upstream Mach cone. For this reason, unlike subsonic flow, even for a wing with infinitely long span we have to consider the three dimensional problem. This means, two dimensional analysis of a supersonic external flow cannot take into consideration the true physical behavior of a lifting surface while calculating its aerodynamic coefficients. However, for academic purposes, we are going to treat a wing section as a two dimensional flow case and obtain the lifting pressure coefficient for the sake of demonstrating the integrations involved in the process.

In a two dimensional flow the source strength remains constant in spanwise direction. For this reason the velocity potential Eq. 5.67 can be expressed as follows.

$$\phi(x,z,t) = \frac{1}{\beta}\int_0^{\xi_1} A(\xi)\int_0^{\pi}(e^{i\omega\tau_1}+e^{i\omega\tau_2})d\theta d\xi, \eta_0 = -\eta_1 = \eta_2 = \sqrt{(\frac{x-\xi}{\beta})^2 - z^2} \tag{5.68}$$

We can find the downwash expression by differentiating Eq. 5.68 with respect to z using the Leibnitz rule as follows.

$$\frac{\partial\phi}{\partial z} = \frac{1}{\beta}\int_0^{\xi_1} A(\xi)\int_0^{\pi}\frac{\partial}{\partial z}(e^{i\omega\tau_1}+e^{i\omega\tau_2})d\theta d\xi + \frac{1}{\beta}\frac{\partial\xi_1}{\partial z}A(\xi_1)\int_0^{\pi}(e^{i\omega\tau_1}+e^{i\omega\tau_2})d\theta \tag{5.69}$$

First term of right hand side of Eq. 5.69 as inner integral becomes

$$\frac{\partial}{\partial z}(e^{i\omega\tau_1}+e^{i\omega\tau_2}) = \frac{i\omega}{a\beta}\sin\theta\frac{\partial\eta_0}{\partial z}(e^{i\omega\tau_1}-e^{i\omega\tau_2}) = \frac{i\omega z}{a\beta\eta_0}\sin\theta(e^{i\omega\tau_1}-e^{i\omega\tau_2}) \tag{5.70}$$

The second term of Eq. 5.69, using the fact that, for $\xi = \xi_1$, $\eta_o = 0$ and $\tau_1 = \tau_2$,

$$(e^{i\omega\tau_1}+e^{i\omega\tau_2})_{\xi=\xi_1} = 2e^{i\omega\left[t-\frac{M}{a\beta^2}(x-\xi_1)\right]} \tag{5.71}$$

The relation Eq. 5.64 stated that depending on (x, y, z) being at the upper or lower surface of the wing we have

$$(\xi_1)_{u,l} = x \mp \beta z.$$

By differentiation we obtain

$$(\frac{\partial\xi_1}{\partial z})_{u,l} = \mp\beta \quad \text{and} \quad (e^{i\omega\tau_1}+e^{i\omega\tau_2})^{u,l}_{\xi=\xi_1} = 2e^{i\omega(t-\frac{Mz}{a\beta})} \tag{5.72}$$

We observe from Eq. 5.72 that the second integral of Eq. 5.69 is independent of θ to give

$$(\frac{\partial\phi}{\partial z})_{u,l} = \frac{i\omega z}{a\beta^2}\int_0^{\xi_1}\frac{A(\xi)}{\eta_o}\int_0^{\pi}(e^{i\omega\tau_1} - e^{i\omega\tau_2})\sin\theta d\theta d\xi \mp 2\pi A(\xi_1)e^{i\omega(t\mp\frac{Mz}{a\beta})} \tag{5.73}$$

Taking the limits for Eq. 5.73 for lower and upper surfaces the downwash value becomes

$$w_{u,l}(x,t) = \lim_{z\to 0^{\pm}}[(\frac{\partial\phi}{\partial z})_{u,l}] = \mp 2\pi A(x)e^{i\omega t} = \bar{w}_{u,l}(x)e^{i\omega t}. \tag{5.74}$$

Using Eq. 5.74 in Eq. 5.73 gives us the velocity potential value as follows

$$\phi_{u,l}(x,t) = \mp\frac{1}{2\pi\beta}\int_0^{\xi_1}\bar{w}_{u,l}(\xi)\int_0^{\pi}(e^{i\omega\tau_1} + e^{i\omega\tau_2})d\theta d\xi \tag{5.75}$$

In order to integrate Eq. 5.75, the integrand of the inner integral must be function of θ only. We can write the integrand as follows

$$\int_0^{\pi}(e^{i\omega\tau_1} + e^{i\omega\tau_2})d\theta = 2e^{i\omega[t-\frac{M}{a\beta^2}(x-\xi)]}\int_0^{\pi}\cos(\frac{\omega\eta_0}{a\beta}\sin\theta)d\theta \tag{5.76}$$

Integral Eq. 5.76 in terms of Bessel function reads as

$$J_0(z) = \frac{1}{\pi}\int_0^{\pi}\cos(z\sin\theta)d\theta$$

and it gives us the velocity potential as

$$\phi_{u,l}(x,t) = \mp\frac{1}{\beta}\int_0^{\xi_1}\bar{w}_{u,l}(\xi)e^{i\omega[t-\frac{M}{a\beta^2}(x-\xi)]}J_0(\frac{\omega\eta_0}{a\beta})d\xi. \tag{5.77}$$

We need to express Eq. 5.77 in nondimensional form while obtaining the lifting pressure coefficient for a thin airfoil. Here, we use the following nondimensional quantities

$$\xi^* = \xi/b, \eta^* = \eta/b, k = b\omega/U, \varpi = kM^2/\beta^2$$

This results in following expression for the velocity potential

$$\bar{\phi}_{u,l}(x,t) = \mp \frac{b}{\beta} \int_{0}^{\xi_1^*} \bar{w}_{u,l}(\xi) e^{-\varpi(x^*-\xi^*)} J_0(\frac{\varpi \beta \eta_0}{M}) d\xi^* \tag{5.78}$$

We know the pressure coefficient in terms of the velocity potential. For a simple harmonic motion the amplitude of the pressure coefficient in terms of the velocity potential reads as

$$\bar{c}_p = \frac{\bar{p} - p_\infty}{1/2\rho_\infty U^2} = -2(ik + \frac{\partial}{\partial x^*})\bar{\phi}_{u,l}/(bU)$$

Since the difference between the lower upper surface pressures gives the lifting pressure if we use Eq. 5.78 without the thickness effect, the lifting pressure coefficient in integral form becomes

$$\bar{c}_{p_a}(x^*) = \bar{c}_{p_l} - \bar{c}_{p_u} = -\frac{4}{\beta}(ik + \frac{\partial}{\partial x^*}) \int_{0}^{x^*} \frac{\bar{w}(\xi^*)}{U} e^{-i\varpi(x^*-\xi^*)} J_0[\frac{\varpi}{M}(x^* - \xi^*)] d\xi^*. \tag{5.79}$$

Let us give an example for unsteady motion of an airfoil in supersonic flow.

Example 1 For the free stream Mach number of 1.5 and the reduced frequency of 0.20 obtain the amplitude of lifting pressure coefficient for simple harmonically heaving airfoil with amplitude $\bar{h}^*$.

Solution With the aid of Mathematica the real and the imaginary parts of Eq. 5.79 can be evaluated along the chord using numerical integration. The graphs of real and imaginary parts of the lifting pressure are plotted below.

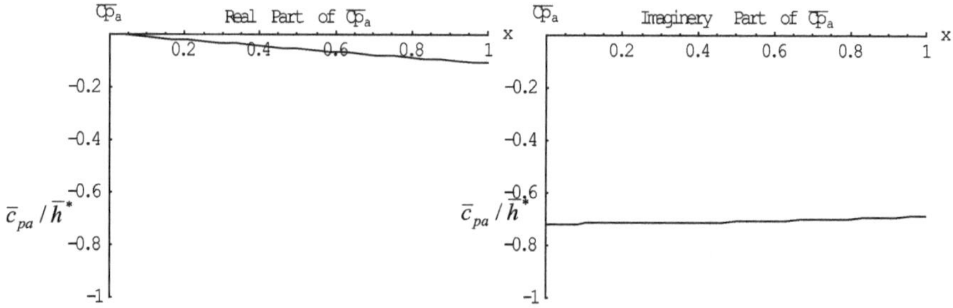

a) real part b) imaginary part

The integral of the real and imaginary parts give the amplitude of the sectional lift coefficient in terms of the heave amplitude as follows.

$$\bar{c}_l = (-0.055883 - 0.705385i)\bar{h}^*.$$

5.10 Supersonic Flow About Thin Wings

Equation 5.67 is given for the three dimensional velocity potential of a finite wing in a supersonic flow. Let us use this expression and the experiences we have had in handling the two dimensional supersonic flow to obtain the downwash expression. For this purpose we take the derivative of the integral expression Eq. 5.67 with respect to z using Leibnitz rule.

$$\begin{aligned}\frac{\partial\phi}{\partial z} &= \frac{1}{\beta}\int_0^{\xi_1}\int_0^{\pi}(e^{i\omega\tau_1}+e^{i\omega\tau_2})\frac{\partial}{\partial z}A(\xi, y+\eta_o\cos\theta)d\xi d\theta \\ &+ \frac{1}{\beta}\int_0^{\xi_1}\int_0^{\pi}A(\xi, y+\eta_o\cos\theta)\frac{\partial}{\partial z}(e^{i\omega\tau_1}+e^{i\omega\tau_2})d\xi d\theta + \frac{1}{\beta}\frac{\partial\xi_1}{\partial z} \\ &\int_0^{\pi}[A(\xi, y+\eta_o\cos\theta)(e^{i\omega\tau_1}+e^{i\omega\tau_2})]_{\xi=\xi}d\theta\end{aligned} \tag{5.80}$$

The derivative of the first term of Eq. 5.80 reads as

$$\frac{\partial}{\partial z}A(\xi, y+\eta_o\cos\theta) = -\frac{z}{\eta_0}\cos\theta\frac{\partial A(\xi,\eta)}{\partial\eta}$$

The second term is given by Eq. 5.70. For the multipliers of the third term we take $\xi = \xi_1$, $\tau_1 = \tau_2$ to obtain

$$\left(\frac{\partial\xi_1}{\partial z}\right)_{u,l} = \mp\beta, A(\xi, y+\eta_o\cos\theta)_{\xi=\xi_1} = A(\xi_1, y), (e^{i\omega\tau_1}+e^{i\omega\tau_2})^{u,l}_{\xi=\xi_1} = 2e^{i\omega(t\mp\frac{Mz}{a\beta})}$$

The third term now becomes enterable, and we finally have for the downwash

$$(\frac{\partial \phi}{\partial z})_{u,l} = -\frac{z}{\beta}\int_0^{\xi_1}\frac{1}{\eta_0}\int_0^{\pi}(e^{i\omega\tau_1}+e^{i\omega\tau_2})\frac{\partial}{\partial \eta}A(\xi,\eta)\cos\theta d\theta d\xi$$
$$+\frac{i\omega z}{a\beta^2}\int_0^{\xi_1}\frac{1}{\eta_0}\int_0^{\pi}(e^{i\omega\tau_1}-e^{i\omega\tau_2})A(\xi,\eta)\sin\theta d\theta d\xi \mp 2\pi A(\xi_1,y)e^{i\omega(t\mp\frac{Mz}{a\beta})} \tag{5.81}$$

Taking the limit of Eq. 5.81 while z approaches zero will give us the upper and lower downwash values as follows

$$w_{u,l}(x,y,t) = \lim_{z\to 0^{\pm}}[(\frac{\partial \phi}{\partial z})_{u,l}] = \mp 2\pi A(x,y)e^{i\omega t} = \bar{w}_{u,l}(x,y)e^{i\omega t} \tag{5.82}$$

In Eq. 5.67, the unknown source strength was A(x, y). If we use the downwash expression, Eq. 5.82, in Eq. 5.67 we obtain an integral expression for the unknown potential in terms of the boundary condition

$$\phi_{u,l}(x,y,z,t) = \mp\frac{1}{2\pi\beta}\int_0^{\xi_1}\int_0^{\pi}\bar{w}_{u,l}(\xi, y+\eta_o\cos\theta)(e^{i\omega\tau_1}+e^{i\omega\tau_2})d\xi d\theta \tag{5.83}$$

The exponents of Eq. 5.83 can be written Cartesian coordinates to give the following expression for the potential

$$\bar{\phi}_{u,l}(x,y,z) = \mp\frac{1}{\pi}\int_0^{\xi_1}\int_{\eta_1}^{\eta_2}\bar{w}_{u,l}(\xi,\eta)\frac{e^{-i\omega\frac{M}{a\beta^2}(x-\xi)}\cos(\frac{\omega R}{a\beta^2})}{R}d\xi d\eta \tag{5.84}$$

If you use the same nondimensional variables used for the two dimensional case, the amplitude of the velocity potential reads as

$$\bar{\phi}_{u,l}(x^*,y^*,z^*) = \mp\frac{b}{\pi}\int_0^{\xi^*_1 1}\int_{\eta^*_1}^{\eta^*_1}\bar{w}_{u,l}(\xi^*,\eta^*)\frac{e^{-i\omega\frac{M}{a\beta^2}(x^*-\xi^*)}\cos(\frac{\varpi R^*}{M})}{R^*}d\xi^* d\eta^* \tag{5.85}$$

We know the relation between the lifting pressure and the velocity potential. Using that relation we obtain

$$\bar{c}_{p_a}(x^*,y^*) = -\frac{4}{\pi}(ik+\frac{\partial}{\partial x^*})\iint_V \frac{\bar{w}(\xi^*,\eta^*)}{U}\frac{e^{-i\omega\frac{M}{a\beta^2}(x^*-\xi^*)}\cos(\frac{\varpi R^*}{M})}{R^*}d\xi^* d\eta^* \tag{5.86}$$

Here, the area integral V is the intersection between the wing surface and the upstream Mach cone, and R^* reads as

$$R^* = \sqrt{(x^* - \xi^*)^2 - \beta^2(y^* - \xi^*)^2}.$$

In this section, we have obtained the analytical relation for the lifting pressure coefficient as Eq. 5.86. We observe that for two and three dimensional supersonic flows the lifting pressure coefficient can be explicitly obtained from the downwash given at the surface as the boundary condition. This implies, unlike for the case of subsonic flows, determination of the surface pressure discontinuity does not require calculated inversion of an integral equation. Instead, the lifting pressure is computed with direct integration of Eqs. 5.79 and 5.86.

There are two different approaches in calculating the aerodynamic coefficients. The first method uses the analytical approach to calculate these coefficients for special geometries like delta wings. This method is good for rigid body motions like translation and rotation of wings (Miles 1959). The second approach is applicable to wider range of problems and they are approximate numerical methods. There are two different types of numerical methods. The first method is called 'The Supersonic Kernel Function Method', and the second one is the 'Mach Box Method'. Before studying these methods in detail, we have to classify the wing edges as subsonic or supersonic edges. The criteria to decide about the type of the edge depend on the relation between the sweep angle and the Mach angle as indicated in Fig. 5.15.

In Fig. 5.15a for the delta wing lying in the Mach cone of the flow, the component of the Mach number normal to the leading edge is less than 1, therefore, the leading edge shown is subsonic. On the other hand, the normal component of the Mach number at the trailing edge is greater than 1, therefore the trailing edge is supersonic. In Fig. 5.15b, only subsonic edge is the tip because the normal component of the Mach number is zero, whereas the leading and the trailing edges are outside of Mach cone and this makes those edges supersonic.

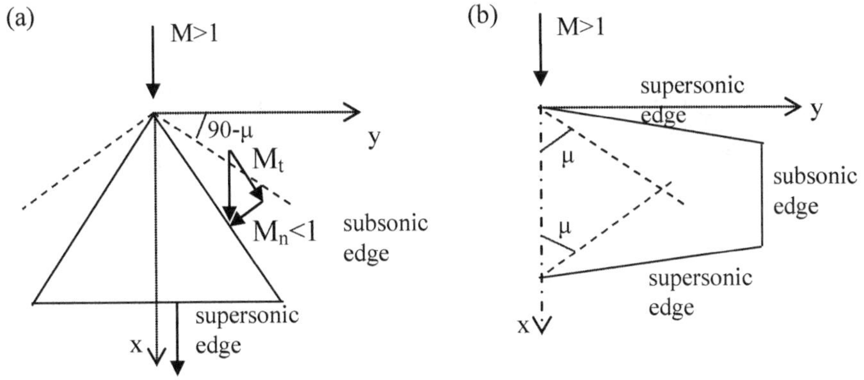

Fig. 5.15 Subsonic and supersonic edges of a wing

5.11 Mach Box Method

Mach Box Method is based on subdivision of the surface of the thin wing in a supersonic flow into small square boxes and summing the effect of all these boxes, as shown in Fig. 5.16, if all the edges of the wing are supersonic (Landahl 1968). In addition, if there are subsonic edges present using the concept of diaphragm, it can be applied to more general cases. First, as shown in Fig. 5.16, let us see its application on a delta wing with all supersonic edges. The boxes in the upstream Mach cone of point (x, y) have influence on the point (x, y) being the center of any square box on the wing surface. Let the jth box with the center coordinates (x_c, y_c) on wing has its vertical displacement as h_j, pitch angle α_j, and roll angle θ_j. The total displacement effect of the jth box on any point (x, y) reads as

$$z_j(x,y,t) = -[h_j(x_c,y_c) + (x - x_c)\alpha_j(x_c,y_c) + (y - y_j)\theta_j(x_c,y_c)]e^{i\omega t}$$

The boundary condition at the surface becomes

$$w(x,y,t) = -[i\omega h_j + i\omega(x - x_c)\alpha_j + i\omega(y - y_j)\theta_j + U\alpha_j]e^{i\omega t}. \quad (5.87)$$

Choosing the equally sized boxes small enough, the downwash for the jth box reads as

$$w_j = -(i\omega h_j + U\alpha_j)e^{i\omega t} = \bar{w}_j e^{i\omega t} \quad (5.88)$$

The amplitude of the lifting pressure coefficient was given in Eq. 5.86 in terms of downwash. If we sum the effect of all boxes influencing any point (x*, y*) on the wing surface the lifting pressure coefficient in summation notation reads as

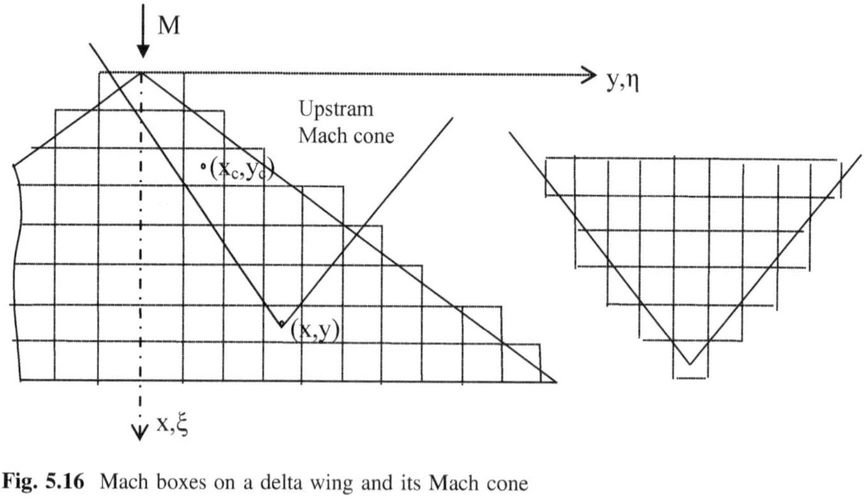

Fig. 5.16 Mach boxes on a delta wing and its Mach cone

$$\bar{c}_{p_a}(x^*,y^*) = -\frac{4}{\pi}\sum_j \bar{w}_j(ik + \frac{\partial}{\partial x^*}) \iint_{V_j} \frac{\bar{w}(\xi^*,\eta^*)}{U} \frac{e^{-i\omega\frac{M}{a\beta^2}(x^*-\xi^*)}\cos(\frac{\varpi R^*}{M})}{R^*} d\xi^* d\eta^* \tag{5.89}$$

Equation 5.89 is an expression with all complex elements. If it is written for all the boxes on the wing surface it will result in the following complex matrix equation.

$$\{\bar{c}_{p_a}\} = -\frac{4}{\pi}[R + iI]\{\bar{w}\} \tag{5.90a}$$

Here, the complex matrix $[R_j + i I_j]$ is the pressure influence coefficient matrix whose elements are the induced velocity at a point (x*, y*) of box V_j.

$$R_j + iI_j = (ik + \frac{\partial}{\partial x^*}) \iint_{V_j} \frac{e^{-i\omega\frac{M}{a\beta^2}(x^*-\xi^*)}\cos(\frac{\varpi R^*}{M})}{R^*} d\xi^* d\eta^*$$

The elements of this complex matrix depend only on the reduced frequency and the Mach number. Once the matrix is composed it can be used for all wing geometry which has supersonic edges. The lifting pressure coefficient at any point (x_i^*, y_i^*) on the surface can be evaluated without inversion of any matrix, i.e. explicitly as follows

$$\bar{c}_{p_a}(x^*,y^*) = -\frac{4}{\pi}\sum_j (R_j + iI_j)\bar{w}_j. \tag{5.90b}$$

While we are developing the Mach box technique we assumed that the lower and the upper surfaces of the wing do not interact with each other. This assumption holds if all edges of the wing is supersonic, since only for the supersonic edges no information can be carried from one surface to another. The picture, on the other hand, is different for the subsonic edges. The surface between the outside of the leading edge of a subsonic edge and the Mach cone is called the dyaphram surface, Fig. 5.17.

On the surface of diaphragm there is downwash with unknown value w_d, but there is no lifting pressure. This gives us the opportunity of partitioning Eq. 5.90-a as follows Hassig et al. (1969).

$$\begin{Bmatrix} \bar{c}_{p_a} \\ 0 \end{Bmatrix} = \begin{bmatrix} P_{ww} & P_{wd} \\ P_{dw} & P_{dd} \end{bmatrix} \begin{Bmatrix} \bar{w} \\ \bar{w}_d \end{Bmatrix} \tag{5.91}$$

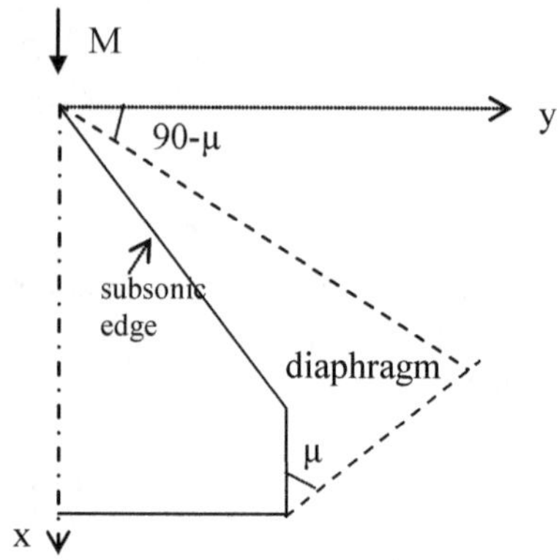

Fig. 5.17 Subsonic edge and the diaphragm

Here,

P_{ww}: Influence coefficient matrix of the wing on itself, square matrix
P_{wd}: Influence coefficient matrix of the diaphragm on the wing,
P_{dw}: Influence coefficient matrix of the wing on the dyaphram,
P_{dd}: Influence coefficient matrix of the diaphragm on itself, square matrix.

We can solve for the downwash value of the diaphragm using the second line of Eq. 5.91 as follows.

$$\{\bar{w}_d\} = -[P_{dd}]^{-1}[P_{dw}]\{\bar{w}\}$$

and substituting the result in 5.91 gives us the desired lifting pressure coefficient.

$$\{\bar{c}_{p_a}\} = \left\langle [P_{ww}] - [P_{wd}][P_{dd}]^{-1}[P_{dw}] \right\rangle \{\bar{w}\}. \quad (5.92)$$

We have mentioned before that there is another method for solving the supersonic flow past thin wings. That was 'Supersonic Kernel Function Method'. We are going to study that method next.

5.12 Supersonic Kernel Method

We have seen the kernel function method for subsonic flows. Now, we are going to use a similar approach to develop supersonic kernel method based on the acceleration potential. The relation between the acceleration potential and the velocity potential for simple harmonic motion was

$$\bar{\psi} = i\omega\bar{\phi} + U\bar{\phi}_x$$

and its solution was

$$\bar{\phi} = \frac{1}{U} e^{-i\omega x/U} \int_{-\infty}^{x} \bar{\psi}(\lambda, y, z) e^{i\omega\lambda/U} d\lambda$$

The acceleration potential was related to the lifting pressure and it was in the following form

$$\psi = A \frac{\partial}{\partial z} e^{i\omega[t - M(x-\xi)/a\beta^2 \pm R/a\beta^2]}/R, R = \sqrt{(x-\xi)^2 - \beta^2[(y-\eta)^2 + z^2]}$$

For the velocity potential we can write

$$\bar{\phi} = A \frac{\partial}{\partial z} e^{-i\omega(x-\xi)/U} \int_{-\infty}^{x-\xi} e^{i\omega(t + \lambda/U - M\lambda/a\beta^2 + R/a\beta^2)} \frac{d\lambda}{R}$$

We can write the downwash expression in terms of the velocity potential as follows

$$\bar{w}(x, y) = \lim_{z \to 0} \iint_S L(\xi, \eta) d\xi d\eta \frac{\partial^2}{\partial z^2} \int_{-\infty}^{x-\xi} \frac{e^{i\varpi(\lambda - M\sqrt{\lambda^2 + r^2})}}{\sqrt{\lambda^2 - r^2}} d\lambda, \varpi = -\omega/U\beta^2, r^2$$
$$= \beta^2[(y-\eta)^2 + z^2]$$

Here, L(ξ, η) is the unknown lifting pressure at the wing surface. In supersonic flow, we know that the acceleration potential is affected by two different wave front to give the solution in following form

$$\psi = A[\frac{\partial}{\partial z} e^{i\omega[t - M(x-\xi)/a\beta^2 - R/a\beta^2]}/R + \frac{\partial}{\partial z} e^{i\omega[t - M(x-\xi)/a\beta^2 + R/a\beta^2]}/R]$$
$$= 2A \frac{\partial}{\partial z} [e^{i(\omega t - M^2 \varpi x)} \frac{\cos(M\varpi)}{R}]$$

The supersonic kernel method is based on the numerical solution of a matrix system formed with finite number of points taken for the integral equation written above. It is applicable to the wing geometries having subsonic leading edges and supersonic trailing edges. The matrix system formed by the numerical integration is difficult to invert, however, with proper choice of pressure modes it becomes invertible (Cunningham 1966).

5.13 Arbitrary Motion of a Profile in Supersonic Flow

In a supersonic flow, the aerodynamic response of a profile to an arbitrary motion is function of the free stream Mach number and is obtained by means of the Wagner function and the Küssner function where the former is the response to a sudden angle of attack change and the latter is the response to an arbitrary gust. Here, we are going to use $\varphi(s)$ for the Wagner function and $\chi(s)$ for the Kussner function as usual.

With Wagner function and α_o:

$$c_L(s) = 2\pi\alpha_o\varphi(s) \tag{5.54}$$

Similarly, for a gust of magnitude w_o with Küssner function:

$$c_{Lg}(s) = 2\pi\alpha_o\frac{w_o}{U}\chi(s) \tag{5.55}$$

Here, s is the reduced time based on the half chord. The approximate exponential forms for the Wagner and Kussner function read as follows.

$$\begin{aligned}\varphi(s) &= b_1e^{-\beta_1 s} + b_2e^{-\beta_2 s} + b_3e^{-\beta_3 s}\\ \chi(s) &= \frac{c_{lg}(s)}{c_{lg}(\infty)} = 1 - b_1e^{-\beta_1 s} - b_2e^{-\beta_2 s}\end{aligned} \tag{5.93a, b}$$

Here, the numerical values for the coefficients and the exponents are given in Table 5.2 as provide by Mozalsky and O'Connell (1962).

Table 5.2 The change of Wagner and Kussner function with respect to the Mach number in supersonic flow

	M	b_1	b_2	b_3	β_1	β_2	β_3
	10/9	36.1	−6.62	−29.6	0.300	0.457	2.87
	10/8	−7.56	14.7	−7.15	0.450	0.530	0.700
$\varphi(s)$	10/6	−0.304	2.60	−2.32	1.00	1.38	1.62
	10/5	−0.170	0.997	−0.828	1.00	1.50	2.00
	M	b_1	b_2	β_1	β_2		
$\chi(s)$	$\sqrt{2}$	1.03	−0.0315	0.95	0.15		
	2.0	2.70	−1.70	1.15	0.969		

5.14 Slender Body Theory

The potential theory is also a tool for studying the flow about slender bodies. Here the aim is to calculate the aerodynamic forces and moments exerted on the body by the flowfield. Since the body is slender and the angle of attack considered is small, the small perturbation method is applicable for the case of compressible flows. For this reason, the equation we are going to use the linearized compressible small perturbation equation given with Eq. 2.24. Let us express Eq. 2.24 in terms of Mach number explicitly for the perturbation potential ϕ.

$$\frac{1}{a^2}\phi_{tt} + \frac{2M}{a}\phi_{xt} + (M^2 - 1)\phi_{xx} = \phi_{yy} + \phi_{zz} \tag{5.94}$$

For simple harmonic motion in terms of the amplitude and the frequency the perturbation potential becomes

$$\phi(x, y, z, t) = \bar{\phi}(x, y, z)e^{i\omega t}.$$

Equation 5.94 in terms of the amplitude and the frequency reads as

$$-\frac{\omega^2}{a^2}\bar{\phi} + i\omega\frac{2M}{a}\bar{\phi}_x + (M^2 - 1)\bar{\phi}_{xx} = \bar{\phi}_{yy} + \bar{\phi}_{zz}. \tag{5.95}$$

The nondimensional form of Eq. 5.94 is more convenient to interpret each term in terms of its order of magnitude. We can perform nondimensionalization using Fig. 5.18.

Let l be the length of the body and s be the maximum lateral dimension of the body. The nondimensional coordinates than defined as $\xi = x/l$, $\eta = y/s$, $\zeta = z/s$. If we write Eq. 5.95 for the nondimensional potential defined as $F(\xi, \eta, \zeta) = \bar{\phi}(x, y, z)/(Ul)$, the equation becomes

$$-Ul\frac{\omega^2}{a^2}F + i\omega\frac{2MU}{a}F_\xi + (M^2 - 1)\frac{U}{l}F_{\xi\xi} = \frac{Ul}{s^2}(F_{\eta\eta} + F_{\varsigma\varsigma}).$$

The right hand side of the equation above contains the Laplacian in transformed coordinates. Rearranging it gives

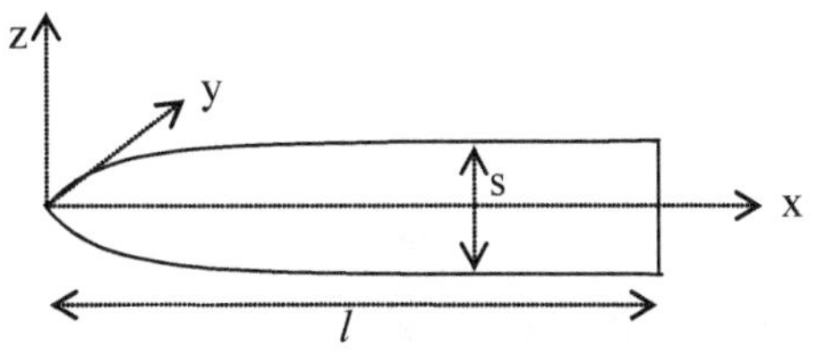

Fig. 5.18 Pertinent dimensions of a slender body

$$F_{\eta\eta} + F_{\varsigma\varsigma} = \frac{s^2}{l^2}\left[-l^2\frac{\omega^2}{a^2}F + i\omega\frac{2Ml}{a}F_\xi + (M^2-1)F_{\xi\xi}\right] \tag{5.96}$$

If we redefine the reduced frequency as $k = \omega l/U$, Eq. 5.96 becomes

$$F_{\eta\eta} + F_{\varsigma\varsigma} = (\frac{s}{l})^2\left[-k^2M^2F + 2ikM^2F_\xi + (M^2-1)F_{\xi\xi}\right] \tag{5.97}$$

For the slender bodies by definition ratio of the lateral dimension to its length is small, therefore $(s/l)^2 < < 1$. In addition if the following parameters k^2 M^2, k M^2 and M^2 is not too large, Eq. 5.97 becomes

$$F_{\eta\eta} + F_{\varsigma\varsigma} = 0. \tag{5.98}$$

Equation 5.98 is the Laplace's equation written involving lateral coordinates only. That is according to Eq. 5.98, the cross flow created by the slender body is incompressible. Equation 5.98 does not contain any term related to ξ, meaning that it seems the potential is independent of flow direction. However, we have to specify the boundary condition along the body surface which makes our potential depend on the low direction. The cross flow behaving incompressible enables us to use Munk's airship theory. This theory is used for finding the aerodynamic forces created by the momentum of the air parcel displaced by the body itself. Let us see the application of Munk's airship theory for steady and unsteady flows.

5.15 Munk's Airship Theory

The aerodynamic forces acting on the airships first was given in a study made by M. M. Munk during the first quarter of the last century (Munk 1924). As we know from the physics of the problem, the rate of change of the momentum of the airflow normal to the freestream is equal to the force crated in that direction. Using this principle, first we can calculate the change of the momentum of the air displaced vertically by an arbitrarily shaped body as shown in Fig. 5.19. Let w_a be the vertical velocity of the air parcel whose density is ρ, and S is the cross sectional area at a

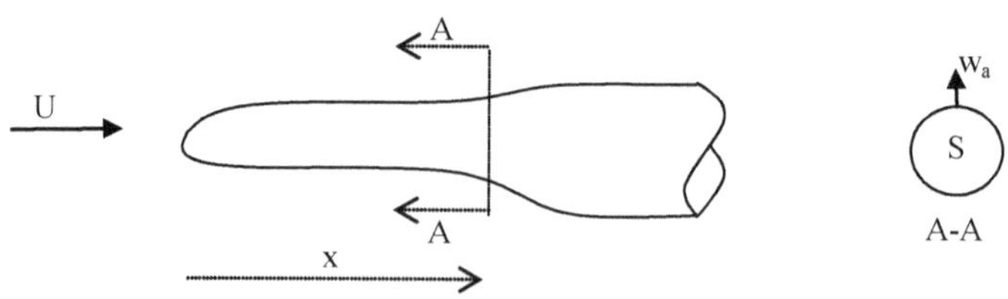

Fig. 5.19 Airship geometry

given station on the body. Since the cross sectional area changes with x, we denote the variables in their differential form as follows.

Let (ρS dx) be the differential air mass displaced vertically with a mean velocity w_a. Its differential momentum in z direction reads as

$$dI_z = (\rho S dx) w_a$$

Here, we assume the body is slender, therefore, we can take the mean vertical velocity value as the velocity of the axis of the body. Accordingly, if the position of the axis of the body is given with $z_a = z_a(x, t)$ the vertical velocity reads as

$$w_a(x,t) = \frac{\partial z_a}{\partial t} + U\frac{\partial z_a}{\partial x}$$

The differential momentum of the air parcel then becomes

$$dI_z = (\rho S dx)(\frac{\partial z_a}{\partial t} + U\frac{\partial z_a}{\partial x}) \qquad (5.99)$$

The rate of change of the momentum of the air is equal, but opposite in sign, to the vertical force acting in unit body length, i.e.,

$$\frac{dL}{dx} = -\frac{D}{Dt}(\frac{dI_z}{dx})$$

If we use Eq. 5.99 and assume that the air is incompressible for the cross flow, the vertical force for unit length becomes

$$\frac{dL}{dx} = -\rho\frac{DS}{Dt}(\frac{\partial z_a}{\partial t} + U\frac{\partial z_a}{\partial x}) - \rho S\frac{D}{Dt}(\frac{\partial z_a}{\partial t} + U\frac{\partial z_a}{\partial x})$$

If we assume that the cross section of the body does not change with time, and expand the substantial derivative in linearized form we obtain the lift change in x

$$\frac{dL}{dx} = -\rho U\frac{dS}{dx}(\frac{\partial z_a}{\partial t} + U\frac{\partial z_a}{\partial x}) - \rho S(\frac{\partial^2 z_a}{\partial t^2} + 2U\frac{\partial^2 z_a}{\partial x \partial t} + U^2\frac{\partial^2 z_a}{\partial x^2}) \qquad (5.100)$$

Equation 5.100 is in very good with experiments performed for the slender bodies. Writing Eq. 5.100 for steady flow gives

$$\frac{dL}{dx} = -\rho U^2\frac{dS}{dx}\frac{dz_a}{dx} - \rho U^2 S\frac{d^2 z_a}{dx} = -\rho U^2\frac{d}{dx}(S\frac{dz_a}{dx}) \qquad (5.101)$$

According to Eq. 5.101, for steady flows, to have a differential force normal to the free stream we need to have for the body:

(i) an angle of attack and the cross sectional area change,
(ii) if the cross sectional area does not change then there has to be a camber!

Now, let us show the normal force L(x) acting on the slender body with angle of attack and the variable cross sectional area shown in Fig. 5.20 as a typical slender body shape. Using Munk's theory, the aerodynamic force acting normal to the flight direction is at the front and back side of the body where there are cross sectional area changes. The same body does not experience any aerodynamical forces where the cross sectional area remains constant. The vertical forces may affect a short region on the body but the moment arms can be long, therefore effective pitching moments with respect to the center of gravity may be generated. These pitching moments, naturally, affect the stability of the body. One striking example to the unstable body is the one with the pointed nose and the pointed end where the forces create a continuous rotation about the center of gravity. The body shown in Fig. 5.20, on the other hand, has a flaired end to give a stable configuration (See Problem 5.32 for detailed description of stability criterion).

Summary: *Compressible flow past thin wings and slender bodies are studied. First, subsonic flow past thin wings are analyzed by means of potential flow theory. The kernel function method is introduced for arbitrary planforms undergoing simple harmonic oscillations. The spanwise polynomial approximation and the chordwise trigonometric function approximations which automatically satisfy the Kutta condition and inherit the leading edge singularity are used. The Doublet-Lattice method as a more general numerical approach is given for the analysis of the flow past additional surfaces like tail or store surfaces, which are not necessarily in plane with the wing surface or surfaces having spanwise deflections. Airfoil response to the arbitrary unsteady motion is also given for subsonic flows. A brief review of shock waves and Mach waves in a supersonic flow is given. Afterwards, unsteady supersonic potential flow is studied for a simple harmonically oscillating point source. First, unsteady flow past an airfoil is considered. Then, supersonic flow about thin wings are analyzed using the Mach box technique. Introduction to supersonic kernel function method is briefly provided. Arbitrary unsteady motion of an airfoil in a supersonic flow is presented. Slender body theory is introduced to analyze the cross flow past bodies of considerable fineness where the cross flow is shown to be approximately incompressible under certain conditions. In connection with the slender body approximation, the Munk's airship theory is utilized to predict the stability derivatives of missile like bodies.*

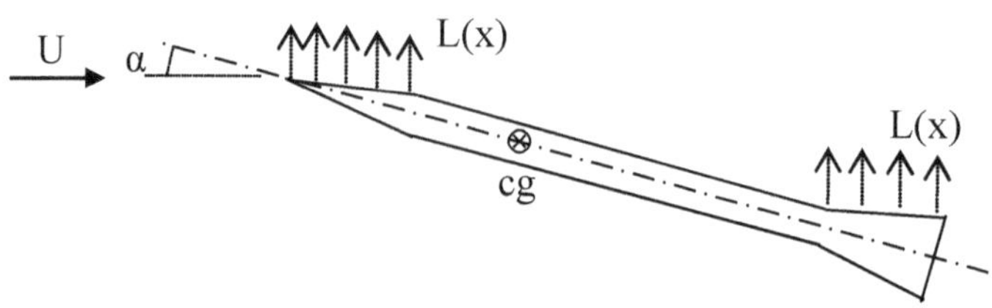

Fig. 5.20 L(x) force acting on the slender body

5.16 Questions and Problems

5.1 Show that Eq. 5.1 given for incompressible source satisfies the Laplace's Equation.

5.2 With Gallilean type transformation show that Equation 5.3 transforms into Laplace's Equation.

5.3 Show that $\bar{g}\frac{e^{\pm ik\bar{R}}}{\bar{R}}$ is a solution for the Helmholtz Equation.

5.4 Show that the perturbation potential equation expressed in moving coordinates for a steady compressible flow, satisfies Eq. 2.24 in original coordinates.

5.5 Find the solution given by Eq. 5.7 for the relation between the time, τ, of initiation of the disturbance and the present time t, in terms of Mach number.

5.6 Show that Eq. 5.10 given for the subsonic unsteady potential satisfies the differential equation given by Eq. 2.24.

5.7 Obtain the subsonic doublet expression Eq. 5.25 with taking the derivative of Eq. 5.24 with respect to z.

5.8 Obtain the relation Eq. 5.26 between the pressure discontinuity and the doublet strength using dimensional analysis.

5.9 Using the general solution for the first order differential equations obtain the relation between the velocity potential and the acceleration potential.

5.10 The choice of cot(θ/2) for the chordwise variation of the pressure coefficient gives the $1/\varepsilon^{1/2}$ type singularity at the leading edge and $\varepsilon^{1/2}$ type distribution at the trailing edge.

5.11 Obtain the amplitude distribution for the lifting pressure of an airfoil plunging simple harmonically with amplitude $\bar{h}$.

5.12 Comment on the term $\sqrt{1-\eta^2}$ written as the coefficient of the polynomial series expressed for the spanwise direction.

5.13 A thin wing with aspect ratio 3 is oscillating simple harmonically in bending mode with the following amplitude distribution along span

$$\bar{h}(y/l)) = 0.18043(y/l) + 1.70255(y/l)^2 - 1.13688(y/l)^3 + 0.25387(y/l)^4$$

For M = 0.24 and k = 0.47, using sufficient number of control points obtain the pressure discontinuity. Integrate the pressure discontinuity for the entire surface to find the amplitude of the lift.

5.14 Solve Problem 5.13 using Doublet-Lattice Method with taking 8 points in spanwise and 8 points in chordwise directions. Compare the results.

5.15 The wing-tail interaction problem for k = 0.75 and M = 0 is shown in Fig. P5.15.
For the simple harmonic plunging obtain the lift amplitude. Take 10 panels in spanwise direction with Multhopp distribution and 5 panels in chordwise direction. The distance between the wing and the tail at the root is 0.6 *x/l*.

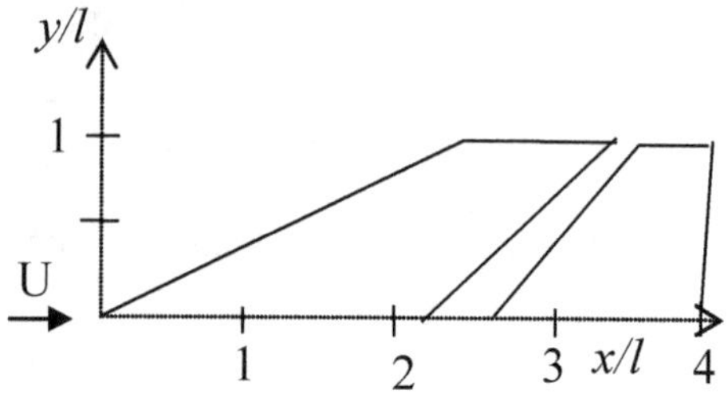

Fig. P5.15 Wing and tail

5.16 A wing with small aspect ratio and a store at the tip is shown in Fig. P5.16. For the simple harmonic plunging of the wing at k = 0.86 and M = 0 find the spanwise lift distribution

(i) wing alone,
(ii) wing plus the store. (take 9 for the wing and 2 panels for the store in spanwise and 5 chordwise panels).

5.17 For Mach number 0.6, obtain the lift response of the airfoil to the arbitrary motion given by Problem 3.10.
5.18 For Mach number 0.7, find the lift response of the airfoil subjected to the gust given by Problem 3.13.
5.19 Using the Lorentz transformation given for the supersonic flow obtain the classical wave equation for the linearized potential.
5.20 Show that the solution for the classical wave equation satisfied by the supersonic potential is given by Eq. 5.28.
5.21 Find the solution to the quadratic Eq. 5.59a,b to obtain two different times for the initiation of the disturbances, 5.59a,b.
5.22 For the supersonic flow, show that the area of the integration over the surface is hyperbola. The chordwise integration process has the upper surface denoted by –z, and the lower surface by +z in 5.64.

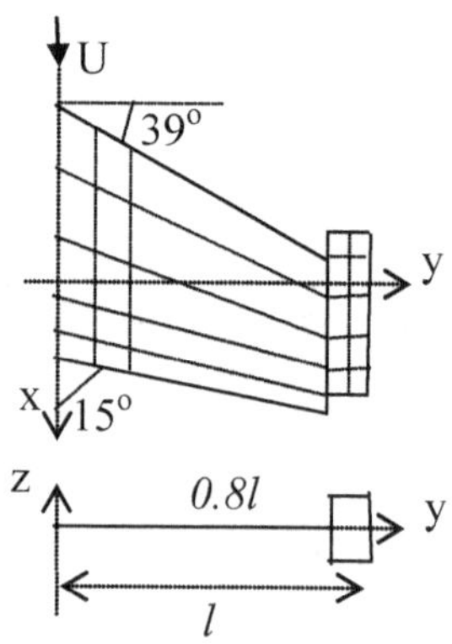

Fig. P5.16 Wing and store

5.23 Using the Euler's formula which gives the relation between the exponential function and the trigonometric functions, obtain the inner integral for the velocity potential expression given by 5.76.

5.24 Find the lifting pressure amplitude for a thin airfoil pitching about its leading edge with $\alpha = \bar{\alpha}e^{i\omega t}$ in M = 1.5. Obtain the lift amplitude also.

5.25 A delta wing with sweep angle of 45° undergoes a simple harmonic plunge oscillations at M = 1.054 at k = 0.525. Find and plot the real part of the velocity potential (i) at the root, (ii) at y/l = 0.6, and (iii) y/l = 0.9. (use 25 Mach box for half of the wing).

5.26 A 45° sweep angle delta wing is in the supersonic flow with M = 1.5.

(i) Divide the half of the wing with equally sized 25 boxes and give identification numbers for the boxes ranging from 1 to 25.

(ii) Use sufficient number of same size boxes to discretize the upstream Mach cone and number them separately with double digits. (use Fig. 5.16 as a sample).

(iii) Form the R + i I influence coefficient matrix as a 25 × 25 matrix which shows the effect of each box in the upstream Mach cone to the box on the wing.

5.27 The wing given by Problem 5.25 oscillates with the reduced frequency of k = 0.2. Obtain the lifting pressure curve for the spanwise change. Find the total lift coefficient.

5.28 Show that for a steady supersonic flow the lifting pressure coefficient becomes

$$C_{pa}(x,y) = -\frac{4}{\pi}\frac{\partial}{\partial x}\iint_{V}\frac{w(\xi,\eta)}{U}\frac{d\xi d\eta}{R}.$$

Find the lift line slope for the wing of Problem 5.27 using Mach box technique. Note that over each box the integral of the Kernel reads as

$\frac{\partial}{\partial x}\iint\limits_{Box}\frac{d\xi d\eta}{R} = \left[\left(-\arccos\frac{\bar{\eta}}{\bar{\xi}}\right)_{\bar{\eta}_l}^{\bar{\eta}_u}\right]_{\bar{\xi}_l}^{\bar{\xi}_u}$, where $\bar{\xi} = x - \xi$, $\bar{\eta} = \beta(y - \eta)$, $\beta = \sqrt{M^2 - 1}$

(If $\bar{\eta}/\bar{\xi} > 1$ then take:$\arccos\frac{\bar{\eta}}{\bar{\xi}} = 0$) and l and u stand for lower and upper integral limits respectively.

5.29 Find the spanwise variation of lift for a delta wing with sweep angle 60° at M = 1.5 and k = 0.2, at simple harmonic plunge. (Use 50 Max box for a half wing).

5.30 Obtain and plot the lift response for the arbitrary motion of the airfoil given in Problem 3.10 at M = 2.

5.31 Find and plot the lift response of the airfoil at Mach number of $\sqrt{2}$ experiencing the gust described in Problem 3.13.

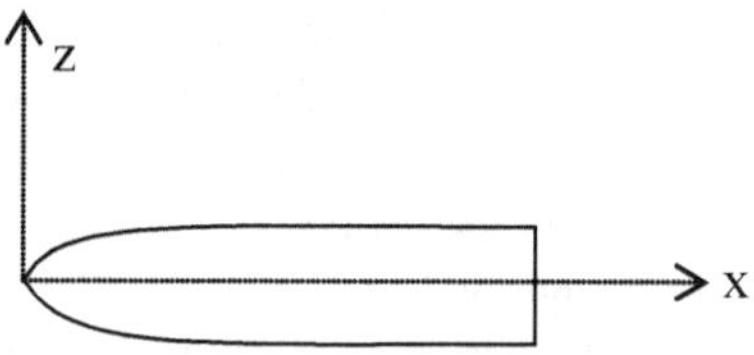

Fig. P5.33 Slender body

5.32 For a supersonic flow under which condition the root pressure distribution of a thin wing can be determined with 2-D analysis. Why?

5.33 The surface equation for a slender body shown in Fig. P5.33 is given by

$$z = 0.05\sqrt{x/l}, 0 \leq x \leq l.$$

Find the following derivatives of the aerodynamic coefficients,

$$i)\partial C_L/\partial\alpha,\ \ ii)\partial C_M/\partial\alpha,\ \ iii)\partial C_M/\partial(\dot{\alpha}l/U).$$

It is known that;

For the static stability: $\partial C_M/\partial\alpha < 0$,
For dynamic stability: $\partial C_M/\partial(\dot{\alpha}l/U) < 0$.

Using the stability criterion given above determine (i) static stability, and (ii) dynamic stability of the body assuming that it is made out of an homogeneous material.

References

Albano, E., Rodden, W.P.: A Doublet-Lattice method for calculating lift distributions on oscillating surfaces in subsonic flows. AIAA J. 279–285 (1969)

Ashley, H., Windall, S., Landahl, M.T.: New directions in lifting surface theory. AIAA J. 3–16 (1965)

Bisplinghoff, R.L., Ashley, H., Halfman, R.L.: *Aeroelasticity*, Dover Publications Inc., New York (1996)

Cunningham, H.J.: Improved numerical procedure for harmonically deforming surfaces from the supersonic kernel function method. AIAA J. 1961–1968 (1966)

Hassig, H.J., Messina, A.F., Twomey, W.J.: Using a partial diaphragm when applying the supersonic mach box method. AIAA J. 356–357 (1969)

Korn, G.A., Korn, T.M.: *Mathematical Handbook for Scientists and Engineers*, 2nd edn. McGraw-Hill, New York (1968)

Landahl, M.T.: Numerical lifting surface theory-problems and progress. AIAA J. 2050–2060 (1968)

Miles, J.W.: *The Potential Theory of Unsteady Supersonic Flow*, Cambridge University Press (1959)

Mozalsky, B., O'Connell, R.F.: Transient Aerodynamics of Wings, Lockheed Aircraft Corporation Report No. 11577 (1962)

Munk, M.M.: The Aerodynamic Forces on Airship Hulls, NACA Rep. 184 (1924)

Pierce, G.: Advanced Potential Flow I, Lecture Notes, Georgia Institute of Technology, School of Aerospace Engineering, Atlanta (1978)

Shapiro, A.H.: The Dynamics and Thermodynamics of Compressible Fluid Flow I. The Ronald Press Company, New York (1953)

Watkins, C.E., Runyan, H.L., Woolston, D.S.: On the kernel function of the integral equation relating the lift and downwash distributions of oscillating finite wings in subsonic flow. NACA Rep. 1234 (1955)

Chapter 6
Transonic Flow

In this chapter we are going to study a special case of an external flow for which the free stream speed of the flow is close to the speed of sound, i.e. the Mach number is about unity. Under this condition the flow is called 'transonic'. In transonic flows, the linearized version of the potential equation is not sufficient to model the flow; therefore, we resort to nonlinear but simplified version of the potential flow. The local linearization concept introduced by Dowell will be implemented for the series solution of the nonlinear transonic velocity potential. The local linearization technique enables us to study some simple steady and unsteady transonic aerodynamic problems analytically. Afterwards, we are going to study the examples for the numerical solution of the nonlinear potential equation introduced by Murman and Cole (1971) in their work which handles the transonic flow region with a suitable numerical scheme. In the rest of the chapter, numerical solutions for transonic flow studies with three dimensional unsteady Euler Equation solutions and the effect of viscosity with thin shear layer approach will be considered. Further unsteady topics of transonic flow will be provided in the chapter for Modern Topics.

6.1 Two Dimensional Transonic Flow, *Local Linearization*

The linearized potential Eq. 2.24b was obtained under the assumption that the difference between the free stream speed of sound and the local speed of sound a' was negligible. When the free stream Mach number M_∞ approaches unity this difference becomes important, therefore, it has to be taken into consideration for transonic flows. Integrating the linearized form of the energy equation, Eq. 2.24b along a streamline from the free stream to the point under consideration, gives the relation between the local speed of sound and the perturbation potential as follows

Ü. Gülçat, *Fundamentals of Modern Unsteady Aerodynamics*,
https://doi.org/10.1007/978-3-030-60777-7_6

$$\frac{a^2 - a_\infty^2}{\gamma - 1} = -(\frac{\partial}{\partial t} + U\frac{\partial}{\partial x})\phi' \tag{6.1}$$

Substituting Eq. 6.1 for the local speed of sound in Eq. 2.24a provides us the perturbation potential for the transonic flow in terms of free stream speed of sound in open form in following manner

$$\begin{aligned} &\phi'_{xx}\left[a_\infty^2 - U^2 - (\gamma - 1)(\frac{\partial}{\partial t} + U\frac{\partial}{\partial x})\phi' - 2U\frac{\partial \phi'}{\partial x}\right] + a_\infty^2(\phi'_{yy} + \phi'_{zz}) \\ &\quad - \frac{\partial}{\partial t}(\frac{\partial}{\partial t} + 2U\frac{\partial}{\partial x})\phi' \\ &\quad = 0 \end{aligned} \tag{6.2}$$

Here, the second derivatives are given in indicial notation. The first term of Eq. 6.2 makes the equation nonlinear. In addition, the difference expression $a_\infty^2 - U_\infty^2$ under the bracket of the first term becomes very small as Mach number goes to zero. This makes the first derivative terms to remain in the bracket. After this simplification of Eq. 6.2, we can divide it by a_∞^2 to obtain

$$\phi'_{xx}(1 - M_N^2) + (\phi'_{yy} + \phi'_{zz}) = \frac{1}{a_\infty^2}\frac{\partial}{\partial t}(\frac{\partial}{\partial t} + 2U\frac{\partial}{\partial x})\phi' \tag{6.3}$$

M_N^2 in Eq. 6.3 is given as

$$M_N^2 = M_\infty^2(1 + \frac{\gamma + 1}{U}\phi'_x).$$

Here, the time derivative in the first term is neglected; however, the equation is still nonlinear. Equation 6.3 still contains time dependent terms at its right hand side and it can be used in studying unsteady transonic flows.

Now, we are going to introduce the local linearization concept for the solution of Eq. 6.3. For this purpose let us separate the perturbation potential into its steady and unsteady components in following manner: $\phi' = \phi'_s + \phi'_d$, where subscript s denotes steady and d denotes the unsteady components. The steady component of the perturbation potential from the left hand side of Eq. 6.3 satisfies the following homogeneous equation.

$$(1 - M_N^2)\phi'_{sxx} + \phi'_{syy} + \phi'_{szz} = 0 \tag{6.4}$$

The unsteady component, on the other hand, reads as

$$\phi'_{dyy} + \phi'_{dzz} - \frac{1}{a_\infty^2}\phi'_{dtt} - 2\frac{U}{a_\infty^2}\phi'_{dxt} - e\phi'_{dxx} - f\phi'_{dx} = 0 \tag{6.5}$$

Here, $e = \left[M_\infty^2 - 1 + (\gamma+1)\frac{\phi'_{sx}}{U}\right], f = (\gamma+1)M_\infty^2\frac{\phi'_{sxx}}{U}$, and subscripts x, y, z denotes partial differentiation, whereas subscripts s and d stand for steady and unsteady components as before.

In Eq. 6.5, coefficients e and f contain the partial derivatives with respect to x only; therefore, they can be expressed in terms of the pressure coefficient for the steady part using Eq. 2.21 as $C_p = -2\phi'_x/U$. Let us remember once more that Eq. 6.5 is non linear because of e and f, Therefore, Eq. 6.5 can be solved either numerically or by means of local linearization (Dowell 1995). For two dimensional studies in x–z coordinates, the local linearization is made as follows. We first decompose the steady perturbation potential into its two different components as follows: $\phi' = \phi^0 + \hat{\phi}$. In addition, let us expand coefficients e and f into (x–x_0) power series as follows

$$e = \sum_{m=0}^{\infty} e_m(x - x_0)^m \quad ve \quad f = \sum_{m=0}^{\infty} f_m(x - x_0)^m$$

Now, for obvious reasons, the homogeneous equation is used with the non homogeneous boundary conditions, i.e.,

$$\phi^0_{zz} - e_0\phi^0_{xx} - f_0\phi^0_x = 0. \tag{6.6}$$

The non homogeneous equation to satisfy the homogeneous boundary conditions, i.e.

$$\hat{\phi}_{zz} - e_0\hat{\phi}_{xx} - f_0\hat{\phi}_x = (\phi^0_{xx} + \hat{\phi}_{xx})\sum_{m=1}^{\infty} e_m(x - x_0)^m + (\phi^0_x + \hat{\phi}_x)\sum_{m=1}^{\infty} f_m(x - x_0)^m \tag{6.7}$$

For a good approximation, we expect: $\phi^0 \gg \hat{\phi}$. In order to satisfy this condition the choice of x_0 plays an important role in determining coefficients e and f. Another important fact here is that the first terms of e and f are independent of x, which means the value of e_0 is a constant. Approaching from supersonic side will make the sign of e_0 positive, and subsonic side will make it negative. Now we know that free stream conditions being slightly supersonic causes the perturbations from a point to be felt only in downstream Mach cone of that point. Utilizing this fact enables us to use unilateral Laplace transform in x direction for Eq. 6.6 with prescribed non homogeneous boundary conditions. The definition of Laplace transform of ϕ^0 in x is given as

$$\overline{\phi}^0 = \int_0^\infty \phi^0(x,z)e^{-sx}dx$$

As is known the Laplace transform of a derivative of a function is given by initial conditions times the powers of s, powers being proportional with the order of the derivative. Applying this property for Eq. 6.6 gives us the following second order ordinary differential equation

$$\overline{\phi}^0_{zz} - (e_0 s^2 + f_0 s)\overline{\phi}^0 = 0 \tag{6.8}$$

Taking $\mu^2 = (e_0 s^2 + f_0 s)$ gives the solution of Eq. 6.8 as follows

$$\overline{\phi}^0(z) = Ae^{-\mu z} + Be^{\mu z} \tag{6.9}$$

In order to satisfy the diminishing radiation condition at infinity in Eq. 6.9 taking B = 0, for upper surface will be consistent with the physics of the problem. On the other hand, the coefficient A can be obtained from the transformed surface boundary condition. For this purpose we match the downwash, w, with the z derivative of the perturbation potential, expression 6.9, at $z = 0^+$ as follows

$$\frac{\partial\overline{\phi}^0}{\partial z}(z = 0^+) = \overline{w} \text{ and, } \overline{\phi}^0(z = 0^+) = \frac{-\overline{w}}{\mu} \tag{6.10a, b}$$

Here, inverse Laplace transform of Eq. 6.10 is performed to express the perturbation potential since the downwash and μ are expressed in s. Here, we have to note that μ^2 is always positive. The variable s is also by definition greater than zero. Hence, the convolution integral (Hildebrand) gives the inverse of Eq. 6.10 as follows

$$\phi^0(z = 0^+) = -\int_0^x e_0^{-1/2}\exp\left(-\frac{f_0\xi}{2e_0}\right)I_0\left(\frac{f_0\xi}{2e_0}\right)w(x-\xi)d\xi \tag{6.11}$$

Here, I_0, is the zeroth order first kind modified Bessel function (Appendix 5). As the surface boundary condition, the downwash w is prescribed; therefore, the pressure coefficient along the chord can be found by integral 6.11. As Mach number goes to 1, the value of e_0 approaches infinity; however, the exponential function of the integrand and the Bessel function simplify the integral 6.11 as follows

$$\phi^0(z = 0^+) = -\int_0^x f_0^{-1/2}(\pi\xi)^{-1/2}w(x-\xi)d\xi \tag{6.12}$$

Let us determine the perturbation potential for a thin airfoil at α = constant angle of attack for the upper surface, $z = 0^+$, using 6.11. The downwash, $w = -U\alpha$, gives the integral 6.11

$$\phi^0(z=0^+)\frac{1}{U\alpha} = \frac{2e_0^{1/2}}{f_0}\xi\exp(-\xi)[I_0(\xi)+I_1(\xi)], \quad \xi = \frac{f_0}{2e_0}x \tag{6.13}$$

The upper surface pressure coefficient reads as

$$\frac{C_p^o}{\alpha} = -2e_0^{-1/2}\exp(-\xi)I_0(\xi) \tag{6.14}$$

The lower surface, on the other hand, has negative z value; therefore, in Eq. 6.9 the radiation condition is applied accordingly to obtain the following lower surface pressure value

$$\frac{C_p^o}{\alpha} = 2e_0^{-1/2}\exp(-\xi)I_0(\xi) \tag{6.15}$$

Equations 6.14–6.15 give the lifting pressure expression as follows

$$\frac{\Delta C_p^o}{\alpha} = 4e_0^{-1/2}\exp(-\xi)I_0(\xi) \tag{6.16}$$

Integrating 6.16 along the chord provides us the sectional lift coefficient as

$$c_l^0 = 4\alpha(f_0 b)^{-1/2}\overline{b}^{1/2}\exp(-\overline{b})\left[I_0(\overline{b})+I_1(\overline{b})\right], \quad \overline{b} = \frac{f_0 b}{e_0}. \tag{6.17}$$

Here, b represents the half chord.

Since we know the perturbation potential component, φ^o related to the non homogeneous boundary conditions, we can now write the following equation for $\hat{\varphi}$ using Eq. 6.5

$$\hat{\phi}_{zz} - f_o\hat{\phi}_z - e_o\hat{\phi}_{xx} = e_1(x-x_o)\phi^o_{xx} \tag{6.18}$$

In order to solve Eq. 6.18, we again take the Laplace transform of it with respect to x coordinate to obtain a non homogeneous second order ordinary differential equation

$$\overline{\hat{\phi}}_{zz} - \mu^2\overline{\hat{\phi}} = -e_1\left(2s\overline{\phi}^o + s^2\frac{d\overline{\phi}^o}{ds} + x_o s^2\overline{\phi}^o\right) \tag{6.19}$$

As a technique, first we solve the homogeneous part of 6.19, and then obtain the non homogeneous solution. The homogeneous part is solved exactly like 6.9. After finding the general solution of 6.19 under homogeneous boundary conditions we can take the inverse Laplace transform of the result to obtain the solution in x coordinates. The downwash expression, $w = -U\alpha$, from the surface boundary condition gives us the perturbation potential and that in turn provides us the surface pressure coefficient as follows

$$\hat{C}_p \cong \alpha e_0^{-3/2} \exp(-\xi)[2\xi(I_1 - I_o) + I_o] \tag{6.20}$$

In Fig. 6.1, plots of C_p^0 given by 6.15, $\hat{C}_p$ given by 6.20, and their summation as the total surface pressure distribution for C_p is shown for Mach number 1. On the same figure also shown is the surface pressure obtained by Dowell using Stahara-Spriter's calculations for 6% thick Guderly airfoil (Appendix 6). In this comparison, coefficients e and f are expanded into the series about $x_o = 2b$. Here, value of x_0 is chosen arbitrarily; therefore, to calculate the pressure coefficient properly calibration with other methods is necessary, (see Problem, 6.9).

6.2 Unsteady Transonic Flow, Supersonic Approach

We have demonstrated before that for the supersonic approach $e_0 > 0$ is the restriction. If we consider simple harmonic motion, two dimensional form of Eq. 6.8, with Laplace transformation of x coordinate, becomes

$$\overline{\phi}^0_{zz} - \mu^2\overline{\phi}^0 = 0, \quad \mu = e_o s^2 + \overline{f}_o s - d \tag{6.21}$$

Here, ω is the angular frequency, and

$$\hat{f}_o = f_o + \frac{2U}{a_\infty^2} i\omega, \quad d = (\omega/a_\infty)^2.$$

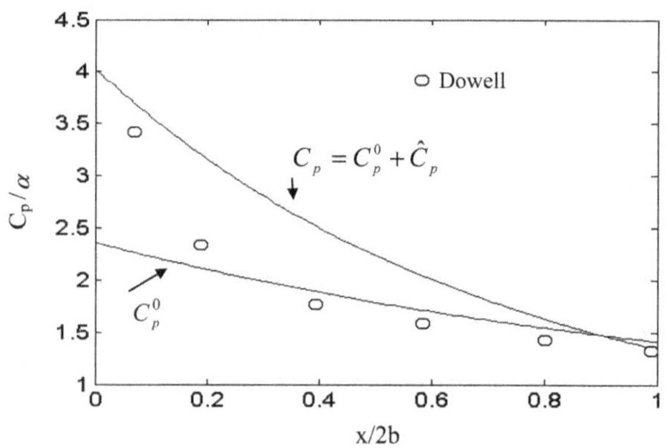

Fig. 6.1 Surface pressure coefficient obtained with local linearization at $M_\infty = 1$ (with $x_0 = 2b, f_o = 2.4/b, e_0 = 0.72$)

The solution of Eq. 6.21 in the Laplace transformed domain can be performed similar to that of 6.8. The inverse transform gives us the solution ϕ^o in x coordinates together with the prescribed boundary conditions as follows

$$\phi^0(z=0) = -\int_0^x e_0^{-1/2}\exp(-\frac{\hat{f}_0\xi}{2e_0})I_0\left\{\left[(\frac{\hat{f}_o}{2e_o})^2+\frac{d}{e_o}\right]^{1/2}\xi\right\}w(x-\xi)d\xi e^{i\omega t} \tag{6.22}$$

For a profile oscillating vertically with $\dot{h}$, Eq. 6.22 becomes

$$\phi^0(z=0) = \dot{h}e_0^{-1/2}\left[(\frac{\hat{f}_o}{2e_o})^2+\frac{d}{e_o}\right]^{1/2}\exp(-e\overline{x})\overline{x}\left[I_o(\overline{x})+\frac{I_1(\overline{x})}{e}\right] \tag{6.23}$$

Here we define:

$$\overline{x} = \left[(\frac{\hat{f}_o}{2e_o})^2+\frac{d}{e_o}\right]^{1/2}x, \quad e = \frac{\hat{f}_o}{2e_o}\left[(\frac{\hat{f}_o}{2e_o})^2+\frac{d}{e_o}\right]^{1/2}$$

We can take the limit of Eq. 6.23 as $M_\infty = 1$ to obtain the expression for the perturbation potential

$$\phi^0(z=0) = 2\dot{h}(\frac{x}{\hat{f}_o\pi})^{1/2} \tag{6.24}$$

Equation 6.24 is used to obtain the pressure coefficient for a simple harmonic vertical oscillation of a thin airfoil

$$C_p = (\pi\hat{f}_o 2b)^{-1/2}\left[-2k^2(x/2b)^{1/2}+2ik(x/2b)^{-1/2}\right]\frac{\overline{h}}{b}e^{i\omega t} \tag{6.25}$$

Here, the motion is prescribed as $h = \overline{h}e^{i\omega t}$, and the reduced frequency is defined as $k = \omega b/U$.

Example 6.1 Find the amplitude of the surface pressure for a profile in simple harmonic vertical oscillation at $M = 1$ and $k = 0.25$.

Solution Let us take $x_0 = b$, $e_0 = 0.12$ and $f_0 = 2.4/b$ to find the surface pressure from Eq. 6.25. The plot of surface pressure coefficient is given in Fig. 6.2 using the complex amplitude given by 6.25. Also shown in Fig. 6.2 is the results of Stahara-Sprieter, given in Dowell, for 6% thick Guderly profile. Comparisons of the graphs indicate that solution with 6.25 is in good agreement with the reference values.

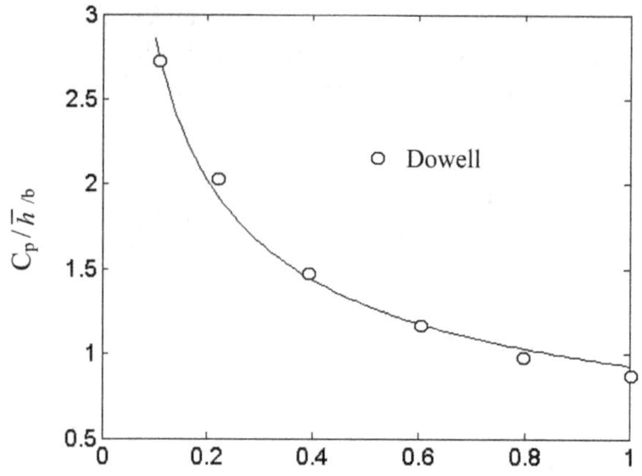

Fig. 6.2 Amplitude of the surface pressure for vertical motion at $M_\infty = 1$

So far we have obtained the transonic steady and unsteady solutions based on local linearization with neglecting the thickness effects. In next section we are going to see the numerical solution of nonlinear transonic flow equation with thickness effects.

6.3 Steady Transonic Flow, Non Linear Approach

Equation 2.15 is the non linear equation which is satisfied by the velocity potential. If we omit the time dependent terms of Eq. 2.15, we obtain the following non linear equation for the velocity potential in two dimensional steady flows

$$(a^2 - \phi_x^2)\phi_{xx} + (a^2 - \phi_y^2)\phi_{yy} - 2\phi_x\phi_y\phi_{xy} = 0 \qquad (6.26)$$

Here, a denotes the local speed of sound and the subscripts denote the partial differentiation with respect to x or y. The character of Eq. 6.26 changes depending on the speed of sound in a transonic flow. If speed of sound is higher than the local flow speed, then the equation is elliptic, and it becomes hyperbolic if the flow speed exceeds the speed of sound. For this reason, the solution of Eq. 2.26 can be obtained either with specific analytical methods for specific profile shapes or numerical methods for arbitrarily shaped airfoils. In general, for a profile with a thickness immersed in a high subsonic free stream the flow speed increases due to thickness effects until reaching the sound line where the flow speed is equal to local speed of sound. After maximum thickness, the supersonic flow expands and speeds up while its pressure drops down. Before reaching the trailing edge there is a sudden increase in the pressure so that the flow pressure eventually reaches the wake pressure. This sudden pressure increase is a normal shock, which is in harmony with the physics of the flow for transition from supersonic to subsonic flow regime. However, for very special geometries it is possible to have shockless transonic flow via inverse design

(Nieuwland and Spee 1968). First shockless transonic flow was studied for symmetrical quasi elliptical profiles (Baurdoux and Boerstoel 1968), and thereafter these techniques were developed for non symmetrical airfoils called supercritical airfoils (Whitcomb 1956; Bauer et al. 1972, 1975).

Now, as an example to the shockless transonic external flow we can obtain the surface pressure coefficient of the symmetrical quasi elliptical profile with finite element solution of Eq. 6.26. Since Eq. 6.26 is a nonlinear equation we have to solve it with an iterative technique. In addition, it has an elliptic-hyperbolic character; therefore, the information in the elliptic region must be carried in all directions. However, in the hyperbolic region the information must travel only in downstream of the node concerned. This forces us to use artificial viscosity with a proper control system while forming the coefficient matrix. This means for the elements in the supersonic region only the information travelling in downstream is permitted, otherwise it is eliminated. This approach gives us a convergent iterative scheme for the solution of the velocity potential (Ecer et al. 1977). In Fig. 6.3 the finite element results with quadrilaterals are compared with the analytical solution for the surface pressure variation C_p of a quasi elliptical Nieuwland profile. Although, a course grid is used, 31 × 11, in computations a good agreement with the analytical solution is achieved in subsonic region and a satisfactory agreement is observed after the critical pressure where local Mach number exceeds unity. Shown in Fig. 6.4a is the discretized flow field for the finite element solution. The 50 step iteration convergence history of the subsonic and the supersonic surface pressure values are given in Fig. 6.4b.

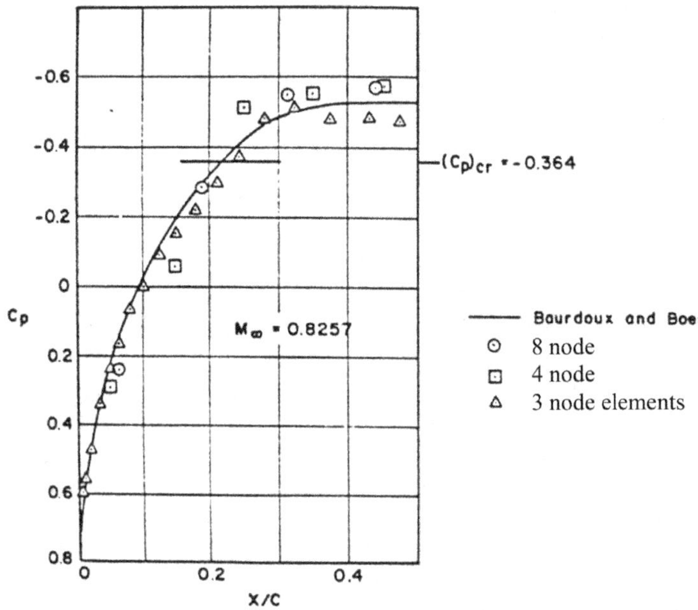

Fig. 6.3 Surface pressure distribution for %10.76 thick Nieuwland profile at M = 0.8257

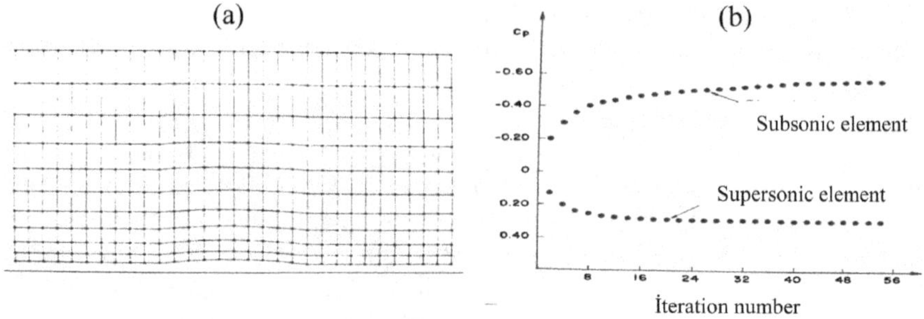

Fig. 6.4 For the symmetrical quasi elliptical airfoil, **a** finite element grid, **b** convergence

The same elliptic-hyperbolic mixed problem was also solved by Murman and Cole using finite difference method with much finer grid on the surface of the airfoil. Their solution agrees well with the analytical solution since they use 50 points on the surface to increase the accuracy. However, their solution required more CPU time. Murman and Cole also considered the off-design behavior of the profile by giving solutions obtained for the free stream Mach numbers slightly different from the design Mach number.

The purpose behind analyzing the transonic flows in detail lies in designing new profiles either without shock or with very weak shock at high subsonic Mach numbers. It is a well known fact that if there is a shock on the surface of the profile at subsonic free stream Mach numbers, the drag coefficient becomes the double of the shockless case. The cause of this drag rise is the shock induced boundary layer separation and the entropy rise across the shock. On the other hand, if the shock occurrence on the airfoil surface is delayed with the increasing of the free stream Mach number, then the lift coefficient will rise while the drag coefficient almost remains the same. Now, we can compare qualitatively the upper and lower surface pressure coefficients and sonic lines for the conventional and supercritical profiles in Fig. 6.5.

As seen in Fig. 6.5a, there is a strong shock present at the upper surface of the conventional airfoil to cause a boundary layer separation whereas at a considerable higher free stream Mach number the supercritical airfoil has weak shocks at the lower and upper surfaces without any flow separation. In conventional airfoils the critical Mach number is reached for lower free stream speeds with lift loss and drag increase as opposed to the supercritical airfoils for which the critical Mach number and the lift is higher and the drag is lower. As an example for a classical NACA airfoil when the free stream Mach number is increased from 0.65 to 0.69 the drag coefficient increases 50%. For a supercritical airfoil, on the other hand, the drag coefficient increases only 10% for the Mach number increase from 0.65 to 0.79 and for M = 0.80 it goes back to the value that was attained at M = 0.65 (Whitcomb and Clark 1956). However, if the free stream speed exceeds the design value of 0.80, the drag coefficient shows a sudden increase. This means one should expect poor performance from the supercritical profiles at off design conditions.

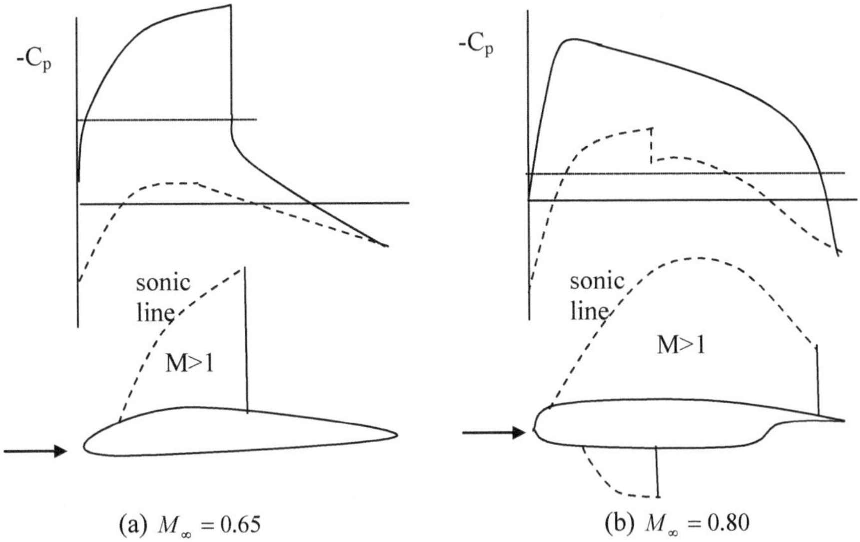

Fig. 6.5 Transonic flow, **a** conventional profile, strong shock, **b** supercritical profile, weak shock

6.4 Unsteady Transonic Flow: General Approach

Previously, we have given the surface pressure coefficient variation for a vertically oscillating thin airfoil with Eq. 6.25 and the amplitude variation along the chord with Fig. 6.2. Now, the real and the imaginary parts of the surface pressure along the chord will be given by Fig. 6.6. Although, Eq. 6.25 which is based on the local linearization does not indicate the presence of the shocks at the leading and trailing edges, it gives agreeable results with experimental pressure measurements. In order to describe the behavior of the surface pressure distribution of a thin airfoil in unsteady transonic flow, the effect of the increase in free stream Mach number must be considered.

In this respect, it is possible to summarize and classify transonic flow conditions for a thin airfoil pitching in oscillatory motion with illustrations similar to that given in Fig. 6.7a–c, based on the experimental and the computational results, obtained for the surface pressure (adapted from McCroskey). The low transonic flow conditions as shown in Fig. 6.7a indicate the presence of a shock on the instantaneous surface pressure distribution C_p^0, and in a periodic motion the real and imaginary parts of the surface pressure C_p^1 depicts a slightly moving shock which is called 'shock doublet' in literature. In addition, the appearance of a strong shock on the upper surface of the airfoil causes boundary layer separation. In Fig. 6.7b shown are the high transonic flow conditions where the free stream Mach number is very close to 1 and the real and imaginary parts are quite similar to that of Fig. 6.6 for which the local linearization technique is implemented. In the flow field for this case, we see the presence of λ shocks around the trailing edges of the upper and

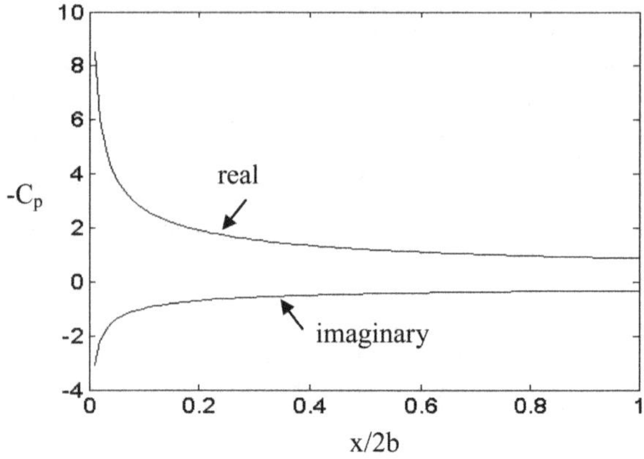

Fig. 6.6 Real and imaginary parts of the C_p distribution along the chord for an oscillating thin airfoil at $M_\infty = 1$

lower surfaces. According to Fig. 6.7c in low supersonic flow regime, the instantaneous surface pressure distribution remains almost constant except around the leading edge which is the same for the unsteady surface pressure distribution. In the flow field of a low supersonic flow, a separated bow shock at the leading edge, and around the trailing edge a fork shaped shock at the upper surface and an expansion fan at the lower surface are present.

In unsteady flows when the viscous effects are negligible i.e. when there is not any shock separated boundary layer flow, the movement of the shock wave is observed for the low and moderate reduced frequencies. In these cases because of shock movement the linearized approach is not suitable. In high reduced frequencies, since the shock movement is not that high, it is possible to use linearized approach (McCroskey 1982). When the linear theory is not applicable either the local linearization or the full non linear potential equation is to be solved. For the cases of strong shocks the presence of vortices forces us to resort to the solution of Euler equations.

In cases of strong viscous effects the presence of flow separation in transonic flows causes some unsteady phenomena such as ‘flutter’, ‘buffeting’ and ‘aileron buzz’ to happen. The flutter phenomenon as a shock induced separation occurring with the shock movement was first observed experimentally with the forced pitching oscillation of profiles. The self induced periodic shock movement on a thick biconvex airfoil in a transonic flow at zero angle of attack was first observed with numerical solutions, and then it was also observed experimentally for certain free stream Mach numbers and frequencies (McCroskey 1982). These observations were useful mostly for the assessment of transonic buzz which indicates the regular response of the structure to the aerodynamic effects. In 1970s, it was possible to predict experimentally the onset of buffeting for a profile with respect to the Mach number and sectional lift coefficient (Kücheman 1978). Shown in Fig. 6.8 is the enveloping curve b) for the onset of buffeting depending on the free stream Mach number and the sectional lift coefficient of a profile. The conditions for the onset of buffet depending on frequency spectrum of the surface pressure oscillations,

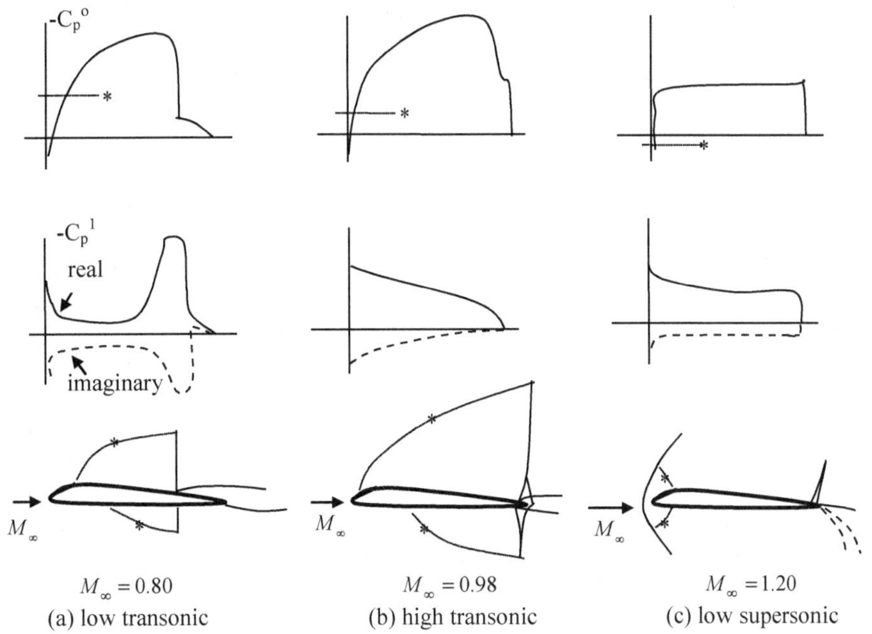

Fig. 6.7 Surface pressure distribution and the flow fields for different transonic Mach numbers (*indicates sonic conditions.)

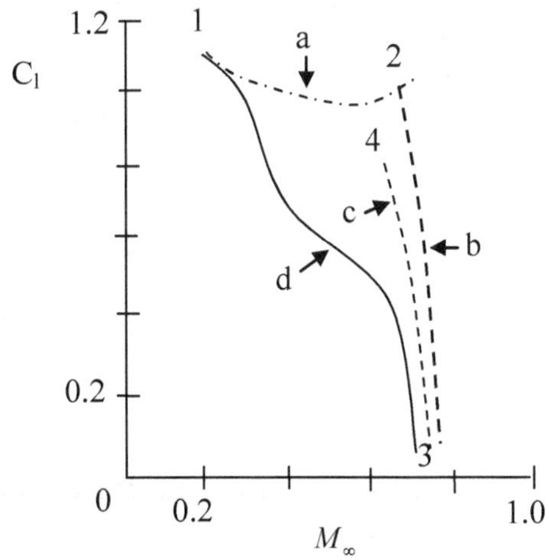

Fig. 6.8 Effect of freestream Mach number and the sectional lift coefficient on **a** flow separation, **b** buffeting, **c** drag divergence and critical Mach number variation

induced frequency and the size of the separation were experimentally determined starting from 1980's. On the other hand, starting from 1990's it has been possible to establish these conditions with numerical solution of Navier-Stokes equations using proper turbulence models. This requires very fine resolution for the computational grids so that the first point away from the surface lies in the viscous sublayer (Isogai 1992). In 2000's the numerical solution obtained with different turbulence models enabled researchers to predict the onset of buffet for NACA0012 at various angles of attack and free stream Mach numbers (Barakos and Drikakis 2000). According to Barakos and Drikakis for the Reynolds number range of 10^6–10^7 the %12 thick symmetric profile at zero angle of attack does not undergo any buffet up to the free stream Mach number of 0.8. On the other hand, at 1° angle of attack and at 0.8 Mach number, and at angle of attack range 2–4° for lower Mach numbers like 0.775 and 0.725 the buffeting starts. Most recent numerical studies on a supercritical airfoil, NLR7301, indicate buffeting at 0.5° angle of attack and in the free stream Mach number range of 0.82–0.83 and Reynolds number range of 1.943×10^6–1.954×10^6! This range is called 'transonic dip' and outside of this range no buffet is encountered (Geistler 2003).

Another unsteady transonic phenomenon is the 'aileron buzz' and it is due to a shock doublet created by the shock movement which causes hinge moments with dissipation at the hinges of aileron. The onset of buzz can happen with weak viscous effects but its maintenance requires strong viscous effects (McCrosky). Both numerical and experimental results enable us to predict the boundaries of buzz with angle of attack and free stream Mach number. Accordingly, the buzz is encountered at lower transonic Mach numbers with increasing angle of attack. In recent years, numerical solution of Navier-Stokes equations performed for designing a 'Supersonic Commercial Plane' by the Japanese National Aerospace Laboratories gives a detailed study of aileron buzz (Yang et al. 2003). In the work of Yang et al., oscillation of an aileron of a wing at a zero angle of attack attached to a fuselage is studied numerically as fluid-structure interaction problem based on an aeroelastic-aerodynamic solution. In their study, a moving deforming grid is employed together with structural damping of the elastic wing. The elastic wing at free stream Mach number of 0.98 indicated undamped aileron buzz to increase the amplitude of oscillations in such a way that eventually the numerical solution diverged. During the diverging of the numerical solution, the amplitude of the oscillation of the angle rises from, 1° to 2° in one cycle. For the case of the rigid wing, however, the same flow conditions caused damping for the aileron oscillations. For the free stream Mach number ranging from 0.95 to 1.02, the aileron oscillations showed damping behavior even for the elastic wing!

6.5 Transonic Flow Around a Finite Wing

Aerodynamically and practically useful three dimensional transonic analyses over finite wings date back to 1940's with implementation of swept wing concept (Polhamus 1984). According to the information given by Polhamus, the first prototype flown with swept wing was Me262 for which the drag rise because of the compressibility of the air was delayed with 40° sweep at the leading edge which enabled the plane to increase its speed. This fact had not been realized by the allied

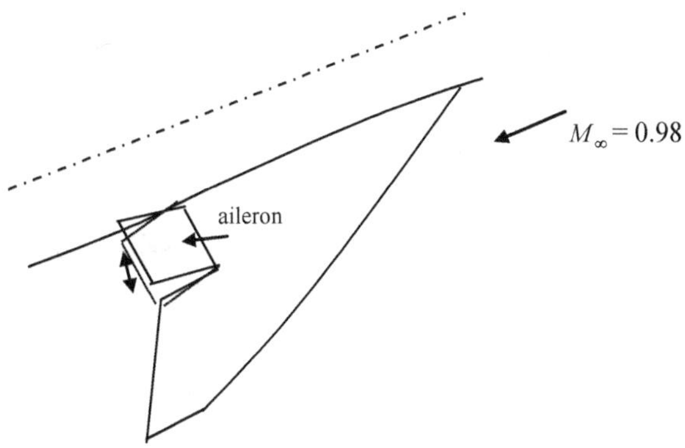

Fig. 6.9 Aileron buzz at transonic flow

forces yet. Busemann's theory on supersonic swept wings was implemented for reducing the effect of compressibility on subsonic wings in wind tunnel testings at 1941. In 1945, however, R.T.Jones was the first, except German aerodynamicists, to start testing swept wings for their aerodynamic utilizations (Jones 1946). A decade after Busemann, Jones' experimental and analytical work independently gave the lift coefficient variations for various wings with respect to the free stream speed, Fig. 6.10. According to Fig. 6.10, the lift coefficient for thin delta wings remains the same except for $M_\infty = 1$. This means, the Jones' theory on thin delta wings state that the compressibility effect at low angles of attack is insignificant at even very high speeds as high as free stream Mach number of 2. That is if we somehow know the lift coefficient of a thin delta wing for incompressible flow, we can safely use that value even for very high speeds. The effect of the aspect ratio of a wing is also shown in Fig. 6.10 for elliptic and rectangular wing forms.

The theory of Jones and the information confiscated from Germans helped the designing of the military and the civilian transonic and supersonic aircrafts having swept wings. In this respect, until 1960's the studies were in general under military contracts; therefore, they were classified. In following years, first the experimental results were presented and/or published in relevant literature (Lock 1962). In his experimental work, Lock designed a 12% thick wing with 55° sweep and a curved leading edge, as shown in Fig. 6.11, at 2.5° angle of attack to give a shockless

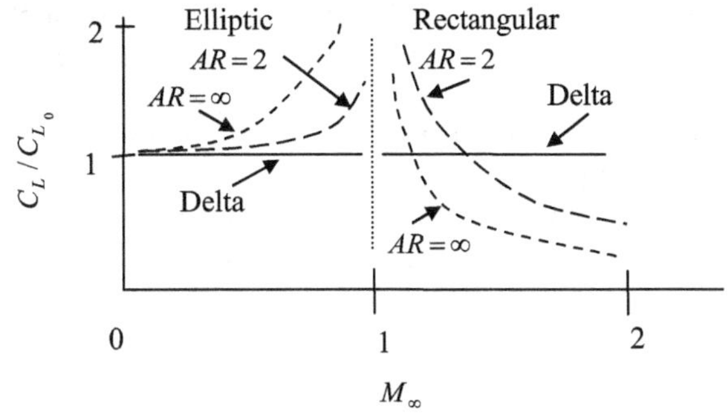

Fig. 6.10 Lift coefficient variation of finite wings with free stream Mach number and the aspect ratio

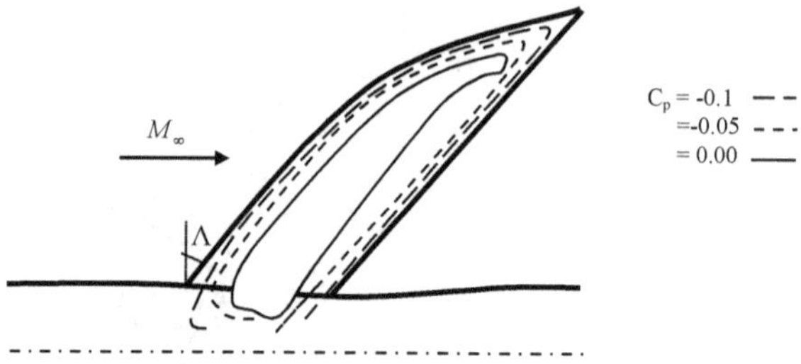

Fig. 6.11 A transonic wing with aspect ratio of 3.54 and the sweep angle of Λ = 55° (The surface pressure plots are for the symmetric non lifting wing)

lifting wing at 0.90, 1.00 and 1.1 free stream Mach numbers. The design lift coefficient of the wing is $C_L = 0.18$. Also shown in Fig. 6.11 are the pressure coefficient contours for the non lifting wing at free stream Mach number of 1. A similar work was performed on a similar wing shape experimentally and numerically by Labrujere and his associates at free stream Mach number $M_\infty = 0.96$ and surface pressure distribution similar to that given in Fig. 6.11 is (Labrujere et al. 1968).

The evolving of the shape of the wing from Fig. 6.11 to its later stages with swept but straight leading edges and with smaller aspect ratios is the given in detail by Kücheman in later years (Kücheman 1978). In later years, with the advent of more sophisticated numerical methods the transonic wing design with higher lift coefficients with small aspect ratios became possible.

So far we have seen the application of the three dimensional potential theory to obtain the surface pressure distribution and the lift coefficient of the wings in transonic flow. The lift coefficient obtained with the potential theory is usually small, i.e. $C_L = 0.18$, since the potential theory is valid for wings at small angles of attack without the presence of shocks and flow separation. In order to obtain higher lift coefficients, we have to increase the angle of attack to the level of partial flow separation, which in turn forces us to perform the viscous flow anlysis via numerical solution of some form of Navier-Stokes equations. As an example, we can give the detailed numerical study over a specially designed swept wing, Wing C, performed by Kaynak implementing a zonal solution technique based on the Euler and Navier-Stokes solutions in mid 1980's (Kaynak 1985). Shown in Fig. 6.12 is Wing C planform and its skin friction lines obtained numerically with the following geometric and flow parameters. The leading edge of the wing has 45° sweep, aspect ratio of 2.6 and spanwise twist of 8.17°. The free stream Mach number is 0.85, the angle of attack is 5.9° and the Reynolds number is 6.8×10^6. In zonal approach, the viscous region around the wing surface is solved with the thin shear layer equation and the outer region is solved with Euler equations (Kaynak 1985). The skin friction lines on the wing surface in Fig. 6.12 indicate that the flow is attached on most of the wing surface, however, only on the tip region there is a local separation due to presence of a shock, and the flow reattaches afterwards in 50–70% spanwise location.

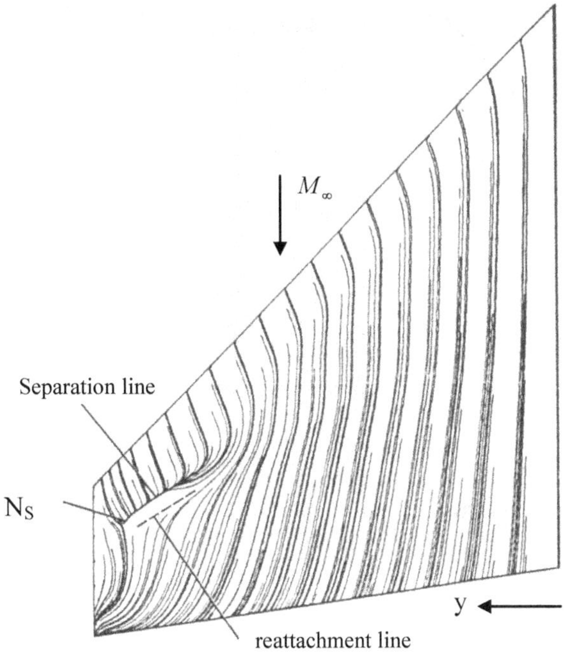

Fig. 6.12 Skin friction lines on Wing C in transonic flow with local separation. (N_s: Nodal point)

In Fig. 6.13, the surface pressure plots are provided at different spanwise locations. From these curves, the presence of a shock at the tip region is visible from 70% spanwise towards the tip itself.

In a transonic regime under the off-design conditions, the increase in the total drag and the decrease in the total lift of a wing show similar drastic changes as shown by a profile in 2-D flow. In this respect a transonic wing has to be re-designed in order to operate in off-design conditions. For this purpose, re-designing process based on the numerical solution of the Navier-Stokes equations is applied on a transonic wing successfully by Jameson. In his work, a transonic wing at buffeting Mach number of 0.86, is re-designed for performance increase under off-design conditions (Jameson 1999).

6.6 Unsteady Transonic Flow Past Finite Wings

In open literature, the linearized potential theory was applied to unsteady transonic lifting surfaces starting with Landahl's zeroth order theory during 1960's (Landahl 1962). In his study, Landahl used a special transformation technique to transform a rectangular wing to a delta wing to implement a previously developed theory for simple harmonic transonic solutions. In this way, the stream function for a simple harmonic motion of a delta wing was expressed, may be not so accurately but

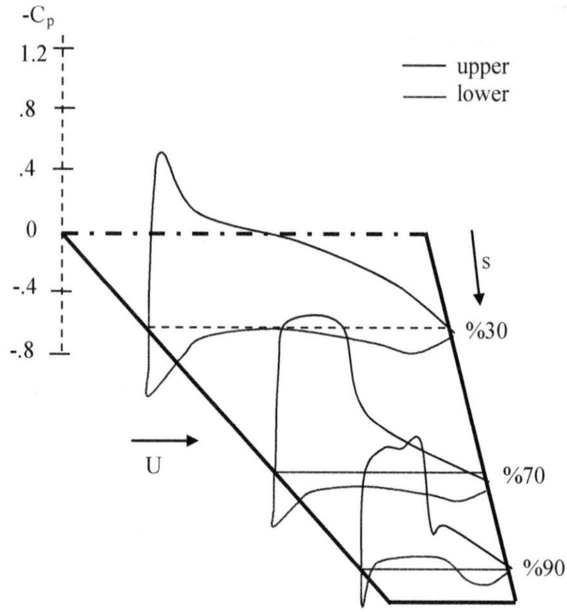

Fig. 6.13 Spanwise variation of surface pressure coefficients along different locations

analytically, for a given reduced frequency. In following years, this approach was used on sub surfaces as transonic panels on the wing in order to increase the accuracy. In addition, the doublet lattice method was successfully applied to the potential flow solutions of transonic flow past swept edged and low aspect ratio wings (Hounjet and Meijer 1985).

The finite difference solution of unsteady three dimensional potential equation, Eq. 6.3, is used for several wings and the results are compared with experiments in a detailed manner by Malone and his associates (Malone et al. 1985). In their study, the numerically computed surface pressure values are in agreement with the experimental results given for the wing of F-5. Shown in Fig. 6.14 are the surface pressure plots at three different spanwise stations of F-5 wing at free stream Mach number of 0.95 and reduced frequency of $k = 0.132$.

A close examination of Fig. 6.14 indicates that a line of shock doublet at both upper and lower surfaces close to the trailing edge appears to grow deeper towards the tip of the wing. When the free stream Mach number is reduced to 0.90, shock doublet is weakened and for further reduction to 0.80 the shock completely disappears from the surface. A similar study performed by Goorjian and Guruswamy (1985) on the F-5 wing gives agreable results with the experiments at free stream Mach number of 0.90 for the wing aspect ratio of 2.98, the taper ratio of 0.71, leading edge sweep angle of 31.9°, and finally trailing edge sweep angle of −5°.

The effect of viscosity on the three dimensional transonic flow solutions to obtain the surface pressure distribution on finite wings was first studied in 1990's with numerical solutions of Navier-Stokes equations (Guruswamy and Obayashi 1992). In their study, a moving grid was used to consider the elastic behavior of the

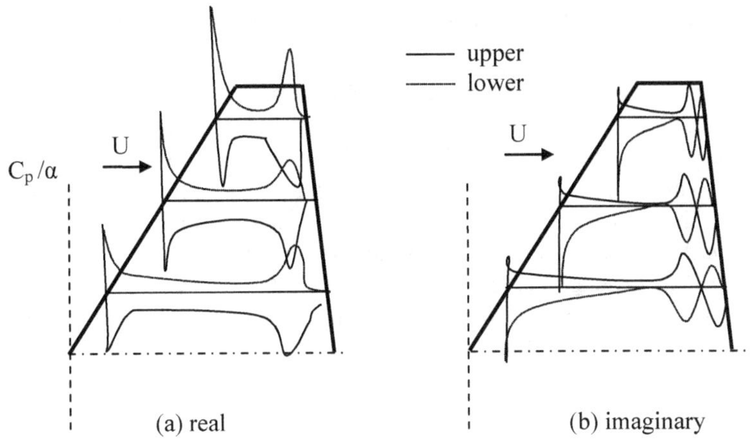

Fig. 6.14 Surface pressure coefficient plots for $M_\infty = 0.95$ of F-5 wing **a** real, **b** imaginary

wing and the root chord was taken as the characteristic length. The viscous effects on F-5 wing at free stream Mach number of 0.90, Reynolds number of 1.2×10^7 and reduced frequency of $k = 0.55$ in pitch oscillations when compared with the experiments show higher pressure rise at the root and lower pressure rise at the tip region because of turbulence modeling. Shown in Fig. 6.15 is the upper surface pressure distribution for a forced pitching oscillations given by $\alpha(t) = 3°–0.5° \sin(\omega t)$ at three different spanwise stations.

The data base published by AGARD (1985) for the unsteady transonic flows past certain profiles and wings gives the surface pressure plots in a detailed report (AGARD-R-702). This report can be used for code validation purposes for the pressure coefficients at various spanwise stations of the swept wings even with high aspect ratios.

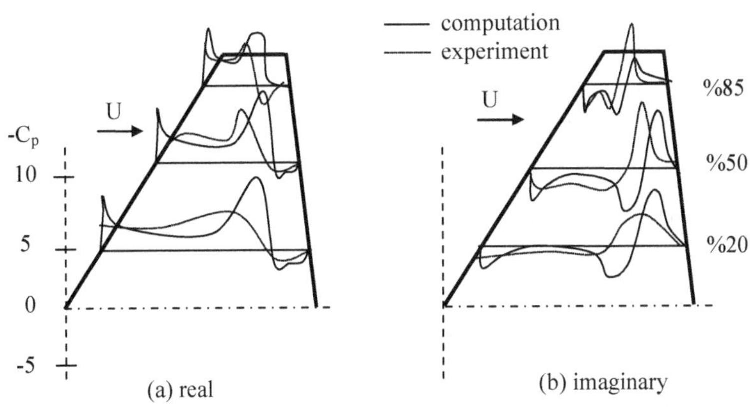

Fig. 6.15 Surface pressure distribution for unsteady transonic flow, viscous solution

6.7 Wing-Fuselage Interactions at Transonic Regimes

The wing-fuselage interaction is always of interest to aerodynamicists since the fuselage effect on the lifting of the wing as well as the drag increase because of the interaction. In transonic regimes this interaction causes almost 50% increase on the drag force when compared with the drag force in low subsonic regimes for the same geometry. The reason for this increase is the creation of a wave drag because of the supersonic flow regime taking place at the intersection of the wing with the body. The experimental determination of the drag and its 50% reduction was possible with the pioneering work of Whitcomb on the delta or swept wings (Whitcomb 1956). In order to reduce drag, the 'area rule' was proposed by Whitcomb as the reduction in the cross sectional area of the fuselage at the intersection with the wing as shown in Fig. 6.16. With the area rule, the sum of the fuselage cross sectional area with the wing area is almost kept constant along the axis of the plane. This enables us to delay the occurrence of Mach waves causing extra drag on the body which in turn reduces the wave drag. The wave drag coefficients C_{Dw} given in Fig. 6.1 are the differences between the measured total drag at zero angle of attack, and the calculated skin friction drag (Whitcomb 1956). As can be seen from Fig. 6.16, the area rule not only reduces the wave drag considerably it also delays

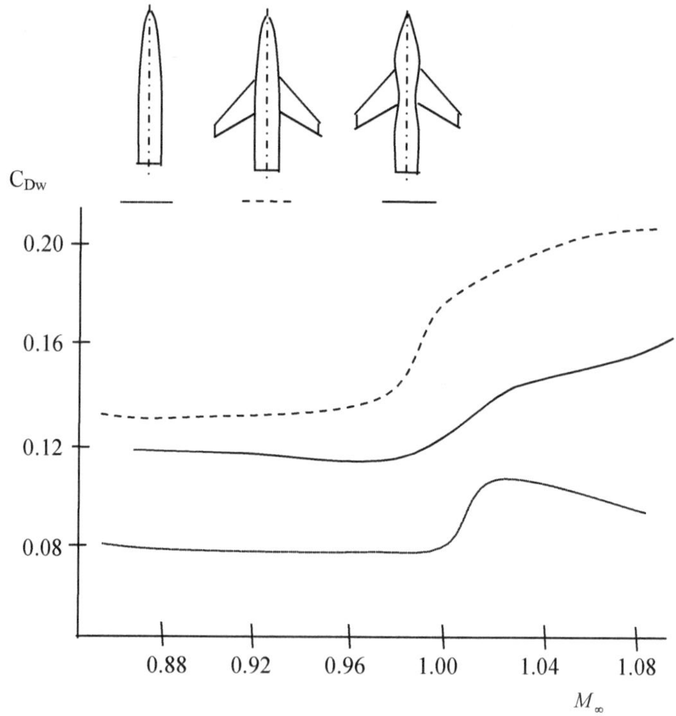

Fig. 6.16 Wave drag variation with free stream Mach number in transonic flows for three different cases

the occurrence of critical Mach number as opposed to the body having no reduction. Theoretical works performed in those years also yielded similar results (Lomax and Heaslet 1956). Lock's previously referred work also can be listed as an example to the experimental study made on wing fuselage interaction. (Lock 1962).

In following years the design criteria for the wing fuselage configuration of the planes having slender bodies with various sweep angles at low supersonic free stream Mach numbers was given by Küchemann (1978). For the wide bodies, however, it is possible to increase the critical Mach number while reducing the wave drag with enlargement of the fuselage cross sectional area at high subsonic cruise (Kuethe and Chow 1998).

In 1970's and 80's with the advances in CFD techniques the analysis and design of wide body wing interactions as well as its unsteady transonic analysis became possible (McCroskey et al. 1985), Nowadays, concurrent with the progress made in computational means the full scale transonic analysis of a full aircraft is possible.

Summary: *Transonic flow, as a special case of compressible flow past two and three dimensional objects is considered. First, local linearization approach for steady thin airfoil theory is introduced to obtain approximate analytical solutions. Then, unsteady analytical solutions are provided for simple harmonically heaving-plunging and/or pitching airfoils. Numerical solutions to full non-linear potential flow past shockless airfoils are also given. Furthermore, comparisons of the surface pressure distributions for the conventional and the supercritical airfoils are provided and the significance of the supercritical airfoils is discussed. The aerodynamic performance of supercritical airfoils at off-design conditions is also given. A general approach to the unsteady transonic aerodynamics is given in terms of low transonic, high transonic and low supersonic considerations. The differences among the three are discussed with emphasis on the real and imaginary parts of the surface pressure distributions. The 'shock doublet' concept and its significance for the low transonic case are also provided. The shock related transonic phenomena such as flutter, buffeting and aileron buzz are briefly mentioned. The viscous effects on the transonic aerodynamics are considered and the concept of 'transonic dip' and its effect on the onset of buffeting is provided. Characteristics for the enveloping curves of the separation points, critical Mach numbers, drag divergence and the onset of buffeting with respect to the lift vs. Mach number are presented. Steady transonic flow over thin wings is provided with its historical development including a pioneering computational work involving the Navier-Stokes solutions over Wing-C. Unsteady transonic considerations for the finite wings are provided for the representative wings for which the measurements are given by AGARD as data bases. Finally, wing-body interactions in terms of the wave drag is studied with introduction of the area rule.*

6.8 Problems and Questions

6.1 Using the energy equation, obtain the linearized form of the relation between the perturbation potential and speed of sound, Eq. 6.1.

6.2 Based on the order of magnitude analysis, show that the time derivative of the perturbation potential is negligible compared to x derivative.

6.3 Obtain Eq. 6.11 from 6.10 by inverse Laplace transform using the convolution integral.

6.4 Show that at constant angle of attack the perturbation potential is given by Eq. 6.13.

6.5 Show that at constant angle of attack the surface pressure coefficient is given by Eq. 6.14.

6.6 Show that lower surface pressure is given by Eq. 6.15.

6.7 Show that the sectional lift coefficient depends on the angle of attack as expressed in Eq. 6.17.

6.8 Obtain the sectional moment coefficient and center of pressure using Eq. 6.16,

6.9 Compare the surface pressure coefficient obtained with Dowell method using $x_0 = \mathrm{b}$ as expansion point and $e_0 = 0.12$ and $f_0 = 2.4/b$ with the pressure coefficient obtained using Stahara-Spreiter method for the 6% thick Guderly airfoil. Discuss the choice of $x_0 = b$.

6.10 Solve Eq. 6.21 in Laplace domain, and obtain the expression 6.22 by inverse transform to give the boundary condition.

6.11 Obtain Eq. 6.23 from 6.24 by the limiting procedure as $M_\infty \rightarrow 1$.

6.12 Show that for simple harmonically heaving plunging thin airfoil the surface pressure expression is given by Eq. 6.25 as the free stream Mach number approaches 1.

6.13 Using the values of Example 6.1, plot the phase lag of the surface pressure coefficient along the chord for a heaving plunging thin airfoil.

6.14 Find the amplitude of the (i) sectional lift coefficient, and (ii) the sectional moment coefficient about the leading edge using the data given in Example 6.1.

6.15 Using Eq. 2.15 expressed for the velocity potential, obtain Eq. 6.26 for compressible steady flows.

6.16 What is a ‘shock doublet’ in unsteady transonic flow?

6.17 Discuss the ‘transonic dip’ phenomenon for the swept wing in a transonic unsteady flow.

6.18 What is the function of transonic dip in unsteady transonic flow?

6.19 Comment on the ‘area rule’ for the wing fuselage in transonic flow.

References

AGARD: Compendium of unsteady aerodynamic measurements, Addendum No.1, AGARD-R-702, May (1985)

Barakos, G., Drikakis, D.: Numerical simulation of transonic buffet flows using various turbulence closures. Int. J. Heat Fluid Flow **21**, 620-626 (2000)

Bauer, F., Garabedian, P., Korn, D.: Supercritical Wing Sections, Lecture Notes in Economics and Mathematical Systems, Springer-Verlag, Berlin (1972)

Bauer, F., Garabedian, P., Korn, D., Jameson, A.: Supercritical Wing Sections II, Lecture Notes in Economics and Mathematical Systems, Springer-Verlag, Berlin (1975)

Baurdoux, H.I., Boerstoel, J.W.: Symmetrical Transonic Potential Flows Around Quasi-Elliptical Aerofoil Sections, Report NLR-TR9007U. National Aerospace Laboratory, NLR, The Netherlands (1968)

Dowell, E.H.(eds.) A Modern Course in Aeroelasticity. Kluwer Academic Publishing Groups (1995)

Ecer, A., Akay, H.U., Gülçat, Ü.:On the solution of hyperbolic equationsusing finite element method. Symposium on applications of computer methods in engineering, Los Angeles, California, August 23–26, 1977

Geistler, W.: Numerical study of buffet and transonic flutter on the NLR7301 airfoil. Aerosp. Sci. Technol. **7**, 540–550 July (2003)

Goorjian, P.M, Guruswamy, G.P.: Unsteady Transonic Aerodynamic and Aeroelastic Calculations about Airfoils and Wings, AGARD-CP-374 (1985)

Guruswamy, G.P., Obayashi, S.: Transonic Aeroelastic Computations on Wings Using Navier-Stokes Equations, AGARD-CP-507, March (1992)

Hounjet, M.H.L., Meijer, J.J.: Application of Time-Linearized Methods to Oscillating Wings in Transonic Flow and Flutter, AGARD-CP-374 (1985)

Jameson, A.: Re-Engineering the design process through computations. J. Aircraft **36**(1), January-February (1999)

Isogai, K.: Numerical Simulation of Shock-Stall Flutter of an Airfoil Using the Navier-Stokes Equations, AGARD CP-507, March (1992)

Jones, R.T.: Properties of Low Aspect Ratio Pointed Wings at Speeds Below and Above the Speed of Sound, NACA TN-1032 (1946)

Kaynak, Ü.: Computation of transonic separated wing flows using an Euler-Navier stokes zonal approach , Ph.D Thesis, Stanford University (1985)

Kuethe, A.M, Chow, C.-Y.: Foundations of Aerodynamics 5th. Ed. J. Wiley, New York (1998)

Küchemann, D.: Aerodynamic Design of Aircraft. Pergamon Press, Oxford (1978)

Labrujere, T.E., Loewe, W., Sloof, J.W.: An approximate method for thedetermination of the pressure distribution on wings in the lower critical speed Range, AGARD CP-35 (1968)

Landahl, M.T.: Linearized theory for unsteady transonic flow, IUTAM Symposium, Aachen, 1962

Lock, R.C.: Some experiments on the design of swept wing body combinations at transonic speeds, IUTAM Symposium, Aachen, 1962

Lomax, H., Heaslet, M.A.: Recent development in the theory of wing-body wave drag. J. Aero. Sci. **23**, 1061–1074 (1956)

Malone, J.B., Ruo, S.Y., Sankar, N.L.: Computation of unsteady transonicflows about two-dimensional and three-dimensional AGARD standard configurations, AGARD-CP-374 (1985)

McCroskey, W.J.: Unsteady airfoils. Ann. Rev. Fluid Mech. (1982)

McCroskey, W.J., Kutler, P., Bridgeman, J.O.: Status and prospects of computational fluid dynamics for unsteady transonic flows, AGARD-CP-374 (1985)

Murman, E.M., Cole, J.D.: Calculation of plane steady transonic flows, AIAA J. **9**(1), 114–121 (1971)

Nieuwland, G.Y., Spee, B.M.: Transonic shock-free flow, fact or fiction? AGARD CP No 35, Transonic Aerodynamics, September 1968

Polhamus, E.C: Applying slender wing benefits to military aircraft. J.Aircraft **21**(8) August (1984)
Whitcomb, R.T.: A study of the zero-lift drag-rise characteristics of wing-body combinations near the speed of sound, NACA Report 1273 (1956)
Whitcomb, R.T., Clark, L.R.: An airfoil shape for efficient flight atsupercritical mach numbers, NASA TMX-1109, July (1965)
Yang, G., Obayashi, S., Nakamichi, J.: Aileron Buzz simulation usingan implicit multiblock aeroelastic solver. J. Aircraft **40**(3), May-June (2003)

Chapter 7
Hypersonic Flow

There exist various criteria to be satisfied by the free stream Mach number M_∞, which makes the flow to be classified hypersonic when it is very high supersonic. Depending on the value of the Mach number, we have hypersonic aerodynamics determined by a predominant parameter with which the flow physics does not change. That is according to some flow parameters the flow is considered to be hypersonic for $M_\infty \geq 3$, and with respect to some other parameter the flow is regarded as hypersonic for $M_\infty \geq 5$ (Anderson 1989). Moreover, the dependence on the Mach number may vary for the same parameter with the body shape. An important parameter of hypersonic flow is the temperature. In the stagnation flow, the temperature may reach some high values which can exceed the durability limits of the materials. For this reason in hypersonic flows heat transfer and thermodynamics play an important role, which forces us to add the concept of aerothermodynamics to aerodynamics (Bertin). In addition, at high temperatures there are considerable changes in the viscosity and specific heats of the air to be accounted in hypersonic flows. The classical approach gives us the thermodynamic properties of the air either in normal temperatures or in very high temperatures approaching infinity. In the temperature ranges which are of interest to us the composition and the properties of the air can be determined by the aid of statistical thermodynamics (Lee et al.). In higher speeds, temperatures and altitudes the chemical composition of the air changes. During the chemical reactions the energy needed is provided by the medium of the air for which the formation energy of each specie must be included in the energy equation. This in turn affects the flow domain about the body and the shock location. The concept of aerothermochemistry is needed to be introduced at this stage. Furthermore, for a ballistic re-entry problem the speeds become so high that ionized flow around the capsule occurs. The ionized flow regions can be studied by plasma flow.

Ü. Gülçat, *Fundamentals of Modern Unsteady Aerodynamics*,
https://doi.org/10.1007/978-3-030-60777-7_7

First studies on hypersonic aerodynamics started after WWII by the design work performed on intercontinental ballistic missiles. The first historic manned hypersonic flight and safe re-entry was made in 1961. Since then, based on the data recorded during re-entry, experiments performed in specially designed hypersonic wind tunnels and the advances made in computational methods considerable progress is achieved in the multi disciplinary field of hypersonic aerodynamics. Naturally, the progress made in this field was not only applicable to re-entry problems and to the flight of intercontinental ballistic missiles but also useful for the design of intercontinental planes with sustainable hypersonic flight in the future.

In this chapter, starting by the Newton's impact theory which predicts the surface pressure coefficient with a simple formula for blunt bodies, various order piston theory to be used for slender bodies, hypersonic similarity based on the Euler equation, viscous hypersonic flow and high temperature gas dynamics knowledge will be provided.

7.1 Newton's Impact Theory

The amount of pressure exerted on a surface with impinging flow is equal to the normal component of the momentum exerted by the impinging fluid particles. The theory proposed by Newton and based on this impact idea was interpreted, until the beginning of 20th century, as the cause of the aerodynamic force acting on flying objects! The impact theory on the other hand was not able to produce sufficient lifting force to balance the weight of flying creatures in nature. Naturally, during the years of the emergence of the impact theory and for the following couple of centuries to come the relation between the air speed and the compressibility of air was not known. Therefore, by shear coincidence the evaluation of surface pressure coefficient with the impact theory, independently from the Mach number itself, at very high Mach numbers was given in the first half of the 20th century (Hayes and Probstein). Now, let us evaluate the pressure exerted by the impact theory on a wall inclined with angle θ_w in a freestream speed of U. The amount of mass per unit time per unit area of the air particles striking the wall, shown in Fig. 7.1, is ρU_n. The momentum exerted on the wall by this air mass is $\rho(U_n^2)$ which is also equal to the pressure exerted.

Expressing the wall pressure in terms of the free stream speed we obtain

$$p = \rho U^2 \sin^2 \theta_w \tag{7.1}$$

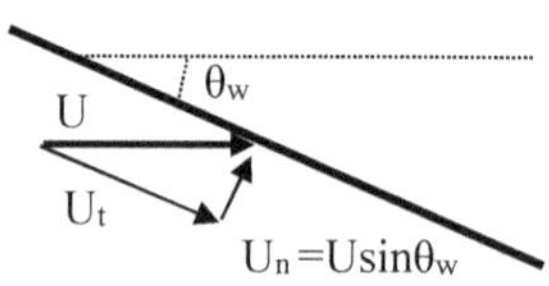

Fig. 7.1 Velocity components of the fluid particles impinging on a wall at a free stream speed of U

Fig. 7.2 Flow about a blunt body at high Mach numbers

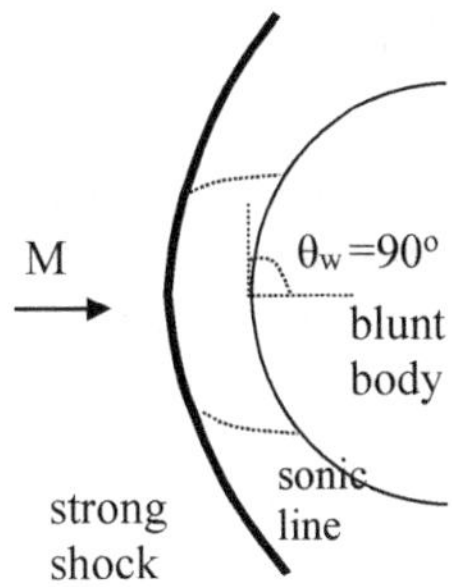

If the area of the wall shown in Fig. 7.1 is S, and the region behind the wall is considered as a vacuum, the normal force acting on the wall becomes

$$N = pS = S\rho U^2 \sin^2 \theta_w \tag{7.2}$$

The normal force coefficient then reads

$$C_N = \frac{N}{1/2\rho U^2 S} = 2\sin^2 \theta_w \tag{7.3}$$

From this normal force coefficient we can obtain the lift coefficient normal to the free stream and the drag coefficient in the direction of free stream as follows

$$C_L = 2\sin^2 \theta_w \cos \theta_w \quad \text{and} \quad C_D = 2\sin^3 \theta_w \tag{7.4a, b}$$

Since the wall angle and the angle of attack is the same for a flat plate, according to the impact theory at small angles of attack the lift coefficient is proportional with the square of the angle of attack, Eq. 7.4 a, b. This lift coefficient is not sufficient even at considerably high speeds. (This was the excuse to explain 'flying is a property of heavenly bodies' for centuries). We can now express the surface pressure coefficient, using the normal force coefficient given by Eq. 7.3, in terms of the free stream Mach number

$$C_p = \frac{p - p_\infty}{1/2\rho U^2} = \frac{p}{1/2\rho U^2} - \frac{p_\infty}{1/2\rho U^2} = 2\sin^2 \theta_w - \frac{2}{\gamma M^2} \tag{7.5}$$

In hypersonic flow the free stream Mach number is high and its square is very high. Therefore, the second term of the right hand side of Eq. 7.5 is negligible compared to the first term. Neglecting the second term gives us the approximate expression which is independent of free stream Mach number as follows:

$$C_p \cong 2\sin^2 \theta_w \tag{7.6}$$

7.2 Improved Newton's Theory

The Newton's impact theory gave us the surface pressure coefficient, Eq. 7.6, for the straight wall. On the other hand, experimental results show that the impact theory is also applicable for the blunt bodies in high Mach numbers. We know that at a high subsonic flow there is a strong detached shock in front of the stagnation point. If we use that stagnation pressure as a coefficient in Eq. 7.6, we get agreeable results between the Newton's impact theory and experiments for hypersonic flow.

If C_{po} denotes the stagnation pressure coefficient of the blunt body, the improve form of Eq. 7.6 reads as

$$C_p \cong C_{po} \sin^2 \theta_w \tag{7.7}$$

At the stagnation point $\theta_w = \pi/2$. This makes Eq. 7.7 yield better results than Eq. 7.6 for the hypersonic flows.

Assuming the detached shock in front of the stagnation point as a normal shock and using the normal shock relations in terms of the Mach number and pressure, it is possible to obtain the stagnation pressure to be used in Eq. 7.7 (see Problem 7.1). Improved Newton formula gives good results for the high angle of attack flows with high wall angles. Shown in Fig. 7.3a is the parabolic surface given by $x = 0.729\, y^2 - 1.0$, and in Eq. 7.3b is the surface pressure coefficient plots obtained by Newton and improved Newton formula at $M = 8$.

The surface curvature effect is not taken into account in Eq. 7.7 when applied to a surface given in Fig. 7.2 while computing the surface pressure coefficient. For this reason it seems it is necessary to add the centripetal force effect in Eq. 7.7. In order to express the relation between the centripetal force and the normal component of the pressure gradient, it is convenient to use s-n coordinate system where s is the tangential and n is the normal coordinates. The centripetal force acting in a unit volume is $\rho V^{2}/R$ where V shows the flow speed and 1/R shows the radius of curvature. The pressure force gradient which balances the centripetal force can be written as

$$\frac{\partial p}{\partial n} = \frac{\rho V^2}{R} \tag{7.8}$$

The continuity equation applied for a tube across the shock to equate the mass flux before and after the shock gives $\rho V dn = \rho_\infty U dy$ where U is the free stream speed and ρ_∞ density, Fig. 7.4.

Integrating Eq. 7.8 from point 1 at the surface to the point 2 without crossing the shock along Δn gives

$$\int_{p_1}^{p_2} dp = \int_{0}^{\Delta n} \frac{\rho V^2}{R} dn \tag{7.9}$$

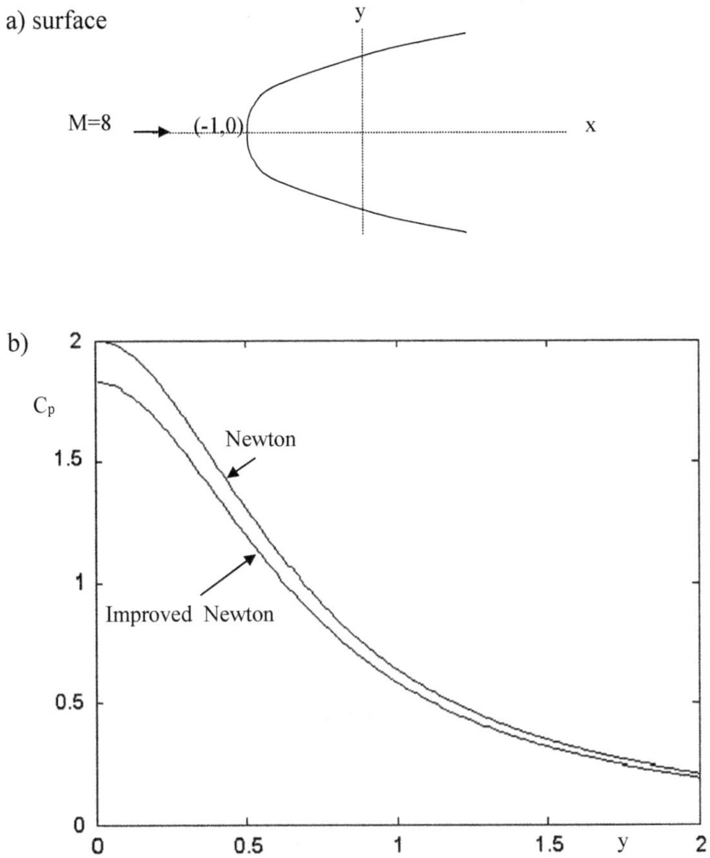

Fig. 7.3 **a**, **b** Surface pressure coefficient of a parabolic surface at M = 8

Fig. 7.4 Balancing of the centripetal force with pressure gradient

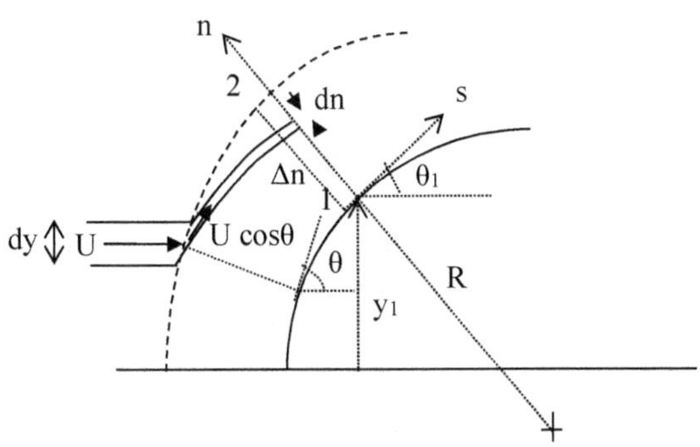

If we use the continuity equation in 7.9 we obtain

$$p_2 - p_1 = \int_0^{y_1 + \Delta n \cos\theta_1} \frac{\rho_\infty UV}{R} dy \tag{7.10}$$

Taking the limit as Δn approaches to zero in Fig. 7.4, and expressing the radius R in terms of tangent angle θ to the surface with V = U cosθ gives

$$p_1 = p_2 + \rho_\infty U^2 \left(\frac{d\theta}{dy}\right)_1 \sin\theta_1 \int_0^{y_1} \cos\theta dy \tag{7.11}$$

Equation 7.11 in terms of pressure coefficient reads as

$$C_{p1} = C_{p2} + 2\left(\frac{d\theta}{dy}\right)_1 \sin\theta_1 \int_0^{y_1} \cos\theta dy \tag{7.12}$$

Now, we can express Eq. 7.12 in terms of the pressure coefficient at point 2, shown in Fig. 7.4 at which the curvature effect is no longer exists, as follows.

$$C_{p1} = 2\sin^2\theta_1 + 2\left(\frac{d\theta}{dy}\right)_1 \sin\theta_1 \int_0^{y_1} \cos\theta dy \tag{7.13}$$

For axially symmetric bodies the expression reads as

$$C_{p1} = 2\sin^2\theta_1 + 2\left(\frac{d\theta}{dr}\right)_1 \sin\theta_1/r_1 \int_0^{r_1} r\cos\theta dr \tag{7.14}$$

The theory developed by the centripetal force inclusion in evaluation of surface pressure is name the Newton-Busemann theory. Although inclusion of the centripetal force seems correct because of physical considerations, it makes the pressure coefficient more disagreeable with the experimental results. The second term of the right hand side of Eq. 7.13 is just a theoretical term, i.e. it is not related to effect of the pressure waves increasing compressibility in the flow direction (Anderson), which makes its implementation rare in engineering applications.

7.3 Unsteady Newtonian Flow

During the re-entry and hypersonic cruise the unsteady motion of the hypersonic vehicle is unavoidable. For this reason we have to determine the unsteady hypersonic aerodynamic coefficients in terms of time dependent pressure variation. Now, let us obtain the pressure coefficient expression of a body whose equation is given by B(x,y,z,t) = 0, has an instantaneous velocity of $\mathbf{q}_B$ and is immersed in a free stream U as shown in Fig. 7.5.

If we denote the unit normal vector into the body with **n** then the normal velocity of the fluid particle that striking the surface of the body reads $(U\mathbf{i} - \mathbf{q}_B).\mathbf{n}$. Afterwards, we can write the mass of the fluid particle which strikes onto the unit surface area in unit time becomes

$\rho\,(U\mathbf{i} - \mathbf{q}_B).\mathbf{n}$. According to the Newtonian impact theory, the momentum acting onto the surface of the body then becomes

$$p = \rho[(\mathrm{Ui} - \mathrm{q_B}).\mathrm{n}]^2 \tag{7.15}$$

The surface pressure coefficient at high Mach number is expressed from Eq. 7.15 as

$$C_p = \frac{p - p_\infty}{1/2\rho U^2} \cong \frac{p}{1/2\rho U^2} = 2[(\mathrm{i} - \mathrm{q_B}/\mathrm{U}).\mathrm{n}]^2 \tag{7.16}$$

Let us now write the unit normal vector $\mathbf{n} = l\ \mathbf{i} + m\ \mathbf{j} + n\ \mathbf{k}$ in terms of the direction cosines l,m,n of the surface equation B(x,y,z,t) = 0 as follows

$$l = -\frac{\partial B}{\partial x}/|grad\,B|, \quad m = -\frac{\partial B}{\partial y}/|grad\,B|, \quad n = -\frac{\partial B}{\partial z}/|grad\,B|$$

Here,

$$|grad\,B| = \left(\frac{\partial B}{\partial x} + \frac{\partial B}{\partial y} + \frac{\partial B}{\partial x}\right)^{1/2},$$

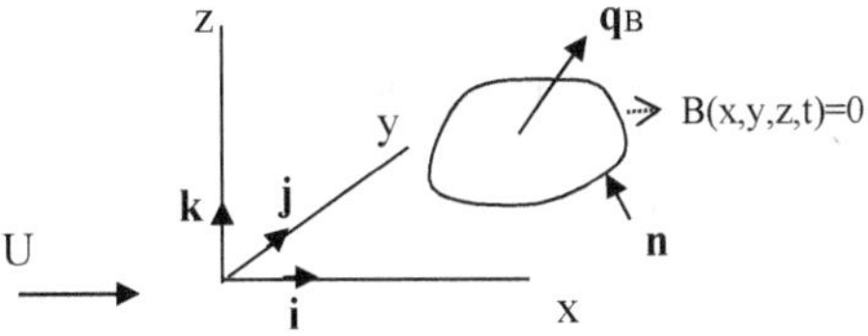

Fig. 7.5 Body B(x, y, z, t) = 0 moving by velocity q_B in a freestream of U

Unit normal **n** reads as

$$\mathbf{n} = -\frac{grad\,B}{|grad\,B|} \quad \text{and} \quad \mathbf{i.n} = -\frac{\partial B}{\partial x}/|grad\,B|$$

Let us express the instantaneous velocity $\mathbf{q}_B$ in terms of its rectangular components

$\mathbf{q}_B = q_x\,\mathbf{i} + q_y\,\mathbf{j} + q_z\,\mathbf{k}$. The normal component of the instantaneous velocity then reads as

$$\vec{q}_B.\vec{n} = -\left(q_x\frac{\partial B}{\partial x} + q_y\frac{\partial B}{\partial y} + q_z\frac{\partial B}{\partial z}\right)/|grad\,B| \tag{7.17}$$

If we assume that the body is rigid, then the shape of the body does not change but its position does in time. Therefore, dB/dt = 0 can be written as

$$\mathrm{dB/dt} = \frac{\partial B}{\partial t} + q_x\frac{\partial B}{\partial x} + q_y\frac{\partial B}{\partial y} + q_z\frac{\partial B}{\partial z} = 0 \tag{7.18}$$

If we substitute Eq. 7.17 and 7.18 into 7.16, the pressure coefficient then becomes

$$C_p = 2\frac{\left[-\frac{\partial B}{\partial x} - \frac{1}{U}\frac{\partial B}{\partial t}\right]^2}{|grad\,B|^2} \tag{7.19}$$

Writing the surface equation in terms of z, the equation becomes B = z - z_1 (x,y, t) = 0.

$$\frac{\partial B}{\partial x} = -\frac{\partial z_1}{\partial x}, \quad \frac{\partial B}{\partial y} = -\frac{\partial z_1}{\partial y}, \quad \frac{\partial B}{\partial t} = -\frac{\partial z_1}{\partial t}.$$

The surface pressure coefficient then reads

$$C_p = 2\frac{\left[\frac{\partial z_1}{\partial x} + \frac{1}{U}\frac{\partial z_1}{\partial t}\right]^2}{1 + \left(\frac{\partial z_1}{\partial x}\right)^2 + \left(\frac{\partial z_1}{\partial y}\right)^2} \tag{7.20}$$

Example 7.1 The surface equation of a body is given by z = $a(x/l)^{1/2,}$ $0 \le x \le l$. If this body pitches by $\alpha = \bar{\alpha}\sin\omega t$ about its nose in small amplitudes, find the surface pressure variation with time.

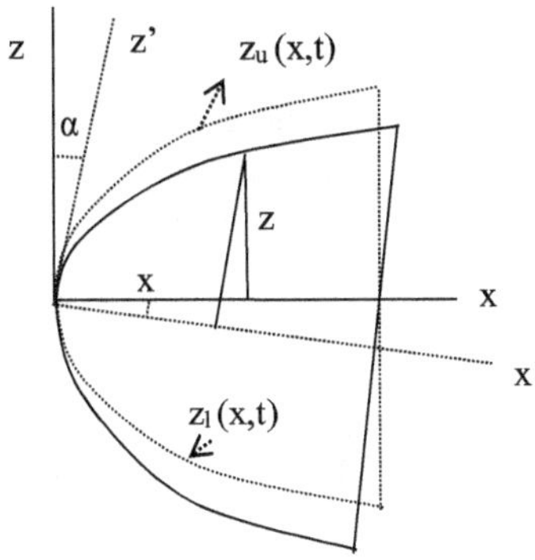

$x' = x \cos\alpha - z \sin\alpha$

$z' = x \sin\alpha + z \cos\alpha$

Solution Let x,z be the fixed and the x',z' be the rotating coordinate system. If we write the surface equation in terms of the fixed coordinates x,z then we have

$$z'_u = a(x'/l)^{1/2} = a\left(\frac{x\cos\alpha - z\sin\alpha}{l}\right)^{1/2}.$$

Now, we can write B(x,z,t) = $z' - z'_u = 0$ to give

$$B(x, z, t = x\sin\alpha + z\cos\alpha - a\left(\frac{x\cos\alpha - z\sin\alpha}{l}\right)^{1/2} = 0.$$

Since the pitching oscillations are small, $\sin\alpha = \alpha$ and $\cos\alpha = 1$ yields

$$B(x, z, t) = x\alpha + z - a\left(\frac{x - z\alpha}{l}\right)^{1/2}$$

Now, we can use the last expression for B to be used in Eq. 7.20 to obtain the unsteady surface pressure coefficient. Here, we have to make a note that even α is small the Newton theory is non linear because of the last term in surface equation..

The Newton and the modified Newton theories are widely used for blunt bodies in engineering applications. For slender bodies, however, the Piston Analogy gives good results for the surface pressure expression in a hypersonic flow.

7.4 The Piston Analogy

Airfoils and control surfaces are thin and/or slender objects which cause free stream to make small angles with leading edge shocks and expansion waves when the flow fields expand for Mach numbers grater than 4. For this reason, i) the changes of speed in free stream directions are negligible compared to the speed changes in normal to free stream, and ii) similarly, the gradients of the flow parameters in the direction of the free stream are negligible compared to the gradients in normal direction. These two facts leads to following 'in the unsteady flow around thin bodies a fluid column normal to the free stream moves in the flow direction while continuing its unsteady motion.' This fact was first observed by W.D. Hayes. Therefore, it is named 'Hayes' Hypersonic Analogy.'

Using the Hayes' Hypersonic Analogy, the pressure variation of two dimensional flows past an airfoil can be expressed in terms of the pressure change due to the waves created by unsteady motion of a piston in one dimensional cylinder (Lighthill). Now, let us study in detail the approach given by Lighthill (Liepmann and Roshko).

According to the similarity law deduced from Fig. 7.6, the surface pressure ratio to the free stream pressure can be obtained using the concept of simple wave created by compression as follows (Liepmann and Roshko)

$$\frac{p}{p_\infty} = \left(1 + \frac{\gamma - 1}{2}\frac{w}{a}\right)^{\frac{2\gamma}{\gamma-1}} \tag{7.21}$$

Here, w shows both the piston velocity and the vertical velocity at the wedge surface, and a indicates the sound of speed at the free stream. For Eq. 7.21 to be valid the flow must be isentropic which is possible only for the case of small perturbations. In addition, the vertical velocity w given by Eq. 7.21 is also valid for the unsteady flows. Let $z = z_a$ (x,t) be the equation of the surface. The associated vertical velocity is then given by

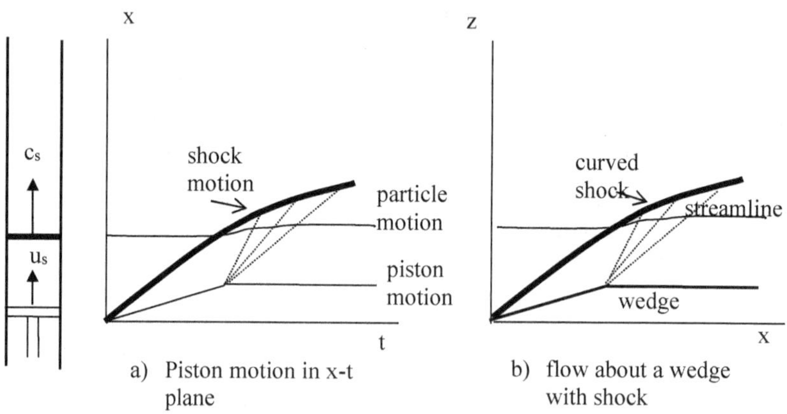

Fig. 7.6 Piston analogy: **a** motion of the piston with speed u_s in a, **b** flow over a wedge in high Mach number

$$w = \frac{\partial z_a}{\partial t} + U \frac{\partial z_a}{\partial x} \tag{7.22}$$

Now, we can obtain the application areas of Eq. 7.21 to a profile with a thickness ratio of τ with the free stream Mach number. Let θ be the turning angle of the flow and w be the vertical velocity because of presence of the airfoil. The approximate value of the vertical velocity then becomes

$$w \cong U\theta = aM\theta$$

This can be written as

$$\frac{w}{a} \cong M\theta \tag{7.23}$$

The product Mθ of Eq. 7.23 is called the hypersonic similarity parameter. The validity of the piston theory depends on this parameter. For piston theory to be valid, the hypersonic flow parameter must be very much less than 1. That is 'the non dimensional turning speed in piston theory is restricted by the product of Mach number and the turning angle.'

Now, we can expand the right hand side of Eq. 7.21 into the series expressed in terms of w/a

$$\frac{p}{p_\infty} \cong 1 + \gamma \frac{w}{a} + \frac{\gamma(\gamma+1)}{4}\left(\frac{w}{a}\right)^2 + \frac{\gamma(\gamma+1)}{12}\left(\frac{w}{a}\right)^3 \tag{7.24}$$

Equation 7.24 is applicable for the hypersonic similarity parameter range $0 < M\tau < 0.6$. The lower order approximation can be made for $0 < M\tau < 0.3$ by neglecting the third term as

$$\frac{p}{p_\infty} \cong 1 + \gamma \frac{w}{a} + \frac{\gamma(\gamma+1)}{4}\left(\frac{w}{a}\right)^2 \tag{7.25}$$

Finally, the linear approach is valid for $0 < M\tau < 0.15$, and reads as

$$\frac{p}{p_\infty} \cong 1 + \gamma \frac{w}{a} \tag{7.26}$$

Including the thickness effect of the airfoil gives better results for the piston theory when compared with the experimental results (Ashley and Zartarian). At high Mach numbers the high flow speeds make the local density low. The low density, on the other hand lowers the value of the Reynolds number which in turn causes the boundary layer thickness to be high. This forces us to consider the profile thickness during the application of the piston theory on hypersonic flows. In addition, in unsteady hypersonic flows there must also be restriction on the free stream Mach number M and the reduce frequency k as follows (Ashley and Zartarian):

$$kM^2 \gg 1 \quad \text{or} \quad (kM)^2 \gg 1 \tag{7.27}$$

The criterion given by Eq. 7.27 show that the piston theory is safely applicable at high reduced frequencies.

The piston theory introduced so far is applicable for $M\tau < 1$ range. In real life, however, even for thin profiles we encounter some cases with $M\tau \geq 1$, which will be examined in the next section.

7.5 Improved Piston Theory: $M^2 \tau^2 = O(1)$

The leading edge shock of a thin airfoil is not weak even if it is attached. Therefore, it creates non negligible turning in the flow field. When this turning angle is multiplied with free stream Mach number, the product is usually in the order of unity, i.e., $M\theta \sim 1$, similarly $M\tau \sim 1$. For these cases we need a valid theory especially for the lifting surfaces in hypersonic aerodynamics, Fig. 7.7.

The development of the pertinent theory is based on the order of magnitude analysis performed on the unsteady Euler equations in terms of the Hayes' similarity parameter. In this way we can obtain an expression for the unsteady lifting pressure for thin airfoils in hypersonic flow. (Van Dyke). Using the Van Dyke's small perturbation theory it is possible to obtain the lifting pressure for arbitrary motion of an airfoil at high Mach numbers with small turning angles with utilization of shock formulae (Pierce).

Let $z_v(x,t) = z_o(x) f(t)$ be the equation of the chord motion of the profile shown in Fig. 7.7. The equation of the upper surface for small angle of attack α becomes

$$z - [z_u(x) - \alpha x + z_v(x,t)] = 0 \tag{7.28}$$

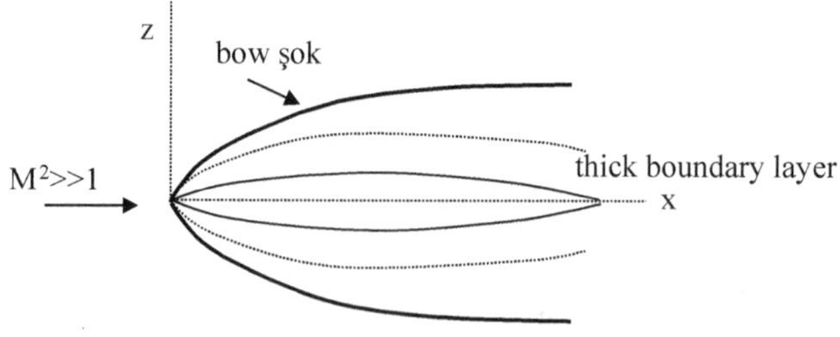

Fig. 7.7 The attached bow shock and a thick boundary layer about a thin airfoil in hypersonic flow

If the turning angle at the leading edge is θ_H, then for the upper and lower surfaces

$$\theta_H = \left.\frac{dz_u}{dx}\right)_{x=0} - \alpha \quad \text{and} \quad \theta_l = \left.\frac{dz_l}{dx}\right)_{x=0} + \alpha \tag{7.29}$$

respectively. The ratio between the value of the pressure right after the leading edge shock and the free stream pressure in terms of the shock angle β reads as (Liepmann and Roshko),

$$\frac{p_H}{p_\infty} = \frac{2\gamma}{\gamma+1}\left[(\beta M)^2 - 1\right] + 1 \tag{7.30}$$

Here, β is the shock angle created by the leading edge turning angle δ_H and δ(x) is the inclination of the surface given as

$$\delta(x) = \theta(x) + \left[\frac{dz_0}{dx} f(t' + x/U) + z_0(x)\frac{\partial f(t' + x/U)}{\partial x}\right] \tag{7.31}$$

Here, t is time and t' is the transformed time with coordinate x and free stream speed U as it appears in the argument of function f. We can now write the averaged flow values in terms of flow parameters as follows

$$\text{Shock's turning angle} \quad M\overline{\beta} = \frac{\gamma+1}{4}(M\theta_H) + \sqrt{\left(\frac{\gamma+1}{4}M\theta_H\right)^2 + 1} \tag{7.32a}$$

$$\text{Leading edge pressure:} \quad \frac{\overline{p}_H}{p_\infty} = \frac{2\gamma}{\gamma+1}(M\overline{\beta})^2 - \frac{\gamma-1}{\gamma+1} \tag{7.32b}$$

$$\text{Mach number:} \quad \frac{\overline{M}_H}{M} = \frac{\gamma+1}{2}\frac{M\overline{\beta}}{\left[\gamma(M\overline{\beta})^2 - \frac{\gamma+1}{2}\right]^{1/2}\left[\frac{\gamma-1}{2}(M\overline{\beta})^2 + 1\right]^{1/2}} \tag{7.32c}$$

$$\text{In terms of Mach number :} \quad \frac{\eta}{M} = \frac{\gamma(\gamma+1)(M\overline{\beta})^3}{\left[\gamma(M\overline{\beta})^2 - \frac{\gamma-1}{2}\right]\left[(M\overline{\beta})^2 + 1\right]} \tag{7.32d}$$

$$\frac{m}{M} = -\frac{\eta}{2M} + \frac{\gamma+1}{2}\frac{M\overline{\beta}}{\left[\gamma(M\overline{\beta})^2 - \frac{\gamma+1}{2}\right]\left[(M\overline{\beta})^2 + 1\right]} \tag{7.32e}$$

At this stage, we substitute Eq. 7.32 into Prandtl-Meyer expansion formula and write Δ(x) = θ(x)- θ_H and E(x,t') = ε(x,t') – ε_H (t'). The surface pressure p = p(x) at any location then reads as

$$\frac{p(x)}{p_\infty} = \frac{p_H}{p_\infty}(1+\eta\varepsilon_H)\left[1+\frac{\gamma-1}{2}\overline{M}_H(!+m\varepsilon_H)\Delta(x)+E(x,t')\right]^{\frac{2\gamma}{\gamma-1}} \tag{7.33}$$

Since E(x,t') is small we can expand Eq. 7.33 into series and neglect the second order terms the lifting pressure expression for a symmetric airfoil at 0° angle of attack reads as follows

$$\frac{p_l - p_u}{p_\infty} = -2\frac{p_H}{p_\infty}\left[1+\frac{\gamma-1}{2}\overline{M}_H\Delta(x)\right]^{\frac{\gamma+1}{\gamma-1}}$$
$$\left[\eta\varepsilon_H\left(1+\frac{\gamma-1}{2}\overline{M}_H\Delta(x)\right)+\gamma\overline{M}_H(m\Delta(x)\varepsilon_H+E(x,t'))\right]$$

The final form of the lifting pressure for general usage by inverse transformation t' = t – x/U becomes

$$\frac{p_l - p_u}{p_\infty} = -2\frac{p_H}{p_\infty}\left[1+\frac{\gamma-1}{2}\overline{M}_H\Delta(x)\right]^{\frac{\gamma+1}{\gamma-1}}\left\langle\left[\eta\left(1+\frac{\gamma-1}{2}\overline{M}_H\Delta(x)\right)+\gamma\overline{M}_H(m\Delta(x)-1)\right]\right.$$
$$\left.\left[\left.\frac{dz_0}{dx}\right)_{x=0} f(t-x/U)+\left.\frac{z_0(0)}{U}\frac{df}{dt'}\right)_{t'=t-x/U}\right]+\gamma\overline{M}_H\left[\frac{dz_0}{dx}f(t)+\frac{z_{0(x)}}{U}\frac{df}{dt}\right]\right\rangle \tag{7.34}$$

Now, we can enumerate the application areas of Eq. 7.34 as follows (Pierce).

(i) As long as the angle of attack is small, the hypersonic similarity parameter satisfies the criterion and Eq. 7.34 is applicable for the flow past profiles with sharp trailing edges.
(ii) The Prandtl-Meyer expansion formula is used to obtain Eq. 7.34 for hypersonic flows with small turning angles. If the turning angle increases, the error also increases, therefore, it has to be corrected by exact shock formulae.
(iii) Eq. 7.34 can be used for finite wings with the concept of strip theory. However, the strip theory is not valid at the sharp wing tips, and it is not capable of modeling the the low aspect ratio wings with high sweep angle.
(iv) Profiles with round leading edges have detached shocks which makes the application of piston theory and the Prandtl-Meyer theory non valid. Therefore, we need to use some other formula for profiles with blunt noses.

The next step is studying the effect of the detached shock created by the round leading edge and the hypersonic flow field after the shock to numerically evaluate the surface pressure of some special geometry.

7.6 Inviscid Hypersonic Flow: *Numerical Solutions*

The detached shock at the leading edge of a blunt body has a slowing effect for a vehicle re-entering to the earth atmosphere. The further away the shock from the body the less will be the heating effect at the stagnation region of the flow because of the high temperature rises after the strong shock. For this reason the location of the detached shock and the flow conditions behind the shock should be known for that high Mach number flows. Wind tunnel and shock tube testings are not capable of giving sufficient data at those high Mach numbers, therefore, the real flight recordings are used in studies when available. On the other hand, the inverse method which is based on determining the body shape and the flow around it for an assumed shock shape was an efficient method at late 1950s (Van Dyke 1958). For two dimensional and axisymmetric shocks a simplified inverse method is used to find the coordinates of the body surface (Maslen). Let us choose the x coordinate parallel to the shock direction and y coordinate normal to that direction to form the local coordinates, Fig. 7.8, and non dimensionalize the flow parameters with free stream conditions and the shock radius. The velocity component u in the flow direction is much larger than the normal v component of the velocity. We can now write the equations

In nondimensional form in terms of χ $\quad \frac{\partial \chi}{\partial x} = (1 - y/R) r^{\sigma} \rho v, \quad \frac{\partial \chi}{\partial y} = r^{\sigma} \rho v$

$$\text{Continuity}: \frac{\partial}{\partial x}[r^{\sigma}\rho u] + \frac{\partial}{\partial y}[(1 - y/R) r^{\sigma} \rho v] = 0 \quad \rightarrow \quad r^{\sigma}\frac{\partial y}{\partial \chi} = -\frac{1}{\rho u} \qquad (7.35\text{a, b})$$

$$\text{y} - \text{momentum}: u\frac{\partial v}{\partial x} + (1 - y/R) v \frac{\partial v}{\partial y} + \frac{u^2}{R} + \frac{1}{\rho}(1 - y/R)\frac{\partial p}{\partial y} = 0, \rightarrow r^{\sigma}\frac{\partial p}{\partial \chi} \cong \frac{u}{R - y} \qquad (7.36\text{a, b})$$

Here, σ = 0 for two dimensional, and σ = 1 for axially symmetric flows. If we nondimensionalize enthalpy with ½ $U^{2,}$ and show the stagnation streamline conditions with s the energy equation reads as: $h + u^2 + v^2 = h_{so} + v_{so}^2$. Then for the energy equation we write

$$u^2 = h_{so} + v_{so}^2 - h - v^2 \qquad (7.37)$$

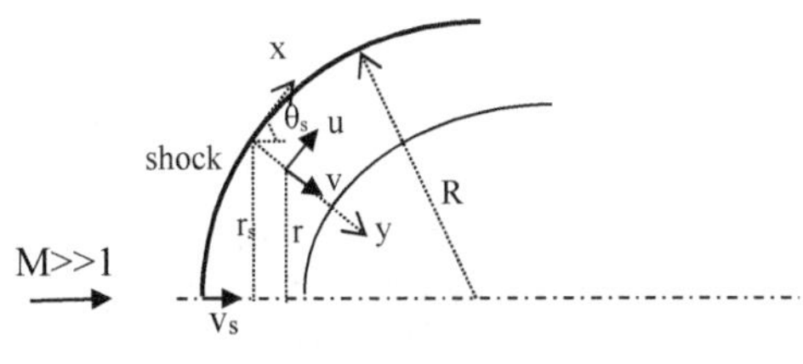

Fig. 7.8 Local coordinates of a spherical shock

$$\text{Conservation of entropy}: u\frac{\partial S}{\partial x} + (1 - y/R)v\frac{\partial S}{\partial y} = 0.$$

After the shock on a streamline we have p/ρ^{γ} = constant.

Simplifying Eq. 7.36b with the assumption of $y/R << 1$, the relation between the pressure and the stream function gives

$$\frac{\partial p}{\partial \chi} \cong \frac{u_s}{r^{\sigma} R_s} \tag{7.38}$$

Integration of Eq. 7.38 with respect to the stream function starting from the stagnation conditions χ_s we obtain

$$p - p_s = \frac{u_s}{r_s^{\sigma} R_s}(\chi - \chi_s) \tag{7.39}$$

Neglecting the term $(v_{so}{}^2 - v^2)$ in energy equation the approximate form of Eq. 7.37 reads as

$$u^2 \cong h_{so} - h \tag{7.40}$$

In order to determine the r coordinate of the body surface, we can write: $r = r_s - y \cos\theta_s$ in terms of the tangent angle θ_s of the shock in Eq. 7.35b to obtain

$$(r_s - y\cos\theta_s)^{\sigma}\frac{\partial y}{\partial \chi} = -\frac{1}{\rho u} \tag{7.41}$$

and for $\sigma = 0,1$ we get

$$(r_s - y\cos\theta_s)\frac{\partial y}{\partial \chi} = -\frac{1}{\rho u} \tag{7.42}$$

Integrating Eq. 7.42 with respect to the stream function gives

$$y - \frac{\sigma y^2 \cos\theta_s}{2r_s^{\sigma}} = \frac{1}{r_s^{\sigma}}\int_{\chi}^{\chi_s}\frac{d\chi}{\rho u} \tag{7.43}$$

The second term at the right hand side of Eq. 7.43 is second order, therefore can be neglected to yield

$$y \cong \frac{1}{r_s^{\sigma}}\int_{\chi}^{\chi_s}\frac{d\chi}{\rho u} \tag{7.44}$$

Since we know the initial conditions, we can March one step to determine the new coordinate from the known stream function value as follows: (i) obtain the pressure p from 7.39, (ii) find the density ρ on the streamline using isentropic relation, (iii) The h enthalpy from equation of state, (iv) the velocity component u from energy equation, and finally (v) y coordinate from Eq. 7.44. Now, let us apply the procedure described above to a spherical shock created by a body whose shape is to be determined.

Example 7.2 For a spherical shock with radius of 1 immersed in $M = \infty$ and $\gamma = 1.4$, find the flow conditions and the first body surface coordinate which has $\phi = 30^o$ at the center of the shock as shown in the figure below.

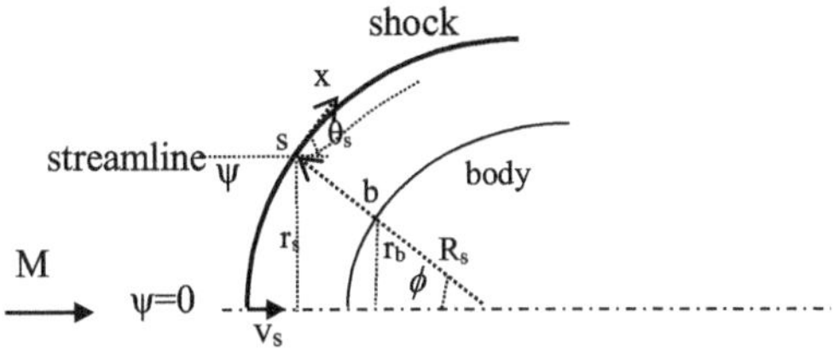

Solution In order to simplify the solution, we assume that the streamline entering into the shock is oblique at with an angle θ_s with existence of oblique shock relations. The pressure behind the shock then becomes: $p_s = 2/(\gamma + 1)\sin^2\theta_s = 0.625$, density: $\rho_s = 6$, nondimensional speed in x: $u_s = \cos\theta_s = \cos(90^o - 30^{o)} = 0.5$. The stream function, since we have a uniform free stream before the shock, at s is for two dimensional flow: $\psi_s = r_{s,}$ and axially symmetric flow: $\psi_s = r_s^{2/}2$. Hence: we find $r_s = \cos\theta_s = 0.5$ and $\psi_s = r_s^{2/}2 = 0.125$. Using Eq. 7.39 we can find the pressure at point b on the surface if we take the stream function value 0 as follows

$$p_b - p_s = u_s(\psi_b - \psi_s)/R_s r_s = 0.5(0 - 0.125)/(1x0.5) \; then \; p_b = 0.5.$$

The entropy remains the same on a streamline. This gives $p_b/(\rho_b)^\gamma = c = p_{so}/(\rho^\gamma_{so)}$ The stagnation pressure is: $p_{so} = 2/(\gamma + 1) = 0.8333$ gives us the value of c. For $\psi = 0$
the density is

$$\rho_b = 4.166.$$

The enthalpy then becomes $h = 2\gamma/(\gamma\text{-}1)(p/\rho) = 0.840$. Similarly, $h_o = 0.968$. Using the energy equation we find the surface velocity as

$$u_b = (h_0 - h)^{1/2} = 0.358.$$

Finally, we can find y_b coordinate of point b, that is the distance of point b to the shock, with 3 point accurate numerical integration as follows

$$y_b \cong \frac{1}{r_s}\int_0^{\chi_s}\frac{d\chi}{\rho u} \cong \frac{1}{r_s}\frac{\Delta\chi}{3}\left[\left(\frac{1}{\rho u}\right)_{\chi=0}+\left(\frac{4}{\rho u}\right)_{\chi=0.0625}+\left(\frac{1}{\rho u}\right)_{\chi=0.125}\right]=0.128$$

This result show that the distance between the shock and the body at point b is the 13% of the shock surface. As a further practice take $\chi = 0.0625$ to find the flow conditions to determine approximate value of $y_{b.}$

Example 7.3 Analyzing the conic shock about a slender cone at M = ∞.

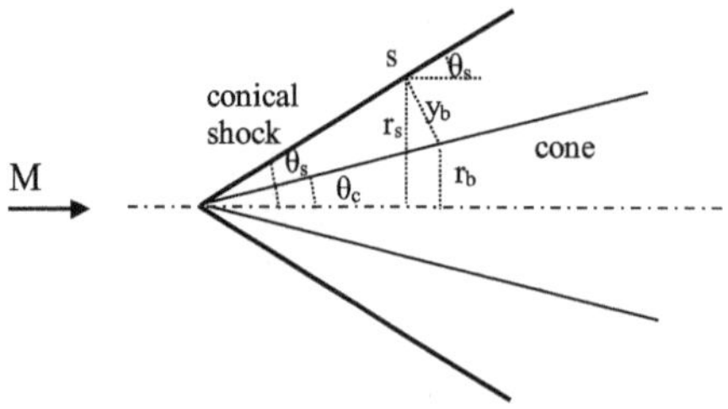

Solution The conical shock shown in the figure above can be considered as an oblique shock with radius $R_s = \infty$. With the method of Maslen the flow conditions behind the shock can be examined in a simple manner as follows:

The pressure $p = p_s + u_s(\psi - \psi_s)(R_s r_s) = p_s$.

The density: $\rho = p^{1/\gamma}\left[\rho_s/\left(p_s^{1/\gamma}\right)\right] = \rho_s$ and the enthalpy: $h = h_s = 2\gamma/(\gamma-1)(p_s/\rho_s)$

From the energy equation: $u = u_s = (h_{so} - h)^{1/2}$.

The distance between the shock and the surface: $y_b = \psi_s/(r_s\rho_s u_s) = r_s/(2\rho_s\cos\theta_s)$

The distance between the axis and the surface: $r_b = r_s - y_b\cos\theta_s = r_s\left[1-(2\rho_s)^{-1}\right]$

The angle between the shock and the surface: $\tan(\theta_s - \theta_{c)} = y_{b/}x = \sin\theta_{s/}(2\rho_s \cos\theta_{s)}$,

With which the cone angle θ_c can be found.

Using the formulae above for γ = 1.4, free Mach number = ∞, and shock angle 15° for a conical axisymmetric shock the cone angle of the body is $c_o = 13.7^{o.}$ We also find $p_s = 0.0558$ and $\rho_s = 6$.

Next, we are going to study and analyze the numerical solution of a shock created by the body of known shape at an angle of attack in hypersonic flow. The 1960s were the years of the space capsules which were sent for manned missions and came back safely to earth. The numerical solutions for the preliminary studies were performed for the Euler equations expressed in spherical coordinates with a method similar to that of Godunov (Bohachevski and Mates 1966). In the numerical solutions the finite difference method was used and the shock was captured within

4–5 discrete points across which the value of the density doubles itself. Shown in Fig. 7.9 is the numerically computed shock location obtained for a space capsule having a half cone angle of 35° with a spherical nose of radius of 4.74 m., immersed in a free stream Mach number of 29 with angle of attack of 20° at altitude of 66 km.

According to Fig. 7.9, the shock location is as close as 25 cm to the nose of the body, and at the maximum 45 cm away. The computed values of the density are at the most 5 times and at the least 2.5 times of the free stream density. For the specific heat ratio of $\gamma = 1.4$ the Rankine–Hugoniot relations also give 5 fold increase of the density after the shock. Another interesting result of this study shows that the streamline having the maximum entropy is the line crossing the shock and staying between the shock and the surface which is different from the stagnation line as shown Fig. 7.9. For a symmetric hypersonic flow, on the other hand, the maximum entropy line coincides with the stagnation streamline (Anderson 1991).

In the early 1970s, flow analysis around space shuttle type configurations started to appear in literature, by numerical solutions of Euler equations (Kutler et al. 1971). The aim of this type of detailed study was to numerically predict the hypersonic inviscid flow with multi shocks about the shuttle re-entering earth's atmosphere at high angles of attack. These studies included calculating the effect of the interaction of the bow shock of the fuselage with the shocks created by the canopy and wings. Since the Rankine-Hugoniot relations are not capable of analyzing this type of complex interactions of shocks, shock capturing methods like MacCormack's is resorted to march in the direction of the flow as a low memory

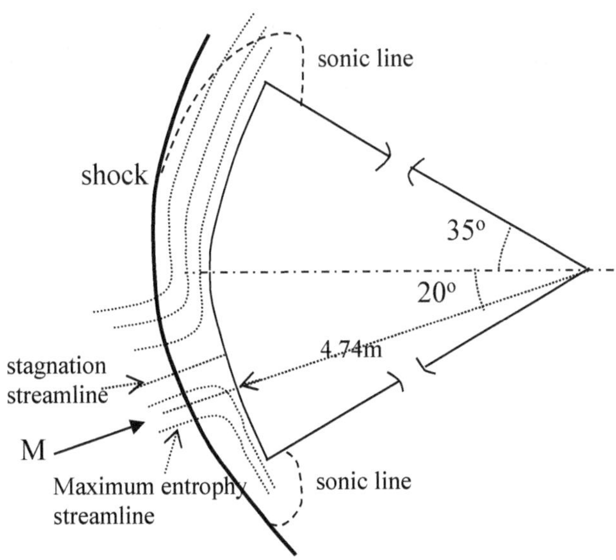

Fig. 7.9 Hypersonic flow about the space capsule with M = 29, $\alpha = 20°$. "Reprinted with permission of the American Institute of Aeronautics and Astronautics"

requiring technique (Kutler and Lomax). In their study, Kutler and the others cast the steady Euler equations in curvilinear coordinates in conservative form and marched in the flow direction step by step in accordance with the CFL (Courant Friedrichs Levy) using the finite difference method. Their solutions were 2nd and 3rd order accurate, and were compared for convergence concerning the resolution. The shuttle like geometry considered in their study had a spherical and a conical surface at the nose, and tangent to that two half expanding ellipses with their major axis coinciding; the flat bottom ellipse and the not so flat top ellipses as shown in Fig. 7.10.

The numerical solution is first made for spherical-conical nose section shown in Fig. 7.10 as the region between the cross Sects. 0–1. The solution obtained at cross Sect. 1 is given as the initial condition for the hyperbolic equation which is then solved by marching by MacCormack scheme until the desired cross section is reached. The numerical results for the position of shocks are provided in Fig. 7.11a, b at two different angles of attack for the free stream Mach number of M = 7.4.

As seen in Fig. 7.11a, at zero angle of attack the bow shock interacts with the weaker canopy shock to generate a slip surface across which there is no pressure difference but velocity difference. The slip surface is only apparent for 3rd order accurate solution. In addition, the bow shock and the weaker wing shock interacts above the wing surface to make the bow shock almost tangent to the tail.

Shown in Fig. 7.11b is the flow field for 15.3° angle of attack at which the bow shock and the canopy shock do not interact and the bow shock is located way over the tail. At zero angle of attack, the surface pressure coefficient is 8 times the free stream pressure right after the canopy shock.

In the beginning of 1990s, ESA (European Space Agency) also sponsored hypersonic flow studies past shuttle like double ellipsoid shapes solved with unsteady Euler (Molina and Huot). This time the finite element method is employed in discretizing the domain of flow for which even viscous effects are also accounted for (Zienkiewicz and Taylor 2000). We can cast the time dependent Euler equations

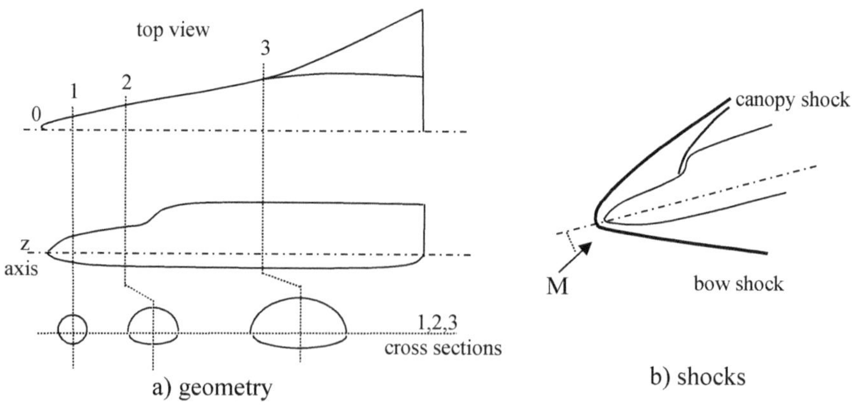

Fig. 7.10 Shuttle s like, **a** geometry, **b** bow shock and canopy shock

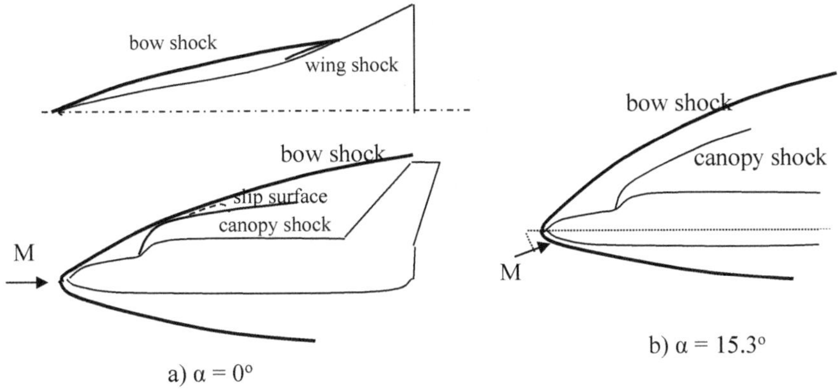

Fig. 7.11 Shuttle like geometry at M = 7.4 shocks at, **a** α = 0°, **b** α = 15.3°

in appropriate form applicable to two dimensional hypersonic flows by defining the energy as $e = h - p/\rho + \frac{1}{2}(u^2 + v^2)$ to give continuity, energy and x and y momentum respectively as follows

$$\begin{aligned} \rho_{,t} + (\rho u)_{,x} + (\rho v)_{,y} &= 0, \\ (\rho e)_{,t} + [(\rho e + p)u]_{,x} + [(\rho e + p)v]_{,y} &= 0, \end{aligned} \tag{7.45a, b}$$

$$\begin{aligned} (\rho u)_{,t} + (\rho u^2 + p)_{,x} + (\rho uv)_{,y} &= 0, \\ (\rho v)_{,t} + (\rho uv)_{,x} + (\rho v^2 + p)_{,y} &= 0 \end{aligned} \tag{7.46c, d}$$

Shown in Fig. 7.12a is the flow field in terms of the Mach equal Mach lines around the double ellipse obtained by the finite element method at M = 8 and 30° angle of attack. The results are obtained by an adaptive scheme using the coarse grid first as shown in Fig. 7.12b. Afterwards, automatic remeshing is used at high gradient regions to get a finer grid as shown in Fig. 7.13c. In Fig. 7.13a, comparison of surface pressure coefficient is shown for the coarse and fine grids. Fine grid solution indicates that the suction is more at the lower surface and the canopy shock is stronger at the upper surface. Shown in Fig. 7.13b is the density residues before and after the adaptive mesh refinement.

The forward time and the finite element discretization of Eqs. 7.45–7.46 in indicial notation become

$$\int_{\Omega} W(U^{t+\Delta t} - U^{\Delta t})d\Omega = -\Delta t \int_{\Omega} (W + \tau A_i^t W_{,i})F_{i,i} d\Omega, \quad \mathrm{i} = 1, 2 \tag{7.47}$$

Here, W is the weighing function, $U = (\rho, \rho e, \rho u, \rho v)^T$ is the unknown vector, $F_1 = (\rho u, (\rho e + p)u, \rho u^2 + p, \rho uv)^T$ and $F_2 = (\rho v, (\rho e + p)v, \rho uv, \rho v^2 + p)^T$ show the flux vector. In addition, $A_i = \partial F_i/\partial U$ is the Jakobian of the fluxes and τ

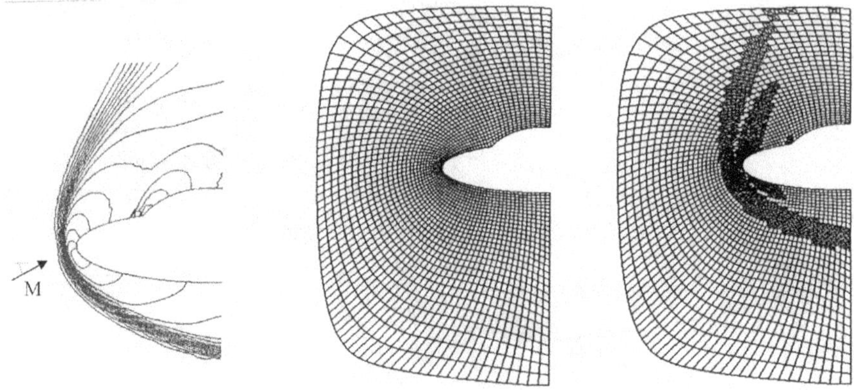

a) Equal Mach lines, b) coarse grid, c) fine grid

Fig. 7.12 At 30° angle of attack and M = 8, **a** Equal Mach lines, **b** coarse, **c** fine grid

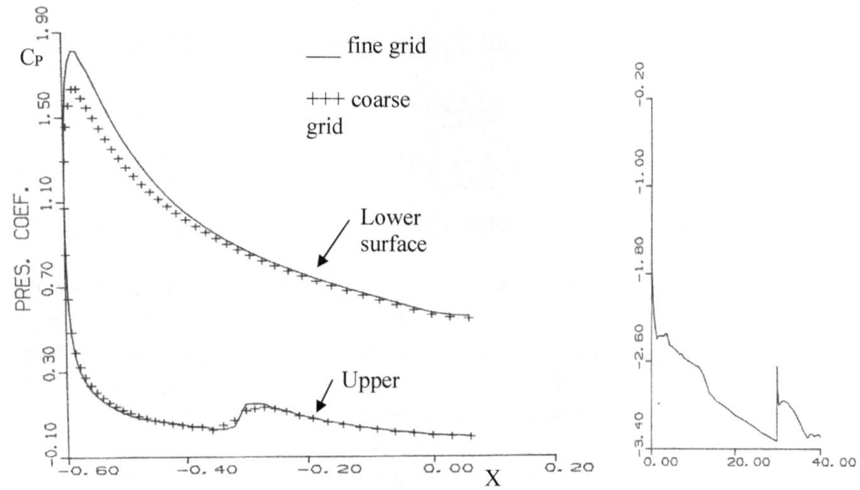

a) Upper and lower surface pressures b) residues

Fig. 7.13 Double ellipse solution **a** surface pressure coefficient, **b** residues

$A_i\ W_{,i}$ is the artificial viscosity. In terms of the element length h, local velocity magnitude$|u|$, and the local speed of sound c we can write $\tau = \frac{1}{2}\,h/(|u| + c)$. On the other hand, Eq. 7.47 can not handle the numerical oscillations around the shock. Therefore, we add the term $\lambda \sum_{i=1}^{2} (W_{,i}U_{,i})$ into the integral at the right hand side. Here, λ is proportional with the square of the element length h and gradient of the local velocity magnitude (see Problem 7.26).

So far, we have seen the hypersonic flow analysis based on the inviscid theory which does not take real gas effects into consideration. From now on in our analysis both the viscous effects and the disassociation caused by the heating of the air because of high speeds will be considered together with comparisons with the inviscid solutions.

7.7 Viscous Hypersonic Flow and Aerodynamic Heating

There are two important reasons to consider the real gas effects in hypersonic flows. The first reason is to predict the viscous drag on the body, and the second is to determine the heating caused by the high velocity gradients of the very high free stream speeds which is to stagnate on the body. The heating effect is much more at the stagnation points of the slender bodies as compared with that of blunt bodies. For example, the heating at the nose of a re-entering slender body is three times more than that of space shuttle (Anderson 1989). On the other hand, the drag caused by the strong detached shock is very high. However the skin friction drag becomes higher for the slender bodies because of having thinner boundary layers as opposed to the boundary layers of blunt bodies. In addition, we have to keep in mind that because of low density yielding low Reynolds numbers, the boundary layers of hypersonic flow must be thick compared to the low speed flows which is somewhat contrary to the common practice. Moreover, behind the detached shock occurring at the nose of a blunt body there exist a layer in which the entropy change is strong. This layer is thicker than the boundary layer and the entropy gradient in this causes extra vorticity even outside of the boundary layer as shown in Fig. 7.14. The Crocco theorem gives the relation between the entropy gradient and the vorticity generated by this gradient as follows (Liepmann and Roshko),

$$V\,x\,\omega = -T\;grad\,S \tag{7.48}$$

For this reason, while studying the hypersonic flow about a blunt body, the entropy change must be considered.

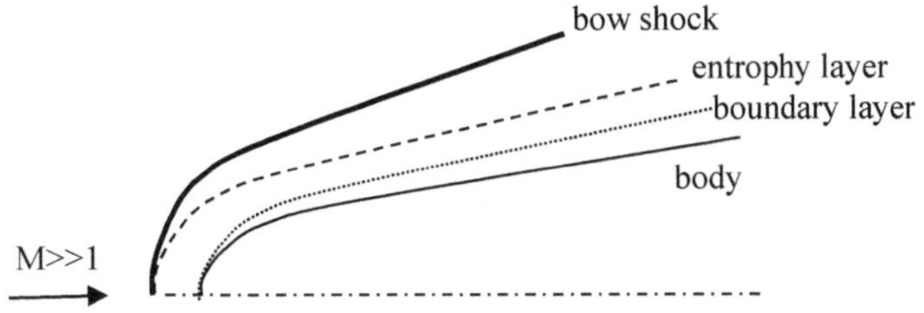

Fig. 7.14 Bow shock, entropy layer and the boundary layer around a blunt body in hypersonic flow

Now, it becomes necessary to derive the formula for the boundary layer thickness in terms of the Reynolds and the Mach numbers of the flow. As we recall from the incompressible viscous theory, the boundary layer thickness δ is inversely proportional with the square root of the Reynolds number, i.e. at station x the boundary layer thickness: $\delta \propto x/\sqrt{R_e}$. In the compressible boundary layer both the density and the viscosity changes considerably with temperature. Therefore, let us write the Reynolds number in terms of the viscosity and the density at the station x on the surface $\delta \propto x/\sqrt{\rho_w U_e x/\mu_w}$. Here, the subscript e refers to the boundary layer edge and w denotes the wall conditions. Now, we can write the boundary layer thickness as follows:

$$\delta \propto \frac{x}{\sqrt{\rho_e U_e x/\mu_e}} \sqrt{\frac{\rho_e}{\rho_w}} \sqrt{\frac{\mu_w}{\mu_e}} = \frac{x}{\sqrt{R_e}} \sqrt{\frac{\rho_e}{\rho_w}} \sqrt{\frac{\mu_w}{\mu_e}} \tag{7.49}$$

According to the boundary layer assumption, the pressure remains constant at a given station. Therefore, with the perfect gas assumption we have $\rho_e/\rho_w = T_w/T_e$, and by assuming linear dependence of viscosity on absolute temperature Eq. 7.49 becomes

$$\delta \propto \frac{x}{\sqrt{R_e}} \frac{T_w}{T_e} \tag{7.50}$$

If we assume the wall temperature as the adiabatic stagnation temperature at high Mach numbers, the relation between the temperature ratio and the edge Mach number becomes $T_w/T_e \cong (\gamma - 1)/2M_e^2$, and the hypersonic boundary layer thickness reads as

$$\frac{\delta}{x} \propto \frac{M_e^2}{\sqrt{R_e}} \tag{7.51}$$

According to Eq. 7.51 the hypersonic boundary layer thickness is proportional with the square of the Mach number.

We have seen that the boundary layer thickness in hypersonic flows grows very thick. For a flat plate at zero angle of attack, on the other hand, the pressure remains the same along x. Is this still possible for very large Mach numbers? The answer to this question lies in the parameter defined as $\overline{\chi} = \sqrt{C}\frac{M^3}{\sqrt{R_e}}$.

If we assume that the viscosity linearly changes with temperature in the boundary layer, we can write $\mu_e/\mu_w = CT_e/T_w$. Let us denote the displacement thickness of a boundary layer at zero angle of attack with δ^*. The slope of the surface because of the displacement thickness then reads as $\theta_e = d\delta^*/dx$. According to the Piston theory, for $M\theta << 1$ a linear approach gives us the pressure distribution from Eq. 7.26 as follows

$$p_w/p_\infty = 1 + \gamma M\theta = 1 + \gamma M d\delta^*/dx \tag{7.52}$$

Here, at high Mach numbers the slope θ for the adiabatic wall conditions can be approximately written as $d\delta^*/dx \cong (\gamma - 1)/2M^2\sqrt{C}/\sqrt{R_e}$ and when substituted in Eq. 7.52 to give

$$p_w/p_\infty = 1 + \gamma(\gamma - 1)\overline{\chi}/2 \tag{7.53}$$

This interaction is called the weak interaction.

For strong interaction, i.e. for Mθ ≫1 (Hayes and Probstein) we obtain

$$p_w/p_\infty \cong \gamma(\gamma + 1)M^2(d\delta^*/dx)^2/2 \tag{7.54}$$

The slope θ for the flat plate is $d\delta^*/dx \cong [(\gamma - 1)/2]^{1/2}(M^2C/R_e)^{1/4}$ (Anderson). Hence, for the strong interaction we have

$$p_w/p_\infty \cong \gamma(\gamma^2 - 1)/2\overline{\chi} \tag{7.55}$$

For both, weak and strong interaction the surface pressure formula for the specific heat ratios of γ = 1.4

$$p_w/p_\infty \cong 1 + 0.464\overline{\chi} \tag{7.56}$$

The Reynolds number depends on x, and in Eq. 7.56 the Reynolds number is in the denominator of the second term which means around the leading edge that term becomes very large and the wall pressure becomes very large. As a result of viscous interaction, the pressure around the leading edge becomes very high compared to the pressure of the ideal flow. This is the indicative of very high drag and intense heating around the leading edge for the case of hypersonic flows. The depiction of this interaction in terms of the wall to free stream pressure ratio $p_{w/}p_\infty$ for the flow about a flat plate is shown in Fig. 7.15.

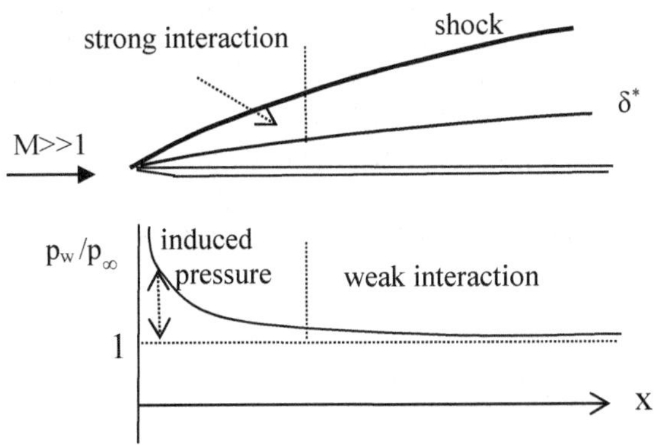

Fig. 7.15 Hypersonic flow interaction around a flat plate

Shown in Fig. 7.15 is the hypersonic strong viscous interaction at the leading edge of the flat plate to create induced pressure zone. Since there is a considerable pressure change at the leading edge region the analysis of the boundary layer in hypersonic flow must be quite different from the classical approach. In addition, since the interaction at the leading edge region is strong, the effect of the leading edge on the stations away from the stagnation point is still felt strongly even in the weak interaction zone. Therefore, the boundary layer solutions obtained in the weak interaction region in hypersonic flow is quite different from the classical boundary layer solutions.

At the leading edge of a flat plate because of the term with $1/(x)^{1/2}$ in Eq. 7.56, the pressure theoretically goes to infinity for finite Mach numbers as sketched in Fig. 7.15. In practice however, reaching to those high Mach numbers only happens during re-entry at high altitudes where the density of the atmosphere is so low. The low density at those altitudes makes the mean free path of the air molecules quite high compared to the dimensions of the leading edge which in turn makes the assumption of continuum no longer valid. For this reason, in order to obtain more realistic results for the pressure about the leading edge, instead of continuum approach, molecular models are preferred with the slip conditions on the surface to replace no-slip conditions as boundary conditions. Using slip conditions helps reducing the pressure values with the Mach number and helps giving results in agreement with the experimental values (Anderson 1989). So far we have seen the viscous interaction for the flat plate in hypersonic flow.

Now, we can further extend the interaction analysis and give a brief summary for different type of bodies in hypersonic flow. If we know the pressure distribution p_c over a conical surface then because of viscous interaction the induced pressure difference approximately reads as follows (Talbot and Hypersonic 1958)

$$\frac{p - p_c}{p_c} = 0.12\bar{\chi}_c, \quad 0 \leq \bar{\chi}_c = M_c^3\sqrt{C/R_{ec}} \leq 4 \tag{7.57}$$

Here, c denotes the potential flow conditions at the conical surface.

In hypersonic flow, the viscous interaction lowers the lift to drag ratio, L/D, by increasing Mach number. This effect is more for the blunt nosed bodies with blunt shapes as opposed to the slender bodies with wings. The maximum L/D ratio is 1.7 for a cone with half cone angle of 9° at M = 9, and it drops down to L/D = 0.5 at M = 19 (Anderson 1989). For a space capsule type of bodies, on the other hand, the L/D ranges between 0.4 and 0.2.

The serious problem of the viscous interaction in Hypersonic flow is the aerodynamic heating. First, let us study the heat transfer problem for a flat plate in high speed flow. If we let q_w be the heat transfer for a unit area in a unit time, we have

$$q_w = S_t \rho_e U_e (h_{ad} - h_w) \tag{7.58}$$

Here, S_t is the dimensionless Stanton number, h_{ad} and h_w are the adiabatic wall temperature and the value of the enthalpy at the wall respectively. The famous

Reynolds analogy states that the Stanton number is related to the skin friction coefficient $c_f = \tau_w/(1/2\ \rho_e\ U_e{}^{2})$ with $S_t \cong c_f/2$ (Schlichting). The enthalpy difference $h_{ad} - h_w$ in Eq. 7.58 is the main factor of the surface heating in high speed flows. The value of the enthalpy at the wall is the value obtained by solving the energy equation with adiabatic wall conditions. Using the engineering approach, the recovery factor r is employed to define the relations between the adiabatic wall enthalpy, boundary layer edge conditions e and the stagnation enthalpy h_0 to give the following

$$h_{ad} = h_e + rU_e^2/2 \quad \text{and} \quad h_0 = h_e + U_e^2/2 \qquad (7.59-\text{a, b})$$

From Eq. 7.59-a,b we obtain

$$r = \frac{h_{ad} - h_e}{h_0 - h_e} \qquad (7.60)$$

Here, the free stream stagnation enthalpy is always greater than the adiabatic enthalpy of the wall to make r is always less than unity.

Defining the Prandtl number as the ratio between the viscous energy loss and the heat conduction, i.e., $P_r = \mu\ c_p/k$, for the flat plate in hypersonic flow conditions the Blasius solution gives (White)

$$r \cong \sqrt{P_r} \qquad (7.61)$$

Equation 7.61 is valid for a wide range of Mach numbers with 2% accuracy. This gives a relation between the Stanton to surface friction ratio in terms of the Prandtl number as follows

$$S_t/c_f = 1/2P_r^{-2/3} \qquad (7.62)$$

Equation 7.62 is also valid with 2% accuracy for a wide range of Mach numbers. Although Eq. 7.62 is obtained for flows around the flat plate, it has been applied to determine the aerodynamic heating caused by three dimensional slender bodies (Anderson).

In case of turbulent hypersonic flow past flat plate with increase in Mach number, there is a considerable decrease in Stanton number. According to the Van Driest's turbulent flow data while Mach number is ranging from 0 to 10, the Stanton number decreases to 0.1 of its incompressible value.

So far we have seen the aerodynamic heating for the slender bodies with sharp leading edges. The solution for stagnation point of the flat plate was singular. Now, we can analyze the aerodynamic heating for the bodies with blunt noses. The analysis of heat transfer at the stagnation point of the circular cylinder and sphere was made by Van Driest, and the following formula for the heat transfer was provided (Anderson):

$$\text{Cylinder}: \qquad q_w = 0.57\,P_r^{-0.6}(\rho_e\mu_e)^{1/2}\sqrt{dU_e/dx}(h_{aw} - h_w) \qquad (7.63-\text{a, b})$$

$$\text{Sphere}: \qquad q_w = 0.763P_r^{-0.6}(\rho_e\mu_e)^{1/2}\sqrt{dU_e/dx}(h_{aw} - h_w)$$

Here, U_e is the boundary layer edge velocity and it naturally takes the value of zero at the stagnation point. The derivative of the edge velocity with respect to x at the stagnation point is non zero, and it is inversely proportional with the local radius of curvature. Assuming Newtonian flow, the derivative of the edge velocity reads as

$$\frac{dU_e}{dx} = \frac{1}{R}\sqrt{\frac{2(p_e - p_\infty)}{\rho_e}} \qquad (7.64)$$

Here R, is the radius of curvature at the stagnation point (see Problem 7.34). If we substitute Eq. 7.64 in Eq. 7.63-a, the heat transfer at the stagnation point of the circular cylinder for hypersonic flow reads as

$$q_w = 0.57P_r^{-0.6}(\rho_e\mu_e)^{1/2}\frac{1}{R^{1/2}}\left[\frac{2(p_e - p_\infty}{\rho_\infty}\right]^{1/4}(h_{aw} - h_w) \qquad (7.65)$$

The aerodynamic heating is inversely proportional with the radius of curvature of the stagnation point. This fact forces the hypersonic vehicles to have a round nose as shown in Fig. 7.9 of the re-entering space capsule. The change in the Stanton number at the stagnation point of a cylinder is experimentally observed to be proportional with the inverse square root of the radius of curvature as given in Eq. 7.65. Naturally, as we move away from the stagnation point, the aerodynamic heating reduces considerably. At $\psi = 45^\circ$ on the surface, halfway between the stagnation point and the shoulder, this reduction goes down to half the value of the stagnation heat transfer, and at the shoulder where $\psi = 90^\circ$, the heating becomes the one tenth of the stagnation value (Anderson).

We have stressed the role of viscous effect concerning aerodynamic heating. In the no slip condition which causes high velocity gradients, i.e. high vorticity value is due to viscosity. In addition, the high entropy gradient occurring behind the strong bow shock generates a vorticity field as given by Eq. 7.48 and creates an entropy layer as shown in Fig. 7.14. The vorticity generated by the entropy gradient also creates non negligible aerodynamic heating starting from the stagnation point. Shown in Fig. 7.16 are the maximum aerodynamic heating values on the line, which is the symmetry line of the bottom surface of the space shuttle, with and without entropy change considered.

According to Fig. 7.16, the difference between the results obtained with entropy and without entropy is small near the stagnation region, and it increases monotonically in the flow direction as the entropy layer increases (Anderson). The results obtained by considering the entropy gradient are agreeable with the experimental measurements, and therefore, they should be preferred.

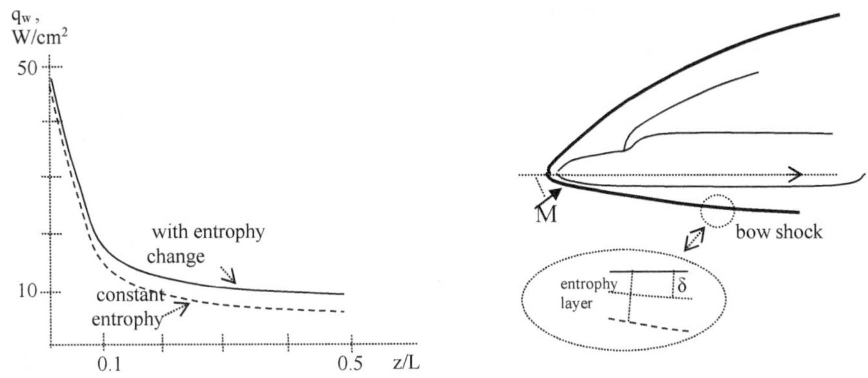

Fig. 7.16 Maximum aerodynamic heating with and without the entropy change line of the shuttle at 40° angle of attack

The last subject to be studied in hypersonic heating is related to the interaction of the strong shock and the boundary layer. This type of heating is usually generated by the oblique shock which is created at a slender leading edge of an external body of hypersonic air vehicle. Since this shock is inclined, it strikes another external part which is located at downstream, and reflects from the boundary layer to generate heat. For the first time, this type of heating is encountered because of the oblique shock created at the intake of an engine interacting with the boundary layer on the engine-body junction of a hypersonic plane in a test flight (Neumann). The sketch and the effect of the λ shock are shown in Fig. 7.17a, b. The variation of the Stanton number and the surface pressure are given in Fig. 7.17b around the region where the shock boundary layer interaction is taking place.

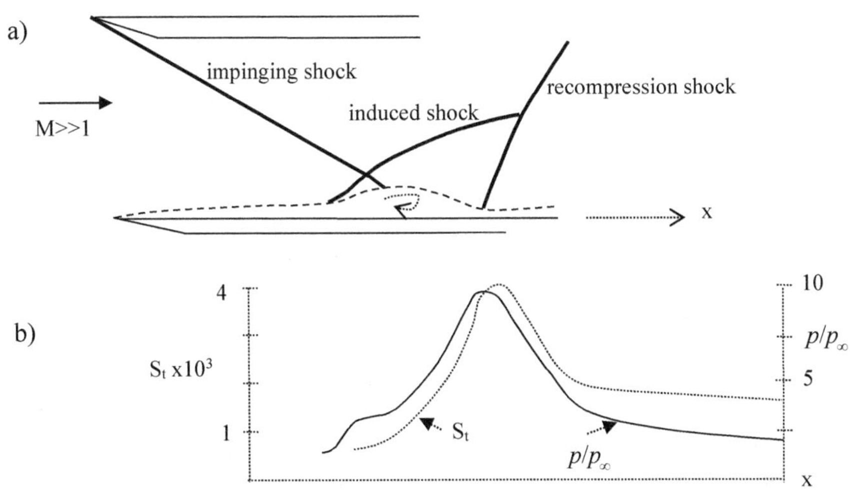

Fig. 7.17 Shock boundary layer interaction: **a** Physics, **b** variation of S_t and pressure

As seen in Fig. 7.17b, the variation of Stanton number is closely related to the surface pressure variation. Right after the λ shock, the value of Stanton number at the surface increases eight times (Marvin et al.). This increase in the Stanton number is the indicative of the aerodynamic heating due to shock boundary layer interaction. The surface pressure increase, on the other hand, is 10 fold across the shock. Now, we can obtain the relation between the pressure change and the aerodynamic heating using the variation of the Stanton number. In laminar flow, the skin friction coefficient c_f is inversely proportional with the Reynolds number. Since the Stanton number is proportional with the skin friction coefficient, the following relation can be deduced

$$S_t \propto 1/\sqrt{R_e} \propto 1/\sqrt{\rho_e} \tag{7.66}$$

Since the pressure is proportional with the density, the definition of the surface heating, q_w becomes

$$q_w \propto \sqrt{p_e} \tag{7.67}$$

to give the proportionality of aerodynamic heating with pressure.

In turbulent flows there is a relation between the surface friction and the Reynolds number as follows: $c_f \propto 1/(R_e)^{1/5}$ (Schlichting). The relation between the surface heating and surface pressure for turbulent flow then becomes

$$q_w \propto (p_e)^{4/5} \tag{7.68}$$

7.8 High Temperature Effects in Hypersonic Flow

We have studied, so far, how high the temperatures can get at the stagnation point of a hypersonically flying vehicle because of a strong shock forming before the nose. At very high temperatures and low pressures of high altitudes, the chemical composition of air can not remain the same. Before the chemical composition change, first the change in the specific heats with temperature occurs. Therefore, the air is no longer a calorically perfect gas. Since the change in the specific heats depends on the temperature, we can assume the air as a thermally perfect gas, and use the temperature dependent specific heat ratios instead of constant $\gamma = 1.4$. On the other hand, the gas constant R has the same dependence on the specific heats both for the calorically or thermally perfect gas, i.e. $c_p - c_v = R$ still holds.

At high temperatures, the temperature dependent behavior of the specific heats of the air can be determined with the aid of 'gas kinetics'. The chemical composition of the air at normal conditions contains 79% molecular nitrogen N_2, 20% molecular oxygen O_2, and 1% other gases. In this composition, neglecting the other gases the air is mainly composed of molecular nitrogen and oxygen. For di-atomic gases, the internal energy of the molecule is composed of the translational and the rotational energies. This internal energy increases linearly with temperature, and is expressed as: $e = e_{tr} + e_{rot.}$ Using statistical methods for di-atomic gases the translational energy depending on temperature T reads as $e_{tr} = 3/2$ RT, and the rotational energy becomes: $e_{rot} =$ RT (Lee et al. 1973). This gives the total internal energy in terms of temperature, and the specific heat constant at constant volume as follows

$$e = \frac{5}{2}RT, \quad c_v = \frac{\partial e}{\partial T} = \frac{5}{2}R \tag{7.69a, b}$$

At higher temperatures, temperatures above 800 K, the bond between the di-atomic molecules starts to vibrate to further increase the internal energy. This increase in the internal energy is called the vibration energy of a molecule. The change in the vibration energy of the molecule is non-linear with the temperature, and the classical thermodynamics is insufficient to calculate the temperature dependence of vibration energy. The quantum mechanical approach with the concept of partition function is necessary to express the vibration energy as follows (Appendix 8):

$$e_{vib} = \frac{h\nu/kT}{e^{h\nu/kT} - 1}RT \tag{7.70}$$

Here, h = 6.625 $\times$ 10^{-34} Js is the Planck constant, ν is the fundamental frequency of the molecule, and k = 1.38 $\times$ 10^{-23} J/K is the Boltzmann constant. The fundamental frequencies for the nitrogen and the oxygen molecules are different, and they are for N_2: $\nu_{N_2} = 7.06x10^{13}$ s^{-1} and for O_2: $\nu_{O_2} = 4.73x10^{13}$ s^{-1} (Anderson 1989). Accordingly, the specific internal energy at high temperatures reads as

$$e = \frac{5}{2}RT + \frac{h\nu/kT}{e^{h\nu/kT} - 1}RT \tag{7.71}$$

Hence, the derivative of the Eq. 7.71 with respect to temperature gives us the specific heat constant at constant volume with the following temperature dependence

$$c_v = \frac{5}{2}R + \frac{(h\nu/kT)^2 e^{h\nu/kT}}{(e^{h\nu/kT} - 1)^2}R \tag{7.72}$$

Shown in Fig. 7.18 is the variation of c_v/R with temperature for the nitrogen and the oxygen molecules having vibrational energies.

According to Fig. 7.18, when 4000 K is reached the value of c_v/R approaches 7/2 as its limit value for di-atomic molecules. The limiting value of the exponential term in Eq. 7.72 approaches unity as temperature goes to infinity.

Fig. 7.18 Variation of c_v/R at high temperatures for O_2 and N_2

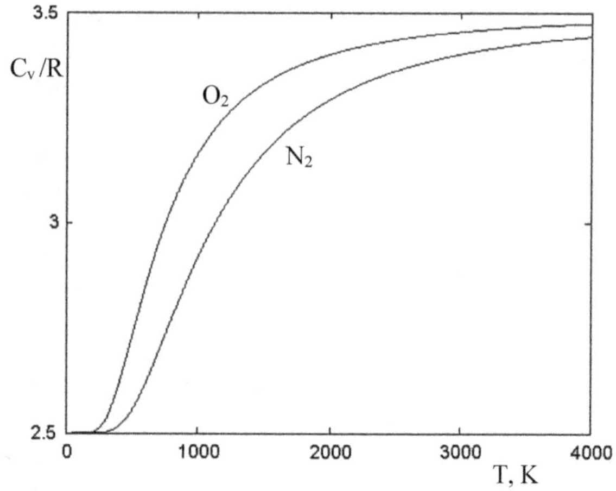

On the other hand, the classical statistical theory gives the value of vibrational energy as $e_{vib} = RT$ which is true only for T approaching infinity. For the values of temperature which are of interest to us, the evaluation of vibration energy with classical theory is not correct. At temperatures above 2000 K the oxygen molecules disassociate and above 4000 K the same thing happens to the nitrogen molecules so that the chemical composition of the air changes, and the relevant chemical reactions must be include at such high temperatures. The necessary reaction energies for the partial or full disassociations of the species are provided by the aerodynamic heating generated by high speeds and the ambient pressure. Under normal room pressure, the full disassociation of oxygen molecules is complete at 4000 K, and the nitrogen molecules are fully disassociated at 9000 K (Anderson). The disassociation of molecules starts at smaller temperatures at low ambient pressures of high altitudes. At higher temperatures than 9000 K, both oxygen and nitrogen atoms start to ionize. For this reason, it becomes necessary to construct a graphical representation for a hypersonic vehicle subjected to aerodynamic heating because of its high speed at different altitudes having different ambient pressure in which the continuum approach still holds, Fig. 7.19 (adapted from Riedelbauch et al. and Anderson).

According to Fig. 7.19, the vibration energy starts before the speed of 1 km/s, and continues up to 2.6 km/s which is indicated by a solid vertical line. Above 3 km/s, the oxygen molecules disassociate at sea level and disassociation starts at 2 km/s at high atmospheric levels. The range of oxygen disassociation, indicated with dashed dotted line, occurs between 3.2–6.5 km/h, at the upper levels it happens at 2.0-5.0 km/h. As the speed increases the nitrogen molecules disassociation range changes in the speed range of 6–10 km/s, whereas this range drops down to 5–8 km/s at higher altitudes as shown with dashed vertical curves. At even higher speeds and altitudes higher than 20 km, the ionization of oxygen and nitrogen, as shown with a solid curve starts. This speed is above 10.5 km/s at the top levels of the atmosphere. Also shown in Fig. 7.19 is the approximate re-entry orbit of the

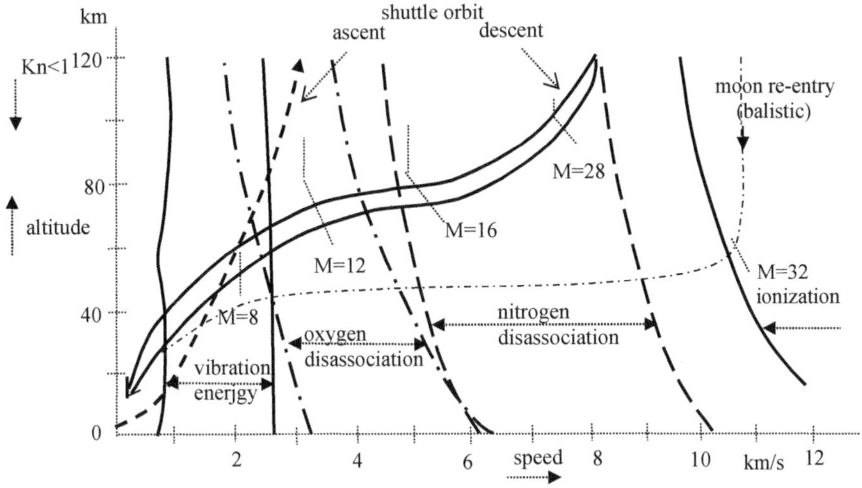

Fig. 7.19 Change in the composition of the air with speed and altitude

space shuttle. According to this orbit, the shuttle cruises aerodynamically and ballistically in disassocited atoms with M = 28 at altitude of 100 km, and descends down to the 50 km altitude as its speed goes down to M = 8 while moving in air molecules full of vibrational energy. At the left side of the Fig. 7.19, shown with a single dashed curve is the approximate ascending path of the space shuttle. On its way up, the shuttle moves through the air molecules put into vibration by its motion, and after the altitude of 70 km it goes through a region of the atmosphere where the oxygen molecules are disassociated. After the level of 80 km the air is no longer continuous medium since the Knudsen number is above 0.1, and no- slip condition no longer prevails.

At the far right side of Fig. 7.19, shown is the path line of a space capsule, like Apollo capsule, making a re-entry at a speed of 11 km/s in an environment that consists of ionized oxygen and nitrogen atoms. The path of the capsule is totally ballistic, i.e., only the drag force is acting to slow it down. At altitude of 52 km, the capsule has a very high Mach number, M = 32, which creates a very strong shock to increase the temperature to the order of 9000 K. Since the temperatures are sufficiently high, we can study the chemical reactions involved in the disassociation of air molecules assuming that the reactions are occurring in equilibrium, and can determine the relations for pressure, density and the temperature of the air which is no longer a perfect gas. Now, we write the law of mass action for each species involved in chemical reactions with equilibrium constants K and the stoichiometric coefficients depending on temperature (Denbigh) as follows

$$O_2 \leftrightarrow 2O \qquad K_{p,O_2}(T) = \frac{(p_O)^2}{p_{O_2}} \tag{7.73a}$$

$$N_2 \leftrightarrow 2N \qquad K_{p,N_2}(T) = \frac{(p_N)^2}{p_{N_2}} \tag{7.73b}$$

$$N + O \leftrightarrow NO \qquad K_{p,NO}(T) = \frac{p_{NO}}{p_N p_O} \tag{7.73c}$$

$$N + O \leftrightarrow NO^+ + e^- \qquad K_{p,NO^+}(T) = \frac{p_{NO^+} p_{e^-}}{p_N p_O} \tag{7.73d}$$

Here, the total pressure p in terms of the partial pressures of 6 species is written as

$$p = p_{O_2} + p_{N_2} + p_N + p_O + p_{NO} + p_{NO^+} + p_{e^-} \tag{7.73e}$$

Since the equilibrium constant for each specie involved in Eq. 7.73 is given in terms of the temperature, we can obtain the total pressure for a given temperature. We know the density from the solution of the continuity equation. Then, we can express the equation of state as a polynomial for the air at high temperatures in terms pressure, density and the temperature. There are various forms of these polynomials in open literature, however, we will be making use of the one which is prepared by Tannehill and Mugge and provided by Anderson for its convenient usage in our next example.

Example 7.4 The re-entry speed of Apollo capsule is 11 km/s which corresponds to M = 32.2 at 52 km altitude. Compute the temperature behind the shock assuming (a) calorically perfect gas, (b) the chemical equilibrium is reached so that new equation of state can be used.

Solution

(a) Assuming calorically perfect gas and $\gamma = 1.4$; from the temperature ratio we get $T_o/T = 1+(\gamma-1)/2\ M^2 = 208$. Here, the temperature before the shock at 52 km altitude $T_1 = T=270.7$ gives us $T_2 = T_o = 208 \times 270.7 = 56305$ K! This temperature is so high and so wrong! Because, under these conditions the air is not calorically perfect and there is a considerable decrease in the temperature behind the shock because of energy used by the formation energy of the new species formed.
(b) The pressure and density ratios for after and before the shock are to be computed as $p_2/p_1 = 1387$ and $\rho_2/\rho_1 = 15.19$ respectively. In order for Tannehill and Munge's polynomial approximations to be used at 52 km altitude we express the pressure and density with respect to sea level values. Accordingly, we find

$$p_2/p_{s1} = p_2/p_1 \times p_1/p_{s1} = 1387 \times 0.0007874 = 1.092 \quad \text{and}$$

$$\rho_2/\rho_{s1} = \rho_2/\rho_1 \times \rho_1/\rho_{s1} = 15.19 \times 0.0008383 = 0.01273$$

From those values and reading from the graph gives us

$T_2/T_{sl} = 40$ and $T_2 = 11520$ K. This gives us a more realistic temperature behind the shock. If we calculate the pressure behind the shock with equilibrium chemistry the result obtained with the calorically perfect gas does not change. However, for the density there is a difference of 2.5 fold!

In the orbit of the re-entering capsule from moon mission shown in Fig. 7.19, argon is the only gas whose concentration, 1%, remains the same. Apart from this ionized nitrogen oxide, NO^{+}, and electron, e^{-}, molal concentrations do not exceed 10^{-3}. This concentration affects the gas dynamic behavior insignificantly. However, the ionization amount is sufficient to destruct the electronical communication between the capsule and earth (Anderson 1989).

Another ballistic orbit is the re-entry path of the Mars mission, which lies outside of Fig. 7.19, and which has the approximate re-entry speed of 15 km/s. The order of aerodynamic heating generated and problems created at this speed, naturally, require special design and analysis.

In order to obtain meaningful lift during the descent of the space shuttle, the flight path inside of the narrow enveloping curve shown in Fig. 7.19 must be followed. In the lower path of the enveloping curve the descent is fast and the lift is high to give $m/(C_L\ S) = 5000\ kg/m^2$, where m is the total mass, C_L is the total lift and S is the lifting surface area of the shuttle. The upper route on the other hand gives $m/(C_L\ S) = 50\ kg/m^2$. The actual intermediate path followed by the shuttle provides us approximately $m/(C_L\ S) = 500$.

Obviously, the addition of chemical reactions and the equilibrium chemistry to the equations we consider further complicates the analysis, however, at high temperatures it provides us with more realistic solutions.

The difference between the solutions based on equilibrium chemistry and the calorically perfect gas assumption at high free stream Mach numbers at high altitudes not only yields different values for temperature and density but also it gives very different flow features and characteristics. Shown in Fig. 7.20 are the different shock locations in front of a circular cylinder which are obtained with the assumption of calorically perfect gas and with chemical equilibrium assumptions.

According to Fig. 7.20, the shock location obtained with the equilibrium chemistry is much closer to the body. Naturally, with the chemical reactions behind the shock, the composition of the air also changes. The changing composition of the air alters the total pressure and the gaseous properties of the air; therefore, the analysis needs to be aerothermochemical rather than aerothermodynamic. In addition, in our aerothermochemical analysis fast chemical composition changes may force us to resort to chemical kinetics which considers the non equilibrium chemistry in flow analysis. Now, we can calculate the shock position assuming different flow conditions in front of a blunt body.

So far we have seen the effect of the equilibrium chemistry on the air composition at high temperature in hypersonic flow. The equilibrium chemistry assumes

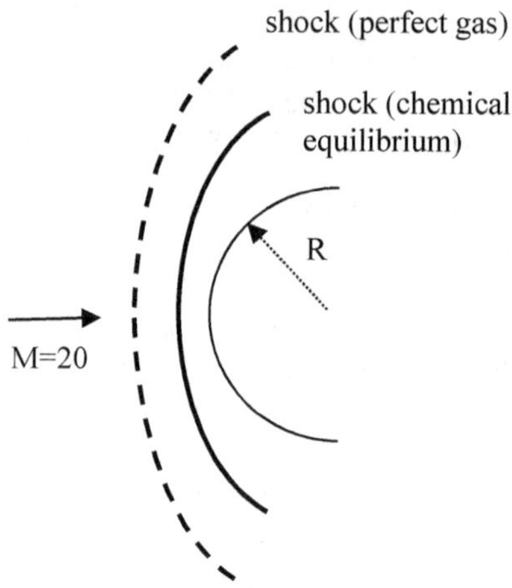

Fig. 7.20 At M = 20 free stream and 20 km altitude shock forming before a circular cylinder with assuming **a** perfect gas **b** chemical equilibrium

that throughout the flow the composition of the air changes instantly under equilibrium. At high flow speeds there are two different procedures to determine the composition of the air. These assumptions are (i) frozen flow, and (ii) flow under finite chemical rates.

In frozen flow, because of high speeds, the air molecules are assumed to continue to flow without undergoing chemical reaction. Because its composition does not change, the flow is assumed to be frozen. In reality, however, the chemical reactions take place with a finite rate, that is the flow composition is between the frozen and the equilibrium chemistry. For this reason we resort to chemical kinetics, which deals with the chemical reactions happening with finite rates and finding the rates of reaction experimentally and/or theoretically. As we have done before, let us start from the vibrational energy of a diatomic gas like oxygen molecule O_2 which becomes effective above 800 K. Equation 7.70 gave us the vibrational energy of a diatomic gas as follows

$$e_{vib}^{eq} = e_{vib} = \frac{hv/kT}{e^{hv/kT} - 1} RT \tag{7.74}$$

The physical model which explains the reason for the vibrational energies of diatomic gas molecules is the collision between the enough number of molecules as the temperature rises to make Eq. 7.74 to become active. In practice, however, the air molecules behind the shock do not start to vibrate immediately. There has to be some time to pass for a molecule to be in complete state of vibration. The previous vibration state of the molecule affects the vibrational energy of the molecule at current time t. Therefore, we need to model the time dependent vibrational energy change for the molecule with a first order differential equation as follows

$$\frac{de_{vib}}{dt} = \frac{1}{\tau}\left(e_{vib}^{eq} - e_{vb}\right) \tag{7.75}$$

Here, τ is the relaxation time which depends on the pressure and the temperature of the air as follows (Vincenti and Kruger).

$$\tau = C\frac{\exp(K_2/T)^{1/3}}{p} \tag{7.76}$$

The values of C and K_2 are determined experimentally. These values are provided for the oxygen, and nitrogen molecules for various temperature ranges by Vincenti and Kruger and summarized in Table 7.1 as follows:

On the other hand, the solution of the first order differential equation, Eq. 7.75, as an initial value problem with initial condition for the vibrational energy: $e_{vib} = e_{vo}$ at $t = 0$ gives us the following

$$e_{vib} = e_{vib}^{eq} + \left(e_{v0} - e_{vib}^{eq}\right)\exp(-t/\tau) \tag{7.77}$$

According to Eq. 7.77, this simple physical model allows us to solve the time dependent vibration energy for the same specie with two different initial conditions. The first possibility is to start with an initial vibration energy which is smaller than the equilibrium vibration energy and to reach the equilibrium vibration energy. In this first case, during acceleration or ascending of the vehicle the temperature increases with the shock strength increase; therefore, the small value of the initial vibration energy increases to reach the equilibrium vibration energy. The second case, on the other hand, is encountered by a re-entering vehicle at very high temperatures, which makes the initial vibration energy higher than the equilibrium vibration energy which is reached by slowing down of the vehicle together with the decrease in temperature.

Now, we can analyze the time dependence of vibration energy given by Eq. 7.77 as a graph provided in Fig. 7.21. According to Fig. 7.21, in the case of re-entry very high speeds create very high temperatures with high initial vibration energy reaching its equilibrium by time. On the other hand, during ascend and before leaving the atmosphere the species around the vehicle first have low vibrational energy and with speed up process the ambient temperature rises and the equilibrium

Table 7.1 Relaxation time constants for vibration energies

Specie	C (atm-μs)	K_2, K	Temperature range (K)
O_2	5.42×10^{-5}	2.95×10^{6}	800–3200
N_2	3.58×10^{-4}	1.91×10^{6}	800–6000
NO	4.86×10^{-3}	1.37×10^{5}	1500–3000

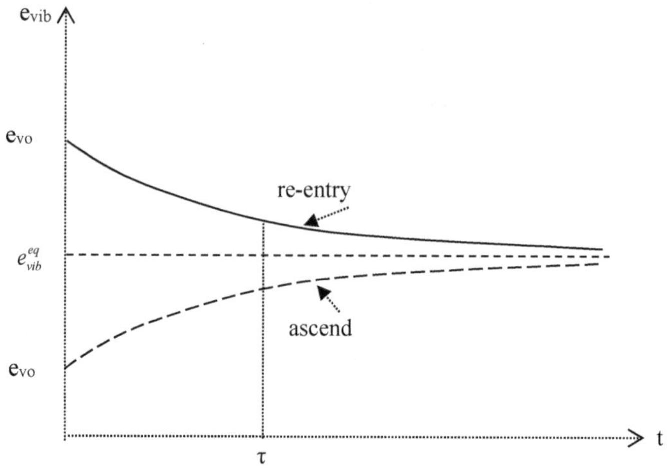

Fig. 7.21 Time dependence of vibration energy

vibration energy is reached in time. Here, in both cases reaching the equilibrium vibration energy happens in micro seconds because of the character of the chemical reactions as demonstrated in the following example.

Example 7.5 Obtain the time dependent expression for the vibration energy of pure oxygen molecules at 1 atmosphere pressure and 3000 K. Find the time elapsed for the difference between the initial vibration energy and the equilibrium vibration energy to drop to value of 1%.

Solution From Table 7.1 the relaxation time $\tau = 1.13 \times 10^{-6}$ s. Equation 7.77 gives

$$e^{-t/\tau} = 0.01$$

and,

$$t = 1.13 \times 10^{-6} \times 4.60 = 5.20 \times 10^{-6} \text{s}. \ .$$

As seen in the example, the passage from higher energy levels to equilibrium energy level takes place in micro seconds. This means non equilibrium vibration energy changes take place 2 or 3 order of magnitude faster than the solution of unsteady flow equations. This in practice means that the one step time resolution of the unsteady flow solvers is more than enough to reach equilibrium in time dependent treatment of non equilibrium chemical reactions using chemical kinetics.

Now, we can look at the rate of chemical reactions taking place at high temperatures, and compare the differences between the results obtained for the flow field with equilibrium chemistry.

Previously, we have given the equilibrium chemistry equations, Eq. 7.73a, d used for air at high speeds. Now, we can construct a new set of equations in which catalyzing effects of M molecules will be used. In the reactions considered here the catalysis is possible either with the reacting molecules themselves, or molecules from outside can act as catalisors. When we give the chemical reactions for the air molecules, we also provide the relevant catalisors in the form of a table. Naturally, with the aid of different catalisors the chemical reaction speeds will be quite different. The following four reactions will take place with different catalyzing agents as follows

$$(1) \qquad O_2 + M \leftrightarrow_{k_{b1}}^{k_{f1}} 2O + M; \quad \frac{d[O]}{dt} = 2k_{f1}[O_2][M] - 2k_{b1}[O]^2[M] \tag{7.78a}$$

$$(2) \qquad N_2 + M \leftrightarrow_{k_{b2}}^{k_{f2}} 2N + M; \quad \frac{d[N]}{dt} = 2k_{f2}[N_2][M] - 2k_{b2}[N]^2[M] \tag{7.78b}$$

$$(3) \qquad N + O + M \leftrightarrow_{k_{b3}}^{k_{f3}} NO + M; \quad \frac{d[NO]}{dt} = k_{f3}[N][O][M] - k_{b3}[NO][M] \tag{7.78c}$$

$$(4) \qquad N + O \leftrightarrow_{k_{b4}}^{k_{f4}} NO^+ + e^-; \quad \frac{d[NO^+]}{dt} = k_{f4}[N][O] - k_{b4}[NO^+] \tag{7.78d}$$

The coefficients k_f and k_b, the forward and backward reaction rate coefficients of Eq. 7.78 a–d are obtained experimentally. In addition, in reactions ranging from (1) to (3), the different M atoms working as catalisor provide different reaction rate coefficients. For only Eq. (1), we can write 5 different equations. That is in order to obtain 2O; N, NO, O, O_2 and N_2 get into chemical reaction with O_2. Similarly, in order to obtain 2 N, reaction (2) has to undergo 5 different sub reactions, and to get NO, reaction (3) must undergo 3 different sub reactions.

Now, let us see how the reaction rate coefficients are calculated. The proper Arrhenius equation for the forward reactions depending on temperature T can be written (Vincenti and Kruger) as follows

$$k_f = C_f T^{\eta_f} \exp(-K_f/kT) \tag{7.79}$$

Using the Vincenti and Kruger's data, the reaction constants for reaction (1) to (4) are provided in Table 7.2

Now, we can plot the forward reaction rates for chemical reactions (1) and (3) in terms of temperature given at Table 7.2, Fig. 7.22. According to the graphs in Fig. 7.22, formation speed of 2O from O_2 becomes different from the formation speed of nitrogen oxide from oxygen and nitrate above 3500 K. At 5000 K, this difference is as high as one order of magnitude. This means, at high temperatures the disassociation of oxygen molecules is much faster compared to the production of nitrogen oxide from disassociated oxygen and nitrogen if reactions 1) and 3) are considered alone.

Table 7.2 Coefficient of reaction speeds at high temperatures

Reaction	Catalisor	C_f	η_f	K_f/k (K)	Temperature range (K)
(1)	O_2	1.9×10^{21}	−1.5	59500	2800–5000
(2)	O_2	1.9×10^{17}	−0.5	113000	6000–9000
(3)	NO	7.9×10^{21}	−1.5	75500	3000–8000
(4)	–	6.5×10^{11}	0	31900	4000–5000

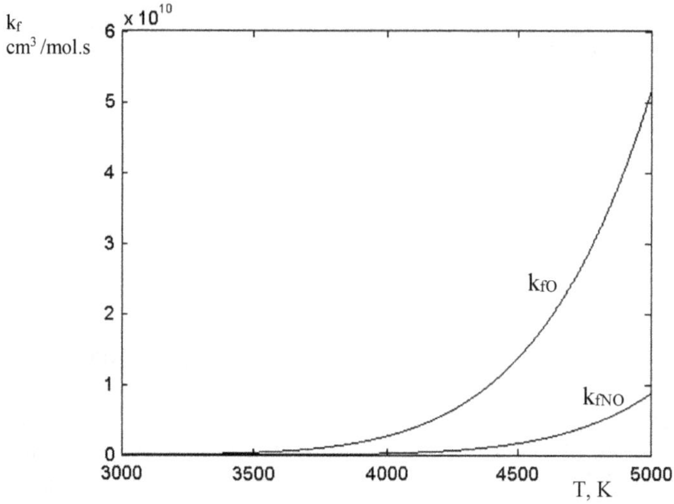

Fig. 7.22 Comparison of the forward reaction rates of disassociation of oxygen with formation of nitrogen oxide

As we give the type of reactions with 7.78 a–d, we also give the relevant forward and backward reaction rates. The Arhenius equation, Eq. 7.79, is the formula to be used to calculate the forward reaction rate. Backward reaction rates, k_b, on the other hand, can be found from equilibrium chemistry. If we write the concentrations at chemical equilibrium using Eq. 7.78 a, in terms K_p we obtain

$$K_c = \frac{k_f}{k_b} = \frac{[O]_{eq}^2}{[O_2]_{eq}} = \frac{1}{RT} K_p \tag{7.80}$$

Now, we can use Eq. 7.80 in 7.78 a to get

$$\frac{d[O]}{dt} = 2k_{f1}\left\{[O_2] - \frac{1}{K_c}[O]^2\right\}[M] \tag{7.81}$$

For the remaining reactions the backward reaction rates can be obtained similarly.

Now, let us formulate the one dimensional reacting flow with chemical kinetics as the simplest flow case. Let M_i be the molal weight of specie in chemical reaction. The mass rate of the same specie will be given by the aid of Eq. 7.81 as $\dot{w}_i = M_i \frac{d[.]_i}{dt}$. The same specie has its partial density, ρ_i, change with the flow velocity in terms of the continuity equation as follows

$$\partial \rho_i / \partial t + \partial(\rho_i u)/\partial x = \dot{w}_i \tag{7.82a}$$

The total sum of partial densities of the species gives the average density of air $\sum_i \rho_i = \rho$. In terms of the mass fraction of the species $c_i = \rho_i/\rho$, Eq. 7.82b becomes

$$\partial c_i / \partial t + u \partial c_i / \partial x = \dot{w}_i / \rho \tag{7.82b}$$

In addition, the vibration energy in terms of mass fraction reads as

$$\partial c_i e_{vib} / \partial t + u \partial (c_i e_{vib}) / \partial x = c_i (e_{vib}^{eq} - e_{vib}) / \tau \tag{7.83}$$

The governing equations: continuity, momentum, and energy equations then respectively read as

$$\partial \rho / \partial t + \partial(\rho u)/\partial x = 0 \tag{7.84}$$

$$\rho \partial u / \partial t + \partial p / \partial x + \rho u \partial u / \partial x = 0 \tag{7.85}$$

$$\rho \partial e / \partial t + p \partial u / \partial x + \rho u \partial e / \partial x = 0 \tag{7.86}$$

Here, e is the total internal energy and, p is the total pressure. The relaxation time τ of the chemical kinetics is used as a constraint in time discretization. During numerical solution, the time step must be one order of magnitude less than the value of τ predicted for that time step.

In this way, we can model the fastest changing energy in the time dependent reacting flow. Here, the time step taken can be very small, the reactions may take place very fast, and the reaction rate coefficients can be very large to make the differential equations 'stiff'. We have to resort to special numerical solution techniques for the stiff differential equations (Hoffman).

The system of differential Eq. 7.82–7.86 solved with the proper initial conditions gives us ρ, u and e values and also c_i and the vibration energy of ith specie. The temperature T of the medium can be found from $\sum_i c_i e = e$ relation as follows

$$e = \sum_i c_i \left[\frac{5}{2} R_i T + \frac{h\nu_i/kT}{e^{h\nu_i/kT} - 1} R_i T + (\Delta h_f)_i^o \right] \tag{7.87}$$

Table 7.3 Shock distance for different air models

Non dimensional shock distance, d/R_s						
Speed, km/s	Alt. km	Shock radius, m.	Chem kin.	Equilb. chem.	Frozen Flw.	Ideal Flw
5	66	0.33	0.0676	0.0625	0.089	0.113
7.5	66	0.21	0.0666	0.0461	0.0805	0.108

Here, $(\Delta h_f)_i^o$ is the effective reference energy of the specie *i*, and is known as the heat of formation of specie *i*. The heat of formation for each specie can be obtained from JANAF (**J**oint **A**rmy **N**avy **A**ir **F**orce) tables.

Now, we can see the two dimensional application of the theory on a blunt body with addition of 16 species equation added to Eq. 7.78 for calculating the flow field interacting with chemical kinetics (Hall et al.). Hall and associates used the theory on flow about the sphere, and compared the solutions with the ideal flow solutions obtained at different altitudes and free stream conditions. The results obtained are compared in Table 7.3 for two different free stream speeds at the same altitude.

According to Table 7.3, chemical kinetics give the shortest shock distance. The time dependent reactions put the shock a little further from the body, the frozen flow pushes the shock even further. The ideal flow solution gives the maximum shock-body distance.

7.9 Hypersonic Viscous Flow: *Numerical Solutions*

In previous chapters we have studied the analytical and experimental work related to viscous hypersonic flow past flat plate at zero angle of attack. Now, we can extend those to further study hypersonic flows with the aid of computational fluid dynamics.

Let us start with the solution of viscous hypersonic boundary layer equations. If we take x coordinate as the direction parallel to the surface and z coordinate as the direction normal to the surface the set of equations reads as

$$\text{Continuity}: \quad \frac{\partial \rho}{\partial t} + \frac{\partial \rho u}{\partial x} + \frac{\partial \rho w}{\partial z} = 0 \tag{7.88a}$$

$$\text{Momentum}: \quad \rho\left(\frac{\partial u}{\partial t} + u\frac{\partial u}{\partial x} + w\frac{\partial u}{\partial z}\right) = -\frac{\partial p}{\partial x} + \frac{\partial}{\partial z}\left(\mu\frac{\partial u}{\partial z}\right) \tag{7.88b}$$

$$\text{Energy}: \quad \rho\left(\frac{\partial h}{\partial t} + u\frac{\partial h}{\partial x} + w\frac{\partial h}{\partial z}\right) = \frac{\partial p}{\partial t} + u\frac{\partial p}{\partial x} + \frac{\partial}{\partial z}\left(k\frac{\partial T}{\partial z}\right) + \mu\left(\frac{\partial u}{\partial z}\right)^2 \tag{7.88c}$$

for unsteady flow. Here, in addition to velocity components u and v, h shows the enthalpy and T denotes temperature where pressure p is as usual provided from outside. In order to solve the boundary layer equations about a blunt body, we start with the stagnation flow solution. Equations 7.88b, c, are partial differential equations. They can be transformed to ordinary differential equations with Illingworth transformation, i.e. $\xi = \int_0^x \rho_e U_e \mu_e dx$ and $\eta = U_e/\sqrt{2\xi}\int_0^z \rho dz$, and the non dimensional $f'(\eta) = u/U_e$ and non dimensional enthalpy then reads $g = h/h_e$

$$(Cf'')' + ff'' = (f')^2 - g \tag{7.89a}$$

$$(Cg')' + P_r fg' = 0 \tag{7.89b}$$

Here, P_r is Prandtl number, and $C = \rho\mu/(\rho_e\ \mu_e)$ (White). If we use the power law for the viscosity variation with temperature for the perfect gas we obtain $C = g^{(n-1)}$.

For air, taking n = 2/3, makes Eq. 7.89a, b in terms of η as the independent variable

$$f''' - \frac{f''}{3g} + g^{1/3}ff'' + g^{4/3} - g^{1/3}(f')^2 = 0 \tag{7.90a}$$

$$g'' - \frac{g'}{3g} + P_r g^{1/3} fg' = 0 \tag{7.90b}$$

Equations 7.90a, b are solved with an iterative method to obtain the stagnation speed and temperature which satisfiy the wall and the edge conditions. These velocity and the temperature values are used as the initial station to start solution of the parabolic Eq. 7.89a, b with a marching technique. Equations 7.88a, b are solved with finite difference as explained in Appendix 10. The calibration of the scheme is made with the hypersonic flow past a flat plate at free stream Mach number of 8. In this bench mark test, the wall to edge temperature ratio is considered 4 as shown in Fig. 7.23a, b (Oksuzoglu and Gulcat 2018).

After observing the agreement with the present solution and the Van Driest solution Fig. 7.22, the numerical procedure is applied to the hypersonic flow past circular cylinder. The pressure distribution for the circular cylinder is provided by the Newtonian Flow.

Shown in Fig. 7.24 is the hypersonic boundary layer flow about the circular cylinder at M = 8, R_e = 10000, and P_r = 0.75. The stagnation flow solution is done with Eq. 7.90a, b at x/R = 0, and the boundary layer edge to wall velocity ratio is taken as 2 for temperature profiles where the wall appears to be cold.

The studies and the related examples for the viscous hypersonic flow are so far for the flows having free stream Mach numbers 8 or smaller. During re-entry when the free stream Mach number is greater than 8, the oxygen and the nitrogen molecules disassociate with speed and altitude as shown in Fig. 7.19. For this reason, as done before, we have to include the chemical reactions in our computations. Since the viscous effects are to be seen near the wall, we have to see

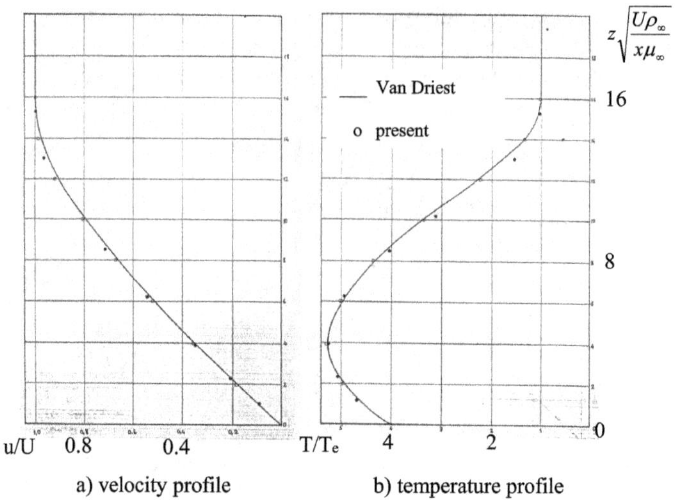

Fig. 7.23 Flow around a flat plate M = 8. **a** Velocity, **b** temperature profiles

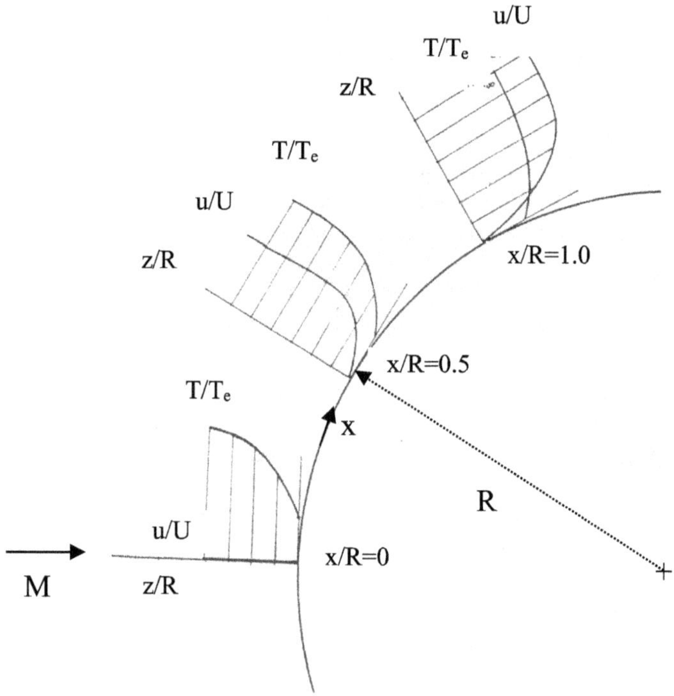

Fig. 7.24 Velocity and temperature profiles at M = 8 and R_e = 10000

whether the wall is a catalytic or not, and add the diffusion terms to the equations together with altering the boundary conditions at the wall in terms of heat conduction. For this reason, let us first determine the mass diffusion coefficient in terms of Fick's law. The Fick's law gives the amount diffusion between the species 1 and 2 as follows: $-\boldsymbol{\rho D_{12}}$ **grad** $\mathbf{c_1}$, where c_1 is the mass fraction of the of the specie 1. Here, D_{12} is the coefficient of diffusion whose values are given in tables in literature to be used in computation in low density flows. The viscous solutions based on the mass diffusion require adding the relevant terms to Navier-Stokes equations. Now, let us re-write the boundary layer equations, Eq. 7.88, by adding the diffusion terms. The continuity and the x momentum equations written for the total species, Eq. 7.88a, b, retain their original forms. The diffusion terms need to be added to the continuity of the species and the total energy equation. The continuity of specie i then reads as

$$\rho u\frac{\partial c_i}{\partial x}+\rho w\frac{\partial c_i}{\partial z}=\frac{\partial}{\partial z}\left(\rho D_{12}\frac{\partial c_i}{\partial z}\right)+\dot{w}_i \tag{7.91}$$

The energy equation then becomes

$$\rho\left(u\frac{\partial h}{\partial x}+w\frac{\partial h}{\partial z}\right)=u\frac{\partial p}{\partial x}+\frac{\partial}{\partial z}\left(\rho D_{12}\sum_i h_i\frac{\partial c_i}{\partial z}\right)+\frac{\partial}{\partial z}\left(k\frac{\partial T}{\partial z}\right)+\mu\left(\frac{\partial u}{\partial z}\right)^2 \tag{7.92}$$

Here, k is the time dependent heat conduction coefficient. Since there are more than one specie, the values of k and μ are computed with a special averaging method (Anderson).

The boundary layer equations, given by Eq. 7.91–7.92 together with the total continuity and the momentum equations, 7.88a, b, with chemical reactions and the wall boundary conditions were seen in the literature starting from mid 1980s. An example to this sort of elaborate work is the study performed by Aupoix and associates. In their study, the two dimensional boundary layer equations at the symmetry plane of the lower surface of the space shuttle are solved at free stream Mach number of 23.4 considering a catalytic and non catalytic wall. Five species, disassociation of O_2, N_2 into O and N, and formation of NO are involved in reacting flow together with catalytic wall conditions. The conclusions of the afore mentioned work is as follows: (i) wall catalysity and the wall temperature have an effect on the heat flux and the displacement thickness, (ii) the surface friction is independent of real gas effects, (iii) chemical reactions for a few number of species are sufficient, and (iv) simple diffusion models yield accurate solutions. In addition, the heat flux q_w at the wall is given by diffusion of 2 species as follows

$$q_w=\left(-k\frac{\partial T}{\partial z}-\rho D_{12}\sum_i h_i\frac{\partial c_i}{\partial z}\right)_w \tag{7.93}$$

For the case of adiabatic wall equating 7.93 to zero is sufficient. In reacting flows, on the other hand, the mass fraction of species is unknown in the equations; therefore, while solving the continuity of the species, the catalytic effects which determine rates of reaction at the wall are to be considered. The expression which gives the catalytic formation of species and their diffusion into the flow is as follows

$$(\dot{w}_c)_i = \rho D_{12}\left(\frac{\partial c_i}{\partial z}\right)_w \tag{7.94}$$

For a non catalytic wall, Eq. 7.94 is set to zero to get $\left(\frac{\partial c_i}{\partial y}\right)_w = 0$. On the other hand, for fully catalytic walls the rate of chemical reactions is infinite; therefore, the mass fractions of the species take their equilibrium chemistry values, i.e. $c_i = (c_{i)eq}$. For partially catalytic walls, since the chemical reactions take place with finite rates, the specie production rate in Eq. 7.94 is given in terms of the gradient of c_i.

Aupoix and his associates also discovered in their study with catalytic and non catalytic wall at 1500 K, M = 23.4, and 71 km altitude using 5 and 10 species that the heat transfer rate at the stagnation point is 0.4 MW/m^2 for catalytic wall and 0.1 MW/m^2 for non catalytic wall. Naturally, reducing the Mach number and the wall temperature reduces the heat transfer.

The non dimensional boundary layer equations for hypersonic flow help us to predict the flow parameters as it was the case for original boundary layer equations. In non dimensionalization we take the velocity, density, enthalpy and the diffusion coefficient at the boundary layer edge as the characteristic values. If we write the non dimensional stagnation enthalpy $h_o = h + u^2/2$ in the energy equation then we have

$$\rho u\frac{\partial h_o}{\partial x} + \rho w\frac{\partial h_o}{\partial z} = \frac{\partial}{\partial z}\left[\frac{\mu}{P_r}\frac{\partial h_o}{\partial z} + (1 - 1/P_r)\mu u\frac{\partial h_o}{\partial z} + (1 - 1/L_e)\rho D_{12}\sum_i h_i\frac{\partial c_i}{\partial z}\right] \tag{7.95}$$

Here, $L_e = \rho D_{12}\, c_{pf}/k$, is the Lewis number and $c_{pf} = \Sigma\, c_i\, c_{pi}$, is the summation of specific heat constants at the frozen flow. According to Eq. 7.95, if the Lewis number is 1 in the boundary layer then the effect of diffusion disappears.

The stagnation point solutions are to be known in order to start the boundary layer solution around blunt bodies. We need to add the term with the Lewis number to Eq. 7.89a, b. With these terms added, the heat transfer rates at the stagnation point of a wall, catalytic or non catalytic, are provided in terms of the Lewis number by Anderson.

Other than boundary layer equations, the hypersonic viscous reacting flow numerical solutions for the space shuttle at 71 km of altitude with free stream Mach number of 23 were published by Prabhu and Tannehill in mid 1980s. In their study, comparison of equilibrium chemistry solution with the perfect gas solution for the value of $\gamma = 1.2$ shows that they match with each other. This indicates that at the upper level of atmosphere flight conditions are better simulated using $\gamma = 1.2$.

Also in mid 1980s, gas kinetic equations were employed to predict the three dimensional flow simulations at the upper level of atmosphere with some degree of slip instead of no-slip condition at the surface (Riedelbauch et al. 1987). In 1990s unsteady three dimensional Navier-Stokes equations were employed for solution of hypersonic aerodynamic problems (Edwards and Flores). In their work, numerical solutions obtained for chemically reacting flow past a sphere-cone configuration at high Mach numbers and high altitudes are in good agreement with the previous studies in terms of surface pressure and heat transfer rates. The specific heat ratio calculation obtained for the reacting flow gives $\gamma = 1.4$ before the shock, and its value reaches minimum of 1.2 after the shock.

7.10 Hypersonic Plane: *Wave Rider*

Hitherto, we have seen the various properties of hypersonic flow past different type of body shapes. Now, we can apply this information to analyze the hypersonic aerodynamics of an aerospace plane. The concept of sustainable hypersonic flight of a vehicle having most of the features described in previous sections and riding on the pressure created by the shock surface under its body formally was introduced in late 1950s (Nonweiler 1959). In his work, Nonweiler proposes inverted W and V shaped cross sections for delta type wings with surface streamlines and lower surface shocks as shown in Fig. 7.25. The weak shocks appearing at the lower

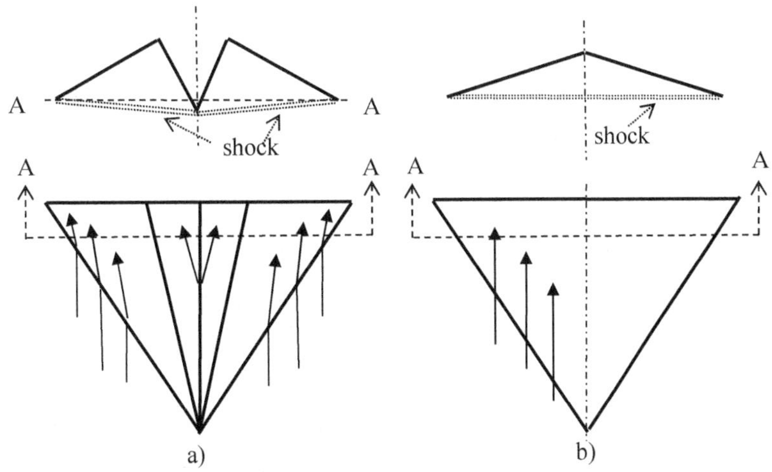

Fig. 7.25 Inverted, **a** W, and **b** V shaped delta wings and shocks and surface streamlines

surface create sufficient lower surface pressure so that the lift generated is adequate for a sustainable flight.

This type of aerospace planes which makes use of the lower surface pressure is also called 'waverider'. According to Nonweiler's analysis an aerospace plane which was named 'karet' then, has stagnation temperature of 1150 °C, at the mid chord region the temperature drops down to 500 °C at 80 km altitude with free stream speed of 6.5 km/s.

The waverider concept was quite popular until 1970s, and lost its popularity for some time because of its low lift to drag ratio, L/D. Thanks to numerical optimization techniques, towards the end of 1980s, the waverider gained popularity again (Bowcutt et al. 1987). In waverider studies, the viscous effects were also considered for the evaluation of optimum $(L/D)_{max}$ for various free stream Mach numbers, at different altitudes and even for planets whose atmospheric properties are known. The previously known $(L/D)_{max}$ barriers were broken with numerical optimization techniques. The free stream Mach number figures given by Küchemann for $5 < M < 10$, can be improved approximately 1.5. As shown in Fig. 7.25, for a wide range of free stream Mach numbers ($2 < M < 25$), Küchemann's previously given $(L/D)_{max} = 4(M + 3)/M$ barrier curve has been changed to $(L/D)_{max} = 6(M + 2)/M$ because of optimization. Bowcutt and his associates even exceed the second curve shown in Fig. 7.26 for M = 20. However, for M = 25 the optimum L/D value is below the given curve. The reason for this behavior is attributed to the Reynolds number effect at high altitudes. The Reynolds number for both free stream Mach numbers is laminar; therefore, for M = 20 viscous drag becomes low to give higher $(L/D)_{max}$. This gives deviations from the averaged behavior for predicting $(L/D)_{max}$ values.

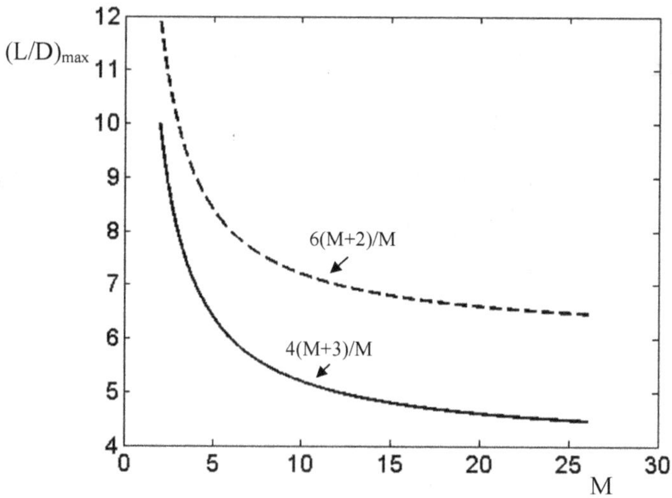

Fig. 7.26 $(L/D)_{max}$ for a waverider: Küchemann—Bowcutt

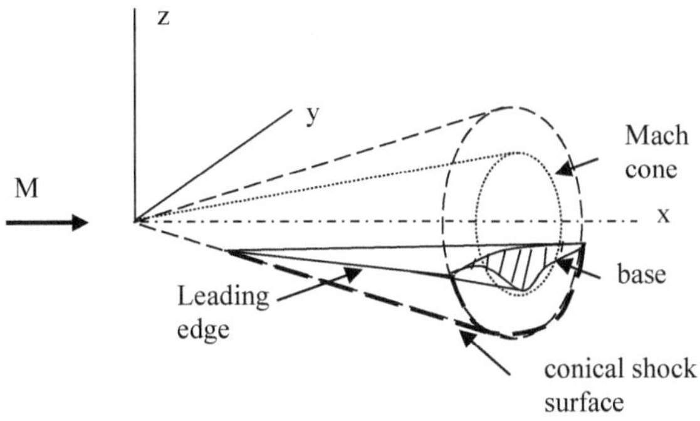

Fig. 7.27 Waverider geometry placed in shock cone

In order to find the optimum shape of a waverider at a given free stream and altitude, first an axially symmetric flow with a weak attached shock of a cone whose angle is the Mach cone angle of the given free stream Mach number is considered (Example 7.3). The leading edges of the waverider are placed inside the conical shock as shown in Fig. 7.27. The upper surface of the waverider is configured as a cylindrical shell surface parallel to free stream. The bottom surface, on the other hand, consists of a base curve tangent to the Mach cone surface, and it joins to the leading edge with a bell shaped curve as shown in Fig. 7.27.

After determining the lower surface pressure created by the shock, and the corresponding lifting pressure, we can calculate the total lift by integrating the lifting pressure over the surface. The total drag then is computed by the wave drag induced by the thickness of the wing and the viscous effects. The surface shape which makes the total drag minimum gives us the maximum (L/D) ratio. Calculating viscous drag with reference temperature method, because of its simplicity, is preferred over the boundary layer integral methods (Anderson 1991). The reference temperature method can give the Reynolds number in transition to turbulence based on the free stream Mach number together with the viscous drag in turbulent flow (Anderson 1991).

The waverider shapes based on the ideal flow analysis as given above were also analyzed by solving the Euler Equations (Jones and Dougherty 1992). In that study, the surface pressure distribution is found to be in good agreement with Bowcutt solution and the experimental results for free stream Mach numbers of 4 and 6. In order to find the matching results, a special grid generation based on the adaptive meshing around the sharp leading edges was used.

The optimum shape and the flow field around a waverider was also studied together with Navier-Stokes and Euler solvers results compared (Takashima and Lewis 1994). In their study, Takashima and Lewis found agreement with the surface pressure solution obtained with Navier–Stokes and Euler solvers at free stream Mach number of 6, Fig. 7.28. Here, in that study, the difference between the lower and upper surface pressures gives us the lifting pressure quite similar to that of ideal solution except around the leading edges. While obtaining the viscous solution the flow field is discretized with a fine grid near the surface where the first point above

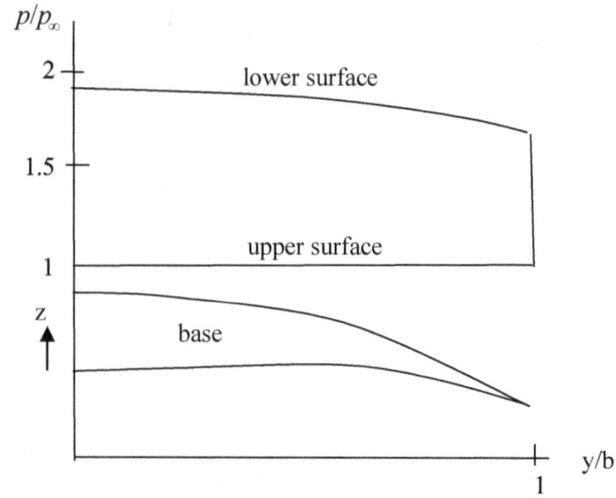

Fig. 7.28 Upper and lower pressure of a waverider at M = 6

the surface has its $y^{+}=0.1$ and the sharp leading edge is a little bit smoothed to have a stable solution. The rounding of a leading edge has an effect on the $(L/D)_{max}$ value. It has been observed that if rounding is made with leading edge radius ratio to length of the waverider less than 0.1% then L/D remains constant, if it is larger than 1% than L/D value decreases (Lewis and McRonald 1992).

In their study, Lewis and McRonald extend their work to the aerodynamic analyses of such waveriders which exceed Mach numbers of 50 while passing through the atmosphere of other planets to make use of a gravity assists in their sustainable hypersonic flight.

Approximately 100 km above sea level in the Knudsen number range of $0.05 < K_n < 1$ the continuum hypothesis is no longer valid. Therefore, the optimum waverider analysis is made with Monte Carlo method which gives very low L/D ratios (Rault). In his study, based on the Monte Carlo Method, Raul find L/D = 0.197 at 105 km altitude, free stream Mach number of M = 25, and Knudsen number of $K_n = 0.05$.

The experimental and theoretical work performed so far indicates that towards the end of the first of quarter of 21st century, the nations or union of nations with advanced technology will probably start the manned or unmanned sustainable hypersonic flight with prototype waveriders.

Summary: *General overview of hypersonic flow theory is given starting with Newtonian impact theory based on the momentum transfer of air particles normal to a wall. Afterwards, modified Newtonian theory based on the stagnation pressure correction to the original theory is introduced. In completing the Newtonian flow centripetal force correction is applied to the hypersonic flow past curved walls. In addition, unsteady Newtonian flow is introduced to examine the oscillatory hypersonic flows past blunt bodies. The piston theory based on the Hayes' hypersonic similarity rule is developed and applied to slender wings. There exist 1st, 2nd or 3rd order piston theories depending on the value of the hypersonic* *similarity parameter (M τ). The modified piston theory on the other hand, handles the unsteady hypersonic flows past wings with sharp leading edges but considerable thickness effects especially because of thick boundary layers. Maslen's simple numerical method is given for the inverse solution of the hypersonic flows past simple shapes. In addition numerical solutions of Euler's equation for a space capsule and shuttle like configurations at various high Mach numbers are provided. Viscous flow analysis and aerodynamic heating and viscous interaction concepts are given in terms of heat transfer at the walls in hypersonic flows. In addition, strong heat transfer effects causing structural damages because of the shock boundary layer interaction is briefly mentioned. The high temperature effects on the specific heat constants of air is studied with the introduction of the vibrational effects of the diatomic molecules like oxygen and nitrogen. The concept of aerothermodynamics is briefly introduced. The composition of air with respect to altitude and speed is provided to assess the disassociation of molecules and the chemical reactions among the oxygen and nitrogen atoms. The equilibrium and non-equilibrium reacting flow concepts are utilized to see the effect of air composition on the temperature field and on the shock position involving blunt body flows. The chemical kinetics for a 1-D flow is provided to describe the flow with chemical non-equilibrium. In addition, a comparison for 2-D curved shock in front of a sphere obtained with chemical kinetics, chemical equilibrium, frozen flow and the perfect gas flow is provided. With this the concept of aerothermochemistry is briefly mentioned. The computational solution to viscous hypersonic flow studies are given for the stagnation flow and the multidimensional reacting flows to find the skin friction and the heat transfer rates based on the Reynolds and the Stanton numbers. The concept of catalycity of walls are introduced in terms of binary diffusion. The nondimensional Lewis number is introduced to show this effect on the energy equation. Finally, the waverider concept is given to yield maximum L/D ratio for hypersonic flying vehicles. The inviscid and the viscous aspects of the waverider are also provided with simple cone flow analysis as well as the modern computational techniques involving solutions of Euler and Navier-Stokes equations. The possible realization of manned or unmanned hypersonic flight based on the waverider concept in near future is stressed.*

7.11 Problems and Questions

7.1 For a given free stream Mach number M, obtain the stagnation pressure coefficient in terms of M. Find the limit of the stagnation pressure coefficient for M approaching infinity.

7.2 Use the point at the surface and the first point Δn away from the surface, obtain Eqs. 7.11 from 7.10 which gives the pressure caused by the curvature. (take $R = -1/(d\theta/ds)$).

7.3 Using Newton–Busemann theory obtain the surface pressure distribution of a circular cylinder at free stream speed of M = 8, and compare the results with Newton and improved Newton method.

7.4 Find the surface pressure distribution of a 2-D body whose surface equation is given with $z_u = x^{1/2}$, $0 \leq x \leq 1$, and pitching about its leading edge at angle of attack amplitude of 5° and k = 0.5.

7.5 A body of revolution is given with equation of radius r: $x = 0.79r^2 - 1$, $-1 \leq x \leq 0$, with respect to symmetry axis x. Obtain the surface pressure distribution with Newtonian and improved Newtonian theory.

7.6 Find the surface pressure distribution of the body of Problem 7.5 which oscillates about its nose with angle of attack amplitude of 5° and k = 0.4.

7.7 Obtain the surface pressure distribution of a spherical segment which has an apex angle of 60° and 2 m radius oscillating about its nose with small amplitude.

7.8 Obtain Eq. 7.6 for steady flows using Eq. 7.20 which expresses surface pressure distribution.

7.9 Newtonian impact theory is non linear. Why?

7.10 Find the sectional drag coefficient for a circular cylinder in hypersonic flow using Newton-Busemann theory.

7.11 Obtain Eq. 7.24 from 7.21 with expanding the equation into the series in terms of powers of w/a.

7.12 For a 5% thick airfoil find the free stream Mach number for which: (i) first order, (ii) second order and (iii) third order piston theory is applicable.

7.13 What should be the minimum value of reduced frequency k for a thin airfoil in a free stream Mach number of 6 so that the piston theory is valid?

7.14 Why is it necessary to consider the thickness effect for a thin airfoil in hypersonic aerodynamics?

7.15 Find the surface pressure distribution of a 5% thick parabolic airfoil in a free stream Mach number of 5 at altitude of 50 km.

7.16 The profile given in Problem 7.12. is in plunging motion with k = 0.3. Determine the upper surface pressure distribution in terms of the amplitude of plunging.

7.17 Using the improved piston theory for the profile given in Problem 7.13 find the lifting pressure distribution. What can be the thickness ratio for the same profile at free stream of M = 3?

7.18. Find the lifting pressure distribution for the profile given in Problem 7.16 at free stream Mach number of M = 3.

7.19 Eq. 7.36a is written for the conservation of momentum in y direction. Obtain Eq. 7.36-b wherein the stream function is independent variable.

7.20 Obtain the oblique shock relations in terms of Mach number M, shock angle θ_s, the specific heat ratios γ.

7.21 Show that the enthalpy can be expressed in terms of the pressure and the density as follows: $h = 2\gamma/(\gamma-1)(p/\rho)$.

7.22 Find the pressure, density, enthalpy and the surface velocity of Example 7.2 for the streamline value of $\chi = 0.0625$.

7.23 Use the Maslen method to find the flow conditions at a point with seen with 30° angle from the center of a circular cylinder in free stream Mach number of infinity.

7.24 Show that the Maslen method gives: $\tan\theta_s = [11\text{-}(121\text{-}48 \tan^2 \theta_{c)}^{1/2}]/2\tan\theta_c$ for the the shock angle θ_s of a cone with cone angle θ_c in a free stream Mach number of infinity with $\gamma = 1.4$.

7.25 What is the advantage of the Maslen method over the Newtonian impact theory?

7.26 Using the Maslen method find the change in the shock angle at the nose of a cone pitching with about its nose with a small amplitude. (Use body attached coordinate system as done in Example 7.1).

7.27 Obtain the Jacobian fluxes $A_i = \partial F_i/\partial U$ in Eq. 7.47 for i = 1,2.

7.28 The coefficient $\lambda = ½\, h^2\, |\partial u_l/\partial x_l|$ is used to prevent the numerical oscillations at a shock. Obtain the expression for $|\partial u_l/\partial x_l|$ in terms of the velocity components if x_l is the direction showing the magnitude of the gradient of a velocity vector in an element.

7.29 An empirical way to determine shock shapes based on experiments is given by Billig. A shape of shock created by a blunt body in x-y coordinate system is given as a hyperbola $x = R+\delta\text{-}R_c \cot^2 \beta[(1 + y^2 \tan^2 \beta/R_c\ 2)1/2\ \text{-}1]$. Here, the blunt body is taken as sphere-cone junction with δ: the shock distance, R: the radius of the sphere and $R_{c:}$ the shock curvature to give

$$\delta/R = 0.143\exp\left(3.24/M^2\right), \quad R_c/R = 1.143\exp\left[0.54/\left(M^2 - 1\right)^{1.2}\right]$$

Here, β is the attached shock angle of the cone alone. Plot th shape of the shock generated by the solid surface shown above for Mach numbers of M = 4 and 8.

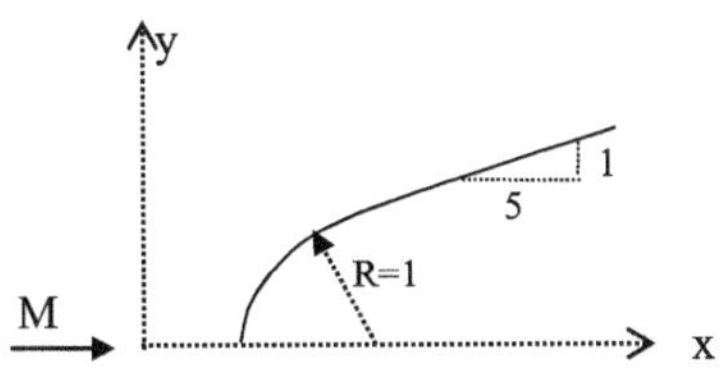

7.30 Using Maslen method, find the approximate value of pressure and density at the junction of the sphere and the cone of Problem 7.29 at Mach number 8.

7.31 A flat plate of 4 m long has the wall temperature of 1200 K at zero angle of attack at free stream Mach number of 25 and 85 km altitude. Using the data given obtain the pressure variation along the plate.

7.32 Find the induced pressure on the surface of cone given in Example 7.3 with the free stream conditions described at Problem 7.31.

7.33 Find the heat transfer rate q_w at x = 10 cm and100cm for the flat plate given in Problem 7.31.

7.34 Show that the derivative of the boundary layer edge velocity is given by Eq. 7.64 for the figure given below.

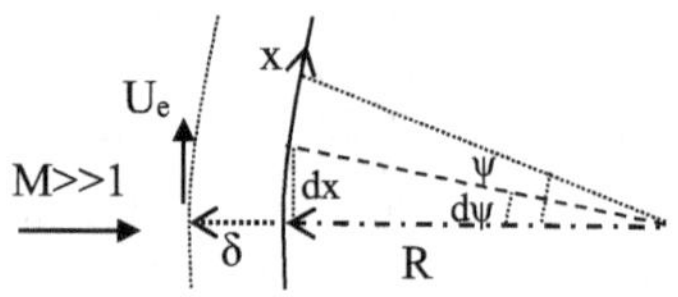

$d\psi/dx = 1/R$

7.35 Obtain aerodynamic heating formula for at the stagnation point of a sphere in terms of radius of curvature. Assuming calorically perfect air, find the aerodynamic heating of the space capsule given in Fig. 7.9. Take the wall temperature as 1200 K. Find also the heating rate at $\psi = 45^{o.}$

7.36 Show that Eq. 7.68 gives the rate of heating in terms of the surface pressure change for the shock boundary layer interaction.

7.37 For a diatomic molecule find the contribution of the vibration energy to the specific heat under constant volume. Neglect the ground level energy effect to the partition function.

7.38 Find the specific internal energy and the enthalpy of the air under 1 atmosphere pressure and 2000 K.

7.39 The surface area of the space shuttle is designed to be 560 000 $in^{2.}$ Determine the lift coefficient of the shuttle during its re-entry.

7. 40 Solve Example 7.5 with equilibrium energy as the initial condition for the Oxygen molecule at 1 atmosphere and 3200 K.

7.41 For $N_2 + O_2 \rightarrow 2\,N + O_2$ plot the graph of forward reaction in 6000 K-9000 K interval and compare with the production of 2O from O_2.

7.42 In pure N_2 flow, find the temperature and density change with respect x coordinate behind the normal shock created with M = 12.28, T = 300 K and p = 1/760 atm.

7.43 Using the energy equation obtain Eq. 7.95 for the stagnation enthalpy h_o in terms of local Prandtl and Lewis numbers.

7.44 Find for a waverider flying at 80 km altitude with a Mach number of 25, find:

(i) approximate lifting pressure, (ii) lift coefficient, and induced drag coefficient. Assume ideal flow.

7.45 If the waverider of the Problem 7.44 has the wall temperature of 1400 K and $Re = 1.371 \times 10^6$ with respect to its length then find the drag coefficient with reference temperature method. Use this result to determine (L/D) with total drag.

References

Anderson, John D.: Hypersonic and High Temperature Gas Dynamics. McGraw-Hill, New York (1989)

Anderson, John D., M.J. Lewis, A.P. Kotari and S. Corda: Hypersonic Waveriders for Planetary Atmospheres, J. Spacecraft, V. 28, No 4, July-August 1991

Ashley, H. and G. Zartarian: Piston Theory-A New Aerodynamic Tool for the Aeroelastician, Journal of Aeronautical Sciences, Dec. 1956

Aupoix, B., C. Eldem and J. Cousteix: Couche Limite Laminare Hypersonique Etude Parametriqeu de la Representation des Effects de Gaz Reel, *Aerodynamics of Hypersonic Lifting Vehicles*, AGARD-CP-428, November, 1987

Bertin, John, J.: *Hypersonic Aerothermodynamics*, AIAA Education Series, Washington, DC, 1994

Billig, F.S.: Shock-Wave Shapes Around Spherical and Cylindirical Nosed Bodies, Journal of Spacecraft and Rockets, V. 4, No 6, June 1967

Bohachevski, Ihor O. and R.E. Mates: A Direct Method for Calculation of the Flow about an Axisymmetric Blunt Body at Angle of Attack, AIAA J., pp 776–782, May 1966

Bowcutt, Kevin G., J.D. Anderson and D. Capriotti: Numerical Optimization of Conical Flow Waveriders Including Detailed Viscous Effects, *Aerodynamics of Hypersonic Lifting Vehicles*, AGARD-CP-428, November, 1987

Denbigh, Kenneth: The Principles of Chemical Equilibrium. Cambridge University Press, Cambridge (1978)

Edwards, Thomas A. And J. Flores: Computational Fluid Dynamics Nose-to-Tail Capability: Hypersonic Unsteady Navier-Stokes Code Validation, J. Spacecraft, V. 27, No 2, March-April 1990

Hall, Gorden J., A.Q. Eschenroeder and P.V. Marrone, Blunt-Nose Inviscid Airflow with Coupled Nonequilibrium Process, Journal of the Aerospace Sciences, Sept. 1962

Hayes, Wallace D. and R.F. Probstein: *Hypersonic Flow Theory Vol.I*, 2nd ed., Academic Press, New York, 1966

Hoffman, Joe D.: Numerical Methods for Engineers and Scientists. McGraw-Hill, New York (1992)

Jones, Kevin D. And F.C. Dougherty: Numerical Simulation of High-Speed Flows About Waveriders with Sharp Leading Edges, Journal of Spacecraft and Rockets, V.29, N. 5, September-October 1992

Kutler, Paul and H. Lomax, Shock Capturing Finite-Difference Approach to Supersonic Flows, Journal of Spacecraft, V.8, No 12, December 1971

Kutler, Paul, R.F Warming and H. Lomax: Computation of Space Shuttle Flow Fields Using Noncenterd Finite-Difference Schemes, AIAA J., V.11, No 2, February, 1973

Lee, John F., F.W. Sears and D.L. Turcotte: *Statistical Thermodynamics*, Addison- Wesley, Reading Mass., 1973

Lewis, Mark J. And A.D. McRonald, Design of Hypersonic Waveriders for Aeroassisted Interplanetory Trajectories, Journal of Spacecraft and Rockets, V. 29 N. 5, September-October 1992

Lighthill, M.J.: Oscillating Airfoils at High Mach Numbers, Journal of Aeronautical Sciences, June 1953

Marvin, J.G., C.G. Hortsman, M.W. Rubesin, T.J. Coakley, M.I. Mussoy: An Experimental and Numerical Investigation of Shock-Wave Induced Turbulent Boundary Layer Separation at Hypersonic Speeds, *Flow Separation,* AGARD-CP- 168, November 1975

Maslen, S.H.: Inviscid Hypersonic Flow Past Smooth Symmetric Bodies, AIAA J, V. 2, No 6, June 1964

Molina, R.C. and J.P. Huot: A One-Point Integration Finite Element Solver for the Fast Solution of the Compressible Euler, Equations, Computer Methods in Applied Mechanics and Engineering, 95, 1992

Neumann, Richard D.: Special Topics in Hypersonic Flow, in Aerodynamic Problems of Hypersonic Vehicles, AGARD-LS 42, 1972

Nonweiler, T.R.F.: Aerodynamic Problems of Manned Space Vehicles, Journal of the Royal Aeronautical Society, September 1959

Oksuzoglu, Hakan: Compressible Boundary Layers, Graduation Thesis. Supervised by U. Gulcat, Faculty of Aeronautics and Astronautics, ITU (1986)

Pierce, Alvin G.: Unsteady Hypersonic Flows About Thin Lifting Surfaces. Lecture Notes, Georgia Institute of Technology (1978)

Rault, Didier F.G.:Aerodynamic Characteristics of a Hypersonic Viscous Optimized Waverider at High Altitudes, Journal of Spacecraft and Rockets, V. 31 N. 5, September-October 1994

Riedelbauch, S., W. Wetzel, M. Kordulla, H. Oertel Jr., On the Numerical Simulation of the Hypersonic Flow in Aerodynamics of Hypersonic Lifting Vehicles, AGARD-CP 428, November 1987

Shapiro, Ascher H.: The Dynamics and Thermodynamics of Compressible Fluid Flow I. The Ronald Press Company, New York (1953)

Talbot, L., T. Koga and P.M. Sharman, Hypersonic Viscous Flow Over Slender Cones, NACA TN 4327, September 1958

Takashima, Naruhisa and M.J. Lewis: Navier-Stokes Computations of a Viscous Optimized Waverider, Journal of Spacecraft and Rockets, V. 31, N. 3, May-June 1994

Van Driest, E.R.: Investigation of Laminar Boundary Layer in Compressible Fluids Using The Crocco Method, NACA TN 2579, January 1952

Van Dyke, M.D.: A Study of Hypersonic Small Disturbance Theory, NACA Report 1194, 1954

Van Dyke, M.D.: The supersonic Blunt Body Problem-Review and Extensions, Journal of Aeronautical Sciences, 1958

Vincenti, Walter G.: And C:H: Kruger. Introduction to Physical Gasdynamics, Wiley, New York, Jr (1965)

White, Frank M.: Viscous Fluid Flow, McGraw-Hill, 1991

Zienkiewicz, O.C., Taylor, R.L.: The Finite Element Method: vol. 3. Fluid Mechanics, Butterworth-Heinemann, Oxford (2000)

Chapter 8
Modern Subjects

Most of the material we have studied so far in general are the topics belonging to classical aerodynamics related to flows past thin or slender objects in small angles of attack for the purpose of generating lift. After the 1970s we see that the boundaries of classical aerodynamics are crossed because of advances made in computational as well as experimental techniques. The flow field analysis of low aspect ratio wings with high swept leading edges at high angles of attack enabled researchers to predict the extra lift generated because of leading edge separation which is exactly the case for some of the biological flows in nature. Utilization of leading edge separation helped aerodynamicists to design highly maneuverable military aircrafts to be used for military purposes. As is known from the classical aerodynamics, the leading edge separation from the wings with a little sweep or no-sweep, on the other hand, causes lift loss. This type of wing must have high lift while cruising at a constant speed and during landing with low speeds must have even higher lift without stalling. Otherwise, unsymmetrical lift loss, either from the left or right wing creates a rolling moment about the axis of the plane, and this causes it to rock. The larger roll moments about the axis of the plane cause spin (Katz and Plotkin 2010).

Since it is not possible to solve the separated flow fields with analytical methods, we have to resort to experimental measurements and visualization techniques or to numerical methods. The numerical and the experimental methods are used together in complementary fashion for the analysis of separated flows to predict the aerodynamic characteristics of relevant configurations. In this chapter, first we are going to study in a detail the flow separation around an airfoil at a constant angle of attack. The sudden lift loss at a constant high angle of attack because of leading edge separation is called static stall. The numerical as well as experimental studies about static stall will be given. Afterwards, the flow separation and the related lift loss at variable angle of attack which is called 'dynamic stall' are going to be analyzed in detail. Finally, three dimensional analyses of swept wings with extra lift created by leading edge separation will be studied. The summary of leading edge flow separation and the vortex sheet formations are shown in Fig. 8.1 for both swept and

Ü. Gülçat, *Fundamentals of Modern Unsteady Aerodynamics*,
https://doi.org/10.1007/978-3-030-60777-7_8

unswept wings. According to Fig. 8.1a (adapted from Katz and Plotkin 2010), for an unswept wing with high aspect ratio there is a two dimensional separation as indicated with A-A cross section. In this two dimensional separation, the clockwise vortices leaving the leading edge and the counter clockwise vortices leaving the bottom surface at the trailing edge form a periodically formed vortex street at the wake of the wing. If h is the vertical distance between the centers of clockwise and counter clockwise vortices, f is the frequency of vortex generation, and U is the free stream speed then the Strouhal number for average Reynolds numbers reads as Katz and Plotkin (2010)

$$S_t = \frac{f\,h}{U} \cong 0.1 - 0.2 \tag{8.1}$$

The wings with moderate sweep at their leading edges have the flow separation with more than one vortex at each side of the wing as shown in Fig. 8.1b. Shown in Fig. 8.1c is the wing with the high sweep, $\Lambda > 70°$, which generates a pair of very strong vortices to roll up immediately after separating from the sharp leading edge, section BB. These high strength vortices generate suction at the upper surface which in turn creates additional lift. The pair of counter rotating vortices generated by the leading edge separation at higher angles of attack of the highly swept wings tends to have their symmetric strength uneven. This causes a rolling moment about the axis of the wing. Initially the rolling moment is small and periodic in nature; therefore, it causes the wing to rock. Further increase in angle of attack causes sudden burst of one of the vortices. This puts the wing in spin. Both wing rock and spin are the unsteady motions induced by the flow.

The periodic heaving and/or pitching motion of an airfoil, as a forced oscillation, is for long known to be the major source of thrust generation for flapping wings. During this type of oscillations the flow separation takes place at a larger angle of attack than the angle at which the static stall occurs. The value of reduced frequency

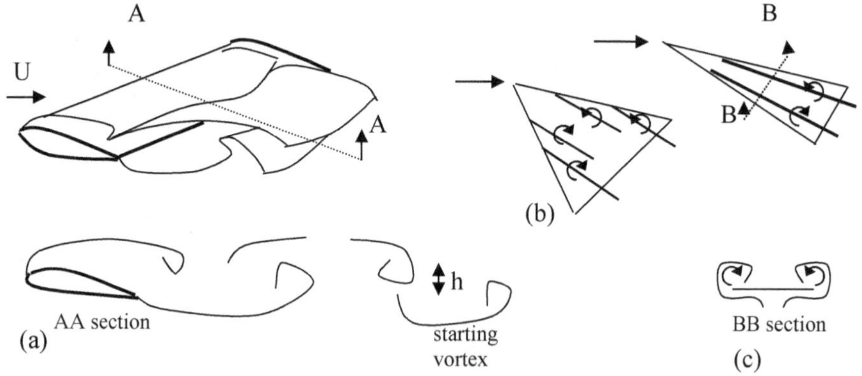

Fig. 8.1 Leading edge separation from a wing with: **a** no sweep, **b** moderate sweep-weak vorticies, **c** high sweep-strong vortices

of oscillation plays an important role in determining the lower limit of the angle of attack at which the dynamic stall takes place. Now, starting from static stall let us see the unsteady aerodynamic aspect of the phenomena which can be included in modern subjects.

8.1 Static Stall

The books on classical aerodynamics depict the picture of static stall as the sudden lift lost after a critical angle of attack. The aerodynamic aspects after the static stall are not usually emphasized (Abbot and Doenhoff 1959). However, even in early 1930s there were some experimental studies performed on different profiles to predict their lifting characteristics beyond stall (Eastman 1931; Anderson 1931). On the other hand, in later years both improvement in visualization and measurement techniques, and numerical solutions of Navier–Stokes equations in advanced computational means and tools enable researchers to study flow separation and corresponding lift lost at least with laminar flow studies (Mehta 1972). The whole flow field was solved by finite difference in Mehta's study which required very large computer memory and time in those days. In addition, solving the attached and separated flow regions with full Navier–Stokes solver caused extra numerical errors. This type of errors and computational time were reduced by means of an integro-differential method (Gulcat, 1981; Wu and Gulcat , 1981). The integro-differential method reduces the computational time with increasing Reynolds number (Wu et al. 1984). Now, we can step by step show the solution domain obtained by integro-differential method for one cycle of flow features by means of instantaneous streamlines of the separated flow past 9% thick Joukowsky airfoil at 15° angle of attack in Fig. 8.2a–j. The laminar flow is studied at Reynolds number of 1000 with the initial conditions given at $t = 0$ as the non circulatory potential flow solution whose streamlines are shown in Fig. 8.2a. Here the non dimensional time is given by the free stream speed and the chord length of airfoil. After the impulsive start, for a short time the flow continues without separation as shown in (b), and as seen in (c) near the leading edge a separation bubble appears. This bubble grows larger to become the main bubble as shown in (d), and forms a large clockwise vortex covering almost the entire upper surface. The main bubble afterwards bursts and separates from the upper surface as shown in (f). The separation of the main vortex from the surface and its movement towards the wake with the main stream generates a counterclockwise rotating vortex at the trailing edge and a secondary weak bubble at the upper surface (f, g). While the secondary surface bubble grows as shown in (g), the trailing edge vortex detaches from the trailing edge and gets carried into the wake (g, h). Meanwhile, recently generated leading edge separation bubble spreads over the entire upper surface in place of the weakening and bursting secondary bubble (h). Thus, one cycle of events becomes complete, starting at $t = 1.89$, and ending at time $t = 7.41$ to have the non dimensional period of $T = 5.52$. The Strouhal number for this flow hence becomes

0.18. The flow separation and one period of vortex formation appear to be unsteady although the boundary conditions of the flow remain the same. Now, we can observe the time variation of the lift and the drag coefficient by examining Fig. 8.3. Impulsively started airfoil at $t = 0^+$ has no circulation and therefore, it has zero lift but very large drag. For this reason, Fig. 8.3 shows the lift and the drag curves together which are started for $t > 0$ just before the flow separation. The lift coefficient C_L reaches its maximum value when the separation bubble covers the entire upper surface at $t = 1.89$. This means the suction created by the separation bubble generates additional lift. However, after the bursting of the main separation bubble, the lift coefficient drops down from 1.3 to its minimum value 0.2 in the time interval $t = 2.93$ to $t = 5.69$. After reaching its minimum value, the lift coefficient increases slightly as the new separation bubble grows and covers the upper surface eventually.

The drag coefficient, given in Fig. 8.3, changes by time similar to the lift coefficient with a phase difference. Since the Reynolds number is low, the drag coefficient values are higher than usual. The drag coefficient takes its maximum value 0.35 when the main separation bubble covers the upper surface causing the largest suction force normal to surface whose streamwise component is quite high.

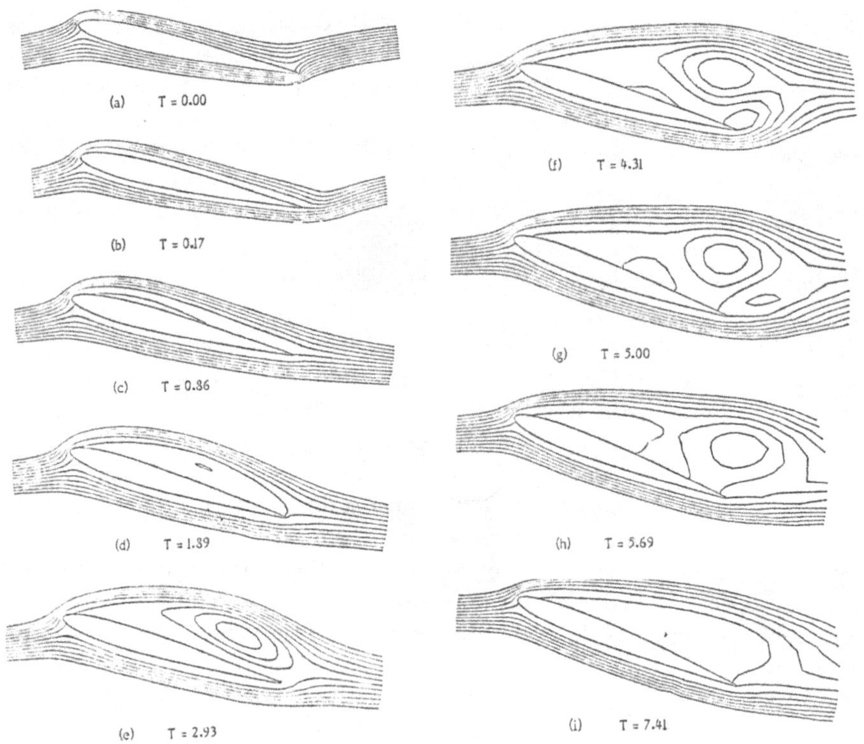

Fig. 8.2 Instantaneous streamline plots at static stall of an airfoil started impulsively from rest

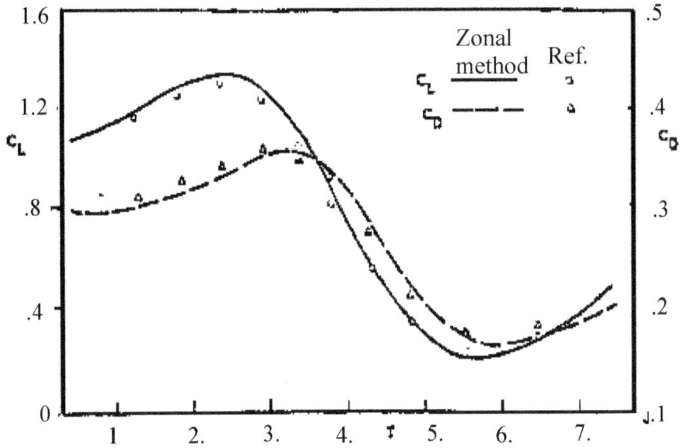

Fig. 8.3 Lift and drag coefficient change by time at static stall

After the bursting of the main bubble, the drag value drops down to its minimum value of 0.15. The growth of new leading edge bubble causes suction to increase, and this in turn makes drag grow to 0.2. As shown in Fig. 8.3, the results of the zonal method is in agreement with the results of the full Navier–Stokes solver given as reference (El-Refaee 1981).

The flow separation causing static stall is a strong separation from the upper surface under constant high angle of attack. The analysis of separated flow regions is possible with numerical solution of Navier–Stokes equations. At the lower surface of the airfoil boundary layer is formed because of favorable pressure gradient. This enables us to divide the entire flow region into two different regions with different flow features, that is attached and detached flow regions which are connected to each other. In the attached flow the boundary layer equation is solved, and Navier–Stokes equations are employed in detached region to give fast and accurate results. These two different regions are interlaced with an integral approach, and the conditions at infinity are satisfied while computing the vortex sheet strength on the surface of airfoil. For this purpose let us express the governing equations in velocity–vorticity formulation in two dimensions. The definition of vorticity: $\omega = \nabla \times \mathbf{V}$ from the velocity field $\mathbf{V}$ gives us the continuity and the momentum equations as follows:

$$\nabla \cdot \mathrm{V} = 0 \tag{8.2}$$

and,

$$\frac{\partial \omega}{\partial t} = -(\mathrm{V} \cdot \nabla)\omega + \nu \nabla^2 \omega \tag{8.3}$$

The boundary layer approximation gives the relation between the vorticity and the velocity component v_s parallel to surface as follows,

$$\omega = -\frac{\partial v_s}{\partial n} \tag{8.4}$$

Here, n is the normal direction to surface, and

$$\frac{\partial \omega}{\partial t} = -(\mathrm{V} \cdot \nabla)\omega + \nu \frac{\partial^2 \omega}{\partial^2 n} \tag{8.5}$$

At a station along the boundary layer, if we know the vorticity value, then integration of (8.5) in the normal direction gives us the velocity component parallel to surface as follows

$$v_s(s,n) = -\int_0^n \omega\,(s,\bar{n})\,dn \tag{8.6}$$

Knowing v_s component in two consecutive stations let us utilize the integral of continuity equation in normal direction to obtain

$$v_n(s,n) = -\int_0^n \frac{\partial v_s(s,\bar{n})}{\partial s} d\bar{n} \tag{8.7}$$

Equation 8.5 is solved to obtain the vorticity values at the new time level t by forward differencing in time, forward differencing in s and central differencing in y direction. This gives us a tri-diagonal system of equations for new time level vorticity values to be found at a given station (Wu and Gülçat). As the boundary conditions of Eq. 8.5, the vorticity at the edge of the boundary layer is taken as zero, and the surface vortex sheet strength computed by integral approach is utilized as the surface boundary condition. We can make use of the continuity of vorticity to obtain an expression for the induced surface velocity by the velocity field at infinity and the vorticity field excluding the surface of the airfoil. This gives us the following integral relation

$$\frac{1}{2\pi}\int_{S^+} \frac{\omega_s x(r-r_s)}{|r-r_s|^2} dS = -V(r_s,t) - \frac{1}{2\pi}\int_{R-S^+} \frac{\omega x(r-r_s)}{|r-r_s|^2} dR + \frac{1}{2\pi}\int_S \frac{(V\cdot n)x(r-r_s)-(Vxn)x(r-r_s)}{|r-r_s|^2} dS \tag{8.8}$$

In Eq. 8.8, S^+ is the neighborhood of the profile surface S, R is the vorticity field, r_s is the point on the surface, ω_s is the vorticity value at the surface, and $V(r_s, t)$ is the time dependent surface velocity vector. Once the surface velocity and the free stream velocity are described, together with the known vorticity field from Eqs. 8.3, 8.5 we can obtain the surface vortex sheet strength from Eq. 8.8. The kinematics of the separated flow region can be formulated in terms of the stream function ψ and the vorticity ω as follows

$$\nabla^2\psi = -\omega \tag{8.9}$$

The kinetics of the separated flow on the other hand is given by Eq. 8.3. The simultaneous solution of Eqs. 8.3 and 8.9 with finite differencing gives vorticity and the velocity fields. The Integro-Differential method applied to a flow past an airfoil at high angle of attack can handle the three different flow regions simultaneously as shown in Fig. 8.4. Theses regions are (i) ideal flow region, (ii) boundary layer region, and (iii) the separated flow region.

The ideal flow region has zero vorticity; therefore, we only need to have the farfield boundary condition effective on the body as the contribution to the flow field from this region. The viscous region, on the other hand, induces velocity according to the Biot-Savart law. The induced velocity can be expressed as the vortex sheet strength on the airfoil surface with the aid of Eq. 8.8. After the impulsive start, the diffused vorticity covers a small flow region around the airfoil surface which is the agent that creates the vorticity. By time this vorticity is convected in the flow direction to increase the vorticity field. We initially take the computational domain small and enlarge it in parallel with the size of the vortex region. This enables us to keep the computational work minimum, and saves us from having spurious errors because of unnecessarily performed computations.

Studying the leading edge separation from certain profiles in two dimensions helps us to understand the nature of the suction generated by the vortex before it bursts. In three dimensional flows because of high sweep we can have the vortex to

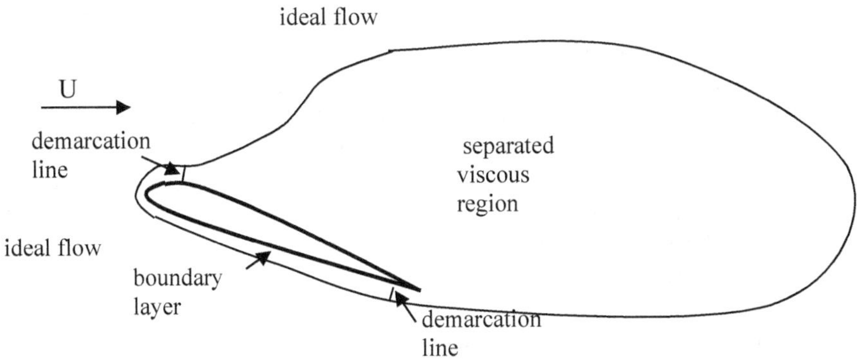

Fig. 8.4 Flow regions around an airfoil at high angle of attack

roll rather than to burst in sustaining the suction force to provide the extra lift at high angles of attack which may cause the 2-D profile to stall.

Now, we can study another stall phenomenon which is dynamic stall of pitching or plunging airfoils in periodic motion wherein the onset of leading edge flow separation is delayed in terms of angles of attack higher than the occurrence of static stall for the same airfoil.

8.2 Dynamic Stall

For an airfoil pitching in oscillatory motion about a given angle of attack, there is a phase lag between the variation of angle of attack, and the lift in time causing a hysteresis for the lift-vs angle attack curve as shown in Fig. 3.9. This phase lag increases and the hysteresis curve becomes more pronounced by increasing as the frequency of oscillation gets larger. This shows us that the response of the airfoil to the angle of attack change is delayed more and more with increasing unsteadiness. This means, while pitching, although the angle of attack can exceed the critical static stall angle the flow may still remain attached. In case of separation, depending on the size and the location of the bubble, as the lift curve returns back from its maximum, the curve deviates from its normal hysteresis behavior until the bubble reattaches itself. The studies related to this behavior were first seen in 1950s during the experiments performed on the helicopter blades in forced pitch oscillation in forward flight (Halfman et al. 1951; Rainey 1957). In later years, a detailed study by Litva gives the following detailed information on a special profile: (i) The effect of Mach number on aerodynamic damping, (ii) the negative damping in large amplitude vertical oscillations, and (iii) The maximum normal force is reported to be significant compared to static case. The wind tunnel used by Litva in his experimental studies had the following operational properties: (i) 0.2–0.6 Mach number range, (ii) $2.2–6.6 \times 10^6$ Reynolds numbers, (iii) pitching reduced frequency range 0.04–0.72, reduced frequency range for heaving-plunging 0.04–0.24, and (iv) $\overline{\alpha} = 0°–25°$ average angle of attack range. For an airfoil which had the 13° angle of attack for static separation and maximum normal force coefficient as $C_N = 1.3$, in dynamic tests, on the other hand, pitching about quarter chord point with $k = 0.062$, $M = 0.4$ and $\overline{\alpha} = 14.92°$ at 17° angle of attack, the maximum normal force coefficient was C_N = 1.s6 (Litva 1969). In Litva's work on dynamic cases an interesting observation was made on the moment coefficient change which starts at 12° angle of attack before the start of lift loss! On the other hand, the increasing frequency delays both the lift loss and the sudden drop of the normal force while reduces the hysteresis effect. Increase in the Mach number affects both the static and dynamic surface pressure before and after the lift loss in such a way that aft of the 10–15% chord the pressure increase is more for the separated flow case. As a result of this, the flow at high subsonic speeds separates at the leading edge and reattaches afterwards; however, the separation occurring near the trailing edge is sustained. The difference between the leading edge separation and the trailing edge separation are given by Ericsson and Reding (1971, 1980). The difference for the dynamic case lift increase is the 50% for the leading edge separation and just 15% for the trailing edge separation. The effect of Mach number on the separation is also observed together with the angle of attack increase (McCroskey 1982). In his work, McCrosky describes the trailing edge separation as 'light stall', and the leading edge separation as 'deep stall'.

According to these definitions, the hysteresis curves for a typical airfoil pitching at small reduce frequencies in a low subsonic free stream are provided in Fig. 8.5. These graphs show the sectional lift C_L, moment C_M, and the drag coefficients C_D changes with respect to angle of attack for one cycle of pitching. (a) At the onset of stall, C_L preserves its elliptic shape accept near high angle of attack, C_M changes in counterclockwise direction with increase in angle of attack, and C_D increases slightly by increasing angle of attack. (b) In light stall, the lift coefficient has lost its elliptic character and as it reaches to its maximum value with a sudden drop it goes down by decreasing angle of attack, the moment coefficient behaves normal with increasing angle of attack, after its maximum with decreasing angle of attack it suddenly drops down, goes down in clockwise and climbs up in counterclockwise manner. The drag coefficient, on the other hand increases by increasing angle of attack to reach its maximum but with decrease in angle of attack the curve even goes down to very small negative region. (c) In the deep stall case, C_L curve climbs up to its maximum value and with a sudden and deep drop goes down to its minimum value in clockwise manner, C_M curve, before its maximum, first in clockwise then in counterclockwise manner completes its cycle, finally the drag coefficient C_D takes quite a large value at the maximum angle of attack and with a sudden drop it goes back to its minimum value as it completes its cycle.

Let us examine the pitching moment change with angle of attack given by a close curve in one cycle. The value of this area, $-\oint C_M d\alpha$, gives us the amount of work, in non dimensional fashion, done on the flow by the profile in one cycle. If the value of the area is positive, then the profile does work on the flow, which reduces the energy of the profile. This has the damping effect on the profile motion. If the area under the $-C_M - \alpha$ closed curve is negative then the flow performs work on the profile to increase its energy, and that in turn increases the amplitude of the pitching oscillations. The increase in the amplitude of the oscillation caused by the negative damping creates the 'flutter'. This type of flutter is called 'stall flutter'. There is a close relation between the stall and the flutter. However, the main difference between the two is that the stall is a more general phenomenon, whereas the flutter is defined as the amplitude of oscillation increase caused by negative damping (McCroskey 1981, 1982). There are some other conclusions that we can draw from Fig. 8.5b, c as follows: (b) in negative damping, the drag has a small propulsive effect because of being negative, and (c) in deep stall, the starting of loss in moment before the loss in lift occurs at the angle of attack at which static separation case loss also occurs.

In case of vertical oscillations, for the angles of attack less than the separation angle, the lift curve preserves its elliptic shape with the vertical coordinate h as

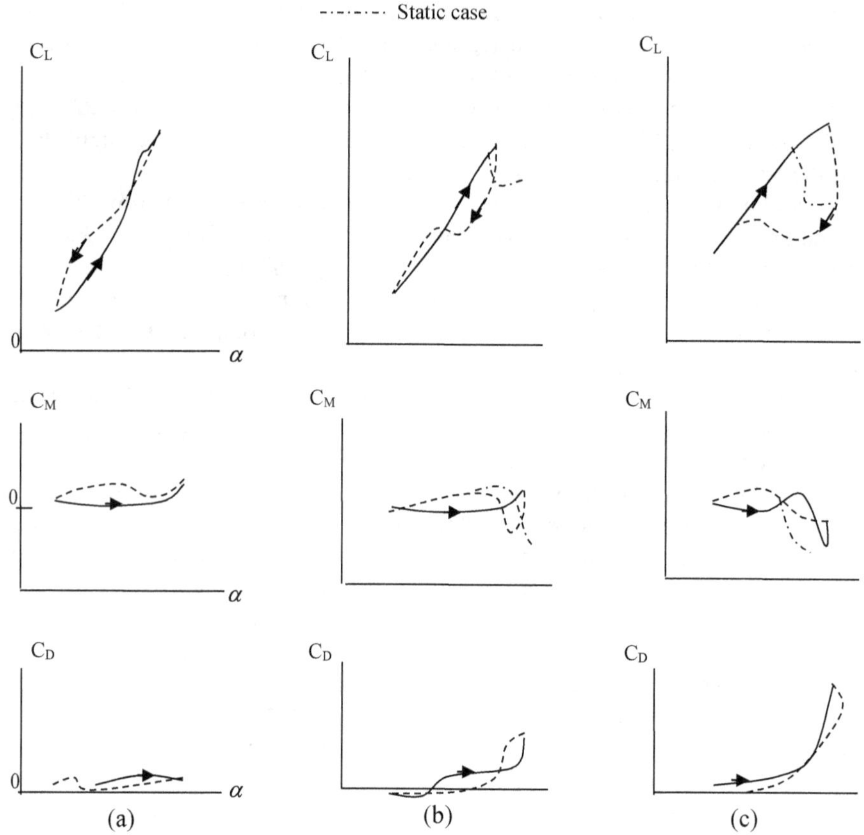

Fig. 8.5 Dynamic stall: **a** on set of stall, **b** light stall, **c** deep stall

indicated by the theory. For the angles of attack larger than the static separation angle, during the downstroke of the profile because of the separation, the lift loss takes its maximum value at the lowest position of the profile, and in upstroke of the airfoil because of reattachment the lift increases until it reaches the highest point (Litva 1969). Here, the area under the close curve of the lift coefficient, $\oint C_L \, dh$, at a moderate Mach number, $M = 0.4$ and at a low reduced frequency $k = 0.068$ yields negative damping in one cycle. In this cycle, the moment coefficient change is about-0.1. The earlier experiments conducted did not reporte any negative damping (Halfman et al. 1951; Rainey 1957). The reason for this is because Litva has worked at higher Mach numbers and at higher amplitudes. A further increase in the Mach number reduces the negative damping eventually to zero because of creation of local shocks at critical Mach number of the profile.

Although the dynamic stall phenomenon experiments and visualizations are helpful for obtaining useful empirical relations, see Problem 8.9, (Ericsson and Reding 1980), it seems more detailed and robust analyses are necessary for engineering applications (McCroskey). The Computational Fluid Dynamics (CFD) as a tool gives this robust and detailed information with numerical solutions of Navier–Stokes equations. The pioneering work on the dynamic stall study of a pitching oscillation of NACA 0012 airfoil at Reynolds numbers of 5000 and 10,000 with reduced frequencies of 0.50 and 0.25 was done by Mehta (1977). The agreement with the experimental work of Werle and the work of Mehta was the early indicative, in those years, of the success of CFD as an analysis tool. The aforementioned work required extensive computational time for one cycle of computations, therefore, especially for the flows with high Reynolds number and the turbulent flows necessitated faster and more efficient codes to reduce the computation times to reasonable levels by means of zonal methods described before (Wu et al. 1984). The Integro-Differential method developed by Wu and Gülçat, is implemented by Tuncet et al. for the dynamic stall analysis of NACA 0012 airfoil at Reynolds number of 10^6 at various reduced frequencies to compare with the experimental work (McCroskey 1981). As a turbulence model Baldwin-Lomax model which is applicable for separated flows also, is implemented. The numerical solution of the Navier–Stokes equations, Eq. 8.2, together with (8.9) was performed in moving coordinates using the effective viscosity. The angle of attack changed as $\alpha(t) = \alpha_{\min} + (\alpha_{\max} - \alpha_{\min})(1 - \cos\omega t)/2$. Shown in Fig. 8.6a is the instantaneous streamlines for k = 0.15, and $\alpha_{\min} = 5°$, $\alpha_{\max} = 25°$, for pitch oscillations of the airfoil. Before the oscillation is started at t = 0, the steady state solution is obtained for a 5° angle of attack. During upstroke, although the static stall angle is exceeded up to angle of attack being 20° flow does not separate. After 20°, however, flow separation starts from the trailing edge and moves toward the leading edge and it reaches to leading edge at 23°. At about 23.9°, at quarter chord a leading edge vortex is formed and it covers the entire upper surface at 24.9°. Going back from the maximum angle of attack of 25°, the vortex leaves the airfoil surface, and moves downstream with a speed of 0.3 U_∞, whereas the experimental value of this speed ranges in 0.35–0.40 U_∞. During downstroke, the vortex starts to separate from the leading edge at 22.8° angle of attack, and with decreasing angle of attack the flow starts to reattach to upper surface similar to that of potential flow until the minimum angle of attack is reached.

Shown in Fig. 8.7a, b are the numerical and experimental values of upper surface pressure coefficient variations during the upstroke and the downstroke motion of the airfoil. As the angle of attack increases, the suction effect of the leading edge vortex and the pressure increase after the separation of vortex from the surface are easily seen, and during the downstroke the flow reattaches, but because of thick boundary layer formation only at the minimum angle of attack the potential surface pressure distribution can be reached, Fig. 8.7.

In Fig. 8.8a shown are the numerical and experimental (i) lift, (ii) drag, and (iii) moment coefficient plots for the same airfoil at k = 0.15 for one cycle of motion at which both experimental and numerical results show the same trend. (i) The lift increases with the increase of angle of attack until reaching maximum, and as the angle of attack becomes smaller the vortex separating from the surface causes lift to drop suddenly. At 22° angle of attack, the new surface vortex forms to increase the

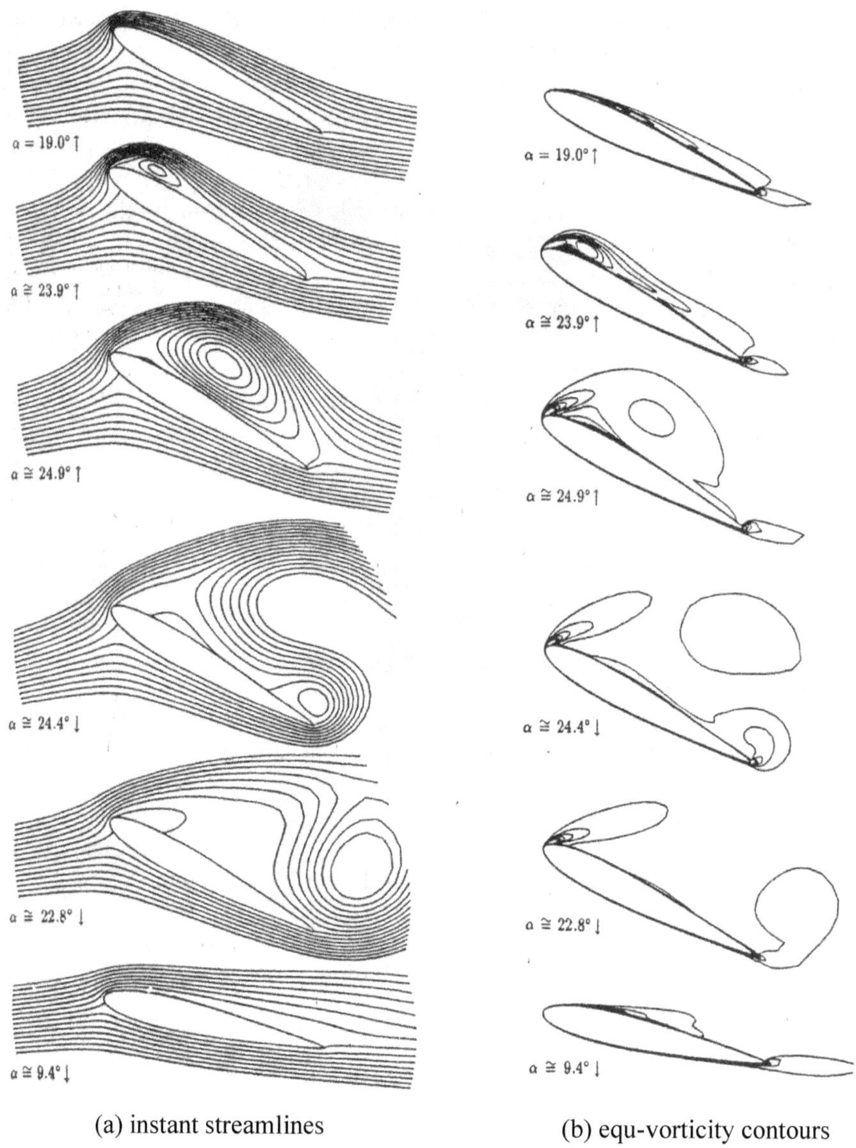

Fig. 8.6 A cycle of pitching motion at Re = 10^6, $k = 0.15$. "Reprinted with permission of the American Institute of Aeronautics and Astronautics"

lift, however, because of the thick boundary layer formation the lift still drops down until 9° angle of attack and it takes its potential value when the angle of attack becomes minimum. (ii) The drag coefficient, on the other hand, increases with increasing angle of attack, after the static separation angle it gradually increases until 23° at which there is a sharp increase to $C_D = 1.0$' at $\alpha = 25°$. During

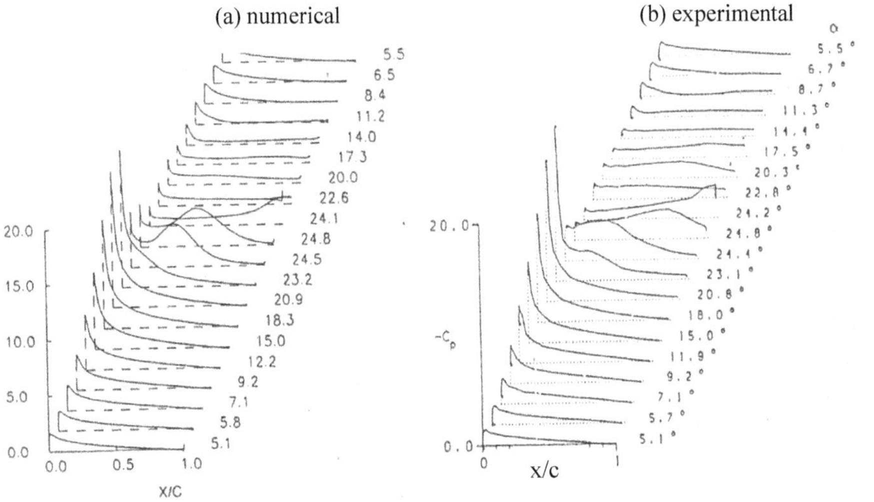

Fig. 8.7 Upper surface pressure coefficient distribution **a** numerical, **b** experimental. "Reprinted with permission of the American Institute of Aeronautics and Astronautics"

downstroke, the drag drops down to values even lower than the values attained at angles of attack equal to that of at upstroke angles, and finally it reaches 0.0 at minimum angle of attack. It is interesting to note that the drag becomes slightly negative when $\alpha < 10°$ during the downstroke, which indicates that there is a slight propulsive force. (iii) The moment coefficient stays at its zero value as the angle of attack increases until the leading edge suction occurs. Afterwards, the moment becomes negative because of growth of the vortex and its streamwise movement which makes $C_M = -0.6$. Returning from maximum angle of attack, moment increases gradually as the angle is in 23°–19° range it decreases again but then it starts to increase to its maximum value of 0.1, and decreases to 0. The moment coefficient is in agreement with the experimental results. However, the area under the $-C_M - \alpha$ curve gives a negative value, larger in magnitude than the value obtained numerically, for the experimental measurements, meaning that the flutter is reached before that predicted numerically.

In Fig. 8.8b shown are the lift and moment coefficient variations with respect to angle of attack in pitching with $k = 0.10$ and $k = 0.25$. For all three reduced frequencies the lift coefficient curves show similar behavior; however the moment coefficients have a tendency to give negative damping with increasing reduced frequencies.

The detailed CFD analysis of the dynamic stall phenomenon has been given here. In Leishman, however, extensive summary of empirical models introduced earlier are given. In addition, the effect of sweep on dynamic stall of a wing is studied with 30° angle of sweep at $M = 0.4$ free stream Mach number (Leishman). In his study Leishman observed that the lift curve slope is not affected with sweep; however, the separation angle increases about 4°, and during the downstroke the

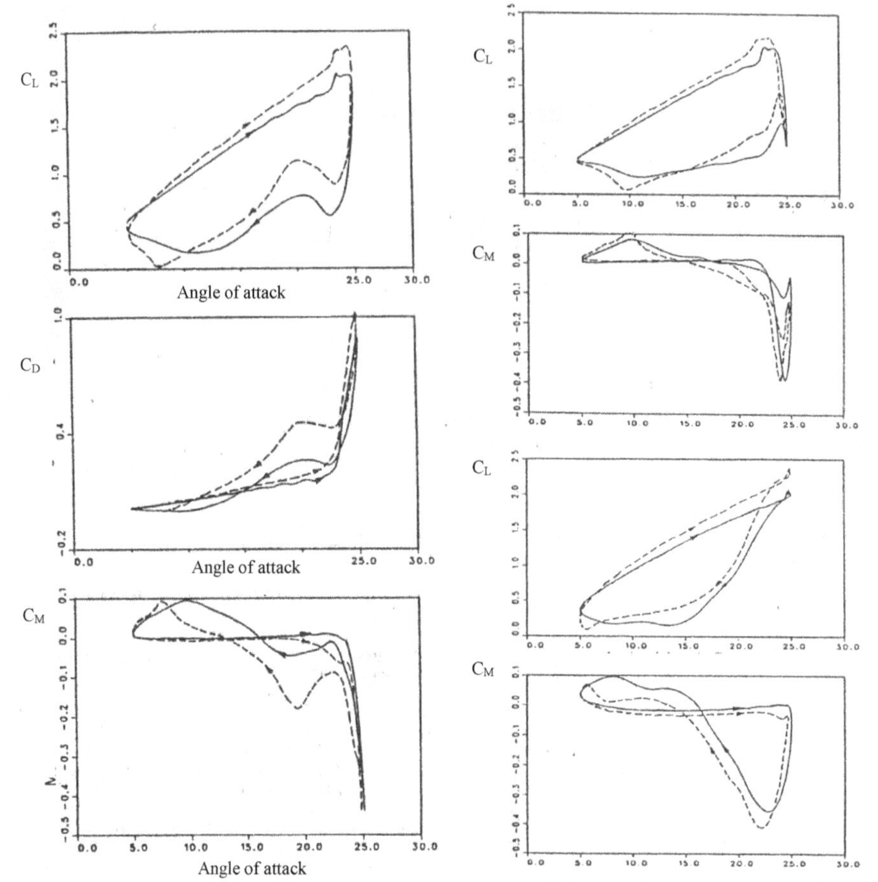

(a) k=0.15

(b) k=0.10 for top two, k=0.25 for bottom two pictures

Fig. 8.8 Lift, drag and moment coefficients at Re = 10^6 for various reduced frequencies. "Reprinted with permission of the American Institute of Aeronautics and Astronautics"

hysteresis curve gets narrower. In the moment diagram, the moment becomes more positive with sweep and the area under the hysteresis curve tends to give more negative damping.

In small sweeps for a finite wing the measurements made at $k = 0.1$ and $M = 0.2$ as the sweep increases:

(i) lift curve slope decreases,
(ii) static separation angle of attack increases,
(iii) the lift curve hysteresis gets narrower,
(iv) the angle of attack at which the moment loss occurs is getting bigger.

A considerably more simple way of studying unsteady airloads at high angles of attack is possible via state-space representation of aerodynamic characteristics based on an input state variable (Goman and Khrabrov 1994). The sectional lift and moment coefficients of an airfoil undergoing an arbitrary unsteady motion can accurately be determined using the static tests separation point movement prescribed as an input state variable (see Problem 8.13).

At large sweep angles the separation phenomenon has a different character than the things happening at small sweeps. Let us next study what happens at high sweeps as the flow separates from the leading edges which may be sharp or round.

8.3 The Vortex Lift (Polhamus Theory)

Classical two dimensional aerodynamics based on the potential theory states that the leading edge suctioncancels the streamwise component of the normal force so that there is no drag force acting on the airfoil. In other words, the leading edge suction obtained by the potential theory can overcome any form of drag except the viscous drag. This makes us wonder if we can make use of the leading edge suction for creating useful aerodynamic forces under special conditions.

It has been observed that the wings with high leading edge sweeps at high angles of attack generate such a high lift that can not be predicted with potential theory. The reason for that is at high angles of attack the vortex generated at the leading edge due to separation merges with the tip vortex to create a strong extra suction force at the upper surface of the wing. This additional lift is called 'vortex lift' and it is predicted with the 'leading edge suction analogy' by Polhamus by early 1970s (Polhamus 1971). This theory is very much in agreement with the experiments, and it is also called Polhamus theory after its validity was proven on delta wings having low aspect ratio. Now, we are ready to study the vortex lift generation with the aid of Fig. 8.9. According to the potential theory the sectional lift coefficient of a thin airfoil in terms of density, freestream speed and the circulation was given by Eq. 1.1 as $l = \rho\ U\ \Gamma$. We can resolve the lifting force l into its components in the direction of the chord and direction normal to the chord. The normal force is denoted by N, and the force in the direction of chord is called the suction force S. The suction force S, here, is the result of the low pressure zone on the upper surface of the airfoil caused by the fast moving flow. Accordingly, if the angle of attack of the airfoil is α, then the suction force will reads as $\mathbf{S} = \boldsymbol{\rho}\ \boldsymbol{U}\ \boldsymbol{\Gamma}\ \mathbf{sin}\ \alpha$. For a profile with a sharp leading edge even a small angle of attack will cause the flow to separate from the leading edge (Fig. 8.9d). The vortex generated by this separation will create a leading edge suction force S which will now be normal to the surface as opposed to the leading edge suction force of the attached flow.

This leading edge suction force creates the extra lift for the wing. For suction force S to be sustainable, the leading edge vortex must merge with wing tip vortex in a stable and steady fashion. The leading edge vortex shown in Fig. 8.10 merges with the tip vortex to form a reattachment line on the upper surface of the wing which has a stable circulation providing continuous extra lifting force.

Let us use the effective circulation Γ and the effective span h of the wing given by Fig. 8.11. If we consider the thrust force T generated by the leading edge vortex, and the adverse effect of the induced downwash w_i we get $\boldsymbol{T = \rho\, \Gamma\, h\, (U \sin \alpha - w_i)}$. Now let us define a non dimensional coefficient K_p related to the potential flow for a wing whose surface area is A as follows

$$K_p = 2\Gamma h/(AU \sin \alpha)$$

As the non dimensional thrust force C_T of the thrust force T we have

$$C_T = \left(1 - \frac{w_i}{U \sin \alpha}\right) K_p \sin^2 \alpha$$

Using the potential lift coefficient K_p we can write non dimensional lift coefficient as $\boldsymbol{C_{L,p} = C_{N,p} \cos \alpha} = K_p \sin \alpha \cos^2 \alpha$. Figure 8.11 gives us the relation between the suction force S in terms of thrust force T as $\boldsymbol{S = T/\cos\Lambda}$, where Λ is the sweep angle of the leading edge. The vortex lift coefficient $C_{L,v}$ after the leading edge separation reads as

$$C_{L,v} = C_{N,v} \cos \alpha = \left(1 - \frac{w_i}{U \sin \alpha}\right) K_p \sin^2 \alpha \frac{\cos \alpha}{\cos \Lambda}$$

The addition of potential and vortex lift gives the total lift coefficient C_L as

$$C_L = K_p \sin \alpha \cos^2 \alpha + K_v \sin^2 \alpha \cos \alpha \tag{8.10}$$

Here: $K_v = (1 - \frac{w_i}{U \sin \alpha}) K_p / \cos \Lambda$.

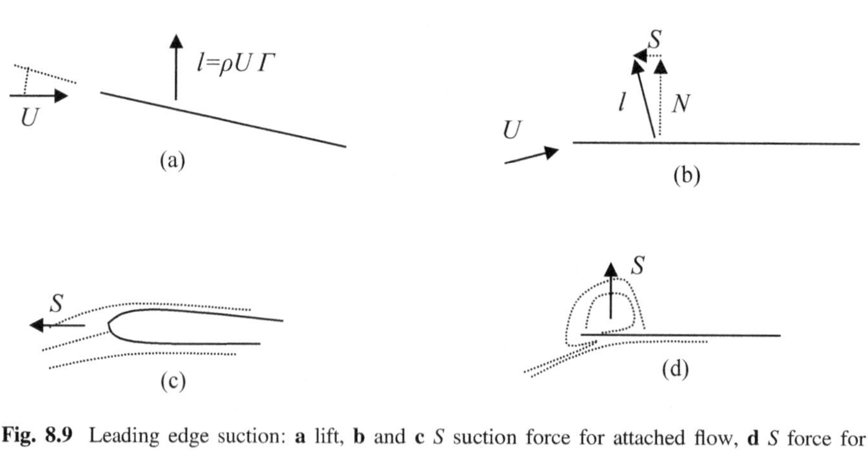

Fig. 8.9 Leading edge suction: **a** lift, **b** and **c** S suction force for attached flow, **d** S force for detached flow

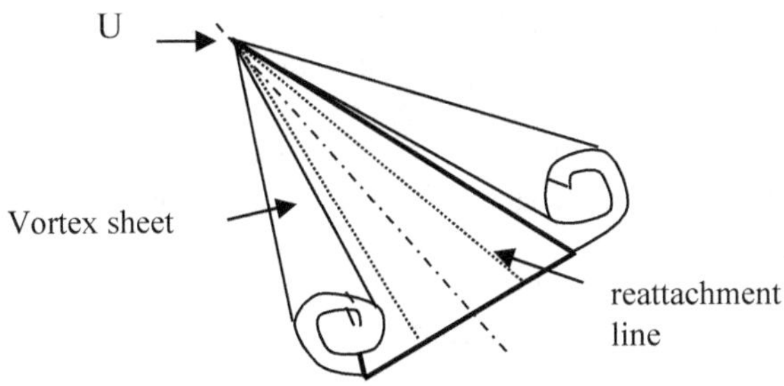

Fig. 8.10 Leading edge vortex sheet and the reattachment line on the upper surface of a delta wing

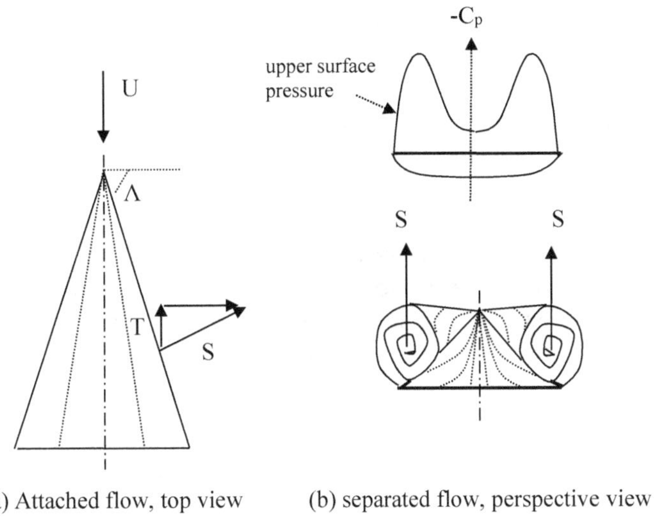

Fig. 8.11 Suction force on the delta wing: **a** attached, **b** separated flow

According to Eq. 8.10, at low angles of attack the potential term, the first term of the right hand side, and at high angles of attack the vortex term which is the second term plays a dominant role in the lift. For the low aspect ratio wings at angles of attack lower than 10° Eq. 1.11 gives a lift proportional with α. Similarly, with assumption of small angle of attack we predict the lift proportional with the angle of attack using Eq. 8.10.

In Fig. 8.11b, shown is the spanwise variation of the upper surface pressure coefficient created by suction force *S* by the separated flow on the upper surface.

In practice, in order to increase the manipulability at high angles of attack using leading edge extension at the root of a wing with moderate sweep creates a new vortex lift (Polhamus 1984; Hoeijmakers 1996). The extra vortex generated by the leading edge extension is shown in Fig. 8.12 with the associated surface pressure coefficient on the upper surface of the wing. Even at moderate angles of attack, the presence of the extra lift contributing to the total is seen in the pressure spanwise distribution. In addition, we observe from the surface pressure curve that the extra vortex creates a strong suction on the upper surface of the wing.

With the aid of Figs. 8.11 and 8.12, we have studied the effects of vortex formation at high leading edge sweep and the leading edge extension of the wings on the spanwise distribution of surface pressure coefficients. Now, we can see the formation of a strong vortex on the total lift coefficient of the wing. Shown in Fig. 8.13 is the total lift coefficient change of a wing with the angle of attack.

The curve in Fig. 8.13 indicated as the potential theory lift curve is obtained by using the first term of Eq. 8.10. In delta wing, on the other hand, both terms of Eq. 8.10 is used in obtaining the lift coefficient curve, where as the leading edge extension lift curve is adopted from Polhamous and Hoeijmaker. In addition, deviation from the Polhamus theory and the experimentally obtained lift loss is shown in (Wentz and Kohlman 1971; Polhamus 1984). Three dimensional wing theory predicts that the induced drag force is proportional with lifting force. The leading edge vortex increases the lifting force, and induces a drag force proportional with the tangent of the angle of attack (Wentz and Kohlman 1971). Accordingly, the total drag coefficient reads as

$$C_D = C_{D_O} + C_L \tan \alpha \tag{8.11}$$

where C_{Do} is the drag at zero angle of attack.

In Chap. 4, the Jones theory provided the lift generated by the cross flow over thin delta wings in incompressible flow. Now, we are going to study the effect of compressibility both in small and large angles of attack. In subsonic flow, depending on Mach number the Prandtl-Glauert formula provides the aerodynamic coefficients (Eq. 1.15a).

In supersonic flows, however, the leading edge sweep angle determines the formula to be used. If the leading edge of the outside of the Mach cone, i.e. supersonic leading edge, the Ackeret formula is sufficient (Eq. 1.20). If it is in the

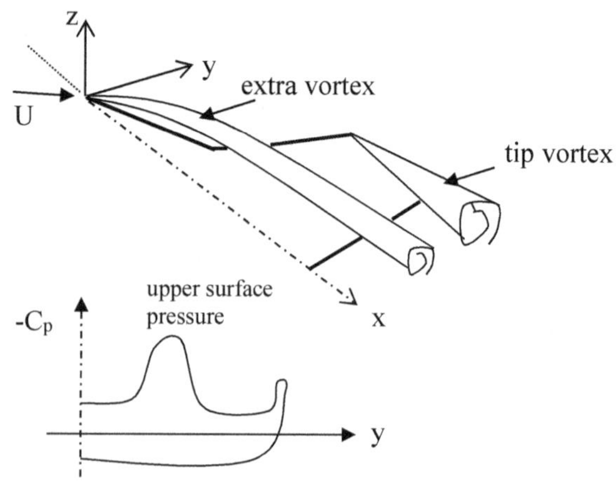

Fig. 8.12 The extra vortex created by the leading edge extension and the associated surface pressure distribution

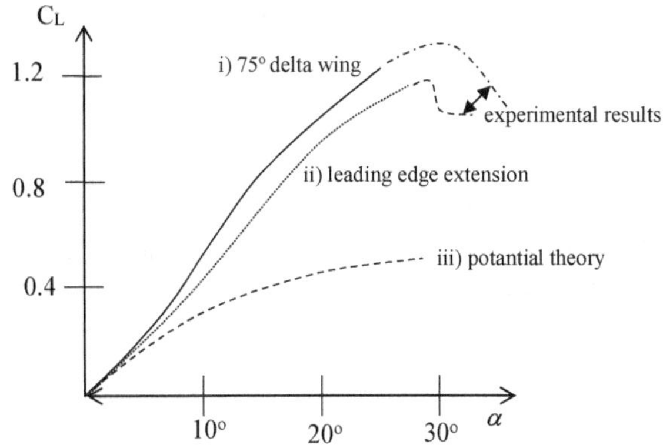

Fig. 8.13 Lift with leading edge vortex: (i) 75° delta wing, (ii) with leading edge extension, (iii) potential theory

Mach cone, subsonic leading edge, the interaction between the upper and lower surfaces has to be taken into account (Puckett and Stewart 1947). In their study, the leading edge sweep angle Λ and the Mach number M are used defining the parameter m as follows:

$$m = (\sqrt{M^2 - 1})\ \cot \Lambda$$

and for $m < 1$.

$$\frac{dC_l}{d\alpha} = 2\,\pi\,\cot \Lambda / E(m') \tag{8.12}$$

Here, $m' = \sqrt{1 - m^2}$ and $E(m')$ is the elliptic integral of second kind (Korn and Korn 1968). If we write Eq. 8.11 suitable for the supersonic flows using Polhamus theory then the potential lift line slope reads as

$$K_p = \pi AR / [2E(m')] \tag{8.13}$$

The vortex lift line slope K_v, in terms of potential lift line K_p with the aid of Fig. 8.11 becomes

$$K_v = \frac{\partial C_{Tp}}{\partial \alpha^2} (\cos \Lambda) \tag{8.14}$$

then for the supersonic

$$K_v = \pi \left[(16 - (AR * \beta)^2)((AR)^2 + 16) \right]^{1/2} / \left[16 * (E(m'))^2 \right] \tag{8.15}$$

Here, $\beta = \sqrt{M^2 - 1}$.

Since the supersonic flow is three dimensional, the induced drag force coefficient for $m < 1$ and small angles of attack becomes (Puckett and Stewart 1947)

$$C_{Di} = \alpha C_L[1 - m'/2E(m')] \tag{8.16}$$

Example 8.1 Euler Equation numerical solver for a delta wing with 75° sweep at $M = 1.95$ and 10° angle of attack gives the normal force coefficient as $C_N = 0.295$ (Murman and Rizzi 1986). Find the normal force coefficient with: (i) incompressible potential flow (ii) supersonic potential flow, and (iii) Polhamus theory.

Solution

(i) For $M = 0$: we find $C_N = \pi/2\ AR\ \alpha \cos\alpha = 0.289$.

(ii) For $M = 1.95$: $K_p = 2\pi\ \cot\Lambda/E(m')$, $m = \sqrt{M^2 - 1}\cot\Lambda = 0.4485$, and $\sin^{-}(1–0.4485^2)^{½} = 63.35$, $E(63.35) = 1.18$ (CRC Tables)

$K_p = 1.446$. This gives $C_{Np} = 0.235$.

(iii) According to the Polhamus theory the extra lift created by the leading edge vortex reads as

$K_v = \pi\ [(16 - 1.1487 \times 2.8025)(1.1487 + 16)]^{½}\ /(\ 16 \times 1.18^2) = 2.087$, and

$C_{Np} = C_{Lv}\cos 10 = 0.061$. Total normal force coefficient:

$C_N = C_{Np} + C_{Nv} = 0.235 + 0.061 = 0.296$.

Accordingly, the closest solution to Euler's result is obtained with the Polhamus theory.

Example 8.2 Find the induced drag coefficient of the wing given in Example 8.1 at 10° angle of attack.

Solution $C_{Di} = 0.033$ from Eq. 8.16.
The viscous and total drag forces acting on the swept wings are given in much detail in Küchemann.

The extra lift created by the leading edge vortex of a delta wing with small aspect ratio is not sustainable, as shown in Fig. 8.11, when the angle of attack increases beyond a critical value because of the spoiling of the symmetry of the vortex pair. Once the symmetry of the vortex pair is broken, the amount of suction on the left and the right side of the wing are no longer equal, therefore, there emerges a non zero moment with respect to the axis of the wing. This moment causes wing to start a rolling motion which is referred to as wing rock.

8.4 Wing Rock

Wings with low aspect ratio, high sweep and sharp leading edges at high angles of attack undergo a self induced unsteady rolling periodic motion called 'wing rock'. In water and wind tunnels, wings having a single degree of freedom, roll only, experiences wing rock beyond a critical angle of attack. The symmetrically formed leading edge vortices, beyond the critical angle of attack can no longer remain symmetric; therefore, their strength changes to create a moment about the axis of the wing. This moment, initially being small, causes wing to roll in one direction. Meanwhile, the vortex on the other side of the wing gets stronger and opposes the roll so that the motion reverses itself. Therefore, a self induced periodic motion is generated. In flight conditions, this rolling takes place together with the side sway and the plunging degrees of freedom. In Fig. 8.14, shown is the three degrees of freedom motion consists of (i) roll, (ii) side sway, and (iii) vertical displacement.

According to Fig. 8.14, the flight direction is in out of y–z plane, and the aircraft rolls about its axis while it moves sideways and descends with high angle of attack. The analysis of this three degrees of freedom motion is possible by evaluating the lifting force, sideways force, and roll moment acting on the body at every instant of the flight in interactive manner. In recent years, the three degrees of freedom problem based on the numerical solution of Euler equations to predict the aerodynamic forces and moment acting on an aircraft in wing rock appeared in literature (Saad and Liebst 2003). On the other hand, the numerical solution of Navier–Stokes

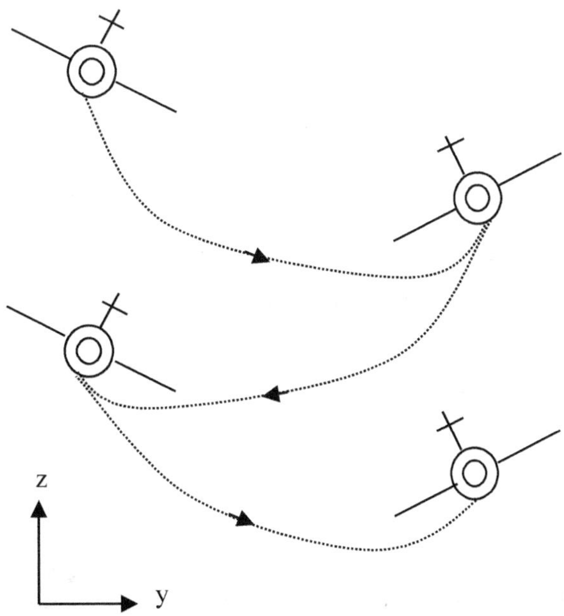

Fig. 8.14 Wing rock with three degrees of freedom: roll, side sway and plunging

equations for delta wings in free or forced rolling oscillations first appeared in the mid 1990s (Chaderjian 1994; Chaderjian and Schiff 1996).

Results of years of experimental as well as numerical studies on unsteady aerodynamics are summarized in Fig. 8.15. On the left side of the graph, where all low aspect ratio wing data was presented, the wing rock occurs above a certain angle of attack for the wings having leading edge sweep more than 74° (Ericsson 1984). If the sweep angle is less than 74°, instead of formation of leading edge vortices we observe their bursting. The bursting of a leading edge vortex causes suction loss on one side of the wing which in turn creates a dynamic instability which is called roll divergence (Ericsson 1984). Starting of roll divergence, however, spoils the periodic rocking motion and causes wing to spin about its own axis. On the right side of the graph, the large aspect ratio effects are visible in terms static and dynamic stall limits.

The numerical studies on wing rock first became possible by modeling the leading edge vortex with unsteady vortex lattice methods (Konstadinopoluos et al. 1985). In that work, the equation of roll motion was based on the conservation of roll moment with roll angle and its time rates. The roll moment equation in terms of the non dimensional moment coefficient C_t as follows

$$I\ddot{\phi} = 1/2\,\rho c A U^2\, C_t(\phi, \dot{\phi}, \alpha) - \mu\dot{\phi} \tag{8.17}$$

Here, I is the roll moment of inertia of the wing, ϕ is the roll angle, A: wing surface area, c: root chord and μ is the bearing resistance to roll. Rearranging the coefficients in Eq. 8.17, and indicating the reduced time by $s = 8Ut/c$ the non dimensional form of Eq. 8.17 reads as

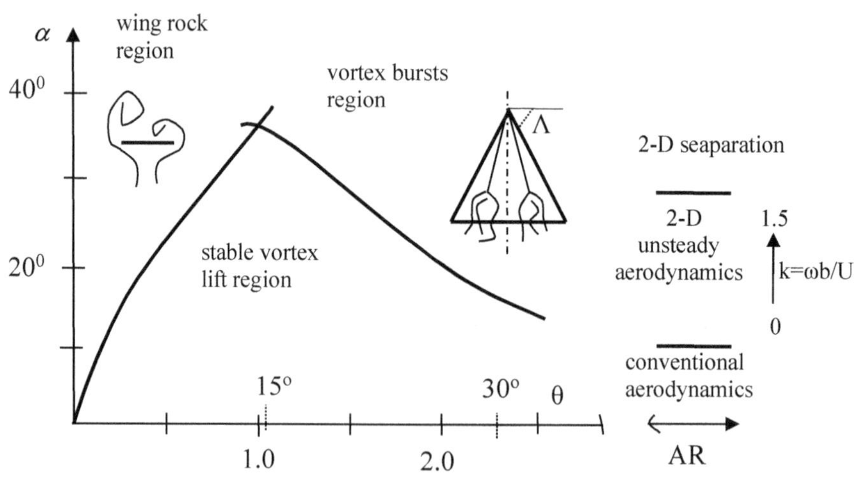

Fig. 8.15 Asymmetric vortex and vortex burst region boundaries depending on AR and $\theta = 90° - \Lambda$

$$\phi''(s) = C_1 C_t(\phi, \phi', \alpha) - C_2 \phi' \quad (8.18)$$

Here, the non dimensional coefficient $C_1 = \rho c^3 S/(128\ I)$ and $C_2 = \mu c\ /(8\ I)$. The roll motion is started with the non zero angle ϕ_o and zero angular velocity. With the sweep angle of 80° and root chord of 42.9 cm the delta wing is set to roll motion at various angles of attack to result in: (i) damped rolling for the angle attack less than 15°, (ii) unstable periodic roll motion for the angle of attack more than 20°.

The simple analytical model construction of roll moment helps to analyze the wing rock phenomenon. Now, we can write a general expression for the non dimensional roll moment coefficient in terms of roll angle ϕ and the time rate of change of that angle ϕ' as follows

$$C_l = a_1 \phi + a_2 \dot{\phi} + a_3 \phi^3 + a_4 \phi^2 \dot{\phi} + a_5 \dot{\phi}^2 \phi + a_6 \dot{\phi}^3 + a_7 \phi^5 + a_8 \phi^4 \dot{\phi} + a_9 \phi^2 \dot{\phi}^3. \quad (8.19)$$

Here, non dimensional coefficients a_i are computed by least square method. The first six terms on the right hand side of the Eq. 8.19 contribute significantly to the value C_l, and the rest of the terms are insignificant. In addition, the terms with odd powers of ϕ constitute the restoring force, and the odd powers of $\dot{\phi}$ are responsible for damping of the roll motion. Accordingly, we can write down the force coefficient C_R which is responsible for the restoring force, the coefficient C_D which is for the damping as follows

$$C_R = a_1 \phi + a_3 \phi^3 + a_5 \dot{\phi}^2 \phi \quad (8.20a)$$

and,

$$C_D = a_2 \dot{\phi} + a_4 \phi^2 \dot{\phi} + a_6 \dot{\phi}^3 \quad (8.20b)$$

Example 8.3 At 25° angle of attack, a wing with 80° angle of sweep rocks with period of 0.39 s. and amplitude of 32°. Obtain the graph of C_R versus ϕ and C_D versus $\dot{\phi}$, and comment on them. Data: $a_1 = -0.0572$, $a_2 = 0.1362$, $a_3 = 0.0514$, $a_4 = -1.403$, $a_5 = -1.943$, $a_6 = 0.075$.

Solution (i) The data is used to express the roll angle in radians $\phi = -32 \sin(2\pi t/0.39)\ \pi/180$ and writing non dimensional $\dot{\phi}$, the restoring force coefficient reads as

$$C_R = -0.0572\phi + 0.0514\phi^3 - 1.943\dot{\phi}^2\phi$$

whose graph is shown Fig. 8.16 which gives the force coefficient in opposite phase with roll angle. These two being in opposite phase make the motion continue in a stable manner. The roll angle changes sinusoidally, however, because expression

(8.20a) being non linear, C_R is periodic but no longer simple harmonic, as seen in Fig. 8.16, especially at the flat peaks of C_R curve.

(ii) The aerodynamic damping coefficient is obtained by subtracting the bearing resistance from Eq. 8.20b as $C^*_D = 0.1362\,\dot{\phi} - 1.403\,\phi^2\dot{\phi} + 0.075\,\dot{\phi}^3 - 0.004\,\dot{\phi}$ Shown in (8.17) is the aerodynamic damping coefficient C^*_D and roll rate $\dot{\phi}$ with time. According to Fig. 8.17 the period of damping coefficient is the half of the period of roll rate (The equations are used with permission of the American Institute of Aeronautics and Astronautics).

According to Figs. 8.16 and 8.17, when the roll angle is approximately zero and the roll rate is near maximum, the damping moment and the roll rate have the same sign, and when the roll angle is maximum and the roll rate is zero, the damping moment and the roll rate have opposite signs. For this reason when the roll rate is maximum, since it has the same sign with the damping moment, there is a loss in damping which means there is a positive feeding of the motion. That is how the wing rock is sustained.

Example 8.4 Plot the hysteresis curve for the roll moments roll angle for the delta wing given in Example 8.3. Indicate the intervals on the hysteresis curve where the motion is damped and where it is fed. Compare the new plot with the comment made on the previous graph.

Solution The total rolling moment coefficient is

$$C_T = C_R + C_D$$

whose graph with respect to roll angle ϕ is plotted in Fig. 8.18. In that Figure when the curve follows the clockwise pattern, there is a negative damping, and the counterclockwise pattern there is a damping. Accordingly, as far as the intervals are concerned, in $-32 \leq \phi \leq -18$ and $18 \leq \phi \leq 32$ there is damping and in -18

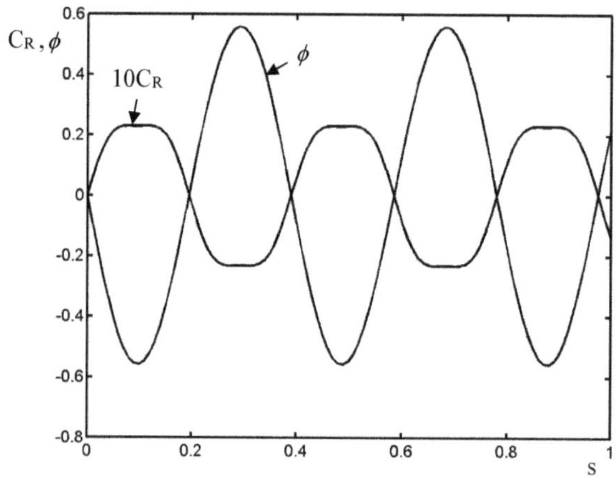

Fig. 8.16 Change in restoring force for the wing rock $10C_R$ and ϕ by time

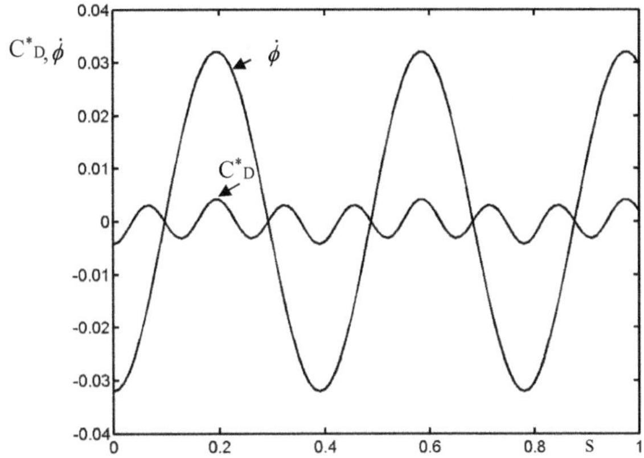

Fig. 8.17 Aerodynamic damping coefficient and the roll angle rate change by time

$\leq \phi \leq 18$ feeding occurs. A similar conclusion is made at high roll angle where there is a negative damping observed.

Examples 8.3 and 8.4 provided us with detailed information about the rolling moment change with roll angle of a wing in a wing rock as a single degree of freedom problem. During the rolling motion of a wing, except at zero yaw angle, while the effective angle of attack changes, the free stream direction also changes with an amount β as the yaw angle as indicated with 3-D representation on Fig. 8.19.

Now, let us express the effective angle of attack α_e , and the effective yaw angle β_e in terms of roll angle $\phi = -\Delta\phi \sin \omega t$. Here, $\Delta\phi$ is the amplitude of roll angle.

$$\alpha_e = \arctan(\tan\alpha_0 \cos\phi) \tag{8.21}$$

$$\beta_e = \arctan(\tan\alpha_0 \sin\phi) \tag{8.22}$$

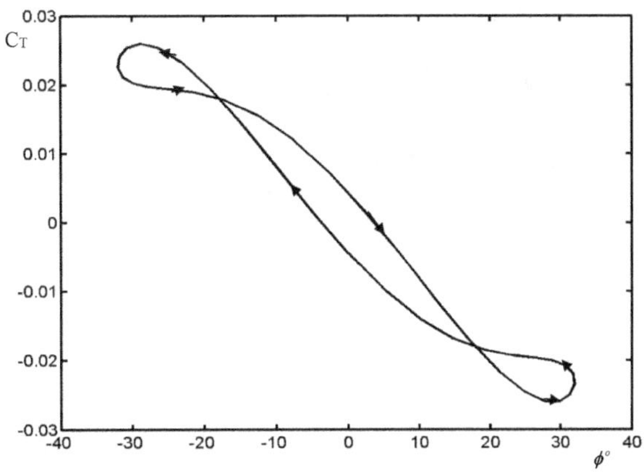

Fig. 8.18 Roll moment coefficent versus roll angle hyterisis curve

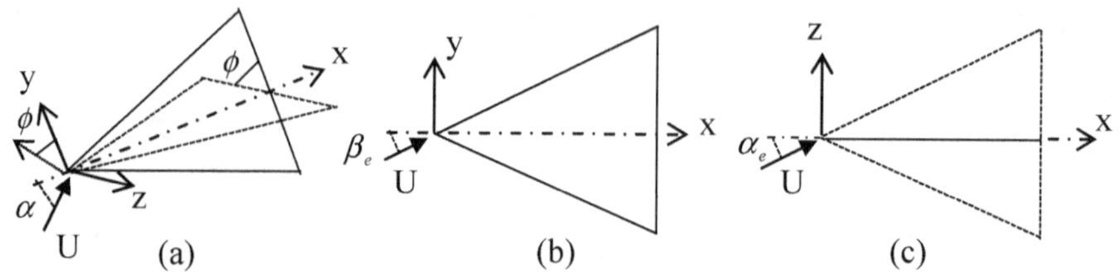

Fig. 8.19 Effective angle of attack and yaw angle of a wing rolling with angle ϕ: **a** angle of attack and yaw angle in 3-D, **b** effective yaw angle β_e, top view, **c** effective angle of attack α_e, side view

For $\phi = 0$ yaw angle we take the angle of attack as α_0. During the rolling motion we consider only the rotational degree of freedom around the root chord of the delta wing. However, Eqs. 8.21 and 8.22 indicate that as the effective angle of attack deceases, the emerging sideways flow causes the flow symmetry to be spoiled. For this reason, the normal force acting on the delta wing changes during rolling and also because of spoiling of symmetry. Shown in Fig. 8.20 is the variation of the effective angle of attack and the yaw angle with the change of roll angle as given in Examples 8.3 and 8.4 for α_0 = 25° and $\Delta\phi$ = 32°. Accordingly, the effective yaw angle changes between −12.5° and 12.5°, while effective angle of attack varies between 22° and 25°. Since the effective angle of attack decreases with rolling, the normal force also decreases. In experiments, however, the static and dynamic cases change in the normal force is found to be different (Levin and Katz 1984). The measurements of Levin and Katz indicate that the time average of the normal force coefficient measured during the roll is smaller than the statically measured values. This difference, for the angles of attack less than 32°, is due to unspoiled vortex symmetry at zero roll angle of for static case, and continuously existing asymmetry for the dynamic case. At higher angles of attack, the vortex bursts occurs earlier for the dynamic case than it happens for the static case which makes the average normal force coefficient for the dynamic case to be 15–20% less than that of the static case.

The measurements made on the delta wing given in Example 8.3 suggest that during rocking, the oscillatory aerodynamic side force acting on the wing has amplitude of 0.5 and a small phase difference between the roll angle (Levin and Katz 1984). About the rolling characteristics of the delta wing of Example 8.3 and 8.4, there is detailed information related to experimental results, conditions and comments at 20, 25, 30 and 35° angles of attack given in Levin and Katz. On the other hand, the frequency of the normal force coefficient is double the frequency of rolling, and the amplitude changes are in minimum 0.3 and maximum 0.7. In their work, the dynamic values of normal and the sideways forces in terms of static and the roll angle values $\phi_{\max}$ and $\phi_{av} \cong \phi_{\max}/2$, as upper and lower limits read,

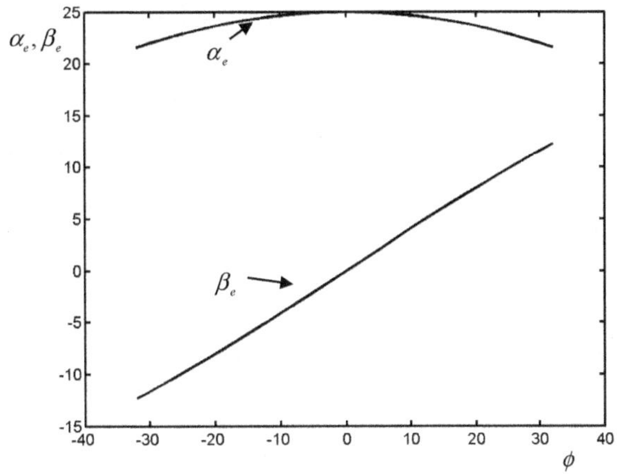

Fig. 8.20 Change in effective angle of attack and yaw angle with roll angle

$$C_{N\,DY} < C_{N_{ST}} \cos \phi_{av} \tag{8.23a}$$

$$C_{Ymaks} \geq C_{N_{ST}} \sin \phi_{maks} \tag{8.23b}$$

During wing rock the maximum dynamic normal force coefficients, as stated before, can not exceed the values attained in static cases as given by inequality (8.23a). On the other hand, according to Expression 8.23b, the lower limit of the side force acting on the wing is proportional with the static value of the normal force. This means, even for a single degree of freedom problem, there is a minimum sidewise force created. For this reason, in real flight condition wing rock analysis we need to consider the sideways and vertical degrees of freedom in addition to the rolling as shown in Fig. 8.14.

Three degrees of freedom simulations in wind tunnels require building of mobile models which are quite expensive to operate, when possible. This forces us to make measurements in flight conditions and/or to perform detailed numerical simulations to compare the results obtained with single degree of freedom problem (Saad and Liebst 2003). In their work, Saad and Liebst use numerical solution of Euler equations to compute the flight path under the aerodynamic forces computed as three degrees of freedom problem. The kinetic yaw angle can be computed if we take the v as the velocity normal to free stream as follows

$$\beta_{kin} = \arcsin (v/U) \tag{8.24}$$

The total yaw angle is determined with addition of angles given by Eqs. 8.21 and 8.23a, b. The geometry of the wing consists of a cone with 30° cone angle and a delta planform with 60° sweep. Two different flow case studies were done about this geometry: first study involves only the roll degree of freedom, and the second study was done with three degrees of freedom. The following was observed:

(i) The angle at which rocking starts is 5° higher for three degrees of freedom.
(ii) Roll angle amplitude is 50% for the three degrees of freedom.
(iii) The sideways motion has 90° phase difference with rolling, which contributes to damping.
(iv) The occurrence of vortex burst causes the amplitude for three degrees of freedom to be less.
(v) The vortex dynamics suggests that the pitching degree of freedom should be included in wing rock analysis.

The comments made above are only based on the numerical solution of Saad and Liebst, and they are not validated by flight measurements.

On the other hand, based on the Navier–Stokes solution, one degree of freedom problem was studied in detail with comparing experimental data given for rolling by Chaderjian and Schiff. Their study is made for the 65° swept delta wing mounted on 8% thick cone-cylinder body. In their study, they consider 15° angle of attack, Reynolds number of 3.67×10^6, Mach number of 0.27, and maximum roll angle of $\phi = 40°$. The dynamic roll motion occurs at a frequency of 7 Hz, and the normal force coefficient shows similar behavior to that of experiments while it is predicted little less than the static force coefficient. Another important conclusion made in their work is the moving of center of pressure towards trailing edge for the dynamic case as compared to the static case. Here, low values of the normal force coefficient could be the reason for the center of pressure to move towards the trailing edge. In addition, the roll moment versus roll angle hysteresis curve is in counter clockwise direction, similar to the case of experimental measures, which indicates that the motion has a damping character. The delta wing left free to roll from the maximum roll angle shows a damping motion experimentally, whereas Navier-Stoke solutions predicts over damping. These computations and experiments are performed at 30° angle of attack, and they converge not to zero roll angles but to half of the maximum roll angle! This shows that if the roll motion is not forced then the delta wing can undergo unsymmetrical damping motion.

The wing rock motion or rolling studied so far is for the slender delta wing whose vortex dynamics is well understood. The non slender wings with round leading edges having about 45° sweep angle at high angles of attack may undergo wing rock for different aerodynamic reasons (Ericsson). The effective angle of attack, as shown in Fig. 8.19b, because of effective yaw angle reads as

$$\Lambda_{efff} = \Lambda \pm \arctan(\tan\sigma \sin\phi) \tag{8.25}$$

Here, σ is the angle between the roll axis and the free stream direction. The effective sweep angle increases on one side of the wing while it decreases on the other side to spoil the symmetry of the vortices. This causes a net roll moment also on the non slender wings, which may lead to the wing rock. Once the wing rock started, one side of the wing goes up and the other side goes down relative to the root chord. The flow separation becomes possible when the effective angle of attack, Eq. 8.20a,b, is near stall angle. For rolling motion to be continuous and periodic wing rock the negative dissipation is necessary. The necessary negative damping is provided by the 'moving wall effect' acting on the boundary layer near the stagnation region. At high angle of attack near the flow separation, the moving wall effect makes the flow to reattach at the uplifting side of the wing and increases the lift at that side while the wing rocks. This way the force increases with the increasing direction of motion to create negative damping. On the other side of the wing which is moving downwards, the moving wall effect increases the separation, which in turn decreases the lift on that side, and naturally reduction in the force and the motion in the same direction create reduction in the damping. Thus, occurrence of negative damping on both sides creates enough energy for wing to rock. This is how the flow induces rocking motion on the non slender wing with round leading edge (Ericsson 2001).

For two different non slender wings both with 45° degree sweep angle as shown in Fig. 8.21a, b, one with the round the other with the sharp leading edges and lower thickness, we observe a completely different rolling behavior at high angles of attack (Ericsson 2003). The planform given in Fig. 8.21a is a 9% thick delta wing with round leading edge and the other planform is 6% thick with sharp leading edge.

For both types of wings given in Fig. 8.21, the experiments performed at high angles of attack give 50% less rolling moments compared to slender delta wings. The sharp leading edged wing at 20°–25° angles of attack is damped at roll angles of 42° and 0° respectively when it was left free at 28° roll angle. On the other hand, for the round leading edged wing at 30°–35° angle of attack range, we observe damped motion at zero roll angle which is left to roll at angles $\phi = 10°$ and 30°, respectively. At 25° and 30° angle of attack, the round leading edged wing when left free to roll from 30° roll angle, rocks with 20° roll amplitude at about $\phi = 50°$. This means that the undamped rolling motion can be observed experimentally only for the non slender wing with the round leading edge starting from certain roll angles. The observed rocking motion is quasi periodic with an approximate period

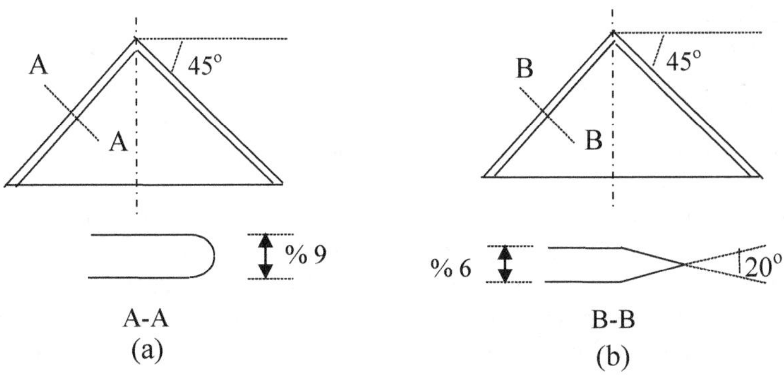

Fig. 8.21 Non slender 45° delta wings: **a** 9% thick with a round leading edge, **b** 6% thick with a sharp leading edge

of 1.5 s. This shows that non slender delta wings with moderate sweep have rocking frequency of one order of magnitude less than that of the slender wings with high sweep. This finally proves that the aerodynamic effects causing the rocking of non slender wings occur slower than that of slender wings.

There is a third kind of wing rock occurring at high angles of attack caused by the periodic shedding of the vortices around the left and right side of a fuselage (Ericsson et al. 1996). For an aircraft having slender fuselage with moderately swept wings at high angles of attack, i.e. $\alpha > 30°$ which exceeds static stall angle, we observe this type of wing rock induced by the shedding of vortices from the part of the fuselage which is ahead of the wing. The occurrence of this kind of rocking motion is caused by the vortex shedding from the separated cross flow about the frontal portion of the fuselage. A cylindrically shaped front body rolls about its axis with an angular velocity while it rocks. During this rolling, there is also a vertical flow because of high angle of attack flow separation. Depending on the value of the Reynolds number based on the cross flow velocity there exists a Magnus force, with known magnitude and direction, acting on the cylinder (Ericsson 1988). The Magnus effect on the cylinder is in the positive direction because of the speed of rotation causing the flow is subcritical and laminar. With the increase in the Reynolds number if the critical flow condition is reached, there emerges a Magnus force which is in opposite direction. In flight conditions the wing rock caused by frontal body is observed experimentally at this critical flow regime. When the Reynolds number based on the free stream speed, body diameter and kinematic viscosity is in the range of 1.0×10^5 and 4.0×10^5, the critical flow conditions are reached. In Fig. 8.22a, b shown is the negative Magnus effect acting on the rotating cylinder in critical flow conditions. The rotational effect on the cylindrical surface causes early transition at the right side of the cylinder, and at the left side the transition is late. The early transition at the right side of the cylinder and reattachment causes a suction force creating negative Magnus force. Meanwhile, from the right side a counter clockwise rotating vortex is shed to the wake. This newly shed strong vortex creates a rolling effect which slows down and stops the clockwise rotation, and causes cylinder to rotate in counter clockwise direction. This time at the left side of the cylinder we observe a suction creating a Magnus force directed towards left. That is how the self induced motion feeds itself in creating sustainable wing rock action. In practice, the wing rock caused by the frontal body is the slowest rocking motion with the period of 3.5 s. Here, the flow separation from the moving body and the vortex shedding play an important role in determining the period of wing rock.

Assuming that an axisymmetric frontal body without a tail wing rocks similar to that shown in Fig. 8.22, we can construct the theoretical hysteresis curve for the roll moment versus roll angle as shown in Fig. 8.23a. The ideal curve given in Fig. 8.23a has the negative damping property for the rolling motion; therefore, the wing rock is self sustainable. The ideal curve indicates that as the body rotates in clockwise direction, the roll angle increases to its maximum value, and when the angular speed is 0, the roll angle reaches its maximum value and changes its direction to counter clockwise rotation. Let us denote the time between two

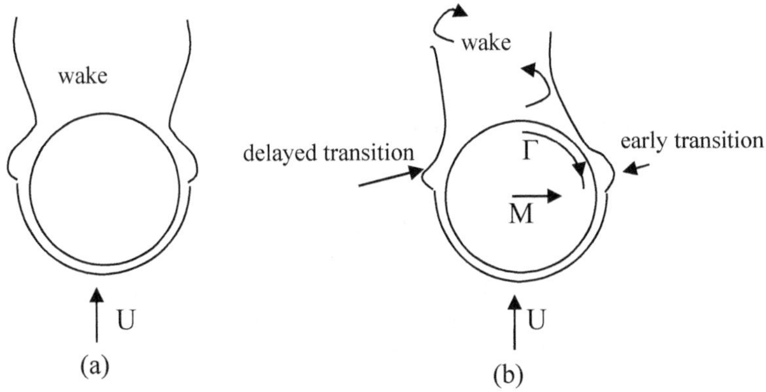

Fig. 8.22 Moving wall effect about rolling in a critical flow: **a** without roll, **b** with roll

successive vortex shedding as Δt. Then the counter clockwise rotating body with the increase of negative roll moment goes back to the zero roll position so that in Δt time interval it starts from 0 roll angle and goes back to 0 roll angle position. In the next Δt time duration it completes its roll to the left side. Finally, in 2 Δt time period it completes one cycle of its motion.

In Fig. 8.23b we observe the real version of the wing rock due to vortex shedding from a frontal portion of a fuselage which rocks in –30° and +30° roll angles. The clockwise direction of the curve near the zero roll region indicates the negative damping while in extreme angles the counter clockwise direction is indicative of positive damping. The difference between the two supplies the necessary energy for rocking.

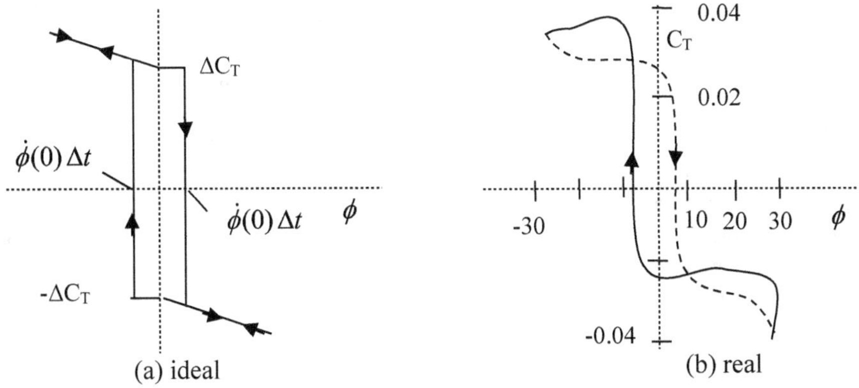

Fig. 8.23 Roll moment versus roll angle hysteresis curves for: **a** ideal, **b** real cases

8.5 Flapping Wing Theory

In recent years, among the subjects of unsteady aerodynamics the flapping wing theory, which is based on the Knoller-Betz effect, has been the most popular one because of ever increasing demands in designing and manufacturing for micro aerial vehicles, MAVs (Platzer et al. 2008; Mueller and DeLaurier 2003).

In order to have sustainable flight with flapping wings, it is necessary to create a sufficient propulsive force to overcome the drag force as well as a sufficient lifting force. In finding the propulsive force we have to evaluate the leading edge suction force created in chordwise direction with pitching-plunging motion of the profile. If we model the profile as a flat plate undergoing unsteady motion, we can obtain the change of the suction force and the lifting force by time using the vortex sheet strength obtained via potential theory (Garrick 1936; Von Karman and Burgers 1935). For the sake of simplicity, let us first analyze the plunging motion of the flat plate undergoing a simple harmonic motion given by $h = \overline{h}e^{i\omega t}$, where $\overline{h}$ is the amplitude of the motion. In terms of reduced frequency k, the Theodorsen function $C(k) = F(k) + iG(k)$, and the non dimensional amplitude $\overline{h}^* = \overline{h}/b$ the sectional lift coefficient reads as

$$\overline{c}_l = -2\pi k \overline{h}^* C(k) i + \pi k^2 \overline{h}^* \tag{8.26}$$

The lifting pressure distribution which provides this lift coefficient also creates a leading edge suction force in the flight direction. The relation between this suction force S and the singular value of the vortex sheet strength at $x = -1$ reads as $\overline{\gamma} = 2P/\sqrt{x^* + 1}$, where $P = \sqrt{2}\, C(k)\, \dot{h}$,

$$S = -\left(\pi \rho P^2 + \alpha L\right) \tag{8.27}$$

Here, α is the angle of attack, and L is the associated lift if there is also pitching. The derivation of (8.27) is given in Appendix 9. The minus sign in front of the suction force indicates that it is in opposite direction with the free stream, which means it provides a force in the direction of flight at pure plunge, and for pure pitching it may give negative propulsion depending on the phase lag between the angle of attack and the associated lift.

As an example using Eqs. 3.25 and 3.26, we can obtain the sectional lift and propulsive force coefficients for a flat plate in vertical unsteady motion given by $h = -0.2 \cos\omega t$ and the reduced frequency of $k = 1.5$. The real part of the sectional lift coefficient is created by the real part of $h(t)$ which corresponds to $\cos\omega t$. Therefore, with small manipulations we obtain for the lift

$$c_l = -2\pi k[(G(k) + k/2)\cos(k s) + F(k)\sin(k s)]\, \overline{h}^* \tag{8.28}$$

and for the suction

$$c_s = -2\,\pi\, k^2\,[G(k)\cos(k\,s) + F(k)\sin(k\,s)]^2 \overline{h}^{*2} \tag{8.29}$$

Here, $s = Ut/b$ shows the reduced time. Shown in Fig. 8.24 is the time variation of the motion of the plate, sectional lift and suction force coefficients with respect to reduced time. During the simple harmonic motion of the plate, since the angle of attack is zero the sectional lift coefficient changes periodically with the amplitude of 1.7 and with the frequency of the motion but with a phase lag. When the profile is at its lowest position, the lift coefficient is negative, and during the early times of upstroke it decreases to its minimum −1.7. While it is still in upstroke motion, the c_l value increases gradually to become positive as the profile reaches the highest position. During early stages of down stroke the lift coefficient starts to increase to reach its maximum value of 1.7, and then its value decreases to become negative as one cycle of motion is completed. In other words, as the bound vortex Γ_a on the plate changes in proportion with the lift, because of the unsteady Kutta condition there is a continuous shedding of vortices with the opposite sign to that of bound vortices into the wake.

During the down stroke of the airfoil, the clockwise rotating bound vortex grows in magnitude for a short time, and after its maximum value it gets smaller while a counter clockwise vortex is shed into the wake from the trailing edge. After the profile passes the midpoint location, the sign of the bound vortex changes to become a counterclockwise rotating vortex while a clockwise rotating vortex is shed into the wake. The schematic representation of the bound vortex formation and the vortex shedding into the wake is shown in Fig. 8.25. In Fig. 8.24, shown is the sectional suction force variation by time which indicates that the propulsive force coefficient remains 0–0.2 in magnitude while its frequency becomes the double of the frequency of the motion. The maximum values of the propulsion occur as the profile passes through the midpoint during its down stroke, and the zero propulsion is observed twice right after the top and bottom points of the profile's trajectory in

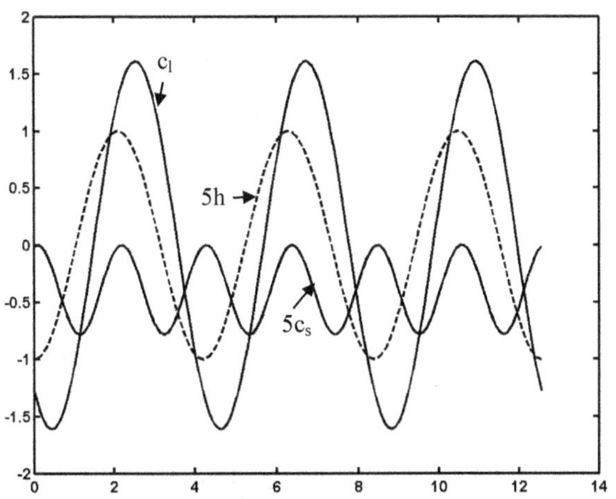

Fig. 8.24 Lift c_l and the suction force c_s coefficient changes with the vertical motion h of the profile

one cycle. This shows us that the creation of the maximum suction force occurs with 90° phase difference with occurrence of maximum or minimum bound vortex. That is when the absolute value of the bound vortex is highest the profile produces zero suction force.

The shedding of vortices in alternating sign from the trailing edge to the wake as described above forms a vortex street. The vortex street in the wake of the oscillating flat plate as shown in Fig. 8.25 indicates that the vortex shed at the top position of the airfoil is in counterclockwise rotation, and the previous vortex shed at the bottom location is in clockwise direction. This means the vortex street has counterclockwise rotating vortices at the top row and clockwise rotating vortices at the bottom row. We note at this point that the vortex street forming at the wake of vertically oscillating flat plate is exactly opposite to the vortex street forming behind the stationary cylinder where the top row of vortices rotate in clockwise and the bottom row vortices rotate in counterclockwise direction. The vortex streets generated behind the circular cylinder and at the wake of the oscillating flat plate have been also observed experimentally (Freymuth 1988). It is a well known fact that the wake formed behind the cylinder creates a drag on the cylinder whereas the wake of the oscillating plate has a structure which is opposite in sign is naturally expected to give a negative drag i.e. propulsion!

Now, let us analyze the physics behind the creation of propulsive force by a vertically oscillating profile using the concept of the force acting on a vortex immersed in a free stream as shown in Fig. 8.26. During the down stroke a clockwise rotating bound vortex is experiencing a vertical velocity component equal to $U_z = \dot{h}$ for the cases (a) the approximate suction force of $S \sim \rho U_z \Gamma_a$, and during the up stroke the counterclockwise rotating bound vortex is under the influence of vertical velocity which is in –z direction to create (b) $S \sim \rho U_z \Gamma_a$ which is the approximate suction force. Here, during (a) down stroke, and (b) up stroke motions the vertical velocity component and the bound vortex change simultaneously so that the suction force S remains in the same direction as a propulsive force. Although the product of the vertical velocity component and the bound circulation $U_z \Gamma_a$ remains the same, its magnitude changes by time as shown in Fig. 8.24.

The time and space variation of the wake vortex sheet strength can be computed in terms of the bound vortex using the potential theory. The relation between the vortex sheet strength γ_w and the bound vortex Γ_a can be established using Eq. 3.13 for a periodic motion of the profile given by $z_a = h \cos ks$ as follows

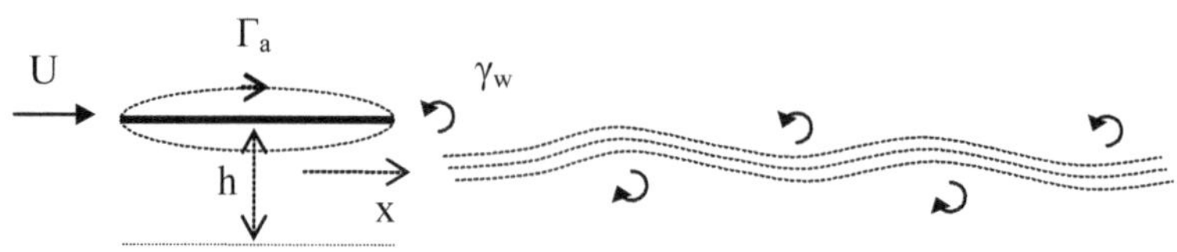

Fig. 8.25 Bound vortex Γ_a and the wake vortices, γ_w, shed from the trailing edge

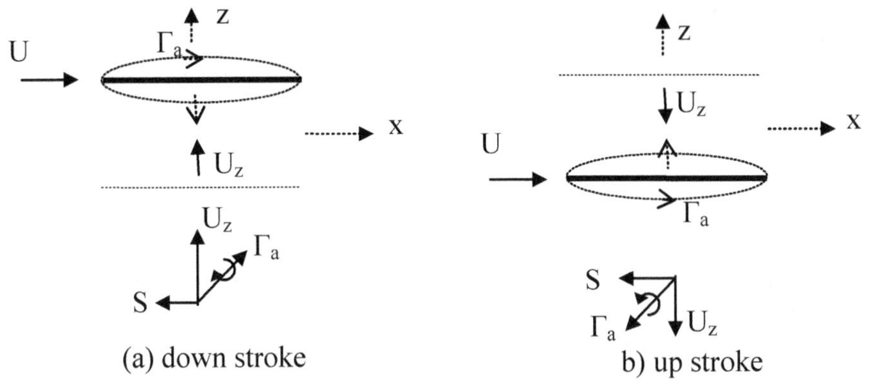

Fig. 8.26 The generation of suction force S during **a** downstroke, **b** upstroke

$$\gamma_w/U = -\frac{k\bar{h}^*}{\left[H_1^{(2)}\right]^2 + \left[H_0^{(2)}\right]^2}\left[(H_1^{(2)}\sin kx + H_0^{(2)}\cos kx)\cos ks + (H_1^{(2)}\cos kx + H_0^{(2)}\sin kx)\sin ks\right] \tag{8.30}$$

Now, with the aid of Eq. 8.30, we can show the spacewise variation of the wake vortex sheet strength at the top and bottom positions of the profile on Fig. 8.27. As shown in Fig. 8.27, at the bottom position of the profile the shed vortex is positive i.e. in clockwise direction, and at the top position it is negative i.e. in counter-clockwise direction. The near wake region vortex signs are in accordance with the signs given in Figs. 8.25 and 8.26 which is indicated in the experimental results of Freymuth.

The propulsive efficiency of the flapping wing is another concern to the aerodynamicist. In order to calculate the average propulsive efficiency in one cycle, we have to know the average energy which is necessary to maintain the propulsion and

Fig. 8.27 Wake vortex sheet when the profile is at **a** bottom, **b** top

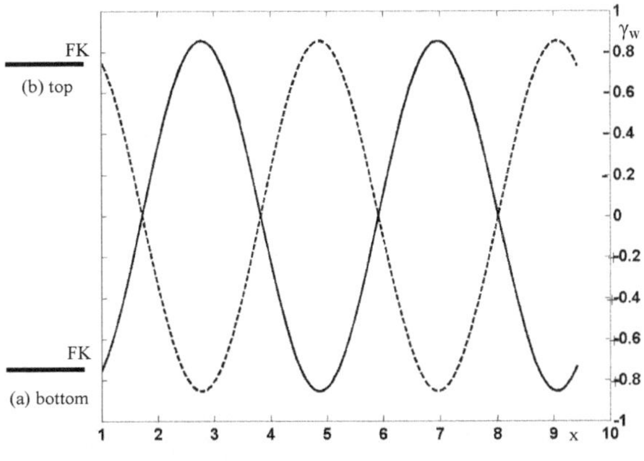

also the average work for the vertical periodic motion. The ratio of the average energy to average work gives as the propulsive efficiency. Accordingly, for a periodic motion given by $z_a = -h\cos\omega t$, with the aid of Eqs. 8.28 and 8.29 we obtain

$$\eta = \frac{\int_0^{2\pi/\omega} S \cdot U\,dt}{\int_0^{2\pi/\omega} L' \cdot \dot{h}\,dt} = \frac{F^2 + G^2}{F} \tag{8.31}$$

Shown in Fig. 8.28 is the variation of the propulsive efficiency with respect to the reduced frequency k.

The theoretical results obtained for the lift and the suction forces of a vertically oscillating thin airfoil at zero angle of attack are in agreement with the solutions obtained using Navier–Stokes equations for NACA 0012 airfoil in plunging motion (Tuncer and Platzer 1996, 2000). Solutions based on the potential flow assumptions and the Navier–Stokes solutions give similar results for the amplitude and the period of both the lift and the suction forces. Naturally, Navier–Stokes solutions also provide viscous and form drags. On the other hand, using the unsteady viscous-inviscid coupling concept and the velocity viscosity formulation the skin friction of the thin airfoil can be determined with numerical solution of the Eqs. 8.5–8.7 with the boundary layer edge velocity $U_e = U_e(t)$ provided by the potential flow as described by Gulcat (Problem 8.28 and Appendix 10). As shown in Fig. 8.28, the difference between the theoretical and the numerical solutions is apparent for the values of propulsive efficiency. The efficiency obtained by the ideal solution is independent of the plunge amplitude, and becomes very high for low frequency oscillations and asymptotically reaches the value of 0.5 for very high frequencies. Obviously, viscous solutions yield lower values of efficiency, and they depend on the amplitude of plunge as shown in Fig. 8.28. The efficiencies obtained with viscous effects indicate that for high plunge amplitude the efficiency values show the tendency to follow the ideal curve. However, for the plunge amplitudes

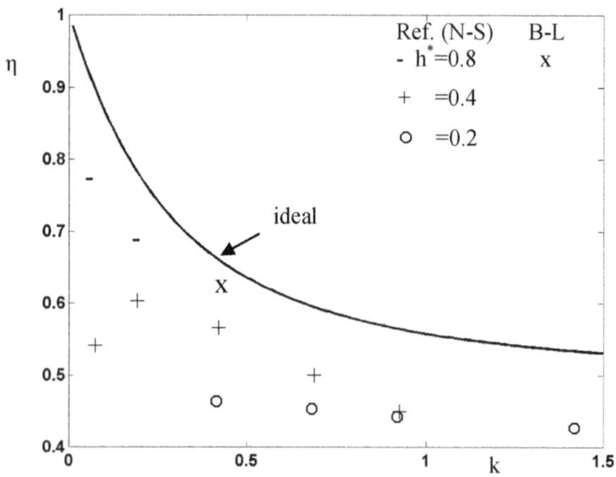

Fig. 8.28 Variation of propulsive efficiency with k

less then 0.4 the efficiencies become small with decreasing of frequencies as opposed to the ideal case, whereas the efficiency obtained with boundary layer approach is in between the ideal and the Navier–Stokes result. According to Navier–Stokes solutions for the efficiency to be more than 0.5, the condition must be $k < 0.6$ and $h > 0.4$.

Previously, we have seen that the dynamic stall takes place at higher angles of attack than the occurrence of static stall depending on the reduced frequency values. The higher the reduced frequency, the more the difference between the static and dynamic stall angles. For a pitching airfoil, the difference between the static and dynamic stall angles $\Delta\alpha$ in terms of the reduced frequency k is given empirically as follows (Prouty 1995)

$$\alpha_{DY} - \alpha_{ST} = \Delta\alpha = \gamma\sqrt{k}, \quad \gamma = 0.3 - 0.5 \tag{8.32}$$

On the other hand, as seen on Fig. 8.6, NACA 0012 profile at the reduced frequency value of $k = 0.15$ can undergo pitching without flow separation up to 20° angle of attach. Above that, between 20° and 23.5° angle of attack, there is a leading edge separation which generates a vortex causing a high lift until the vortex is convected to wake from the trailing edge. Static wind tunnel experiments show that the flow over the profile separates at 13° angle of attack. The 7° difference between the static and dynamic stall angles is slightly higher than the empirically estimated value obtained by Eq. 8.31 using the lower value of coefficient γ. That is to say Eq. 8.31 gives a little bit conservative estimates for the dynamic stall angles of pitching airfoils. The effective angle of attack for the plunging airfoil in a free stream of U becomes zero at the top and bottom locations, and takes its maximum value at the center point. During down stroke the effective angle of attack gives positive lift, and during up stroke it provides negative lift. Now, we can calculate the relation between the effective angle of attack, plunge amplitude and the frequency for an airfoil undergoing vertical oscillations $z_a(t) = -h\cos(\omega t)$ in a free stream as follows. Since he vertical velocity of the airfoil then becomes $\dot{z}_a = h\omega\sin(\omega t)$, the effective angle of attack reads as

$$\tan\alpha_e = \frac{-\dot{z}_a}{U} = -k\,(h/b) \tag{8.33}$$

According to Eq. 8.33, the effective angle of attack is proportional with the product of the free stream and the dimensionless plunge amplitude. The dynamic separation angle as given in Eq. 8.32 depends on only the reduced frequency. The airfoil pitching with reduced frequency of $k = 1.5$ has the dynamic separation angle with $\alpha_{DY} = \alpha_{ST} + \Delta\alpha = 13 + 21 = 34°$. This means at the reduced $k = 1.5$ the profile can undergo plunge oscillation up to the non dimensional plunge amplitude $h/b = 0.45$ without experiencing flow separation if we consider the plunging with the effective angle of attack is equivalent to the pitching with the same angle of attack. This assumption lets us apply the potential flow theory for a wide range of plunge rates with boundary layer coupling to take the viscous effects into account.

Using Eqs. 8.32 and 8.33 we can find the maximum plunge amplitude in terms of the reduced frequency for a profile encountering no flow separation as given in Fig. 8.29. According to this figure, for lower values of the reduced frequency we can safely have high values of plunge amplitude without flow separation.

Now, we are ready to apply the unsteady viscous-inviscid interaction concept to the plunging thin airfoil with $z_a(t) = -h\cos(\omega t)$ to obtain the time variation of the thrust coefficient and its average over one cycle of motion as described in (Gulcat 2009). The leading edge suction force c_s is given in Eq. 8.28. If we calculate the skin friction from the surface vorticity of the boundary layer, B-L, solution then we can obtain the time history of the drag coefficient c_d. The time dependent boundary layer edge velocity is provided from the surface vortex sheet strength of the plunging thin airfoil as follows (Gulcat 2009)

$$\frac{U_e(s)}{U} = 1 \pm [F(k)\sin(ks) + G(k)\cos(ks)]k\bar{h}^* \sqrt{(1-x)/(1+x)} \tag{8.34}$$

The addition of the suction force and the drag gives us the instantaneous propulsive force coefficient as $C_F = c_s + c_d$. The time average of C_F over one period gives us the definition of the average propulsive force C_T as follows:

$$c_T = \frac{1}{T}\int_0^T c_F \, dt \tag{8.35}$$

Table 8.1 gives the averaged propulsive force coefficients obtained for various plunge amplitudes with the viscous-inviscid interaction, and compares with the results obtained with N-S solutions for NACA 0012 airfoil at Reynolds number of 10^5 and the reduced frequency of $k = 0.4$.

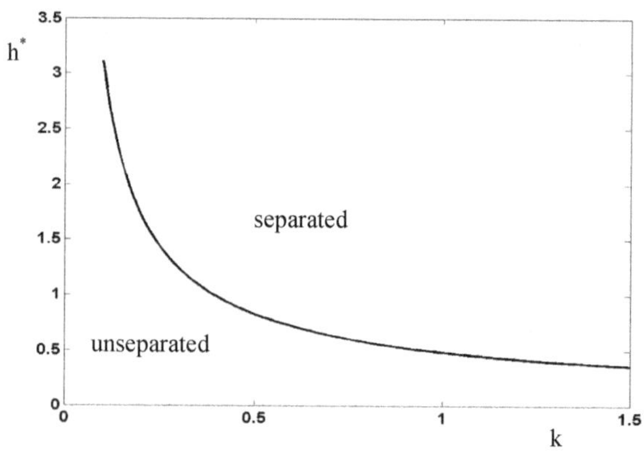

Fig. 8.29 Change of plunge amplitude with the reduced frequency without experiencing dynamic stall

For viscous-inviscid interaction to be applicable at $k = 0.4$, Eq. 8.31 dictates that the effective angle of attack α_e should be less than the dynamic stall angle of NACA 0012, i.e. $\alpha_d = 12° + 0.3\ (0.4^{1/2}) \approx 23°$. For angles of attack larger than 23°, as seen from Table 8.1, the viscous inviscid interaction overestimates considerably the averaged propulsive force. The thickness effect is also important in prediction of drag on an airfoil. If the thickness correction (Van Dyke 1956) is made for the NACA 0012 airfoil, the agreement between the viscous inviscid solution becomes very good for the low effective angles of attack.

We know now the capabilities and limitations of viscous-inviscid interaction approach for plunging thin airfoils. Therefore, we can perform parametric studies to predict the average propulsive force depending on the Reynolds number, plunge amplitude and the reduced frequency. The wind tunnel experiments indicate that to obtain a net propulsive force for a plunging airfoil the product of the reduced frequency and the dimensionless plunge amplitude must be higher than a critical value, i.e. $kh > 0.2$ where the Reynolds number is 17 000 (Platzer et al. 2008). The Reynolds number, however, is also an important parameter to obtain net propulsive force as shown in Fig. 8.30.

The variation of the average propulsive force coefficient, c_T for different dimensionless plunge amplitudes $h/b = 0.2$, 0.4, and 0.6 is given in Fig. 8.30a–c respectively. Figure 8.30a indicates that, for $h/b = 0.2$ to generate a net propulsive force, the Reynolds number must be greater than 10^3 and the reduced frequency must be greater than 1.2. If the plunge amplitude is doubled, that is, for $h/b = 0.4$ according to Fig. 8.30b, for the reduced frequency values greater than 0.5, a net propulsive force is obtained even for a Reynolds number of 10^3. Moreover, increasing the amplitude to 0.6 gives a net propulsive force for a wide range of frequencies, that is, $k > 0.3$, and $\mathrm{Re} > 10^3$ as shown in Fig. 8.30c. A close inspection of Figs. 8.30b–c indicates that when the amplitude is high, the increase in the Reynolds number from 10^4 to 10^5, that is, one order of magnitude increase has very little effect on the propulsive force coefficient. Figure 8.31 shows the Reynolds number dependence with kh variation of the net propulsion generation of a plunging thin airfoil. The region above the line indicates propulsion whereas below the line there is a power extraction area which is of importance to wind engineering when performed as pitching and plunging for significant power extraction for clean energy production (Kinsey and Dumas 2008).

Table. 8.1 Averaged propulsive force coefficient c_T at $k = 0.4$, $\mathrm{Re} = 10^5$

$\bar{h}^*$	c_T, present	c_T, corrected	c_T, Ref.	$\alpha_d = \alpha_s + \Delta\alpha$	$\alpha_e = \tan^{-}(-\dot{h}/U_\infty)$
0.8	–0.129	–0.119	–0.118	23°	18°
1.0	–0.205	–0.195	–0.176	23°	21°
1.2	–0.298	–0.288	–0.134	23°	25°

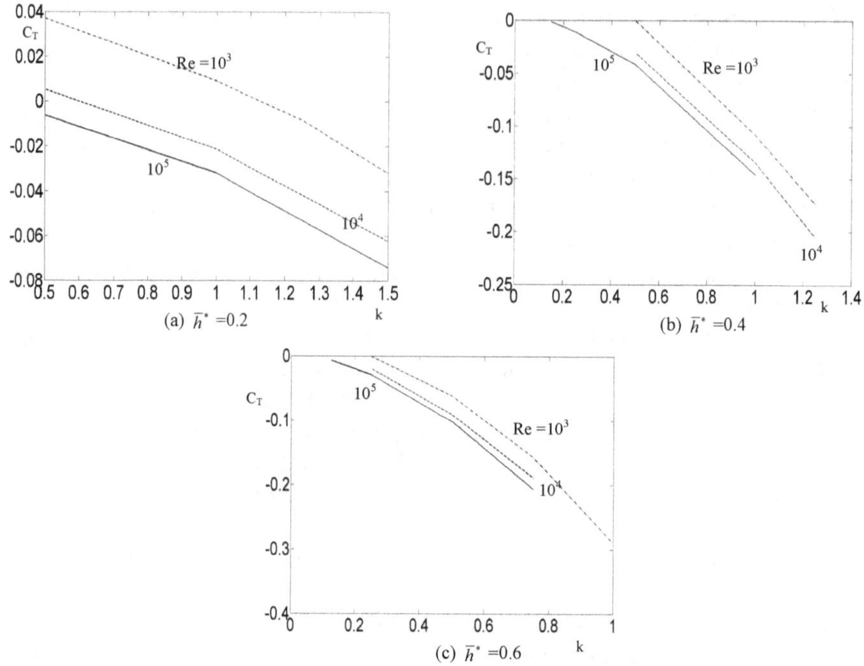

Fig. 8.30 Variation of the averaged propulsive force coefficient c_T with the reduced frequency k and the Reynolds number Re: **a** $\bar{h}^*$ = 0.2, **b** $\bar{h}^*$ = 0.4, **c** $\bar{h}^*$ = 0.6

Fig. 8.31 Reynolds number and *kh* dependence of the propulsion and power extraction

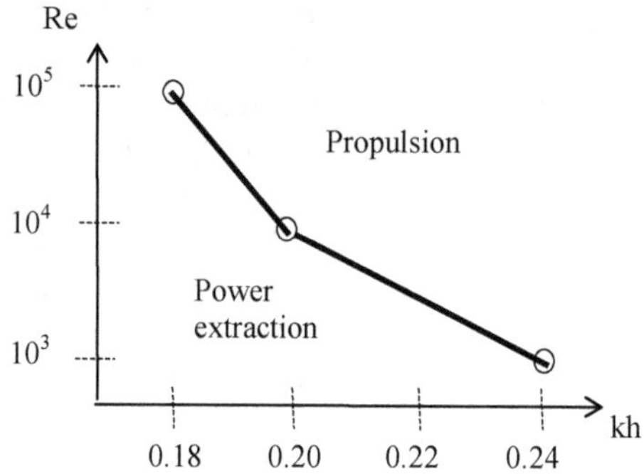

Finally, for the plunging airfoil we can give the propulsive efficiency values obtained with the viscous inviscid interaction. Table 8.2 shows the comparison of the efficiency values with the Navier–Stokes solutions of (Tuncer and Platzer 2000).

According to Table 8.2, there is an 8% difference for the efficiency with the viscous inviscid interaction and the full N-S solution at 80% plunge amplitude with respect to the chord. This discrepancy becomes 18% for 100% plunge amplitude

Table. 8.2 Propulsive efficiency for a plunge at Re = 10^5 and k = 0.4

2 h*	η_{id} [4]	H	η (Ref)	Difference (%)	α_d	α_e = atan(-$\dot{h}/U_\infty$)
0.8	0.668	0.641	0.59	8	23°	18°
1.0	0.668	0.65	0.55	18	23°	21°

because of having high effective angle of attack where N-S solution predicts weak separation at the trailing edge.

So far, we have seen in a detail, lift and propulsive force variations of a plunging airfoil as a one degree of freedom problem. As a result, at zero angle of attack, the lift created is positive during down stroke and negative during up stroke to give zero average value, and the propulsive force is generated for a certain range of *kh* values and Reynolds numbers if we take the viscous effects into consideration. In order to obtain positive lift throughout the flapping motion two degrees of freedom, i.e., pitching and plunging becomes necessary for the airfoil. We can impose a pitching plunging motion on the airfoil for which the lift is always positive because of effective angle of attack if we describe the pitching with α, the plunging with h and the phase difference between the two with φ as follows:

$$\begin{aligned} h &= -\bar{h}\cos\omega t \\ \alpha &= \alpha_0 + \bar{\alpha}\cos(\omega t + \varphi) \end{aligned} \tag{8.36a, b}$$

The unsteady motion of the airfoil given by Eq. 8.36a, b gives the effective angle of attack at the leading edge of the airfoil as

$$\alpha_e = \tan^{-}(-\frac{\dot{h} + d\dot{\alpha}(t)\cos(\alpha(t))}{U - d\dot{\alpha}(t)\sin(\alpha(t))}) + \alpha(t) \tag{8.37}$$

where d is the distance between the leading edge and the pitch axis. If we consider the pitching over a constant angle of attack, during up stroke if we let the angle of attack increase and during down stroke let the angle of attack decrease then we can have an effective angle of attack always positive during the forward flight given by Eq. 8.37, which yields positive lift throughout the pitch and plunge. Now, we can illustrate the whole motion on a simple figure as the superpositioning of Eqs. 8.35a, b, as depicted on Fig. 8.32a, b during (a) down stroke, and (b) up stroke.

According to Fig. 8.32, during (a) down stroke, and (b) up stroke, the effective angle of attack shows very little change. If we can keep the effective angle of attack given by Eq. 8.36a, b lower than the dynamic separation angle, we can use the viscous-inviscid interaction to predict the propulsive and the lifting forces of a pitching plunging airfoil. Problem 8.29. Detailed numerical studies of a pitching plunging airfoil were given in late 1990s as Euler and Navier–Stokes solutions at Re = 10^5 (Isogai et al. 1999), and comparison is made with the Lighthill's potential solution. Isogai et al. studied the motion of NACA 0012 airfoil in dimensionless

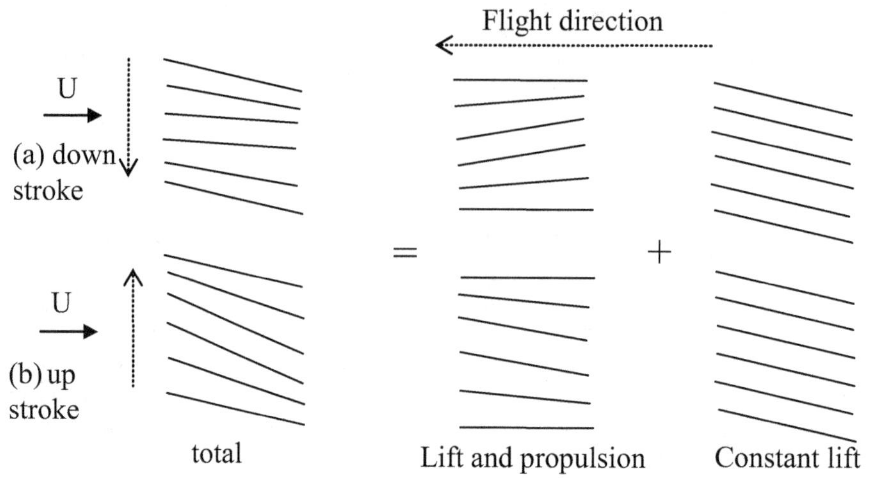

Fig. 8.32 **a** Down stroke, and **b** up stroke motions resolved with Eq. 8.36a, b

plunge amplitude of 1.0, angle of attack amplitude of 20°, pitch axis location as the midchord, and the phase angle as 90° to calculate the propulsive force coefficient and the efficiency in terms of the reduced frequency k. Naturally, the highest efficiency is obtained with potential theory, and the Euler and N-S solutions yield less values of efficiency respectively at k values ranging 0.5–1.0. As it happens for the case of pure plunge the efficiency decreases with increasing k for the pitch-plunge case. As k changes in 0.5–1.0, the efficiency of the potential flow ranges in 0.85–0.75, Euler solution gives 0.8–0.6, and N-S yields 0.7–0.55, respectively. For the Navier–Stokes solutions, there is no significant efficiency variations for laminar and the turbulent cases. At the same range of reduced frequency, the propulsive force coefficients vary between 0.4 and 0.6 for the potential solution, 0.35–0.75 for Euler, and 0.3–0.6 for N-S solutions. These results indicate that the propulsive force coefficient increases with increasing reduced frequency. The Navier–Stokes solutions performed by Tuncer and Platzer under similar flow conditions agree well with the work of Isogai et al. However, at phase difference of 30°, there is a discrepancy between two approaches as far as the leading edge separation of the solution given by the latter is concerned.

As indicated with Eq. 8.26, pure plunging always creates a leading edge suction which yields a propulsive force. However, it is not so for the pure pitching motion of an airfoil because of the phase lag between the angle of attack α and the lifting force L. This phase lag may yield negative average propulsion, i.e. drag even with potential flow analysis, depending on the position of the pitch axis a for all ranges of reduced frequency. Let us consider the pure pitching motion with $\alpha = \overline{\alpha}\cos(\omega t + \varphi)$, wherein only the second term of the right hand side of

Eq. 8.35b is considered. The averaged propulsive force from (8.26) to (8.30) with the aid of (Garrick 1936), and with cosiderable correction, reads as

$$\frac{c_T}{\overline{\alpha}^2}(k,a) = \pi k^2(F^2+G^2)\left[\frac{1}{k^2}+\left(\frac{1}{2}-a\right)^2\right]+\pi k^2\left[\left(\frac{1}{2}-a\right)\left(\frac{1}{2}-F\right)-\frac{F}{k^2}-\left(\frac{1}{2}+a\right)\frac{G}{k}\right] \tag{8.38}$$

Shown on Fig. 8.33 are the curves for the averaged propulsive force coefficients plotted against the inverse of the reduced frequency.

According to Fig. 8.33, by definition, negative values of averaged propulsive force indicate propulsion whereas the positive values mean the fluid extracts power from the pitching airfoil. For the pitch axis at three quarter chord, i.e. $a = 1/2$, at all values of reduced frequency there is not any propulsion predicted. At large values of k the pitching about the leading edge $a = -1$, the quarter chord point $a = -1/2$, the trailing edge $a = 1$, and the mid chord $a = 0$, we observe that it is not possible to generate propulsion. However, for small values of k, i.e. $k < 1$, we see that except for $a = 1/2$, generation of propulsive force is possible. Therefore, according to the ideal theory, if we want to have contribution to the propulsive force from the pitching, it is necessary to choose a proper pitch axis as well as the reduced frequency range for o pitching plunging airfoil. This adverse effect of pitch axis location on the propulsive force naturally alters the propulsive efficiency. The ideal efficiency formula for the pitching plunging airfoil, Eq. 8.35a,b, with the phase difference of φ can be obtained as

$$\eta = \frac{\int_0^{2\pi/\omega} S\cdot U\,dt}{\int_0^{2\pi/\omega}(L'\cdot\dot{h}+M\dot{\alpha})dt} = \frac{a_1 h^{*2}+(a_2+b_2)\overline{\alpha}+2(a_4+b_4)h^*\overline{\alpha}}{c_1 h^{*2}+c_2\overline{\alpha}^2+2c_4 h^*\overline{\alpha}} \tag{8.39}$$

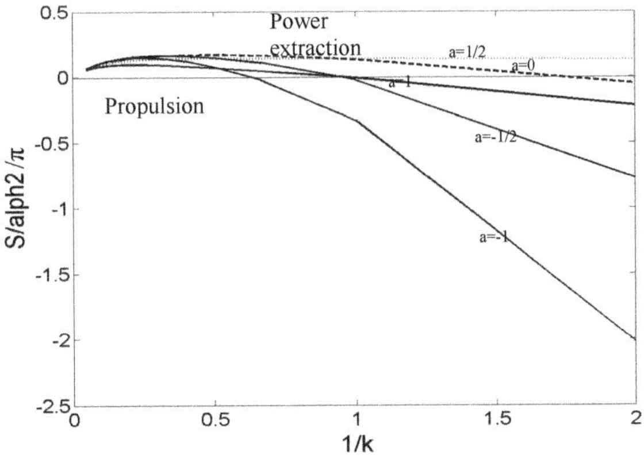

Fig. 8.33 Averaged propulsive force coefficient $c_T/(\alpha^2\pi)$ versus inverse of the reduced frequency for different pitch axis

where:

$$a_1 = F^2 + G^2, \quad a_2 = a_1\left[1/k^2 + (0.5 - a)^2\right] + 0.25 - (.5 - a)F - G/k,$$
$$a_4 = a_1[1/k\ \sin(\varphi) + (0.5 - a)\cos(\varphi)] - 0.5F\cos(\varphi) - 0.5G\sin(\varphi),$$
$$b_2 = -0.5a - F/k^2 + (.5 - a)G/k, \quad b_4 = 0.5(.5 + G/k)\cos(\varphi) - F/k\sin(\varphi),$$
$$c_1 = F, \qquad c_2 = 0.5(0.5 - a) - (a + 0.5)[F(0.5 - a) + G/k],$$
$$c_4 = 0.5(0.5 - 2\,aF + G/k)\cos(\varphi) + 0.5(F/k - G)\sin(\varphi)$$

Knowing that the pitch may hamper the propulsive efficiency we have to choose the pitch axis with caution as well as the phase between the pitch and plunge. Equation 8.38 gives the ideal propulsive efficiency η = 0.87 for a flat plate pitching about mid chord with k = 1, h^* = 1.5, $\overline{\alpha} = 15°$ and φ = 75°, whereas η = 0.54 is computed with N-S solution for NACA 0012 airfoil at Re = 10^4 and with the same flow parameters (Tuncer and Platzer 2000). There exist further studies, based on the N-S solutions, to optimize the efficiency and/or thrust in terms of plunge magnitude, pitch magnitude and the phase lag (Tuncer and Kaya 2005). Table 8.3 shows the comparison of the optimized propulsive efficiency computed using N-S solutions for NACA 0012 airfoil at Re = 10^4 with the ideal efficiency calculated using Eq. 8.38 for an airfoil pitching about its midchord.

According to Table 8.3, there is a 20% difference in the ideal efficiency and the efficiency evaluated with N-S solutions, and the efficiency increases with increasing pitch amplitude. Furthermore, solving for maximum efficiency may not yield a good thrust coefficient as well as searching for maximum thrust may not produce very high efficiency.

Now we are ready to give examples to evaluate the effective angle of attack of a pitching plunging airfoil for various $\overline{h}$, $\overline{\alpha}$ and k values for which dynamic separation angles are larger than the effective angle of attack.

Example 8.5 Assume an airfoil pitching about its leading edge and plunging with k = 0.35 as follows

$$h = 1.1\ \cos(\omega t)$$
$$\alpha = 10° + 10°\cos(\omega t + \pi/2)$$

Table. 8.3 Propulsive efficiency for a plunge at k = 0.5

2 h*	η_{id} (%)	η (Ref)	Difference (%)	$\overline{\alpha}$	Φ
0.45	73	58.5	20	15.4°	82.4°
0.57	79	63.8	20	21°	86.7°

Solution Since the reduced frequency is given we describe the motion in reduced time with following equations:

$$h = 1.1\ \cos(ks)$$
$$\alpha = 10^\circ + 10^\circ\ \cos(ks + \pi/2)$$

Taking $d = 0$ for Eq. 8.36 gives the expression for the effective angle of attack $\alpha_e = \alpha_e(s)$, whose plot for a period of motion is given as follows:

According to Fig. 8.34, the effective angle of attack remains less than 23° which is under the dynamic separation angle given for NACA 0012 profile with Eq. 8.31. That means the profile can undergo high amplitude pitch and plunge without encountering separation. During down stroke, the angle of attack gets smaller but the relative air velocity in vertical direction causes increase in the effective angle of attack. During up stroke, however, the increase in angle of attack makes the effect of the negative vertical air velocity vanish. As a result of this pitch and plunge it becomes possible to have an unseparated flow throughout the motion because of having the effective angle of attack under 20°. At the same time, the angle of attack and the effective angle of attack remains positive to yield a positive lift. It is necessary to make a note here that according to Fig. 8.33 the propulsion due to pitch is also favorable because of pitch axis location and the k being 0.35.

Example 8.6 The NACA 0012 airfoil is pitching and plunging with a reduced frequency of 1 as given below

$$h = 0.65\ \cos(\omega t)$$
$$\alpha = 20^\circ + 20^\circ\ \cos(\omega t + \pi/2)$$

Show that the effective angle of attack remains under the dynamic separation angle of attack.

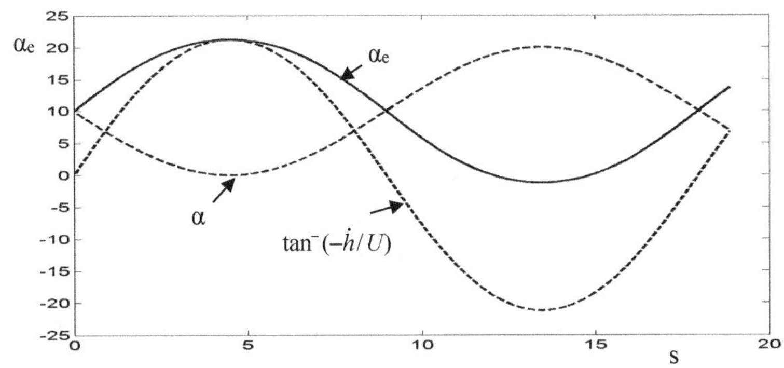

Fig. 8.34 The effective angle of attack α_e variation with reduced time s

Solution The dynamic separation angle of attack is found as
$\alpha_{cr} = 13° + 0.3\sqrt{k} * 180/\pi = 30°$. In terms of reduced time s it reads as

$$h = 0.65 \cos(s)$$
$$\alpha = 20° + 20° \cos(s + \pi/2)$$

The superposition of pitch and plunge gives us the effective angle of attack less than 30° as shown in Fig. 8.35. Since the effective angle of attack remains above 10°, the instantaneous lift is always positive and relatively high. As seen from Fig. 8.35 during the flapping motion relative to free stream the angle of attack changes between 0° and 40°.

The results of Examples 8.5 and 8.6 indicate that: (i) for low reduced frequencies, i.e. $k < 1$, pitching with small angles of attack and plunging with high amplitudes and with 90° phase angle we obtain effective angles attack less than the dynamic separation angle, (ii) for $k > 1$ with small plunge amplitudes and large angles of attack, flapping without exceeding the dynamic separation angle is possible.

So far we have studied the pitch plunge motion of an airfoil prescribed as simple harmonic motion. However, a nonsinusoidal motion of the flapping airfoil is also observed to yield sufficient propulsive force through path optimization (Kaya and Tuncer 2007). In their study Kaya and Tuncer used B splines for the periodic flapping motion. They showed that thrust generation may significantly be increased, compared to the sinusoidal flapping, with the characteristics of the path for optimum thrust generation staying at about constant angle of attack at most of the upstroke and downstroke, while pitching is happening at extremum points of plunge.

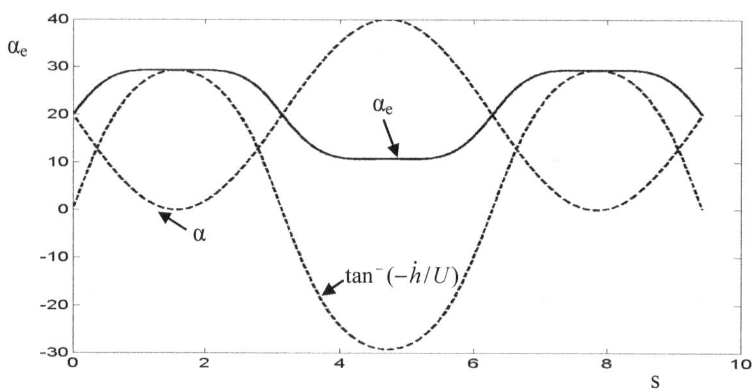

Fig. 8.35 High lift and high propulsion with high reduced frequency pitch and plunge

We know now that in order to create a propulsive force we need to create a reverse Karman vortex street at the wake of the oscillating airfoil. The creation of the reverse Karman street is possible either with attached flow or with flows creating strong leading edge vortices which in turn generate appreciable leading edge suction. If the leading edge vortex formed, because of angle of attack exceeding the dynamic separation angle, does not burst at the trailing edge, it will create considerable suction at the upper surface which will help for propulsion and lift as well. As seen in Figs. 8.6 and 8.7, the N-S solutions which are in agreement with experiments, show increase in lift although the dynamic separation angle is exceeded by 3–4°. Further increase in the angle of attack creates bursting of the vortex at the trailing edge to cause lift lost. However, if the reduced frequency is increased above 0.15 it is possible to go to higher angles of attack without causing vortex burst at the trailing edge. (Isogai et al. 1999). At high Reynolds numbers, laminar or turbulent, it is possible to create a propulsive force without resorting to high angles of attack. On the other hand, at low Reynolds numbers, i.e. $R_e \leq 1000$, the pitching motion may provide propulsion at low frequencies if the angle of attack exceeds 20°. For this case maximum thrust is achieved in 45°–60° angle of attack range (Wang 2000).

The last aspect of the pitching plunging airfoil to be briefly mentioned here is the power extraction from the oscillating airfoil (Kinsey and Dumas 2008). This time rather than having propulsion with the unsteady motion which is provided by the energy of the fluid, the energy will be given to the fluid by the motion of the airfoil to generate power which is useful in harvesting wind energy. The pitch plunge motion here is conventionally defined with $\alpha(t) = \overline{\alpha}\sin(\omega t)$, and $h = \overline{h}\sin(\omega t + \phi)$ with the approximate definition of the feathering parameter (Anderson et al. 1998; Kinsey and Dumas 2008)

$$\chi = \frac{\overline{\alpha}}{\tan^{-}(\omega\overline{h}/U)} \tag{8.40}$$

which is approximately associated with propulsion for $\chi < 1$, whereas $\chi > 1$ corresponds to power extraction, and naturally, $\chi = 1$ yields neutral motion called feathering for which neither propulsion nor power production exist. If the average power extraction coefficient over a cycle due to plunge and pitch combined is denoted with $\overline{C}_P$ then the power extraction efficiency reads as

$$\eta = \frac{\overline{P}}{\overline{P}_d} = \overline{C}_P\frac{b}{\overline{h}} \tag{8.41}$$

where, $\overline{P}$ is the total power produced and $\hat{P}_d$ is the total power of the oncoming flow passing through the swept area during plunge. The power extraction efficiency is theoretically limited by 59% from a steady inviscid stream tube, whereas Kinsey and Dumas report about 33% efficiency and almost 2.82 average total force coefficient for NACA 0015 airfoil pitching about its 1/3 chord with $b/\overline{h} = 2$, $\overline{\alpha} = 76.33°$ and $k = 0.56$ at Reynolds number of 1100.

8.6 Flexible Airfoil Flapping

The flexible wing flapping in oscillating airfoils provides aerodynamic benefits in terms of lift and thrust generation as well as providing inherently light structures (Heatcote and Gursul 2007). The real positive effect of the chordwise flexibility in forward flight is the prevention of the flow separation by means of reducing the effective angle of attack while changing the camber of the airfoil periodically. During plunge motion with large amplitudes, we can keep the effective angle of attack lower than the dynamic separation angle with flexible camber (Gulcat 2009). If we assume a parabolic camber, whose amplitude changes periodically with $z_a(x, t) = -a^* \cos \omega t\ x^2/b^2$ for a thin airfoil as shown in Fig. 8.36, we can obtain the boundary layer edge velocity due to flexible camber as

$$\frac{U_e}{U} = 1 \pm \{[(1+2x+F) - Gk/2]\cos ks - [(G+(x+x^2+F/2)k]\sin ks\}\,\overline{a}^*\sqrt{\frac{1-x}{1+x}} \tag{8.42}$$

and the suction force as

$$c_s = -2\pi[(F-1-Gk/2)\cos(ks) - (G+Fk/2\sin(ks))]^2\overline{a}^{*2} \tag{8.43}$$

where, $\overline{a}^* = a^*/b$ is the non dimensional maximum camber amplitude.

If we give the plunging motion as $h = -h^* \cos(\omega t)$, and the camber motion with 90° phase,

$$z_a(x_2 t) = -a(t)x^2/b^2, a(t) = a^*\cos(\omega t + \pi/2), \quad -b \le x \le b$$

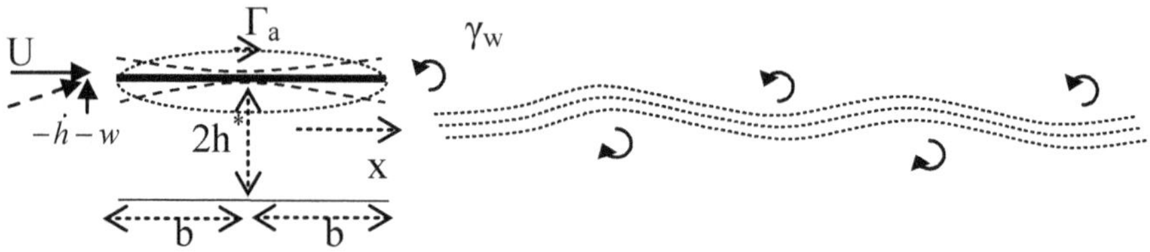

Fig. 8.36 Plunging chordwise flexible thin airfoil and its wake

this provides us with the effective angle of attack which is less than dynamic stall angle. Now, the effective angle of attack for the combined motion at the leading edge is determined as follows

$$\alpha_e = \tan^{-}\left[-\left(\dot{h} + w_{aLE}\right)/U_\infty\right] \tag{8.44}$$

where w_{aLE} is the downwash at the leading edge caused by the time dependent camber change. Shown in Fig. 8.37 is the time variation of the propulsive force coefficient plots obtained including viscous effects for the flexible airfoil at Re = 10^4 and k = 1 for three different camber ratios: (a) $\overline{a}^*$ = 0.05, (b) 0.1, and (c) 0.15. The corresponding average force coefficients are found as (a) C_T = –0.3265, (b) –0.3316, (c) –0.3398, respectively. The ideal average force coefficients and the computed values are compared in Table 8.4 at associated effective angles of attack, all less than the corresponding dynamic stall angle, which is 29°.

According to Table 8.4, tripling the camber ratio from 5 to 15% results in only a 4% increase in the average force coefficient, that is, from —0.3265 to –0.3398. This

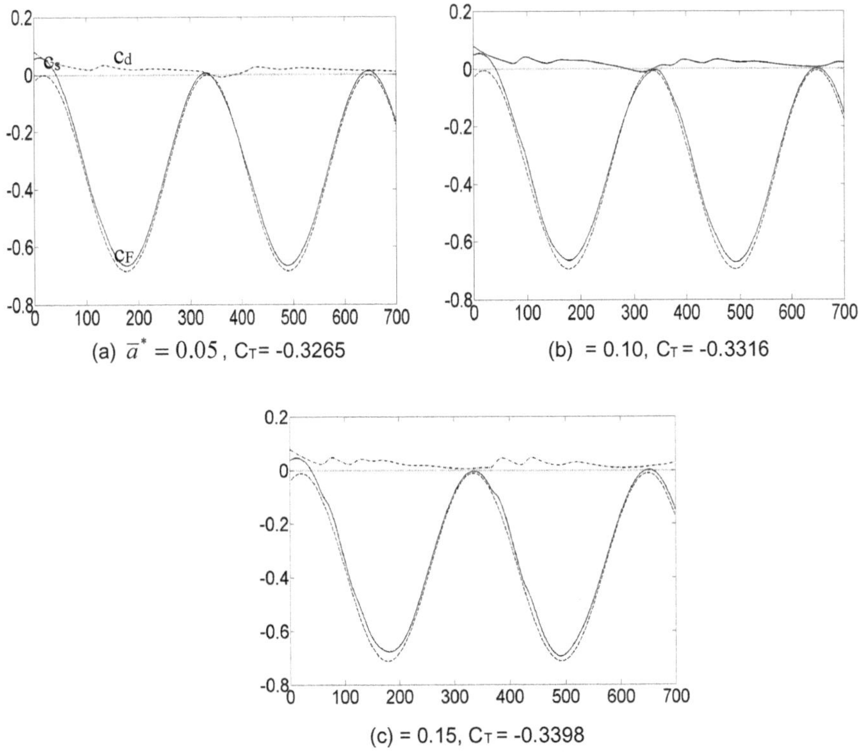

Fig. 8.37 Time variation of propulsive force coefficients for heaving plunging flexible airfoil at Re = 10^4 and k = 1, Δt = 0.01, for **a** $\overline{a}^*$ = 0.05, **b** = 0.10, and **c** = 0.15

Table. 8.4 Thrust coefficients for different $\bar{a}^*$ at Re = 10^4, $\bar{h}^*$ = 0.6, and k = 1

$\bar{a}^*$	C_T	C_{Tid}	α_d, [44]	$\alpha_e = a\tan\left[-\left(\dot{h} + w_{aLE}\right)/U_\infty\right]$
0.05	–0.3265	–0.3433	29°	27°
0.10	–0.3316	–0.3505	29°	23°
0.15	–0.3398	–0.3625	29°	18°

shows that increasing the camber ratio does not produce a significant overall propulsive force increase for the case of a flexibly cambered airfoil undergoing plunge.

The viscous drag acting on the parabolically cambered thin airfoil is also obtained using the boundary layer equations. Equations 8.2 and 8.5 give the inertial values of the velocity vector $\vec{v} = u\vec{i} + v\vec{j}$ and vorticity ω, which is necessarily used in skin friction calculations, in moving deforming coordinates attached to the body as a non inertial frame (Gulcat 2009) as shown in Fig. 8.38. Let x–y be the rectangular coordinates attached to the body, and let ξ-η be the curvilinear local coordinates with surface fitted ξ coordinate's tangent angle with x axis being α_1, and let η be parallel to z axis. At a given point (x, y) this yields $x = \xi\cos\alpha_1$, and $y = x\sin\alpha_1 + \eta$, wherein the continuity and the vorticity transport respectively reads as

$$\frac{1}{\cos\alpha_1}\frac{\partial u}{\partial\xi} - \tan\alpha_1\frac{\partial u}{\partial\eta} + \frac{\partial v}{\partial\eta} = 0 \tag{8.45}$$

and,

$$\frac{\partial\omega}{\partial t} + \frac{u}{\cos\alpha_1}\frac{\partial\omega}{\partial\xi} + (v - u\tan\alpha_1)\frac{\partial\omega}{\partial\eta} = \frac{1}{\text{Re}}\frac{\partial^2\omega}{\partial\eta^2} \tag{8.46}$$

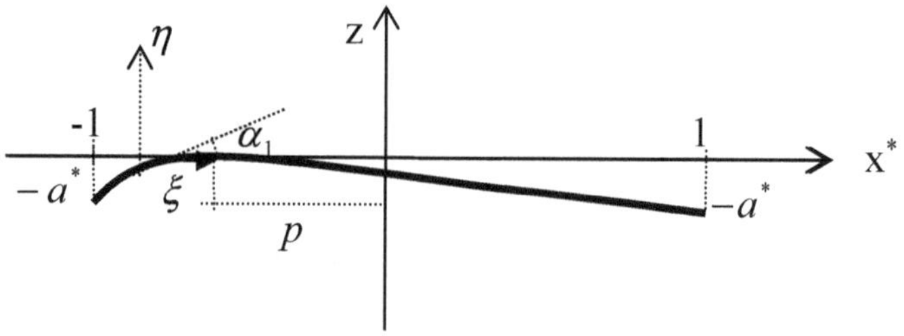

Fig. 8.38 Body fixed x–y coordinates, and body fitted ξ-η coordinates for a parabolically cambered thin airfoil

The discretized form of Eqs. 8.44 and 8.45 for boundary layer solutions can be written in a way similar to those given in Appendix 10 except for new coefficients resulting from the scale factors expressed in terms of the surface angle α_1.

So far, we have seen the aerodynamic benefits of the chordwise flexibility for the case of the periodic camber variation normal to the chord direction. Next, we are going to analyze the flexibility effects as the maximum camber location varying along the chord. Let the camber geometry of the thin airfoil be as shown in Fig. 8.38, and let the maximum camber location vary periodically with time. According to Fig. 8.38, the camberline equation for a piecewise parabolic variation with the maximum camber a located at p reads as

$$z(x) = \begin{cases} -a(x-p)^2/(1+p)^2, & x<p \\ -a(x-p)^2/\,(1-p)^2, & x \geq p \end{cases} \tag{8.47}$$

The time dependent downwash expression, $w(x, t) = \partial z/\partial t + U\,\partial z/\partial x$ with $\dot{p} = \partial p/\partial t$ then becomes (Gulcat 2009a),

$$w(x,t) = \begin{cases} 2a\dot{p}(x-p)^2/(1+p)^3 + 2a(x-p)(\dot{p}-U)/(1+p)^2, & x<p \\ -2a\dot{p}(x-p)^2/(1-p)^3 + 2a(x-p)(\dot{p}-U)/\,(1-p)^2, & x \geq p \end{cases} \tag{8.48}$$

The full unsteady lift coefficient can be calculated for a simple harmonic motion using Eq. 3.32a. However, even if we assume that the periodic movement for the maximum camber location is simple harmonic, according to Eqs. 8.46 and 8.47, both the camber motion and the associated downwash are periodic but they are no longer simple harmonic. Therefore, we have to be cautious while using the formulae derived for unsteady force and moment coefficients. Nevertheless, for oscillations with small frequencies as a first approximation we can use the concept of *steady aerodynamics*, i.e. $\dot{p} = 0$, the piecewise integration of Eq. 8.47 with Eq. 3.31a from –1 to p, and p to 1 gives the sectional lift coefficient as

$$c_l = 2a\left[(-2p(p^2+1)+p^2+1)\pi + 4p(2p-1)\sin^{-}(p) + (8p(1-p)+4p^2)\sqrt{1-p^2}\right]/[(p^2-1)]^2 \tag{8.49}$$

For the maximum camber location at the midchord, i.e. $p = 0$, Eq. 8.48 gives $c_l = 2a\pi$ as expected. The boundary layer edge velocity for the quasi steady case reads as

$$\begin{aligned} \frac{U_e}{U} = 1 \pm \frac{u'}{U} = 1 \pm \frac{ap(p-x)}{\pi}\left[8\ln(abs(p-x)^{-1}\sqrt{(1-x^2)(1-p^2)}+1-xp)\right]/[(1-p)(1+p)]^2 \\ + \frac{a}{\pi}\sqrt{\frac{1-x}{1+x}}\left[(2(1+x)-2p)(p^2+1)\pi + 8(p^2-p(1+x))\sin^{-}(p) - 8p\sqrt{1-p^2}\right]/[(1-p)(1+p)]^2 \end{aligned} \tag{8.50}$$

Here, + is used for upper and – is used for the lower surfaces of the airfoil. As expected, for $p = 0$ which means that the maximum camber at the mid-chord Eq. 8.49 gives

$$\frac{u'}{U} = 2a\sqrt{\frac{1-x}{1+x}}(1+x) = 2a\sqrt{1-x^2} \tag{8.51}$$

The steady sectional moment and lift coefficients obtained for an airfoil having 2% camber with its maximum camber location at p where $-0.5 \leq p \leq 0.5$ are given in Fig. 8.39. As observed in Fig. 8.39, the moment coefficient becomes positive for the p values which are aft of the mid-chord where lift coefficient increases significantly.

Shown in Fig. 8.40 is the steady surface velocity perturbation change with the location of the maximum camber. As expected, the peak value of the perturbation moves toward the mid-chord as the position of the maximum camber point moves the same way. Also shown in Fig. 8.40 is the surface velocity perturbation for a corrugated airfoil, bilinear in nature, with maximum camber location at quarter chord.

For non-negligible frequency values we have to consider $\dot{p} \neq 0$, therefore, the downwash expression, $w = w(t, x)$ must include the relevant terms known as *quasi-steady aerodynamics*, of expression 3.31a. The sectional lift coefficient then reads as

$$\begin{aligned} c_l^{qs} &= a_1(\sin^{-}(p)/2 - p\sqrt{1-p^2}/2 - (p^2+2)\sqrt{1-p^2}/3 + \pi/4) \\ &\quad + b_1(\sin^{-}(p)/2 - p\sqrt{1-p^2}/2 - \sqrt{1-p^2} + \pi/4) \\ &\quad + c_1(\sin^{-}(p) - \sqrt{1-p^2} + \pi/2) + a_2(-\sin^{-}(p)/2 \\ &\quad + p\sqrt{1-p^2}/2 + (p^2+2)\sqrt{1-p^2}/3 + \pi/4) \\ &\quad + b_2(-\sin^{-}(p)/2 + p\sqrt{1-p^2}/2 + \sqrt{1-p^2} + \pi/4) \\ &\quad + c_2(-\sin^{-}(p) + \sqrt{1-p^2} + \pi/2) \end{aligned} \tag{8.52}$$

wherein,

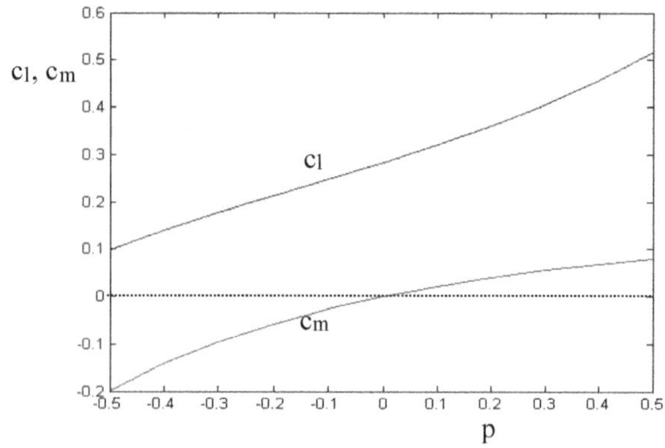

Fig. 8.39 Lift and moment coefficient variations with the maximum camber location p

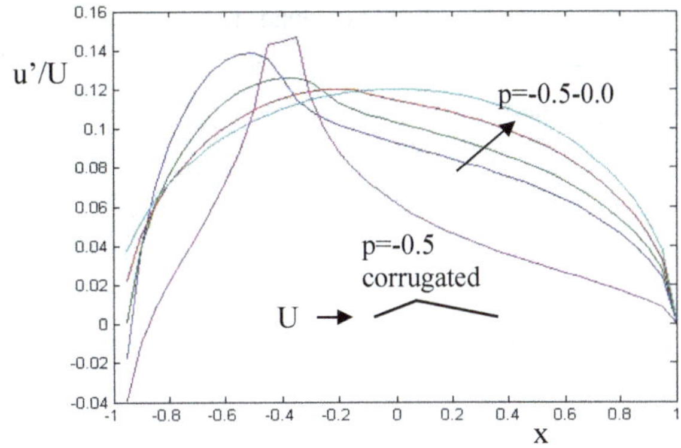

Fig. 8.40 Surface velocity perturbation variation with maximum camber location

$$
\begin{aligned}
& a_1 = 2a\dot{p}/(1+p)^3, \qquad b_1 = -2a_1 p + a_1(1+p) - 2aU/(1+p)^2, \\
& c_1 = a_1 p^2 - a_1(1+p)p + 2aU/(1+p)^2 \\
& a_2 = -2a\dot{p}/(1-p)^3, \qquad b_2 = -2a_2 p + a_2(1-p) - 2aU/(1-p)^2, \\
& c_2 = -a_2 p^2 - a_2(1-p)p + 2aU/(1-p)^2
\end{aligned}
$$

The edge velocity for the quasi steady aerodynamics reads as follows

$$
\begin{aligned}
\frac{U_e}{U} = 1 \pm \frac{u'(x)}{U} &= 1 \pm [c_1(1+x) + b_1 x(1+x) - c_2(1+x) - b_2 x(1+x)] \\
& \sqrt{1-x^2}\ln\left[1/(p-x)\sqrt{(1-x^2)(1-p^2} + 1 - px\right] \\
& - \left[c_1 + b_1(1+x) + a_1(1+2x^2)/2\right](\sin^{-}(p) + \pi/2) \\
& + \left[c_1 + b_1(1+x) + a_1(1+2x^2)/2\right](\sin^{-}(p) - \pi/2) \\
& + [(a_1 - a_2)(p/2 - 2x) - b_1 + b_2]\sqrt{1-p^2}
\end{aligned}
\tag{8.53}
$$

The steady and *the quasi steady aerodynamic* approaches do not consider the effect of the wake as phase lag between the motion and the aerodynamic response such as lift or moment, and the reduction in their amplitudes. As we know, the measure of this lag and the amplitude reduction is the Theodorsen function *C* (*k*) = *F*(*k*) + *iG*(*k*). The amplitude of the lift coefficient for the *quasi unsteady aerodynamics* according to Eq. 3.32a reads as

$$
\overline{c}_l^{qu} = C(k)\, c_l^{qs} \tag{8.54}
$$

The apparent mass term plays no role in quasi unsteady aerodynamics to give a simple relation between the vortex sheet strength and the lifting pressure, i.e. $\gamma_a = \overline{c}_{pa}/2$. The boundary layer edge velocity then is found from the perturbation velocity: $u' = \gamma_a/2$. The leading edge suction velocity P is given as $P = \lim_{x^* \to -1} \left(\frac{1}{2}\gamma_a\sqrt{x^*+1}\right)$.

In expanded form it reads:

$$P = \frac{\sqrt{2}}{\pi}\left\{\begin{array}{c} -(c_1+1.5a_1)(\sin^- p+\pi/2)+(c_2+1.5a_2)(\sin^- p-\pi/2) \\ +[(p/2+2)a_1-b_1-(p/2+2)a_2-b_2]\sqrt{1-p^2} \end{array}\right\} \quad (8.55)$$

which is to be used in Eq. 8.27 to calculate the suction force. Knowing P from Eq. 8.55, the quasi unsteady lift from Eq. 8.54, and the equivalent angle of attack from quasi-steady lift, i.e. $\alpha = c_l/2\pi$, we can obtain the propulsive force S from Eq. 8.27.

The effect of the unsteady motion of the camber location is studied under various conditions for the maximum camber location changing with $p = -0.25[1 - \cos(ks)]$, where $s = Ut/b$ is the reduced time. Shown in Fig. 8.41 are the typical lift and thrust variation plots for the cambered thin airfoil having chordwise flexibility with maximum camber of 3% and reduced frequency of $k = 0.2$. The quasi unsteady lift and thrust coefficients shown with _____ indicates the expected phase lag between the motion and the aerodynamic response. Since the reduced frequency $k = 0.2$ is small, the differences among the steady, quasi steady and the quasi unsteady lift and thrust coefficients are not too large. According to Fig. 8.41, the maximum lift and the zero thrust are obtained for $p = 0$ for which the maximum camber is at the midchord, and the minimum lift and the maximum thrust are achieved when the maximum is at quarter chord.

The averaged suction force coefficients obtained by time integration of the curves over a period given in Fig. 8.41 are represented in Table 8.5 for (i) steady, *st*, (ii) quasi steady, *qs*, and (iii) quasi unsteady cases, *qu*. According to Table 8.5, the force coefficient becomes smaller for quasi-unsteady treatment with increasing reduced frequency.

For a flat plate at Re = 10 000 the drag coefficient according to Blasius is $c_d = 0.0266$. The boundary layer solution obtained with the procedure as described

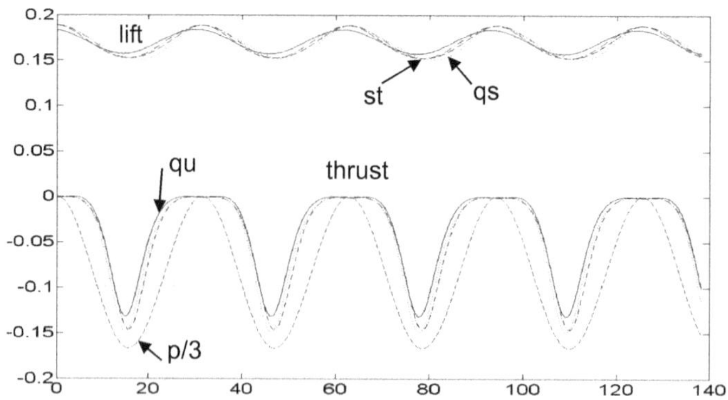

Fig. 8.41 Lift and thrust coefficient variations with time for $k = 0.2$ and $a = 3\%$

Table. 8.5 Averaged thrust coefficients for $a = 3\%$

	$k = 0.1$	$k = 0.2$	$k = 0.4$	$k = 0.8$
Steady	0.0432	0.0432	0.0432	0.0432
Quasi steady	0.0433	0.0434	0.0439	0.0460
Quasi unsteady	0.0384	0.0356	0.0344	0.0341

in Appendix 10, and based on the edge velocity given by Eq. 8.53, gives the viscous drag opposing to the motion as 0.0286 for $k = 0.2$ and 0.0266 for $k = 0.8$. This shows that the smallest propulsive force coefficient 0.0341, obtained with quasi unsteady approach for $k = 0.8$, for an airfoil morphing with a fixed camber ratio of %3, easily overcomes the viscous drag produced by the chordwise flexible airfoil.

The chordwise change in the camber is considered simple harmonic. However, the associated downwash w given by Eq. 8.48 is no longer simple harmonic, especially for motions having high frequencies. Shown in Fig. 8.42 is the quasi steady lift, Eq. 8.51, change with time and the quasi unsteady lift obtained with the FFT applied to the equivalent motion whose angle of attack determined via Eq. 8.48 as an arbitrary motion. Comparison of Figs. 8.41 and 8.42 shows the effect of the reduced frequency, which is low for the small values of k, on the lift coefficient amplitude of the chordwise flexible motion, whereas the time averaged lift coefficient is almost the same for quasi steady and the quasi unsteady approaches as seen in Fig. 8.42.

The full unsteady approach includes the apparent mass term given by the second term of the right hand side of Eq. 3.27. The apparent mass term contributes to lift but makes zero contribution to leading edge suction term.

In this section we have analyzed the active chordwise flexibility of a thin airfoil. There are experimental, in water tunnels (Heatcote and Gursul), as well as

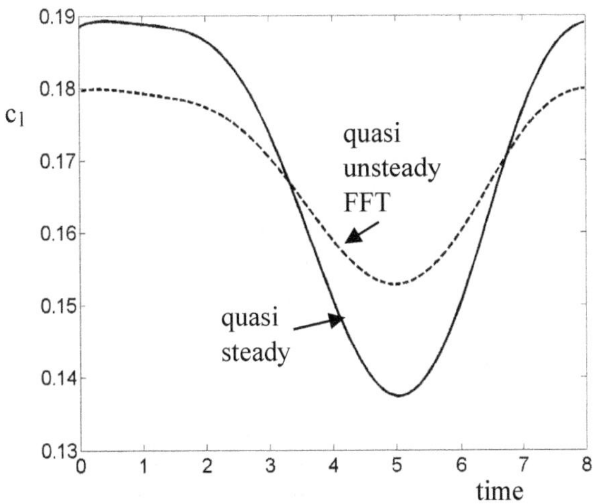

Fig. 8.42 Sectional lift coefficients: — quasi steady, and ____ quasi unsteady with FFT at or $k = 0.8$

numerical studies based on fluid–structure interaction (Zhu) concerning the passive flexibility with known or assumed elastic behavior of the thin hydrofoil flapping in water. The experimental and the numerical results agree well for the deformation of a thin and a thick flexible steel plate undergoing periodic heaving motion. The results obtained for a pitching plunging elastic airfoil by Zhu indicate that with increasing stiffness the thrust coefficient increases while the efficiency decreases. The effect of the maximum angle of attack is, however, opposite i.e., the efficiency increases and the thrust coefficient decreases as the maximum effective angle of attack increases. The behavior of the steel plate in air as inertia driven deformation is somewhat similar at least qualitatively. However, for low stiffness values both the thrust and the efficiency are very small. Furthermore, the thrust becomes negative, which implies drag, for even lower values of stiffness.

8.7 Finite Wing Flapping

The finite wing flapping differs, especially for the low aspect ratio wings, from the 2-D oscillatory motions of airfoils because of the presence of the tip vortex which is likely to interact with the leading edge vortex of the wing. For the large aspect ratio wings, however, the strip theory, based on the quasi 2-D approach, can give the approximate values for the total lift and the propulsive force once the type of motion is described. During the flapping of the wing, since the heaving amplitude changes linearly along the span, the dynamic separation angle also changes from one strip to another as well. Therefore, one has to make sure that each strip does not experience the dynamic stall. If there is a dynamic separation present in any strip then the leading edge vortex must be checked for bursting so that it does not lose its suction force. In case of a lost of suction in any strip, the contribution coming from that strip to the lifting and propulsive force must be reduced from the total accordingly (DeLaurier). Based on their modified strip theory Mueller and DeLaurier give their predicted averaged total thrust coefficient as negative and it agrees well with experimental values for a specific wing at low reduced frequencies, i.e. $k < 0.1$, which indicates power reduction, i.e. windmilling. There is an over estimated positive thrust for $k > 0.1$, and the over estimation is as high as 10%, for the reduced frequency of $k = 0.2$. The theoretical and the measured lift coefficients remain almost constant with respect to reduced frequency, wherein the theory over estimates the lift coefficient about 15% compared to experimental values.

Further experimental studies were conducted to model the 3-D dynamic stall of low aspect ratio wings oscillating in pitch (Tang and Dowell 1995) and (Birch and Lee, 2005). Tang and Dowell modeled a low aspect ratio wing with a NACA 0012 in periodic pitch, and they observed that results of their simple model showed qualitative similarities with the data of corresponding 2-D airfoil. Birch and Lee (2005), on the other hand, investigated the effect of near tip vortex behind the pitching rectangular wing with NACA 0015 airfoil profile having aspect ratio of 2.5 at $\mathrm{Re} = 1.86 \times 10^5$ within the reduce frequency range of 0.09–0.18. Their experimental results indicate small hysteretic behavior during the upstroke and downstroke motions for both the attached and the light stall oscillations. In case of deep stall oscillations, however, during upstroke the lift and the lift induced drag values increased with the airfoil incidence more than during downstroke for which the size of the tip vortex was larger compared to that of upstroke.

More detailed and extended wind tunnel as well as numerical study of oscillating finite wings was given by Spentzos et al. Five different wing geometry, varying from rectangular to highly tapered planforms with swept back tips, whose aspect ratios ranging from 3 to 10 and Reynolds numbers ranging from 1.3×10^4 to 6×10^6, are studied in dynamic stall conditions. The reduced frequencies of pitching oscillations range from 0.06 to 0.17. A light stall study of a rectangular wing with NACA 0015 section and with aspect ratio of 10 at $Re = 2 \times 10^6$ and $M = 0.3$ indicates that hysteresis curves for the lift and the drag narrow down considerably from half span to the tip both for the experimental and the computational results. At the tip region, however, there is a considerable positive shift between the experimental and the numerical results for the coefficients, which is attributed to the flexibility of the wing at the tip region (Spentzos et al. 2007).

The spanwise flexibility is also effective in thrust production of a pitching plunging finite wing (Zhu 2007). For a flexible wing, modeled as a thin foil in air, there is an initially sharp increase in thrust coefficient with increase in the stiffness of the foil, and it remains almost constant after dropping to a certain stiffness value. However, the efficiency shows a small increase with increasing stiffness. The increase in the average pitching angle decreases the amount of thrust but has an increasing effect on the efficiency of the foil. Nevertheless, for hydrofoils, where the calculations are performed for water, the thrust gradually increases with increasing stiffness, and the efficiency decreases slightly. The effect of average pitching angle is the same as it was for the case of air. The effect of spanwise flexibility on the thrust of a finite wing may change with the tip vortex and the leading edge vortex interaction which may enhance or weaken the leading edge suction force created by the foil. For more precise assessment, further investigations for the wings with tip vortex reducing devices become necessary.

The frequency of the flapping plays additional essential role in finite wing flapping because of presence of the tip vortex. As the frequency of the flapping increases, the vortex generation frequency also increases during the creation of lift.

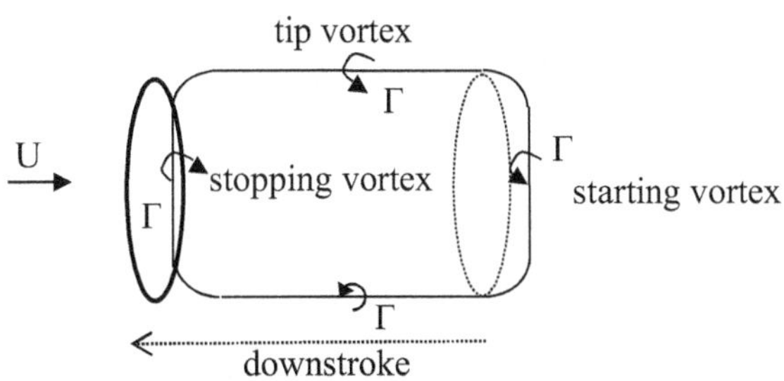

Fig. 8.43 Starting and stopping vortex generated during the downstroke

The starting vortex, the tip vortices shed from the left and the right tips of the wing and the bound vortex on the wing itself altogether form a vortex ring during the downstroke. At the end of the downstroke, since there is no lift on the wing, the bond vortex becomes a *stopping vortex* as shown in Fig. 8.43.

The starting and stopping vortices are equal in magnitude but opposite in sens, and both are normal to the free stream direction. The size of the idealized vortex ring in Fig. 8.43 depends on the wing span and the frequency of the flapping. For the case of high frequency flapping the starting vortex can not move downstream away from the wing, therefore, it affects the lift unfavorably. On the other hand, once the wing is at its lowest position for upstroke, the effective angle of attack must create a lift generating vortex so that another starting vortex, which is in opposite sign with the stopping vortex, forms after a little lag. At the end of the upstroke, when the wing is its top position, a new stopping vortex, which is almost equal to the previously formed stopping vortex, and the new tip vortices are formed to make a new vortex ring. This way, once a cycle of motion is complete with downstroke and upstroke a ladder type wake, which consists of stopping and starting vortices, is generated as shown in Fig. 8.44a.

In the ladder type wake, which is produced by flapping finite rigid wing, the starting vortex having an opposite sign with the bound vortex causes delaying effects on the lift. In order to avoid this delay and not create vortices which are normal to flight direction, the length of span is reduced during upstroke with making use of spanwise flexibility. During downstroke the wing has a full span to give wider gap between the tip vortices whereas this gaps narrows down because of having smaller wing span during upstroke, which makes the strength of the tip vortex to remain the same. Hence, in an alternating manner, we observe one wide and one narrow tip vortex street, which in literature is called concertina type wake as shown in Fig. 8.44b, Lighthill. In concertina type, unlike the ladder type, the periodic occurrence of wake vortices normal to the flight direction which plays a delaying effect in lift generation, disappears. Therefore, the spanwise flexibility, which generates concertina type wake pattern is preferable for man made flapping wings having high aerodynamic efficiencies similar to the efficiencies of the wings exist in nature.

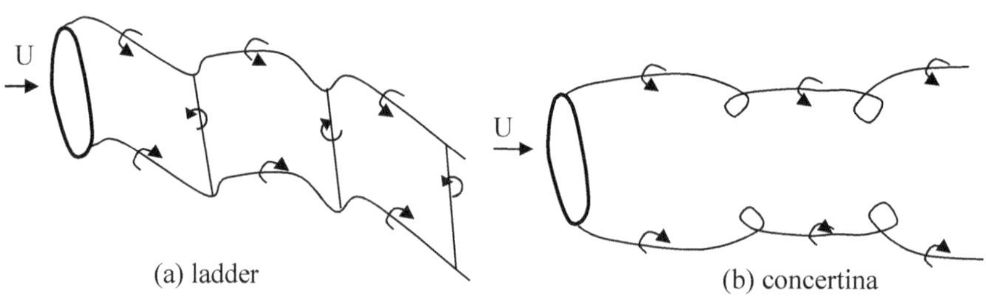

Fig. 8.44 Flapping finite wing vortices: **a** ladder, **b** concertina type

Let us give a numerical example for unsteady calculations based on an example given in Chap. 4.

Example 8.7 The lift and the propulsive force of a heaving plunging rectangular wing: Here, we will make use of the sine series expansion of the sectional circulation amplitude similar to that given in Sect. 4.3.2 which yields, for $y = l\cos\theta$, and the sectional lift $\overline{L}'_i = \rho U \overline{\Gamma}_i$ as

$$\overline{\Gamma}_i(\theta_i) = bU\sum_{n=1}^{N}\overline{K}_n\frac{\sin n\theta_i}{n}\text{in matrix notation}\left\{\overline{\Gamma}\right\} = \left[\frac{\sin n\theta}{n}\right]\left\{\overline{K}\right\}$$

The nondimensional sectional lift amplitude can be expressed as

$$\frac{\overline{L}}{2\rho U^2 b} = \pi\left[-\frac{k^2}{2} + ikC(k) + ik\sigma_h(y)\right]h^*$$

Here, $\sigma_h(y)$ is given in terms of the ratio of the sectional value of the reduced circulation to the 2-D reduced circulation as defined in Sect. 4.3.2. Using the data given in Example 4.3 for a rectangular wing of $AR = 6$ heaving plunging with the reduced frequency of $k = 2/3$, we can find 4 unknown complex coefficients $\overline{K}_n$ by solving the 4×4 matrix equation. For symmetric loading taking $n = 1, 3, 5, 7$ gives these values as

$$\begin{aligned} K_1 &= -0.5465 - 1.3908i, K_3 = -0.2768 - 0.6246i, K_5 \\ &= 0.2831 + 0.1398i, K_7 = 0.3037 + 0.2828i \end{aligned}$$

The total lift can be obtained through the integration of the sectional lift along the whole span which is

$$\overline{L} = \int_{-l}^{l}\overline{L}'(y)dy = h^*\int_{\pi}^{0}2\rho UbU\sum_{1}^{N}\overline{K}_n\frac{\sin n\theta}{n}(-l\sin\theta)d\theta$$

Interchanging the order of integral with the summation and integrating the result, we get the contribution only for $n = 1$ as follows

$$\overline{L}/h^* = \rho U^2 b\, l\pi\,\overline{K}_1$$

The total lift coefficient amplitude then reads as

$$\overline{C}_L/h^* = \frac{\rho U^2 b\, l\pi\,\overline{K}_1}{1/2\,\rho U^2 4b\, l} = \frac{\pi}{2}\overline{K}_1 = (-0.5465 - 1.3908i)\frac{\pi}{2}$$

For $h(t) = \overline{h}\cos(\omega t),\;\; h^* = \overline{h}/b$ the total lift history becomes

$$C_L(s)/h^* = -0.8584\cos(2s/3) - 2.1846\sin(2s/3)$$

The propulsive force on the other hand is calculated similar to that of 2-D case using (8.27) as follows (Gulcat 2011)

$$\overline{P} = -\sqrt{2}b[C(k) + \sigma_h]i\omega\overline{h}$$

Here, let us call $\overline{H}_i = [C(k) + \sigma_h]_i = \sum_{n=1}^{N} \overline{S}_n \frac{\sin n\theta_i}{n}$.

The sectional propulsive force then for $h = \overline{h}\cos(\omega t)$ type motion becomes

$$F_s'(t) = -\pi\rho P^2 = 2b\rho\pi\left[(G + \sigma_h^i)\cos(\omega t) + (F + \sigma_h^r)\sin(\omega t)\right]^2\omega^2\overline{h}^2$$

From (Reissner) the spanwise H values reads as

$y^* =$	0.0	0.4	0.8	1.0
$H =$	$0.569 - 0.123i$	$0.563 - 0.116i$	$0.511 - 0.113i$	$0.11 - 0.313i$

Solution for the S values gives

$$S_1 = 0.6633 + 0.1691i, S_3 = 0.2994 + 0.4358i, S_5 = -0.0676 + 1.0501i, S_7 = -0.1329 + 0.7763i$$

The time dependent total propulsive force then reads in series form as

$$C_{Fs}(t) = \frac{\int_{-l}^{l} F_s' dy}{\rho U^2 A/2} = \pi h^{*2}k^2/2\sum_m\sum_n (S_m^r \sin\omega t + S_m^i \cos\omega t)(S_n^r \sin\omega t + S_n^i \cos\omega t)I_{mn}/mn$$

where,

$$I_{mn} = \int_0^{\pi} \sin(m\theta)\sin(n\theta)\sin\theta\, d\theta$$

and, $I_{nn} = 4n^2/(4n^2 - 1),\ n = 1, 3, 5, 7$ *and* $I_{13} = I_{31} = -4/15,\quad I_{15} = I_{51} = -4/105$.

With above values the averaged propulsive force coefficient over a period gives

$$\overline{C}_{Fs}/h^{*2} = \frac{1}{2\pi/k}\int_0^{2\pi/k} C_{Fs}(s)\, ds = -0.312$$

The propulsive force obtained above is opposed with the induced drag D_i which is calculated as follows

$$\overline{D}_i(y) = -\overline{L}_i'(y)\alpha_i(y), \text{ with } \overline{\alpha}_i(y) = -\frac{1}{4\pi U}\int_{-l}^{l}\frac{d\overline{\Gamma}}{y_i-\eta}, \quad d\overline{\Gamma}$$

$$= 2bUh^*\sum_{1}^{N}\overline{K}_n\cos n\theta(-d\theta)$$

Then the induced angle of attack becomes

$$\alpha_i = -\frac{b/l}{2\pi}h^*\int_0^{\pi}\frac{\sum_1^N\overline{K}_n\cos n\theta\, d\theta/n}{\cos\phi-\cos\theta}$$

After changing the order of integration with summation and making use of the famous formula for the singular integral, which is $-\pi\sin n\phi/\sin\phi$, the results reads as

$$\alpha_i = \frac{b/l}{2}h^*\frac{\sum_1^N\overline{K}_n\sin n\phi}{\sin\phi}$$

The total drag for a heaving plunging wing then reads as

$$D_i t) = -\int_{-l}^{l} L_i(y)\alpha_i(y)dy = \rho U^2\frac{b^2}{l}h^{*2}\int_0^{\pi}\sum_{m=1}^{N}\text{Re}(\overline{K}_m e^{i\omega t})\frac{\sin m\phi}{m}\sum_{n=1}^{N}\text{Re}(\overline{K}_n e^{i\omega t})\sin n\phi(-ld\phi)$$

The contribution to the total drag is different from zero for $m = n$ only, otherwise it is zero. Hence, after interchanging with summation, the integral with respect to y gives

$$D_i(t) = \rho U^2\frac{\pi b^2}{2}h^{*2}\sum_{n=1}^{N}(K_n^{r2}\cos^2\omega t - K_n^{i2}\sin^2\omega t)/n$$

The total drag coefficient then reads as

$$C_D(t) = \frac{D(t)}{\rho U^2 A/2} = \frac{\pi b^2 h^{*2}}{4bl}\sum_{n=1}^{N}(K_n^{r2}\cos^2\omega t - K_n^{i2}\sin^2\omega t)/n$$

The time averaged propulsive force coefficient over a period becomes

$$\overline{C}_D/h^{*2} = \frac{\pi}{2\pi/k}\int_0^{2\pi/k} C_D(s)\,ds = \frac{\pi}{8AR}\sum_{n=1}^{N}(K_n^{r2} - K_n^{i2})/n = 0.113$$

The net propulsive force is the difference between the total propulsive force and the induced drag force. Therefore, the net propulsive force coefficient reads as

$$\overline{C}_F = \overline{C}_{Fs}/h^{*2} + \overline{C}_D/h^{*2} = -0.312 + 0.113 = -0.199$$

On the other hand, the 2-D propulsive force coefficient from (8.29) reads: $\overline{c}_s = -0.486$, compared to –0.312 which is the value of the propulsive force for the finite wing.

3-D Unsteady Boundary Layer: The viscous drag acting on thin wings also play important role on determining the net propulsive force generated during flapping. For determining the viscous drag from the skin friction, the 3-D unsteady boundary layer equations are solved separately at the upper and the lower surfaces of the wing. For this purpose the velocity–vorticity formulation is employed similar to the procedure that applied for 2-D case in Sect. 8.6. This time, the thin wing is represented with a vortex sheet having chord-wise and span-wise vorticity components which satisfy the following equations

$$\text{Chord wise vorticity:}\frac{D\omega_x}{Dt} = \omega_x\frac{\partial u}{\partial x} + \omega_y\frac{\partial u}{\partial y} + \frac{1}{\text{Re}}\frac{\partial^2\omega_x}{\partial z^2} \tag{8.56}$$

$$\text{Span wise vorticity:}\frac{D\omega_y}{Dt} = \omega_x\frac{\partial v}{\partial x} + \omega_y\frac{\partial v}{\partial y} + \frac{1}{\text{Re}}\frac{\partial^2\omega_y}{\partial z^2} \tag{8.57}$$

$$\text{together with the continuity equation: } \vec{\nabla}\cdot\vec{V} = 0. \tag{8.58}$$

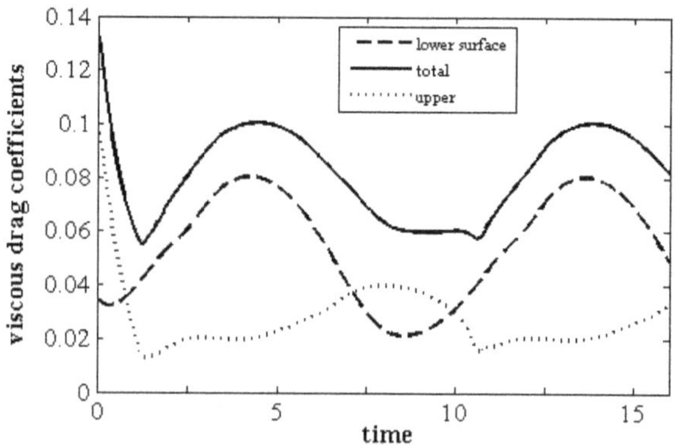

Fig. 8.45 Viscous drag coefficients versus time

The finite difference discretization of Eqs. (8.56) and 8.57) yields penta-diagonal matrix equations as follows

$$d_{jk}(\cdot)_{i,j-1,k} + a_{jk}(\cdot)_{i,j,k-1} + b_{jk}(\cdot)_{ijk} + c_{jk}(\cdot)_{i,j,k+1} + e_{jk}(\cdot)_{i,j+1,k} = r_{jk} \quad (8.59)$$

Values of the coefficients *a*, *b*, *c*, *d* and the right hand side *r* are given in Appendix 11 together with the solution procedure for the velocity field as well. Here, the edge velocity components are obtained from

$$U_e = \int_0^{\partial} \omega_y dz \quad \text{and} \quad V_e = -\int_0^{\delta} \omega_x dz \quad (8.60\text{a, b})$$

For equally spaced discretization in z

$$0.5\omega_{x,i,j,1} + \omega_{x,i,j,2} + \omega_{x,i,j,3} + \cdots + \omega_{x,i,j,bs} = -V_e/\Delta z$$
$$\text{and} \quad (8.61\text{a, b})$$
$$0.5\omega_{y,i,j,1} + \omega_{y,i,j,2} + \omega_{y,i,j,3} + \cdots + \omega_{y,i,j,bs} = U_e/\Delta z$$

Shown in Fig. 8.45 is the viscous drag variation for the rectangular wing of an aspect ratio 3, heaving and plunging simultaneously with root flapping at Re = 1000 by

$$h_{ut} = 1.2b\cos(0.667s)$$
$$h_{lt} = \frac{1.68}{3}\frac{y}{b}\cos(0.667s + \varphi).$$

Here, $\phi = 1.9\pi$, which seems to be the phase difference between the uniform and root flappings which gives the maximum thrust (Gulcat 2011). The time averaged drag coefficient shown in Fig. 8.45 is C_D = 0.0789. The difference between the upper and lower viscous drag history is due to the phase difference between the uniform and

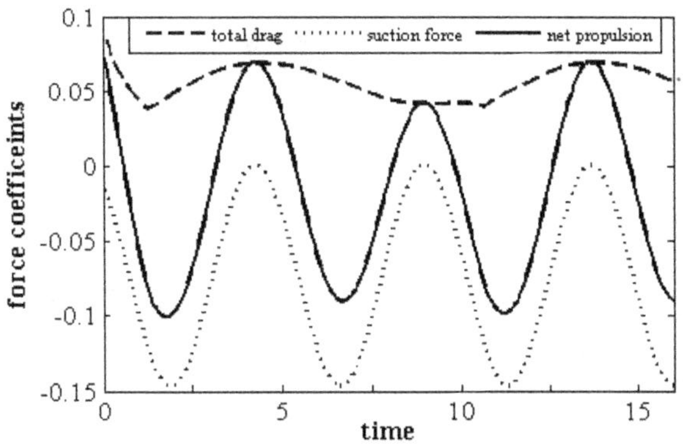

Fig. 8.46 Net propulsive force coefficients versus time

the root flapping.The Reynolds number, here, is low enough for viscous forces to overcome the suction forces evaluated as described in Example 8.7.

For Re = 2000, however, the suction forces overcome the viscous drag to give a net propulsive force as shown in Fig. 8.46. The drag for the Reynolds number of 2000 reduces to 0.0544, which gives the time averaged thrust force coefficient as $\overline{C}_T = -0.0716 + 0.0544 = -0.0172$.

8.8 Ground Effect

Air vehicles performing close to the ground experience the ground effect as changes in their aerodynamic lift and moments. This change is function of two variables: distance to the ground and the angle of attack of the wing. The distance to the ground plays an increasing role on the lift, whereas the angle of attack has a decreasing effect. When the unsteady behavior is concerned, the effect of unsteadyness also comes into picture in predicting the performances of helicopters, Micro Air Vehicles and the high speed ground vehicles. Here, first we study the ground effect on the airfoil performance and then on the finite wing.

Shown in Fig. 8.47 is the vortex system of a thin airfoil modeled as a vortex sheet and its ground image to yield the amplitude of the following downwash

$$\overline{w}(x) + \frac{ik\overline{\Omega}}{2\pi}\int_{-1}^{\infty} e^{-ik\xi}\left(-\frac{1}{x-\xi} + \frac{2h_g\alpha}{(x-\xi)^2+4h_g^2} + \frac{x-\xi}{(x-\xi)^2+4h_g^2}\right)d\xi =$$
$$\frac{1}{2\pi}\int_{-1}^{1} \overline{\gamma}_a(\xi)\left(-\frac{1}{x-\xi} + \frac{2h_g\alpha}{(x-\xi)^2+4h_g^2} + \frac{x-\xi}{(x-\xi)^2+4h_g^2}\right)d\xi \tag{8.62}$$

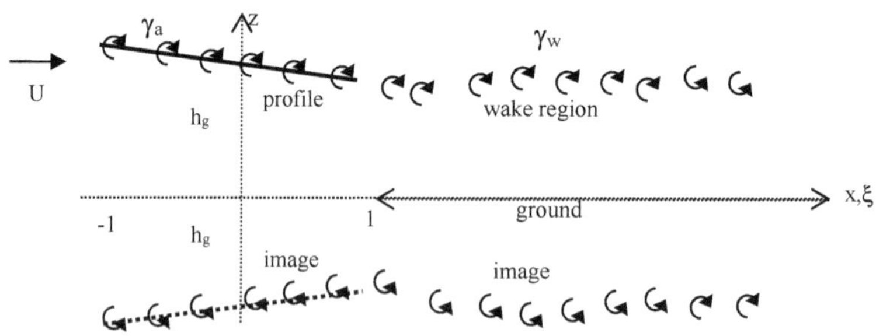

Fig. 8.47 Bound and image vorticies and their wake

which satisfies the unsteady Kutta condition which relates the wake vortex shett strength to the reduced circulation as follows: $\overline{\gamma}_w(x) = -ik\overline{\Omega}\, e^{-ik\,\xi}$. Since the motion of the airfoil is prescribed the associated downwash is known. Implementing the series approach to the kernel of (8.62) lets us to invert it to get an approximate value of the bound vortex sheet which yields for the following lift coefficient amplitute, see Appendix 12,

$$\begin{aligned}\overline{C}_l &= 2\,\pi\, C(k)\left[\overline{\alpha} + ik(\overline{h} + \overline{\alpha}/2)\right] - \pi k^2(\overline{h} + \overline{\alpha}) + \frac{\pi}{2}\left[ik\overline{h} + \left(1 - \frac{3ik}{4} - 4\alpha h_g\right)\overline{\alpha}\right]/h_g^2\\ &+ ik\frac{\overline{\Omega}}{U}\left\{\frac{3}{16}\left(\frac{1}{2k^2} - \frac{1}{4ik} + \frac{\alpha h_g}{\pi}\right)e^{-ik} + \frac{1}{4}\left[\left(-\frac{1}{k^2} + \frac{3}{2ik}\right)e^{-ik} + \frac{3}{2}C_1(k) - C_2(k)\right]\right\}/h_g^2\\ &- k^2\frac{\overline{\Omega}}{U}\left\{\frac{5}{16}\left(\frac{1}{k^2} - \frac{1}{2ik}\right)e^{-ik} + \frac{5}{4}\left(\frac{e^{-ik}}{ik} + C_1(k)\right) - 2\left(\frac{e^{-ik}}{k^2} + C_2(k)\right)\right\}/h_g^2\end{aligned} \tag{8.63}$$

For the steady case, since, $k = 0$ and $C(k) = 1$ Eq. (8.63) becomes

$$C_l = 2\pi\alpha(1 - \alpha/h_g + 0.25/h_g^2) \tag{8.64}$$

This expression (8.64) is in agreement with the formula given in Katz and Plotkin (2010). The second term in (8.64) reduces the lift near ground especialy at high angles of attack.

The propulsive force is obtained from the leading edge suction using (8.27) with

$$\overline{P} = \sqrt{2b}\left\{\begin{aligned} &U\left[C(k)\overline{\alpha} + ikC(k)(\overline{h} + \overline{\alpha}/2) - ik\overline{\alpha}/2 - \frac{\overline{\alpha}\alpha}{h_g}\right] + \frac{U}{4h_g^2}\left[\overline{\alpha} + ik(\overline{h} + \overline{\alpha})\right]\\ &+ \frac{ik\overline{\Omega}}{4\pi h_g^2}\left[(-\frac{3}{4k^2} + \frac{7}{8ik} + \frac{\alpha h_g}{ik})e^{-ik} + C_1(k) - C_2(k)\right]\end{aligned}\right\}$$

As an application, flow past a plunging thin airfoil at zero angle of attack and at Re = 10 000 with the reduced frequency range at $k = 0.5$–0.8 is studied. The time history of the distance to the ground, in terms of the half chord, is cosidered asd

$$h_g(t) = 1.1 - 0.4\cos(ks). \tag{8.65}$$

Shown in Fig. 8.48 are the thin airfoil motion, lift coefficient variaton for out of ground effect, without the wake and the ground effect with distance to the ground changing by (8.65). Figure 8.48 indicate that the ground affects the airfoil at the most as it moves toward to the ground, whereas, as it moves away from the ground the effect is only on the reduction of the amplitude of the lift coefficient. Table 8.6 shows the averaged sectional lift coefficient in OGE for various reduced frequecies. From the table it is observed that as the reduced frequency increases the lift coefficient also increases as expected.

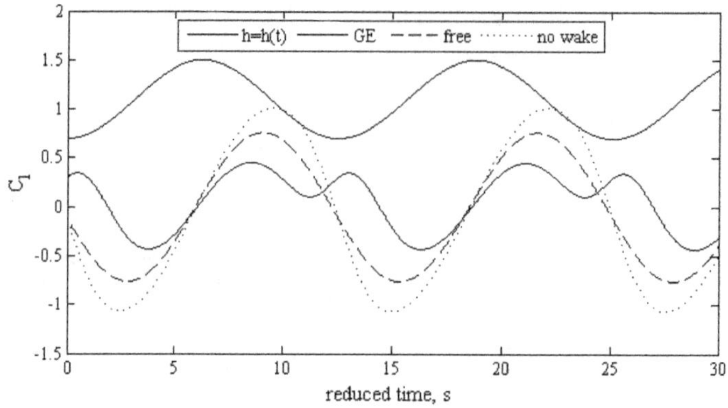

Fig. 8.48 C_l history under the ground effect and its wake for $k = 0.5$ and $h_g(s) = 1.1–0.4\cos(ks)$

Table. 8.6 Averaged sectional lift

	$k = 0.5$	$= 0.6$	$= 0.7$	$= 0.8$
GE	0.0802	0.0987	0.1125	0.1215
OGE	0.0	0.0	0.0	0.0

Fig. 8.49 Lift change for GA (W)-1 profile under the ground effect, ____ present, —o—reference

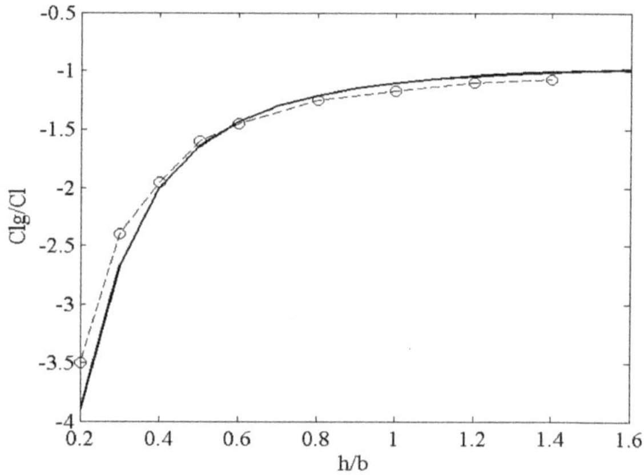

Here, as a non zero angle of attack case, the ground effect on the lift and the propulsive force change for GA(W)-1 profile is studied and compared with the available data of (Moryossef and Levy 2004) which is obtained with RANS. Shown in Fig. 8.49 is the comparison of the present work results on C_l versus h/b and the results given by the referance for the steady case at $\alpha = -3.2°$.

As for the unsteady case the plunging motion of GA(W)-1 at $\alpha = -3.2$ for $k = 0.5$ with $h_g = 1.342–0.2\cos(ks)$ the turbulent velocity, u and the eddy viscosity, eps profiles at the leading and trailing edges are provided in Fig. 8.50a. The viscous

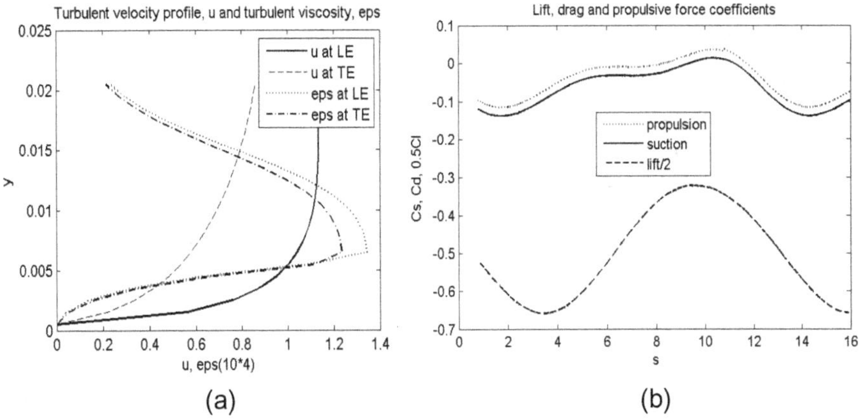

Fig. 8.50 **a** Turbulent velocity and viscosity profiles, **b** lift, drag and propulsive force versus s

drag calculations are made at Re = 460 000 with Cebeci-Smith turbulent modeling to predict the eddy viscosity.

The average propulsive fore coefficient value is –0.0307, and the value given by (Moryossef and Levy 2004) using RANS reads as –0.0285. These two values are in reasonable agreement as far as the engineering computations are concerned.

For visualization of the far wake of the heaving plunging thin airfoil, we make use of the velocity field given by the wake vortex sheet in terms of the reduced circulation

$$\gamma_w(x) = k(\Omega_i \cos x - \Omega_r \sin x) \cos ks + k(\Omega_i \sin x + \Omega_r \cos x) \sin ks, \quad \overline{\Omega} = \Omega_r + \Omega_i i \tag{8.66}$$

The far wake velocity vector components induced by the vortex strength given by (8.66) reads as

$$u(x,z) \cong \frac{z}{2\pi} \int_x^\infty \frac{\gamma_w(\xi) d\xi}{(x-\xi)^2 + z^2} \text{ and } w(x,z) \equiv -\frac{1}{2\pi} \int_x^\infty \frac{(x-\xi)\gamma_w(\xi) d\xi}{(x-\xi)^2 + z^2} \tag{8.67a, b}$$

Using (8.66) in (8.67a, b) with the aid of the transformation $x - \xi = \eta$ the intagrals are obtained for a point x in transformed coordinate η as follows

$$\gamma_w(x,\eta) = (c \sin x - d \cos x) \sin \eta + (c \cos x + d \sin x) \cos \eta, c$$
$$= \Omega_i \cos ks + \Omega_r \sin ks, \;\; d = \Omega_i \sin ks - \Omega_r \cos ks$$

With this transformation the integrals in η coordinates bocome $\int_1^{\infty} \frac{e^{ik\eta}}{k^2\eta^2+z^2} d\eta$, which has closed form solutions in k and z (Mathematica 2008). Accordingly, the reduced circulation values are calculated for (8.66) using (8.67a, b). Shown in Fig. 8.51a, b de the velocity fields are for the weak and strong ground effects respectively. When the ground effect is weak the velocity vectors at the lower and upper region have reverse sense but equal magnitudes, however, for the strong effect, the velocities near the ground get larger as opposed to the velocities away from the ground. The velocity field for Fig. 8.51a, b are obtained for the far wake with $k = 1$ and for $h = 1.1–0.4\cos(ks)$.

Further details for the derivations and the comparison of the far wake with the Navier–Stokes solutions are in Gulcat 2015a. The result is shown on Fig. 8.52a, b is the wake of a heaving-plunging NACA 0012 airfoil at Reynolds number 17 000 and $kh = 0.45$ obtained by a (a) coarse and (b) fine mesh. The fine mesh wake solution is costly but very much agrees with the result shown in Fig. 8.51a wherein the solution is based on ideal flow. It is also observed that if the flowfield near the body is resoved with the same number of discrete points, the fine or coarse mesh resolution of wake does not effect the aerdynamic forces or moments acting on the airfoil (Gulcat 2015a).

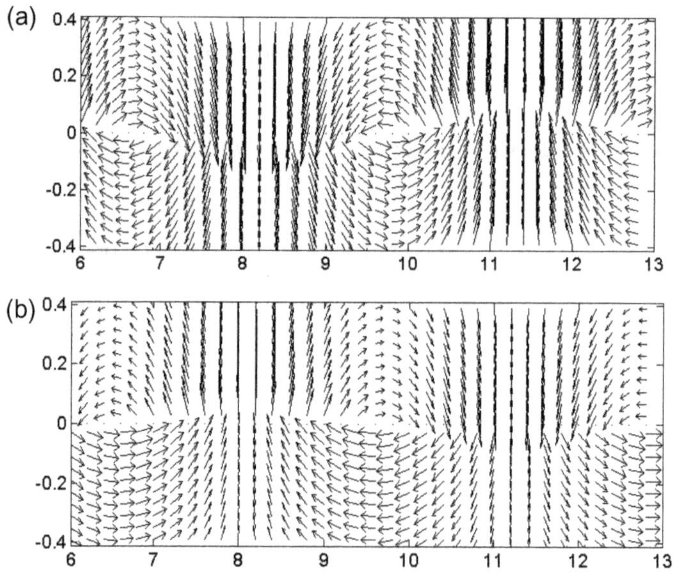

Fig. 8.51 **a** Far from ground, $h_g = 1.5\text{b}$ and **b** near ground, $h_g = 0.7\text{b}$, wake velocity vectors in flow fixed coordintes, phase difference $\varphi = \pi$.

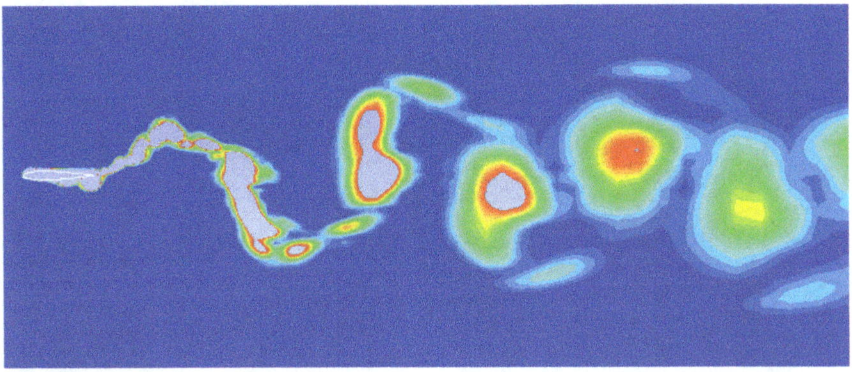

(a)Vortices for coarse mesh solution at Re=17000 k.h=0.45 (44K cell)

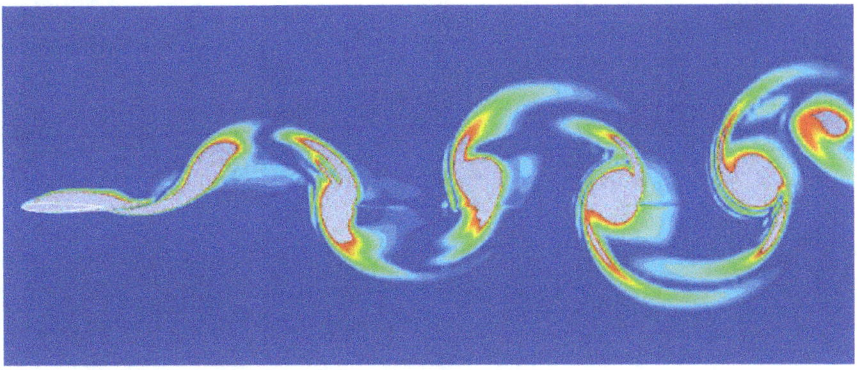

(b)Vortices for fine mesh solution at Re=17000 k.h=0.45 (1.25M cell)

Fig. 8.52 **a** coarse wake, **b** fine wake resolution results

Now, we can study the ground effect on finite wings. Here, concerning the ground effect, two different derivations are involved because of the aspect ratio considerations. First, moderate to high aspect ratio wings are considered, and then the low aspect ratio wings are studied for steady state cases to be utilized as basis for unsteady studies.

Moderate to High Aspect Ratio Wings:

First, we consider the steady state solution to the ground effect using Prandtl's Lifting Line theory based on the concept of a horse shoe vortex and its image due to the presence of ground which is shown in Fig. 8.53.

The downwash induced by this vortex system consist of the w_w, wing induced, and the w_g, image induced. The downwash induced by the ground is formulated as follows

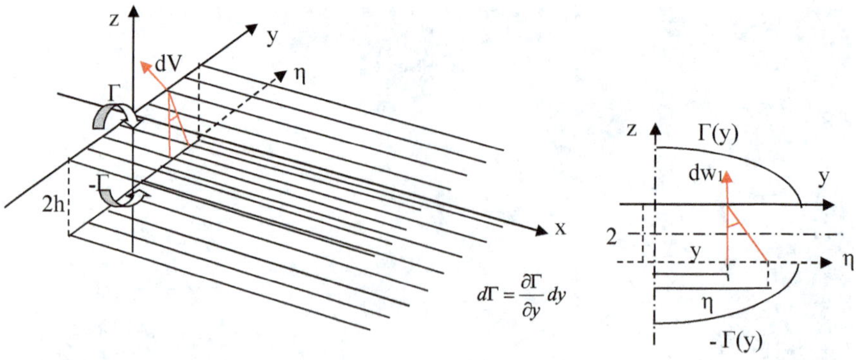

Fig. 8.53 Horse shoe vortex and its image near the ground

$$
\begin{aligned}
w_g =& \frac{1}{4\pi}\int_{-b}^{b}\int_{-l}^{l}\frac{\gamma_a(\xi,\eta)(x-\xi)}{R^3}d\xi\,d\eta + \frac{1}{4\pi}\int_{-b}^{b}\int_{-l}^{l}\frac{\delta_a(\xi,\eta)(y-\eta)}{R^3}d\xi\,d\eta \\
&+\frac{1}{4\pi}\int_{-l}^{l}\delta_w(\xi,\eta)(y-\eta)\int_{b}^{\infty}\frac{d\xi}{R^3}d\eta
\end{aligned}
\tag{8.68}
$$

where, $w_g = (U-u')\frac{\partial z_a}{\partial x} = -(U-u')\alpha, u' = \frac{1}{4\pi}\int_{-b}^{b}\int_{-l}^{l}2\alpha h_g\frac{\gamma_a(\xi,\eta)}{R^3}d\xi\,d\eta,$ and $R = \left[(x-\xi)^2 + (y-\eta)^2 + 4h^2\right]^{1/2}$.

In short: $w = \frac{1}{4\pi}(I_1 + I_2 + I_3)$.

These integrals are evaluated by parts as given in Bisplinghoff et al. (1996) to yield

$$
\begin{aligned}
-U\alpha =& -\frac{1}{2\pi}\int_{-b}^{b}\frac{\gamma_a(\xi,\eta)}{x-\xi}d\xi - \frac{1}{2\pi}\int_{-b}^{b}\frac{2\alpha h_g\gamma_a(\xi,\eta)}{(-b-\xi)^2+4h^2}d\xi \\
&+\frac{1}{2\pi}\int_{-b}^{b}\frac{\gamma_a(\xi,\eta)(x-\xi)}{(x-\xi)^2+4h^2}d\xi + \frac{1}{4\pi}\int_{-l}^{l}\frac{d\Gamma}{d\eta}\left[\frac{-1}{y-\eta}+\frac{-(y-\eta)}{(y-\eta)^2+4h^2}\right]d\eta
\end{aligned}
$$

If we multiply each term with $\sqrt{(b+x)/(b-x)}$ and integrate the result wrt x from $x = -b$ to $x = b$, with changing the order of integration and with noting that $\int_{-b}^{b}\gamma_a(x,y)dx = \Gamma(y)$, we get

$$U\alpha(y)\int_{-b}^{b}\sqrt{\frac{b+x}{b-x}}dx=\left[-\frac{1}{2}+\alpha(y)h\int_{-b}^{b}\frac{d\xi}{(-b-\xi)^2+4h^2}-\frac{1}{2\pi}\int_{-b}^{b}\int_{-b}^{b}\sqrt{\frac{b+x}{b-x}}\frac{(x-\xi)dx}{(x-\xi)^2+4h^2}d\xi\right]\Gamma(y)$$
$$-\frac{b}{4}\int_{-l}^{l}\frac{d\Gamma}{d\eta}\left[\frac{-1}{y-\eta}+\frac{-(y-\eta)}{(y-\eta)^2+4h^2}\right]d\eta \tag{8.69}$$

For $h^{*} = h/b > 1$ that is distance to the ground is higher than the half chord, the Kernels are simplified and integrated to yield

$$\left(1-\frac{\alpha(y)}{h^*}+\frac{1}{4h^{*2}}\right)\Gamma(y)=Uba(y)\left[\alpha(y)-\frac{1}{2a(y)U}\int_{-l}^{l}\frac{d\Gamma}{d\eta}\left[\frac{1}{y-\eta}+\frac{y-\eta}{(y-\eta)^2+4h^2}\right]d\eta\right] \tag{8.70}$$

The Glauert's Fourier series substitution: $\Gamma(\theta)=U\,a_o b_o\sum_{n=1}^{\infty}A_n\sin n\theta$ with the following coordinate transformation $y=l\cos\theta\, and\, \eta=l\cos\varphi$ gives

$$\sum_{n=1}^{\infty}A_n\left\{\sin n\theta\left(1-\frac{\alpha(y)}{h^*}+\frac{1}{4h^{*2}}\right)+\frac{b\pi n}{2l\sin\theta})-\frac{b}{2l}\int_{0}^{\pi}\sin n\phi\left[\frac{4h^{*2}b^2}{l^2r^2}-\frac{(\cos\theta-\cos\phi)^2}{r^2}\right]\sin\phi d\phi\right\}=\frac{ab}{a_o b_o}\alpha \tag{8.71}$$

where, subscript o denotes the root values and $r^2=[(\cos\theta-\cos\phi)^2+4h^{*2}b^2/l^2]^2$. Note that the integral in (8.71) represents the ground effect with h^* in the coefficient of A_n, and needs to be evaluated numerically while obtaining its contribution to the lift. Since the wing is rectangular $l/b = AR$ and $ab/a_o b_o=1$. With these information (8.71) is solved numerically for A_n's after choosing finite number of spanwise stations. As we have the solution for oge, the total lift coefficient for the wing reads as $C_L=\pi\, a_o b_o l A_1\alpha/S=\pi\, a_o\alpha A_1/4$ and the induced drag becomes

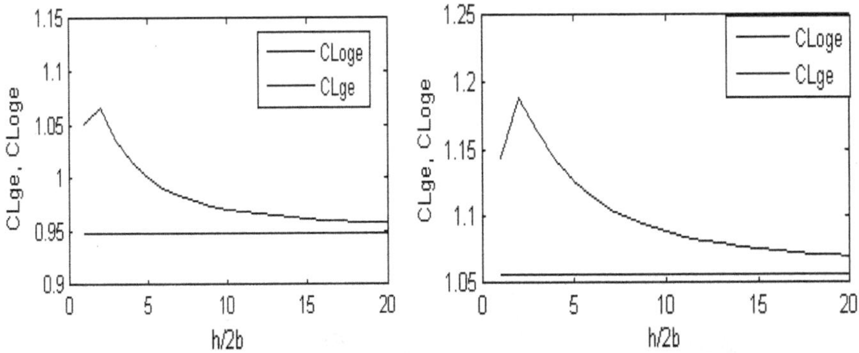

Fig. 8.54 Ground effect on a rectangular wing at $\alpha = 12°$, left: $AR = 6$, right: $AR = 10$

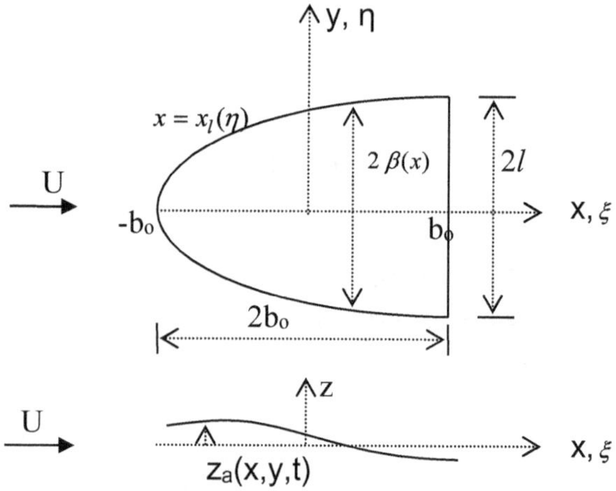

Fig. 8.55 Low aspect ratio wing, top and side views

$$C_{Di} = C_L^2/(\pi AR)\sum_1^{\infty} nA_n/A_1. \tag{8.72}$$

For a thin wing of $AR = 5.18$ at $\alpha = 10°$ and $h = 1$: $C_L = 1.206$. In Traub (2015) this value reads as $CL = 1.20$, wherein rectangular wing with S8036 profile is used which has 5.15 as lift curve slope $CL = 0.5$ at zero AoA for low Reynolds number flows. Calculated $C_{Di} = 0.0271$, whereas in Traub (2015) it reads as 0.037 which includes viscous drag also.

Shown in Fig. 8.54 is the effect of the aspect ratio at ground proximity. For $h/2b < 2$, the effect of the ground is adverse on the airfoil because of the $-\alpha/h$ term. The ratio of C_{Lge}/C_{Loge} almost remains the same for two different aspect ratios, $AR = 6$ and 10.

Low Aspect Ratio Wings:

Shown in Fig. 8.55 is the top and side view of a low aspect ratio wing with root chord being $2b_o$ and maximum span being 2 l at the trailing edge. The leading edge curve is $x = x_l(\eta)$, and the full span changes with $2\beta(x)$ from nose to the trailing edge.

The downwash due to image vortex generated by the ground reads as

$$\begin{aligned} w_g(x,y) &= \frac{1}{4\pi}\int_{-b}^{b}\int_{-l}^{l}\frac{\partial\gamma_a}{\partial\eta}\frac{(x-\xi)(y-\eta)}{\left[(x-\xi)^2+4h^2\right]R}d\xi\, d\eta \\ &+ \frac{1}{4\pi}\int_{-b}^{b}\int_{-l}^{l}\frac{\partial\delta_a}{\partial\xi}\frac{(x-\xi)(y-\eta)}{\left[(y-\eta)^2+4h^2\right]R}d\xi\, d\eta \\ &+ \frac{1}{4\pi}\int_{-l}^{l}\delta_a(b,\eta)\frac{-(y-\eta)d\eta}{\left[(y-\eta)^2+4h^2\right]} \end{aligned} \tag{8.73}$$

Noting that $\delta_a(b,\eta) = d\Gamma/d\eta$ and $\frac{\partial \gamma_a}{\partial \eta} = \frac{\partial \delta_a}{\partial \xi}$ combining first and second terms of the downwash as

$$w_g(x,y) = \frac{1}{4\pi}\int_{-b}^{b}\int_{-l}^{l}\frac{\partial \gamma_a}{\partial \eta}\frac{(x-\xi)(y-\eta)\left[(x-\xi)^2+4h^2+(y-\eta)^2+4h^2\right]}{\left[(x-\xi)^2+4h^2\right]\left[(y-\eta)^2+4h^2\right]R}d\xi\, d\eta$$
$$+\frac{1}{4\pi}\int_{-l}^{l}\frac{d\Gamma}{d\eta}\frac{-(y-\eta)d\eta}{\left[(y-\eta)^2+4h^2\right]} \tag{8.74a}$$

Simplifications of the integrals involved in the downwash expression are quite similar to that of high aspect ratio wings. The evolution of the integrals are provided in Appendix A13, which is based on the work by Gulcat (2015b). Accordingly, we get

$$w_g(x,y) = \frac{-2\,U\alpha}{4\pi\, h^{*2}}\int_{-1}^{1}\frac{y^*-\eta^*}{\sqrt{1-\eta^{*2}}}\eta^* d\eta^* = \frac{U\alpha}{4h^{*2}} \text{ with } \frac{\partial\, z_a}{\partial\, x} = -\alpha \tag{8.74b}$$

The image vorticity induced x component of the perturbation velocity reads as

$$u'(x,y) = \frac{1}{4\pi}\int_{-b}^{b}\int_{-l}^{l}\frac{2h_g\gamma_a}{R^3}d\xi\, d\eta \tag{8.75}$$

and the lifting pressure because of the presence of the x component of the induced velocity obtained in **A.13** reads as

$$C_{pa} = \frac{1}{2h_g}\frac{\partial}{\partial x}\left\{\frac{\partial z_a}{\partial x}\frac{\partial}{\partial x}\left[\sqrt{\beta^2(x)-y^2}\left(-\beta^2(x)\frac{\partial z_a}{\partial x}\right)\right]\right\} \tag{8.76}$$

The sectional lift for the wing is then evaluated as

$$L'(y) = \frac{1}{2}\rho U^2\int_{x_l}^{b_0} C_{pa}dx = -\frac{\rho U^2}{4h_g}\sqrt{l^2-y^2}\left[\frac{\partial z_a}{\partial x}\frac{\partial}{\partial x}\left(\beta^2\frac{\partial z_a}{\partial x}\right)\right]_{TE} \tag{8.77}$$

The total lift then is evaluated for $-u'\frac{\partial z_a}{\partial x}$ as:

$$L_{u'} = \int_{-l}^{l} L'(y)dy = -\frac{\pi\rho U^2 l^2}{4h_g}\left[\left(\frac{\partial z_a}{\partial x}\right)^2\frac{\partial \beta^2}{\partial x}\right]_{TE} \tag{8.78}$$

Here, $\frac{\partial \beta}{\partial x} = 0$ implies no contribution to the lift if the wing leading edge is parallel to the x axis at the tip of the wing.

Finally, the lift coefficient acting on the wing is

$$C_L = \frac{L}{1/2\rho U^2 S} = \frac{\pi\,\alpha}{2} AR\left(1 + \frac{1}{4h^2} - \frac{\alpha}{4h}\frac{\partial \beta^2}{\partial x}\bigg|_{TE}\right) \tag{8.79}$$

At high AoA the sharp leading edge vortex gives (Polhamus 1971),

$$C_L = K_p \sin\alpha\,\cos^2\alpha + \pi \sin^2\alpha\,\cos\alpha \tag{8.80}$$

The image of the leading edge vortex induces up wash $w = \Gamma/(4 \times 2\pi h)$ which in turn gives the extra leading edge thrust as $T_g = 2\;l\;\rho\;\Gamma^{\,2}/(8\pi h)$ with $\Gamma = K_p$ $SU\sin\alpha/4\;l$. The associated image vortex lift then reads as

$$C_{Lvi} = \frac{K_p^2}{32AR\,h/2l}\sin^2\alpha\,\cos\alpha, \quad K_p = \frac{2\pi\,AR\,\cos\Lambda}{AR\sqrt{1 + (2\,\cos\Lambda/AR)^2} + 2\cos\Lambda} \tag{8.81a, b}$$

Using (8.80, 8.81a), the total lift for a delta wing with 65° sweep at 20° degrees angle of attack is found as C_L = 0.474 + 0.345 = 0.819, which is in very good agreement with the value given by (Qu et al. 2015) as 0.818 for h/C_r = 0.5, where C_r is the root chord.

Utilizing steady state values of the lift coefficients obtained for wings with various aspect ratios, their unsteady behaviours are studied in the following section.

8.9 State-Space Representation

The state space representation of aerodynamic characteristics of lifting surfaces is an efficient tool to predict their unsteady performances, especially, at high angles of attack. Here, first let us introduce the general approach to the idea of the state space representation.

The state-space representation of any dynamical system can be represented with a first order ordinary differential equation as an input-state-output dynamical system (Goman and Khrabrow 1994) as follows

$$\frac{dx}{dt} = f(x, h_s) \;\; and \;\; C = g(x, h_s) \tag{8.82a, b}$$

where, h_s is system input, C is output and x is state space internal dynamic variable which is function of time t.

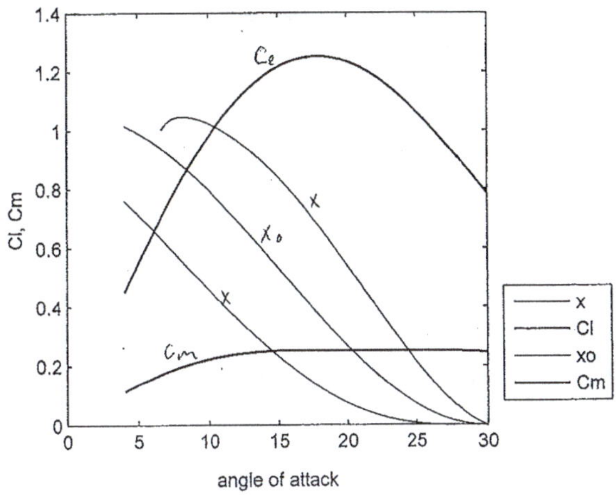

Fig. 8.56 Lift and moment coefficients change and separation point movement versus AoA

For an airfoil unsteady lift and moment coefficient change with time is given in terms of the state variable x which is represented as the position of the separation point on the upper surface of the airfoil. The ordinary differential equation for x reads as

$$\tau_1 \frac{dx}{dt} + x = x_o(\alpha - \tau_2 \dot{\alpha}) \tag{8.83}$$

where, argument $(\alpha - \tau_1 \dot{\alpha})$ indicates shift for the angle of attack rate $\dot{\alpha}$ of the static variation of $0 \leq x_o(\alpha) \leq 1$. Here, τ_1 and τ_2 are the time constants expressed with the chord to free stream speed ratio c/U. The output functions as the force and moment coefficients then become for $0 = LE \leq x \leq 1 = TE$

$$c_l(x, \alpha) = \frac{\pi}{2}(1 + \sqrt{x})^2 \sin \alpha, \quad c_m(x, \alpha) = c_l(x, \alpha) \frac{5(1 - \sqrt{x})^2 + 4\sqrt{x}}{16} \tag{8.84a, b}$$

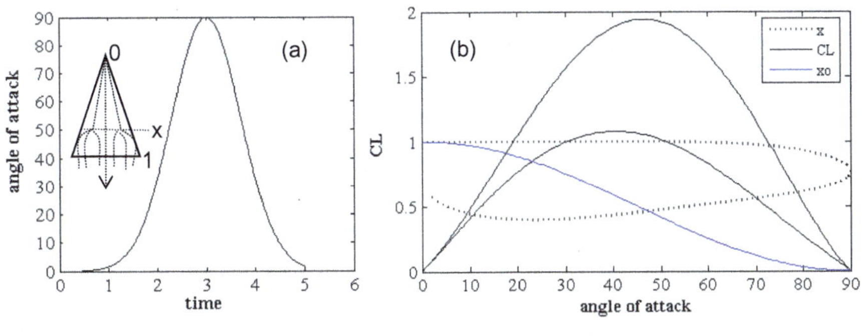

Fig. 8.57 **a** Rapid α change, **b** corresponding variation of lift coefficient for a delta wing.

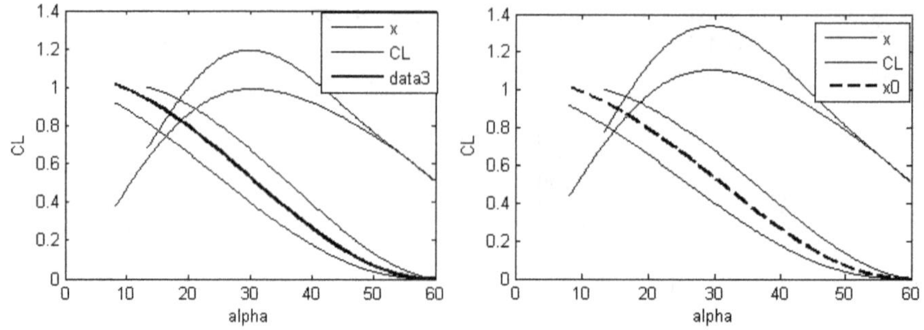

Fig. 8.58 Unsteady ground effect on the lift of a delta wing within $10° < \alpha < 60°$; left: out of ground effect, right: with ground effect $h/b_o = 1.25$ (data3: x_o)

The lift and the moment coefficient variation wrt α is shown in Fig. 8.56 for an airfoil pitching with $\alpha(t) = 30° \sin(\omega t)$ about its quarter chord point. Here: $x_o(\alpha) = \cos^2(3\alpha),\ 0 \le \alpha \le 30°$ and take $\tau_1 = 0.5c/U$ and $\tau_2 = 4.0c/U$ with $k = \omega c/U = 0.05$.

For finite wings of moderate to large aspect ratio equivalent version of (8.85a, b) can safely be used with the change made in the coefficient of (8.85a) as

$$C_L(x,\alpha) = \frac{\pi^2 A_1}{2}(1+\sqrt{x})^2 \sin\alpha, \quad C_M(x,\alpha) = c_l(x,\alpha)\frac{5(1-\sqrt{x})^2 + 4\sqrt{x}}{16} \tag{8.85a, b}$$

where, A_1 is the coefficient of the first term of the Glauert's series given with (4.22a) solution.

Further application of the state space representation on a large aspect ratio wing can be found in (Reich and Albertani 2011) in studying post stall behavior of a MAV in extreme maneuvers.

For the low aspect ratio wings the state space representation is applied with different time constants associated to a different state variable x, which is the vortex bursting point on the upper surface of the wing. New pair of equations is

Table. 8.7 Sensitivity of the lift coefficient with respect to time constants

τ_1	τ_2	Maximum lift (upstroke)	Maximum lift (downstroke)
1.4	0.6	1.80	1.0
1.5	0.5	1.75	1.0
1.6	0.4	1.65	1.0

$$\tau_1 \frac{dx}{dt} + x = x_o(\alpha - \tau_2 \dot{\alpha})$$
$$C_L(\alpha, x) = K_p \sin\alpha \cos^2\alpha + x^2 \pi \sin^2\alpha \cos\alpha \qquad (8.86a, b)$$

Solving (8.86a) numerically method together with (8.86b) for $\tau_1 = 1.5c/V$ and $\tau_2 = 0.5c/V$, we obtain the variation of lift for a delta wing with $AR = 1.5$ as shown in Fig. 8.56b which determined by the rapid angle of attack change as a bell shaped curve as prescribed in Fig. 8.57a, b.

As the final application of the state space representation, the effect of the ground on low aspect ratio wing is considered. Here, including the effect of the image vortex at the leading edge we consider the additional thrust, as described earlier, to yield an extra lift. Accordingly, for the following values of $\tau_1 = 1.5c/U \, and \, \tau_2 = 0.5c/U$ and $x_o = \cos^2(1.5 \arg)$ where arg is given in terms of AoA which is $\alpha(s) = \pi/3 * \sin(0.05s)$, and its rate of change. In Fig. 8.58-left the lift change in OGE is shown; wherein the maximum lift coefficient is 1.2 and it occurs during upstroke, which is 1.0 during down stroke. In Fig. 8.57-right, the ground effect increases the lift coefficient to its maximum values as 1.35 and 1.05 during upstroke and down stroke respectively. This indicates that the presence of the ground increases the maximum lift coefficient more than 12%.

For the dynamical systems modeled with (8.83a,b), an important issue is the sensitivity of the solution to the time constants involved. Table 8.7 is prepared for this purpose and it shows the sensitivity of the lift coefficient, in oge, with changing of time constants. It is observed that the maximum lift, which occurs during upstroke, changes with change in time constants whereas the maximum lift during downstroke almost remains the same. Here, we take AoA arbitrary as $\alpha(s) = \pi/3 * \exp(-(0.05(s-3))^2/2)$, where s is the reduced time, $s = Ut/b$.

So far only single state variable x is involved in examples given above. However, for some unsteady cases there may be more than one state variable. For example: the 'roll motion' of a low aspect ratio wing involves the left and the right side vortex burst points, x_l and x_r respectively, movement on the upper surface of the wing during the roll. This problem can be simplified to the solution of a single state variable with a new variable which is $x = x_l - x_r$ to solve the problem of roll similar to the examples given above. The solution details can be found in (Goman and Khrabrow 1994).

It seems, the state-space representation of the aerodynamics is a nice shortcut for the prediction of the unsteady aerodynamic loads especially at high angles of attack maneuverings (Reich et al. 2011). Based on the state-space, a recent work summarizes applications of various unsteady aerodynamic models, ranging from Theodorsen's to the Navier–Stokes', for studying agile flights at low Reynolds numbers suitable for MAV technology (Brunton 2012).

Summary: *Modern subjects in Aerodynamics involving high angle of attack flows past airfoils and wings are studied with extensive coverage. The strongly separated flow regions at high angle of attack flows are analyzed via the numerical solution of*

Navier–Stokes equation. In this respect, 2-D static and dynamic stall of airfoils at low and high Reynolds numbers are considered. An integro-differential method is implemented for solving the development of strong separation about an airfoil which impulsively starts from the rest. Then to study the dynamic stall of an airfoil in turbulent regime, uniform flow past a simple harmonically oscillating airfoil is considered. The critical angles of attack for the occurrence of dynamic stall are found for the same airfoil at different reduced frequencies. The hysteresis curves for the lift and the moment coefficients with respect to the angle of attack are obtained. The Mach number dependence of these curves is also given. The stall flutter problem is also briefly discussed in connection with the negative damping provided with those hysteresis curves. Polhamus theory based on the leading edge vortex lift is studied for the thin wings having high sweep angles. The leading edge suction force providing additional lift because of a vortex roll up is considered. The effect of compressibility at high angles of attack is also provided for the delta wings at supersonic speeds. The induced drag and the effect of the Mach number on this drag are demonstrated via a numerical example. In addition, three different types of wing rock are analyzed extensively. First, the wing rock range of slender wings is provided with the introduction of an enveloping curve which relates the angle of attack to the aspect ratio. The roll degree only is considered for prediction of the restoring and the damping components of the rocking of the slender wing. The effective angle of attack based on the instant roll angle is also considered for the starting of the wing rock phenomenon. In order to simulate the actual flight conditions, a three degrees of freedom problem which allows the translational planar motion of the wing in addition to the rolling is considered. The effect of the additional degrees of freedom on the wing rock is mainly to increase the angle of attack at which the rolling begins and reduce the amplitude of the roll angle. The second type of wing rock is the wing rock of non-slender wing with round leading edge. In this type of wing rock, the frequency of rocking is one order less in magnitude than the frequency of slender wing rock. The third type of wing rock occurs at high angles of attack and it is due to periodic flow separation from the frontal portion of fuselage. This type of wing rock is called forebode wing rock and is well explained with 'moving wall effect' and has even slower rocking characteristics than the non-slender wing rocking. Finally, flapping wing propulsion and lift generation is studied for the heaving-plunging and/or pitching airfoil. First, purely heaving-plunging airfoil in a free stream at zero angle of attack is considered and the propulsive force generated from the leading edge suction is calculated. The critical angle of attack for the airfoil is usually kept under the dynamic stall angle so that attached flow conditions prevail. Also given is the propulsive efficiency for this airfoil and it is compared with the efficiency calculated with N-S solutions. The average lift generated in this case is zero. As the second case study, the pitching is superimposed on heaving-plunging so that there is always a positive lift present as well as the propulsive force. The flapping wings with low Reynolds numbers require much higher angle of attacks to generate positive thrust while creating reverse Karman vortex street. In pure pitch, the pitch axis location is an important parameter to create thrust, because at high frequencies there is a possibility of creating drag instead of thrust even for potential flows. The effect of chordwise flexibility is also considered, and it is observed that the periodic camber change can produce sufficient thrust to overcome the viscous effects even with low maximum camber ratios. At last, finite wing flapping is considered via strip theory. The 3-D vortex wake picture indicates the suitability of the flapping pattern. The constant span down stroke and up stroke creates a ladder type wake vortex where there is the danger of having Wagner effect that delays the lift generation. Whereas, the down stroke with a long span and up stroke with a short span creates a concertina type wake without any cross vortex present. In concertina type wakes, as it occurs in nature, there is no Wagner effect present. Therefore, this type of spanwise flexible wing flapping is recommended for ornithopter technology or micro air vehicle design which are currently of interest.

8.10 Problems and Questions

8.1 What is the effect of flow separation at (i) swept, and (ii) unswept wings at high angle of attack.

8.2 Obtain a pseudo tri-diagonal matrix equation to solve the vorticity transport equation in a boundary layer with forward differencing in time and with appropriate differencing in space suitable for marching along the surface starting from the leading edge (Appendix 10). Write a subprogram first for the solution of pseudo tri-diagonal matrix solution.

8.3 Find the velocity component, in the direction parallel to the surface, by integrating the discrete vorticity values, obtained in Problem 8.2, in the normal direction starting from the wall.

8.4 Obtain an explicit expression for the vertical velocity component using a finite difference scheme prescribed in Appendix 10.

8.5 Derive the 2-D vorticity transport equation, and discretize this equation to obtain the vorticity field at time level $n + 1$ using SLUR (Successive Line Underrelaxation).

8.6 Obtain the relation between the stream function and the vorticity as the kinematic relation of the 2-D flow. Apply SOR (Successive Overrelaxation) technique to solve the elliptic equation.

8.7 What are the differences between the light stall and the deep stall. Comment on the differences as regards the sectional lift and the moment coefficients.

8.8 Comment on the effect of the (i) separation, and (ii) Mach number on the negative drag for a plunging airfoil.

8.9 At high angles of attack, the empirical formulae for the lift and moment coefficients for airfoils pitching at high frequencies are given in terms of maximum dynamic moment coefficient $(C_{M\ \max})_{\mathrm{DYN}}$ and the normal force coefficient ΔC_{nv} due to vortex as follows.

$$(C_{M\,\max})_{\mathrm{DNN}} = -0.75\Delta C_{nv}, \quad \Delta C_{nr} = 1.5\pi \sin^2(\alpha_{vs})_{\mathrm{eff}}$$

and $(\alpha_{vs})_{\mathrm{eff}} = \alpha_o + \Delta\theta \sin[(\omega t)_{vs} + 0.45\,k]$.
Here, $(\omega t)_{vs}$ and $\Delta\theta$ is the pitch amplitude:
if $\omega\Delta\theta\cos(\omega t)_{vs} < 0.02$ then: $(\omega \mathrm{t})_{vs} = 2\tan^{-}\left[\frac{\cos(0.995)}{1.5k + \sin(0.995) + (\alpha_0 - \alpha_s)}\right]$

$$x\left\{\sqrt{1 + \frac{(1.5k + \sin(0.995))^2 - ((\alpha_0 - \alpha_s)/\Delta\theta)^2}{\cos^2(0.995)}}\right\}$$

and if, $\omega\Delta\theta\,\cos(\omega t)_{vs} > 0.02$ then: $(\omega t)_{vs} = 0.995 + \sin^{-}(\frac{0.995 + \alpha_s - \alpha_o}{\Delta\theta})$.
Wherein: α_s is the static stall angle, α_o is the average amplitude for the angle of attack. Using these formulae, find the normal force and the moment coefficients for NACA0012 airfoil, whose separation angle is 14.5° pitching with $k = 0.25$ at 15° average angle of attack with 10° pitch amplitude.

8.9 During dynamic stall, the drag coefficient is less for pitch-up than for pitch-down, whereas the lift coefficient is larger for pitch-up than for pitch-down. Why?

8.10 The indicative of the stall flutter is the sign of the integral under the curve of (i) lift vs vertical displacement for plunging, and (ii) moment vs angle of attack for pitching. Why? In obtaining the closed integral for a complete cycle take the clockwise line integral positive, and determine a criterion for stall flutter.

8.11 Using the potential theory obtain the damping for a cycle of (i) plunge, and (ii) pitch oscillations.

8.12 The state-space representation is based on a state function x satisfying the first order ODE $\tau_1\dot{x} = x_o(\alpha - \tau_2\dot{\alpha}) - x$, where argument $(\alpha - \tau_1\dot{\alpha})$ indicates shift for the angle of attack rate $\dot{\alpha}$ of the static variation of $0 \le x_o(\alpha) \le 1$. Here, τ_1 and τ_2 are the time constants expressed with the chord to free stream speed ratio c/U. The output functions as the force and moment coefficients then become

$$c_l(x,\alpha) = \frac{\pi}{2}(1+\sqrt{x})^2 \sin\alpha, \quad c_m(x,\alpha) = c_l(x,\alpha)\frac{5(1-\sqrt{x})^2 + 4\sqrt{x}}{16}$$

Obtain the lift and the moment coefficient variation wrt α for an airfoil pitching with $\alpha(t) = 30° \sin(\omega t)$ about its quarter chord point. Assume: $x_o(\alpha) = \cos^2(3\alpha)$, $0 \le \alpha \le 30°$ and take $\tau_1 = 0.5c/U$ and $\tau_2 = 4.0c/U$ with $k = \omega c/U = 0.05$.

8.14 Consider a delta wing with sweep angle Λ. Show that the expressions 8.11 and 8.12 give the same lift line slope for the delta wing.

8.15 Using the Polhamus theory obtain the drag polar for a delta wing with sweep angle 75°.

8.16 Obtain the vortex lift line slope, given by Eq. 8.14, for a supersonic delta wing.

8.17 A delta wing has an aspect ratio of 1. (i) Plot the coefficients K_p and K_v with respect to Mach number and (ii) for $M = 2$, plot the lift coefficient wrt angle of attack. Comment on the limiting values involved in the graph.

8.18 The delta wing given in Fig. 8.59 has the supersonic lift line slope, according to Puckett and Stewart, as follows

$$\frac{dC_l}{d\alpha} = (2\,\pi\,\cot\Lambda\;/E(m'))\,H(a),$$

$$H(a) = \frac{2}{\pi}\left[\frac{a}{1+a} + \frac{1-a}{(1-a^2)^{3/2}}\cos^{-}(-a)\right]$$

What is the lift coefficient of the wing given in Example 8.1 having a leading edge with 35°, and a trailing edge with 75° sweep?

8.19 The induced drag coefficient of the delta wing given in Fig. 8.59, according to Puckett and Stewart, reads as

$$C_{Di} = \alpha\,C_L[1 - m'/\,(2\,(1-a)H(a)\,E\,(m'))]$$

Find the induced drag of the wing given in Problem 8.17.

8.20 The delta wing given in Example 8.3 is in yaw oscillating with 35° amplitude and 0.40 s period. Using the coefficients given for yawing moment and $C_2 = 0.003$ obtain: (i) the restoring moment coefficient, (ii) yaw angle change with time, (iii) damping moment coefficient, and (iv) rate of change of yaw angle with time. Plot the total yaw moment-yaw angle hysteresis curve, and indicate the feeding and the damping zones on the curve.

8.21 Derive the formulae 8.20a,b and 8.21 which give the effective angle of attack and the effective yaw angle in terms of the yaw angle φ.

8.22 For the wing given in Example 8.3, evaluate (i) the maximum normal force Coefficient, and (ii) minimum side force coefficient.

8.23 Comment on the aerodynamic mechanisms causing the wing rock of the round leading edged non-slender wings.

8.24 Comment on the causes of different types of wing rock and the differences of the period durations involved.

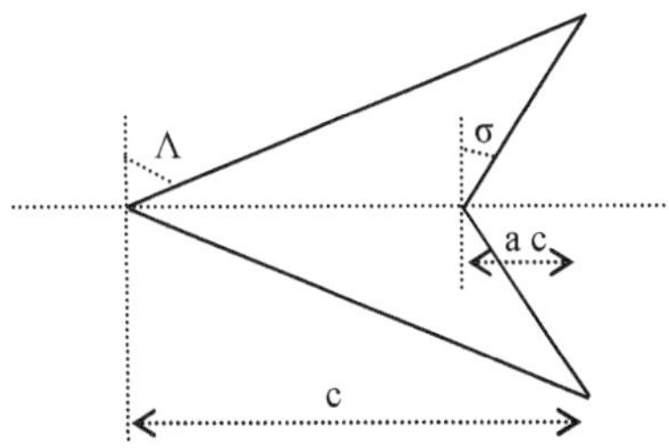

Fig. 8.59 Delta wing

8.25 Using the 'Moving wall effect', comment on the negative damping for (i) the plunging profile, (ii) the pitching profile, and (iii) the periodically rotating cylinder in a free stream.

8.26 Obtain the sectional leading edge suction force coefficient for a profile plunging with $z_a = h$ cos*ks* in a free stream at zero angle of attack.

8.27 Obtain the expression which gives the wake vortex sheet strength, Eq. 8.29, for Problem 6.25,

8.28 Show that for an heaving-plunging airfoil the aerodynamic propulsion efficiency is $\eta = \frac{F^2+G^2}{F}$.

8.29 The unsteady boundary layer solution based on the edge velocity values gives us the skin friction distribution for a body. Obtain the upper and lower boundary layer edge velocity expressions for a thin airfoil plunging with $h = -0.2\cos(1.5s)$. Using the edge velocity expression, obtain the time dependent surface vorticity values with Eqs. 8.5–8.7 (Appendix 10).

8.30 Derive Eq. 8.37 for the thrust coefficient of an airfoil in pure pitching about the point a with reduced frequency k.

8.31 A thin airfoil is plunging with $h = \overline{h}e^{iwt}$, and pitching with $\alpha = \overline{\alpha}e^{i(\omega t + \phi)}$ about a point a. Obtain the general expression for the leading edge suction force for this airfoil in two degrees of freedom problem. Here, take $P = \sqrt{2}C(k)(\dot{h} + U\alpha + b(1/2 + a)\dot{\alpha})$.

8.32 Derive the thrust efficiency formula, Eq. 8.38, for a pitching plunging airfoil. Comment on the effect of the ratio of the plunge to pitch amplitude on the efficiency.

8.33 Obtain the time variation of the lift and propulsive force coefficients and their plots for the airfoil given by Example 8.5. Assume that the profile pitches about quarter chord point.

8.34 Obtain the lift and propulsive force coefficients of an airfoil given in Example 8.6, and compare the results with Problem 8.30. Assume the profile pitches about midchord.

8.35 What are the values of the feathering parameters for the airfoils given by Examples 8.5 and 8.6?

8.36 For a chordwise flexible airfoil obtain the quasi unsteady edge velocity, Eq. 8.41, and the suction force coefficient, Eq. 8.42, formulae assuming that the parabolic camber of the airfoil, whose maximum camber is at the midchord, changes simple harmonically.

8.37 Derive the equations of continuity, Eq. 8.44, and the vorticity transport, Eq. 8.45, for skewed coordinates as shown in Fig. 8.38.

8.38 Obtain the time dependent but steady lift coefficient, Eq. 8.48, and the boundary layer edge velocity, Eq. 8.49 for a chordwise flexible parabolically cambered thin airfoil whose equation is given by Eq. 8.46 and maximum camber location along the chord is given by p.

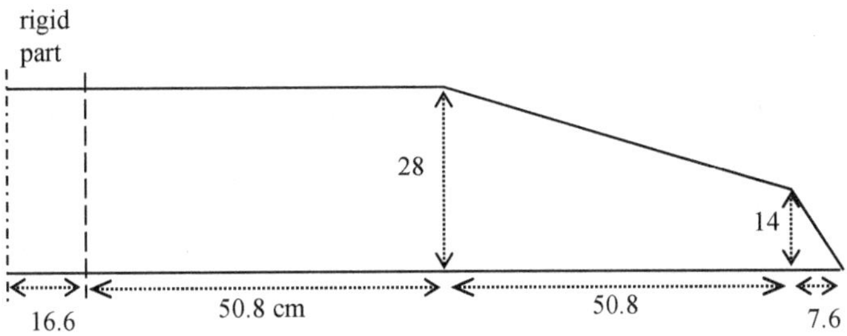

Fig. 8.60 Ornithopter wing geometry

8.39 Obtain the quasi steady lift coefficient, Eq. 8.51, and the boundary layer edge velocity, Eq. 8.52 for a chordwise flexible and parabolically cambered thin airfoil whose equation is given by Eq. 8.46, where the maximum camber location p along the chord changes by SHM.

8.40 Obtain the quasi unsteady lift coefficient using FFT and the arbitrary angle of attack change associated with the equivalent quasi steady lift given by Eq. 8.51 for the reduced frequency of 0.8 and 1.0. Comment on the differences of both lift coefficient curves.

8.41 The wing shown in Fig. 8.60 pitching and plunging with 3 Hz in a free stream of 15 m/s. Using the strip theory, obtain the total lift and the propulsive force coefficients change by time. The wing is undergoing a motion having the dihedral angle $h = y \sin\Gamma$, starting from the end of the rigid part with maximum of $\Gamma = 20°$. The phase difference between the plunge and the pitch is 90°, and the average pitch angle is 6°. (Use 10 equally spaced strips for the strip theory).

8.42 Which type of spanwise flexibility is preferred for a finite wing?

8.43 Find the numerical values for calculating the net propulsive force of a thin wing given in Example 8.7.

8.44 Write down a numerical solution algorithm for the LU decomposition solution of the pseudo penta diagonal matrix equation given by (A11.4)

8.45 Obtain Eq. 8.66 as a relation between the wake vorticity and the reduced circulation.

8.46 Write a numerical algorithm to solve Eq. 8.71 for A_n. Obtain lift coefficients for $h/2b = 2$ and AR = 6, 10 at AoA = 12.

8.47 Using (8.76) find the chord wise variation of the lift as $L'(x)$.

8.48 Using the time constants given for the solution of (8.86a, b) for the delta wing of $AR = 1.5$, obtain the lift coefficient variation for $\alpha(s) = \pi/3 * \exp(-(0.05(s-3))^2/2)$ and $x_o = \cos^2(1.5\,\arg)$

References

Abbott, I.H., Von Doenhoff, A.E.: Theory of Wing Sections. Dover Publications Inc., New York (1959)

Anderson, R.F.: Aerodynamic characteristics of six commonly used airfoils over a large range of positive and negative angles of attack. Naca TN-397 (1931)

Anderson, J.M., Streitien, K., Barrett, D.S., Triantafyllou, M.S.: Oscillating foils at high propulsive efficiency. JFM **360** (1998)

Baldwin, B.S., Lomax, H.: Thin layer approximation and algebraic model for separated flows. In: AIAA Paper 78-0257, January 1978

Birch, D., Lee, T.: Investigation of the near-field tip vortex behind an oscillatingwing. JFM **544** (2005)

Bisplinghoff, R.L., Ashley, H., Halfman, R.L.: Aeroelasticity, Dower Publications Inc., New York (1996)

Brunton, S.L.: Unsteady aerodynamic models for agile flight at low reynolds numbers. Ph.D. dissertaion, Aerospace Engineering Department, Virgina Polytechnique Institute and State University (2012)

Chaderjian, N.M.: Navier-stokes prediction of large–amplitude delta-wing roll oscillations. J.Aircr. **31**(6) (1994)

Chaderjian, N.M., Schiff, L.B.: Numerical simulation of forced and free-to-roll delta-wing motions. J. Aircr. **33**(1) (1996)

CRC: Standard Mathematical Tables, 24th edn. CRC Press (1974)

Eastman, N.T.: Test of six symmetrical airfoil in the variable density wind tunnel. NACA TN-385 (1931)

El-Refaee, M.M.: A numerical study of laminar unsteady compressible viscous flow over airfoils. Ph.D. thesis, Georgia Institute of Technology (1981)

Ericsson, L.E., Reding, P.: Unsteady airfoil stall, review and extension. J. Aircr. (1971)

Ericsson, L.E., Reding, P.: Dynamic stall at high frequency and large amplitude. J. Aircr. (1980)

Ericsson, L.E.: The fluid mechanics of slender wing rock. J. Aircr. (1984)

Ericsson, L.E.: Moving wall effects in unsteady flow. J. Aircr. **25**(11) (1988)

Ericsson, L.E: Wing rock of nonslender delta wings. J. Aircr. (2001)

Ericsson, L.E.: Effect of leading-edge cross-sectional shape on nonslender wing rock. J. Aircr. **40** (2), 407–410 (2003)

Ericsson, L.E., Mendenhall, M.R., Perkins, S.C.: Review of forebody-induced-wing-rock. J. Aircr. **33**(2) (1996)

Freymuth, P.: Propulsive vortical signature of plunging and pitching airfoils. AIAA J. **26**(7) (1988)

Garrick,I.E.: Propulsion of a flapping and oscillating airfoil. NACA Report 567 (1936)

Goman, M., Khrabrov, A.: State-space representation of aerodynamic characteristics of an aircraft at high angles of attack. J. Aircr. **31**(5) (1994)

Gulcat, U.: Separate numerical treatment of attached and detached flow regions in general viscous flows. Ph.D. dissertation, Georgia Institute of Technology, School of Aerospace Engineering, Atlanta (1981)

Gulcat, U.: Propulsive force of a flexible flapping thin airfoil. J. Aircr. **46**(2) (2009a)

Gulcat, U.: Effect of maximum camber location on the aerodynamic performance of a thin airfoil. In: 5th Ankara International Aerospace Conference Ankara, Turkey, 17–19 August 2009 (2009b). http//aiac.ac.metu.edu.tr. ISBN: 978-975-1656-4-1

Gulcat, U.: Minimization of induced drag for the low aspect ratio wings in flapping. In: CFD and Optimization 2011, An ECCOMAS Thematic Conference, 23–25 May 2011 (2011). ISBN: 978-605-61427-4-1

Gulcat, U.: The ground effect on a plunging airfoil at a constant angle of attack. In: 8th Ankara International Aerospace Conference, METU, Ankara, Turkey, 10–12 September 2015 (2015a). https://aiac.ae.metu.edu.tr/paper.php?No=AIAC-2015-101

Gulcat, U.: The ground effect on thin wings with various aspect ratios. In: 8th Ankara International Aerospace Conference, METU, Ankara, Turkey, 10–12 September 2015 (2015b). https://aiac.ae.metu.edu.tr/paper.php?No=AIAC-2015-051

Halfman, R.L., Johnson, H.C., Haley, S.M.: Evaluation of high angle of attack aerodynamic derivative data and stall-flutter prediction techniques. NACA TN 2533 (1951)

Heatcote, S., Gursul, I.: Flexible flapping airfoil propulsion at low reynolds numbers. AIAA J. **45** (5) (2007)

Hoeijmakers, H.W.M.: Vortex wakes in aerodynamics. AGARD CP-584 (1996)

Isogai, K., Shinmoto, Y., Watanabe, Y.: Effects of dynamic stall on propulsive efficiency and thrust of flapping airfoils. AIAA J. **37**(10) (1999)

Katz, J., Plotkin, A.: Low-Speed Aerodynamics. Cambridge University Press, New York (2010)

Kaya, M., Tuncer, I.H.: Nonsinusoidal path optimization of a flapping airfoil. AIAA J. **45**(8) (2007)

Kinsey, T., Dumas, G.: Parametric study of an oscillating airfoil in a power-extraction regime. AIAA J. **46**(6) (2008)

Konstadinopoulos, P., Mook, D.T., Nayfeh, A.H.: Subsonic wing rock of slender delta wings. J. Aircr. (1985)

Korn, G.A., Korn, T.M.: Mathematical Handbook for Scientists and Engineers, 2nd edn. McGraw-Hill, New York (1968)

Küchemann, D.: Aerodynamic Design of Aircraft. Pergamon Press, Oxford (1978)

Levin, D., Katz, J.: Dynamic load measurements with delta wings undergoing self-induced roll oscillations. J. Aircr. **21**, 30–36 (1984)

Lighthill, J.: The inaugural goldstein memorial lecture – some challenging new applications for basic mathematical methods in the mechanics of fluids that were originally perused with aeronautical aims. Aeronaut. J. (1990)

Litva, J.: Unsteady aerodynamic and stall effects on helicopter rotor blade airfoil sections. J.Aircr. (1969)

Mathematica-7: Wolfrom (2008)

McCroskey, W.J.: The phenomenon of dynamic stall. NASA TM-81264, March 1981 (1981)

McCroskey, W.J.: Unsteady airfoils. Ann. Rev. Fluid Mech. (1982)

Mehta, U.B.: Starting vortex, separation bubbles and stall- a numerical study of laminar unsteady flow around an airfoil. Ph.D. thesis, Illinois Institute of Technology (1972)

Mehta, U.B.: Dynamic stall of an oscillating airfoil. In: Paper 23, Unsteady Aerodynamics, AGARD CP-227, September 1977 (1977)

Moryossef, Y., Levy, Y.: Effect of oscillations on airfoils in close proximity to the ground. AIAA J. **42**(9), 1755–1764 (2004)

Mueller, T.J., DeLaurier, J.D.: Aerodynamics of small vehicles. Ann. Rev. Fluid Mech. (2003)

Murman, E.M., Rizzi, A.: Application of Euler equations to sharp edged wings with leading edge vortices. NATO, AGARD, CP-412 (1986)

Platzer, M.F., Jones, K.D., Young, J., Lai, J.S.: Flapping-wing aerodynamics: progress and challenges. AIAA J. **46**(9) (2008)

Polhamus, E.C.: Predictions of vortex-lift characteristics by a leading-edge suction analogy. J. Aircr. **8**, 193–199 (1971)

Polhamus, E.C.: Applying slender wing benefits to aircraft. J. Aircr. **21**, 545–559 (1984)

Prouty, R.W.: Helicopter Performance Stability and Control. Krieger Publishing Company, Malabar (1995)

Puckett, A.E., Stewart, H.J.: Aerodynamic performance of delta wings at supersonic speeds. J. Aeronaut. Sci. **14** (1947)

Qu, Q., Lu, Z., Guo, H., Liu, P., Agarwal, R.K.: Numerical investigation of the aerodynamics of a delta wing, in ground effect. J. Aircr. **52**(1), 329–339 (2015)

Rainey, A.G.: Measurement of aerodynamic forces for various mean angle of attack on an airfoil oscillating in pitch and on two finite-span wings oscillating in bending with emphasis on damping. NACA Report 1305 (1957)

Reich, G.W., Eastep, F.E., Altman, A., Albertani, R.: Transient poststall aerodynamic modeling for extreme maneuvers in micro air vehicles. J. Aircr. **48**(2), 403–411 (2011)

Saad, A.A., Liebst, B.S.: Computational simulation of wing rock in three-degrees-of-freedom problem for a generic fighter with chine-shaped forebody. Aeronaut. J. (2003)

Spentzos, A., Barakos, G.N., Badcock, K.J., Richards, B.E., Cotton, F.N., Galbraith, R.A.McD., Berton, E., Favier, D.: Computational fluid dynamics study of three-dimensional stall of various planform shapes. J. Aircr. **44**(4) (2007)

Tang, D.M., Dowell, E.H.: Experimental investigation of three-dimensional dynamic stall model oscillating in pitch. J. Aircr. **32**(5) (1995)

Traub, L.W.: Experimental and analytic investigation of ground effect. J. Aircr. **52**(1), 235–243 (2015)

Tuncer, I.H., Kaya, M.: Optimization of flapping airfoils for maximum thrust and propulsive efficiency. AIAA J. **43**(11) (2005)

Tuncer, I.H., Platzer, M.F.: Thrust generation due to airfoil flapping.AIAA J. **34**(2) (1996)

Tuncer, I.H., Platzer, M.F.: Computational study of flapping airfoil aerodynamics. J. Aircr. **37**(3) (2000)

Tuncer, I.H., Wu, J.C., Wang, C.M.: Theoretical and numerical studiesofoscillating airfoils. AIAA J. (1990)

Van Dyke, M.D.: Second order subsonic airfoil theory including edge effects. NACA TR-1274 (1956)

von Karman, Th., Burgers, J.M.: Durand, W.F. (ed.) General Aerodynamics Theory-Perfect Fluids, Aerodynamic Theory, vol. II. Julius Springer, Berlin (1935)

Wang, J.Z.: Vortex shedding and frequency selection in flapping flight. JFM **410**, 323–341 (2000)

Wentz, W.H. Jr., Kohlman, D.L.: Vortex breakdown on slender sharp-edged wings. J. Aircr. **8**, 156–161 (1971)

Werle, H.: Hydrodynamic flow visualization. Ann. Rev. Fluid Mech. (1973)

Wu, J.C., Gulcat, U.: Separate treatment of attached and detached flow regions in general viscous flows. AIAA J. **19**(1) (1981)

Wu, J.C., Wang, C.M., Gulcat, U.: Zonal solution of unsteady viscous flows. In: AIAA Paper 84-1637, June 1984 (1984)

Zhu, Q.: Numerical simulation of flapping foil with chordwise and spanwise flexibility. AIAA J. **45**(10) (2007)

Chapter 9
Unsteady Applications: *Thrust Optimization, Stability and Trim*

The unsteady aerodynamics related material developed in the previous chapter for the practical applications is going to be utilized in this chapter for the thrust optimization and the flight stability of bodies having flapping wings. The flapping wing technology demands the optimum values of the amplitudes of the pitch and plunge and the phase difference in between them so that the maximum thrust or the most efficient performance can be achieved during the flapping period.

Flight dynamics require the fast prediction of the unsteady aerodynamic loads as forcing functions in the equations of motion. For this purpose the state-space representation of aerodynamic loadings are implemented with taking into account the circulatory and the non-circulatory terms generated with the pitch and the pitch rate. Letting the state variables interact with the equation of the motion, we obtain the coupled version for the stability derivatives matrix of the perturbation equations. The time averaged stability derivative matrix is readily applied for the prediction of the longitudinal stability and the trimming of a flapping body in hover. For the full unsteady treatment, the symmetric flapping cannot be trimmed whereas the trim is possible with the anti-symmetric wing flapping. The details of the formulations and applications are going to be provided in the following sub-sections.

9.1 Thrust Optimization

The objective of this section is to obtain and implement a formulation for the maximum thrust for a simple harmonically pitching-plunging and morphing airfoil in finite and zero freestreams. The method is also extended to the path optimization for non-sinusoidal pitch-plunge motions, where doubling the thrust generation is possible compared to the results previously obtained by sinusoidal path optimization. The formulations are based on ideal flow therefore some constraints are

Ü. Gülçat, *Fundamentals of Modern Unsteady Aerodynamics*,
https://doi.org/10.1007/978-3-030-60777-7_9

applied on the amplitude of the motion to satisfy the ideal flow limitations. This way, the problem is solved as an eigenvalue problem which gives the maximum thrust as its maximum eigenvalue and the amplitude of the motion as the corresponding eigenvector. In addition, emposing an efficiency constraint results in with an iterative solution to achieve the maximum thrust and corresponding motion at that efficiency or iterating for the maximum efficiency as well.

Optimization of flapping airfoils for maximum thrust and efficiency, based on Navier Stokes solutions, first appeared in Tuncer and Kaya (2005). Furthermore, the nonsinusoidal path optimization of a periodically flapping airfoil was studied by implementing the nonuniform rational B-splines (Kaya and Tuncer 2007). Recently, the optimum thrust for a pitching-plunging airfoil at zero free stream is studied theoretically and experimentally in Bulut et al. (2016), wherein the theoretical part is based on the Garrick's approach (Garrick 1936) and its extended version in Walker (2012) and Walker and Patil (2014). Furthermore, Gulcat applied a similar optimization technique with both linear and nonlinear modeling of the leading edge vortex effect on the optimum thrust in Gulcat (2017a, 2019a) respectively.

The leading edge suction force and the unsteady lift force contribute to the propulsive force generated by the flapping of a rigid airfoil is given as follows (Garrick 1936):

$$S = -(\pi \rho P^2 + \alpha L) \tag{9.1}$$

where, $P = \lim_{x \to -1} (\gamma_a \sqrt{x^* + 1})/2$ is the leading edge suction velocity, α is the angle of attack and the lift amplitude from (3.30-a)

$$\bar{L} = \rho U^2 b \left[-2C(k) \int_{-1}^{1} \sqrt{\frac{1+\xi^*}{1-\xi^*}} \frac{\bar{w}(\xi^*)\, d\xi^*}{U} - 2ik \int_{-1}^{1} \sqrt{1-\xi^{*2}} \frac{\bar{w}(\xi^*)\, d\xi^*}{U} \right] \tag{9.2}$$

wherein $\bar{w} = i\omega \bar{z}_a + U\, \partial \bar{z}_a / \partial x$ is the amplitude of the downwash w.obtained from the airfoil motion $z_a = z_a(x, t)$, and C(k) = F(k) + iG(k) is the Theodorsen function.

The amplitude of the vortex sheet strength, obtained from (3.16) for simple harmonic motion is given in terms of the circulation $\bar{\Gamma}_a$, as

$$\bar{\gamma}_a(x^*) = \frac{2}{\pi} \sqrt{\frac{1-x^*}{1+x^*}} \int_{-1}^{1} \sqrt{\frac{1+\xi^*}{1-\xi^*}} \frac{\bar{w}(\xi^*)\, d\xi^*}{x^* - \xi^*} + \sqrt{\frac{1-x^*}{1+x^*}} \frac{ik\bar{\Gamma}_a}{b\pi} e^{ik} \int_{1}^{\infty} \sqrt{\frac{\lambda+1}{\lambda-1}} \frac{e^{-ik\lambda}}{x^* - \lambda} d\lambda \tag{9.3}$$

where: $\frac{ik\bar{\Gamma}_a}{b}e^{ik} = \frac{4/\pi}{H_1^{(2)}(k)+H_0^{(2)}(k)} \int_{-1}^{1} \sqrt{\frac{1+\xi^*}{1-\xi^*}}\bar{w}(\xi^*)\,d\xi^*$

The improper integral in the second term of the right hand side can be expanded into the series in terms of $\frac{x}{\lambda}$ is given by (A12.8) of appendix as follows

$$I(k,x) = \int_1^\infty \sqrt{\frac{\lambda+1}{\lambda-1}}\frac{e^{-ik\lambda}}{x-\lambda}d\lambda = -\int_1^\infty \frac{\lambda+1}{\sqrt{\lambda^2-1}}\frac{1}{\lambda}\left(1+\frac{x}{\lambda}+\frac{x^2}{\lambda^2}+\frac{x^3}{\lambda^3}+\cdots\right)e^{-ik\lambda}d\lambda$$

$$= -\int_1^\infty \frac{1}{\sqrt{\lambda^2-1}}\left[1+\frac{1}{\lambda}(x+1)+\frac{1}{\lambda^2}(x^2+x)+\frac{1}{\lambda^3}(x^3+x^2)+\cdots\right]e^{-ik\lambda}d\lambda,$$

$$= i\frac{\pi}{2}H_0^{(2)}(k) - \int_1^\infty \frac{1}{\sqrt{\lambda^2-1}}\left[\frac{1}{\lambda}(x+1)+\frac{1}{\lambda^2}(x^2+x)+\frac{1}{\lambda^3}(x^3+x^2)+\cdots\right]e^{-ik\lambda}d\lambda$$

As x approaches -1, the second term vanishes to leave us with the definition of the leading edge suction force amplitude as

$$\bar{P}_i = \frac{\sqrt{2}}{\pi}\int_{-1}^{1}\sqrt{\frac{1+\xi^*}{1-\xi^*}}\frac{\bar{w}_i(\xi^*)\,d\xi^*}{-1-\xi^*} + \frac{\sqrt{2}}{\pi}[1-C(k)]\int_{-1}^{1}\sqrt{\frac{1+\xi^*}{1-\xi^*}}\bar{w}_i(\xi^*)\,d\xi \quad (9.4)$$

We can now obtain formulae for the lift and the leading edge suction velocity amplitudes for basic type of motions of a thin airfoil implementing (9.2 and 4) as follows:

(i) For a heaving-plunging thin airfoil in SHM $z_a(x,t) = -he^{i\omega t}$ *with* $\bar{w} = -hi\omega$, Hence,

$$\bar{P}_0 = \sqrt{2}C(k)ik\bar{h}U, \ \ \bar{h} = h/b$$

and $\bar{L}_0 = \rho U^2 b[2C(k)ik - k^2]\pi\,\bar{h}$

(ii) For pitching plunging airfoil with pitch axis at *ab*:

$$z_a(x,t) = -\bar{\alpha}\,e^{i\omega t}(x-ab) \text{ with } \bar{w}(x) = -\bar{\alpha}\,i\omega(x-ab) - U\bar{\alpha}$$

The leading edge suction velocity then becames

$$\bar{P}_1 = \sqrt{2}[C(k) + (C(k)-1)ik/2 + (1-C(k))a - a\,ik]U\bar{\alpha}$$

and the lift $\bar{L}_1 = \rho U^2 b[2C(k)(ik/2+1) + ik - 2C(k)ika - ika + k^2 a]\,\pi\,\bar{\alpha}$

(iii) For oscillating flap

$$z_a(x,t) = -\bar{\beta}\,e^{i\omega t}(x-eb) \text{ with } \bar{w}(x) = -\bar{\beta}\,i\omega(x-eb), \text{ for } x \geq eb$$

The leading edge suction velocity and the lift due to flap oscillations then read as Garrick (1936)

$$\bar{P}_{1\beta} = \sqrt{2}\left(-2U\sqrt{1-e^2}\beta + T_4 b\dot{\beta}\right)\Big/2\pi$$

and,

$$\bar{L}_{1\beta} = \rho U b^2\left(UT_4\dot{\beta} + bT_1\ddot{\beta}\right) + 2\pi\rho UbC(k)\left(UT_{10}\beta + bT_{11}\dot{\beta}\Big/2\right)\Big/\pi$$

where,

$$T_1 = -(2+e^2)\sqrt{1-e^2}/2 + e\ \cos^{-} e, \quad T_4 = e\sqrt{1-e^2} - \cos^{-} e$$

$$T_{10} = \sqrt{1-e^2} + \cos^{-} e, \quad T_{11} = (2+e)\sqrt{1-e^2} + (1-2e)\cos^{-} e$$

(iv) For parabolic camber oscillations:

$$z_a(x,t) = -qx^2/b^2 e^{i\omega t} \ with\ \bar{w} = -(i\omega x^2/b^2 + 2x/b^2 U)\bar{q}$$

The leading edge suction force reads as

$$\bar{P}_2 = \sqrt{2}[C(k)ik/2 + C(k) - 1]\,Uq/b$$

and the lift contribution reads as

$$(\bar{\alpha}\bar{L}_2) = \int_{-b}^{b} (p_L - p_U)\frac{dz_A}{dx}dx$$

Using the expression for the lifting pressure the above integral reads as

$$\frac{\bar{\alpha}\bar{L}_2}{\rho U^2 b} = -3\pi\, ik\, q^{*2} + 4\pi\,(1-C)(-1-ik/2)q^{*2} + \pi k^2 q^{*2}$$

Now we can write down a general expression for the thrust generated by a having-plunging, pitching and morphing thin airfoil. That is

$$T = T_{LES} + \int_{-b}^{b} (p_L - p_U)\frac{dz_a}{dx}dx$$

$$T_{LES} = \rho\pi U^2 b\,(\sum_{i=0}^{2} P_i)\,(\sum_{j=0}^{2} P_j)$$

and

$$\sum_{i=0}^{2}\alpha\bar{L}_i = \sum_{i=0}^{2}\int_{-b}^{b}(\bar{p}_L - \bar{p}_{U_i})\frac{dz_{ai}}{dx}dx = -\bar{\alpha}\bar{L}_0 - \bar{\alpha}\bar{L}_1 - \bar{\alpha}\bar{L}_2$$

Now, we can form a table showing the sectional suction and lift coefficients as follows:

Let us form a vector **Q** with elements $\vec{Q} = \left|\bar{h}\,\bar{\alpha}\,\bar{q}\bar{\beta}\right|$, and consider h = h_r as real and $\bar{\alpha} = \alpha_r + i\alpha_i$ and $\bar{q} = q_r + iq_i$ and $\bar{\beta} = \beta_r + i\beta_i$ are complex to indicate that these are out of phase with h. Hence we have the complex quantities represented as a real vector $\bar{Q} = \begin{Bmatrix} h_r \\ \alpha_r \\ \alpha_i \\ q_r \\ q_i \end{Bmatrix}$ or $\bar{Q} = \begin{Bmatrix} h_r \\ \alpha_r \\ \alpha_i \\ \beta_r \\ \beta_i \end{Bmatrix}$

This enables us to write a quadratic form for the average thrust as follows

$$T_{avg} = \frac{1}{2}\text{Re}\left[\pi\rho\bar{P}conj(\bar{P}) + \sum_{i=0}^{4}\bar{\alpha}\,\bar{L}_i\right] \tag{9.5}$$

Hence, only the real part of the equation above contributes to the average thrust. The imaginary part contains product of sine and cosine which integrates to zero over one period to give

$$T_{avg} = \frac{1}{2}\pi\rho U^2 b\{Q\}^T\left[H_{avg}\right]\{Q\} \tag{9.6}$$

Wherein, H_{avg} is a 5 × 5 matrix which constitutes the quadratic form for the thrust function which has all real entries as functions of the reduced frequency k and the real and imaginary parts of the Theodorsen function, i.e. F and G.

Optimization: The average thrust function given with (9.6) can be maximized if its gradient is set equal to zero, i.e.

$$\vec{\nabla}T_{avg} = \vec{0} \quad \rightarrow \quad \pi\rho U^2 b\left[H_{avg}\right]\{Q\}_{\max T} = \vec{0}$$

where, $\{Q\}_{\max T}$ is the vector which gives the maximum value of the average thrust. The solution of this equation is trivial since the H_{avg} is not singular. On the other hand, the derivation of the equation is based on the linear aerodynamic theory, therefore, any increase in the elements of Q causes increase in thrust. Hence, there is not any maximum as the problem posed. However, we can find a maximum thrust if we impose restrictions on the motion as constraint. There are several ways to impose the constraint for the optimization; one of them is the magnitude constraint.

Magnitude constraint: Both aerodynamic and the mechanical restrictions can justify the magnitude constraint in the following form

$$\sum_{i=1}^{5} Q_i^2 \leq 1 \text{ or } \{Q\}^T\{Q\} \leq 1 \tag{9.7}$$

With this restriction, we stop the increase in maximum thrust with setting a limit to the magnitude of the motion. In the design space this constraint will put a peak to the maximum thrust. Hence, we can write the constraint as an equality, i.e.

$$f(Q) = \{Q\}^T\{Q\} - 1$$

This enables us to write a Lagrangian composed of the average thrust function and the constraint without altering the value of the average thrust to be maximized as follows:

$$L(Q, \lambda) = \frac{1}{2}\pi\rho U^2 b\{Q\}^T \left[H_{avg}\right]\{Q\} - \lambda f(Q) \tag{9.8a}$$

Wherein, λ is the Lagrange multiplier for the constraint. Now, we can set the gradient of the Lagrangian (9.8a) to zero to obtain

$$\pi\rho U^2 b \left[H_{avg}\right]\{Q\} - \lambda\{Q\} = 0 \tag{9.8b}$$

$$\{Q\}^T\{Q\} - 1 = 0$$

The first equation is nothing but an eigenvalue problem. It has n principle directions and corresponding n eigenvalues in general. The largest eigenvalue is the maximum average thrust and the corresponding eigenvector gives the associated motion and the deformation vector. With the aid of the second equation we can prove that the Lagrangian is the maximum. If we rewrite the Lagrangian with $\{Q\}^T_{\max}\{Q\}_{\max} - 1 = 0$ we then obtain

$$L(Q_{\max}, \lambda_{\max}) = \frac{1}{2}\pi\rho U^2 b\{Q\}^T_{\max}\lambda_{\max}\{Q\}_{\max} = \frac{1}{2}\pi\rho U^2 b\lambda_{\max}$$

Example Let us consider a thin airfoil undergoing plunge-pitch and morfhing simultaneously at k = 0.5. Let us find the maximum value of the aerodynamic thrust and the corresponding motion and the deformation.

Using above the formulation given, with pitch point changing, we find the values given in the following Table 9.2, where $T_{\max} = L(Q_{\max}, \lambda_{\max})/(\rho U^2 b)$

As seen from Table 9.2, the minimum of the maximum thrust occurs for the pitch point located at midchord, $a = 0$. The pitch point location towards the leading edge increases the value of the thrust within the limits of the linear theory. The choice of pitch point location away from the midchord towards the trailing edge increases the maximum thrust but the motion associated with exceeds the limits of linear theory.

Efficiency constraint: The thrust optimization with magnitude constraint does not necessarily result in most efficient solution. Therefore, we need to consider the optimization aiming a desired efficiency. For this purpose, the efficiency can be forced to a certain value while the thrust is made maximum. Hence, generating a Pareto front for thrust and efficiency is the proper way of optimizing the thrust with efficiency constraint. (On the other hand, the problem of efficiency constraint without a magnitude constraint is not a well posed problem and is not considered.) The thrust efficiency is defined as the ratio of the work generated by thrust to the work required for the motion of the airfoil. Thus it is

$$\eta = \frac{W_T}{W_M} = \frac{T_{avg} U}{W_M} \tag{9.9}$$

The work required for the motion of the airfoil for one period reads as $W_M = \frac{1}{2}\rho U^2 \int\limits_0^T \int\limits_{-b}^{b} C_{pa}(x,t) \frac{\partial z_a(x,t)}{\partial t} dx\, dt$, where $\frac{\partial z_a}{\partial t} = i\omega\, \bar{z}_a(x) e^{i\omega t}$ *and* $C_{pa} = \bar{C}_{pa} e^{i\omega t}$

With proper non-dimensionalization of the integral the average of this work becomes

$$W_{Mav} = \rho U^3 b \frac{1}{2} \mathrm{Re}\,(\int\limits_{-1}^{1} \bar{C}_{pa} ik\, \bar{z}_a^* dx^*)$$

Substituting the motion and the deformation parameters of the airfoil, this result can be written in matrix form as follows

$$W_{Mav} = \frac{1}{2} \pi \rho U^3 b \{Q\}^T [H_{avg}] \{Q\}$$

The efficiency constraint can be written in terms of chosen reference efficiency as follows

$$\eta(\{Q\}, k) - \eta_{ref} = 0$$

Then we have

$$\eta(Q) = \{Q\}^T \langle [H_{avg}] - \eta_{ref} [H_M] \rangle \{Q\} = 0 \tag{9.10}$$

This enables us to write the Lagrangian

$$L(Q, \lambda, \lambda_\eta) = \frac{1}{2}\pi\rho U^2 b\{Q\}^T[H_{avg}]\{Q\} - \lambda f(Q) - \lambda_\eta \eta(Q) \tag{9.11}$$

Here, λ_η is the Lagrange multiplier for the efficiency constraint. Now, we can find the maximum of it by setting the gradient of the Lagrangian (9.11) to zero vector to get

$$\begin{gathered}\pi\rho U^2 b[H_{avg}]\{Q\} - \lambda\{Q\} - \lambda_\eta\langle[H_{avg}] - \eta_{ref}[H_M]\rangle\{Q\} = 0 \\ \{Q\}^T\{Q\} - 1 = 0 \\ \{Q\}^T\langle[H_{avg}] - \eta_{ref}[H_M]\rangle\{Q\} = 0\end{gathered} \tag{9.12a, b, c}$$

This is no longer an eigenvalue problem. It involves N unknowns for the motion and the deformation amplitudes and also 2 Lagrange multipliers to result into solve N + 2 nonlinear equations. For the solution we use an iterative method for systems. Sole efficiency constraint problem without a magnitude constraint looks like a generalized eigenvalue problem, however, it turns out to be an ill-posed problem, and therefore it is not looked into. For any given efficiency value, we can solve the double eigenvalue problem with an iterative procedure as described below. (On the other hand, the solution for the maximum thrust case provides a reasonably good estimate to start the iterations for the numerical method to create a specific front for the optimization.)

Let $f(x_i) = 0$ be the non-linear system of equations to be solved iteratively. Expanding f_i about x_i and setting it to zero, in indicial notation, gives

$$f_{i+1} = f_i + \frac{\partial f_i}{\partial x_j}\delta x_j \cong 0 \quad \rightarrow \quad \frac{\partial f_i}{\partial x_j}\delta x_j = -f_i$$

For each iteration, the following matrix equation is solved for δx_j

$$\sum_j c_{ij}\delta x_j = -f_i \;\; with \;\; c_{ij} = \frac{\partial f_i}{\partial x_j}$$

Hence, $x_i^{new} = x_i^{old} + \delta x_i$ until convergence!

The elements of the matrix c_{ij} and the load vector f_i are given in the Appendix 14.

As an application to the efficiency constraint the pitching-plunging airfoil with k = 0.5 having various efficiencies the following results from (9.12-a,b,c) are obtained and provided at Table 9.3.

The convergence is quite fast. Above results are obtained with 3 decimal place accuracy in 4 iterations and with 6th decimal place accuracy in 10 iterations. If one compares the results for the efficiencies for 0.30 and 0.60, one finds that the efficiency is doubled whereas the maximum thrust is reduced only 10%. The last row of Table 9.3 shows the results of the case for which only the magnitude constraint is considered, wherein $T_{max} = 0.2158$ is found as the highest eigenvalue with the lowest efficiency.

Efficiency optimization: Using multi-objective optimization, it is possible to optimize the motion to yield the maximum efficiency. The multi-objective optimization requires an iterative procedure, because of nonlinear dependence of the efficiency on the motion. based on the creation of a Pareto front while maximizing the average thrust. The previously defined efficiency in terms of the motion Q reads

$$\eta = \frac{W_T}{W_M} = \frac{\{Q\}^T [H_{avg}] \{Q\}}{\{Q\}^T [H_M] \{Q\}}$$

Now, we want to maximize the efficiency as well as the average thrust function with magnitude constraint. This requires maximization of (9.8a) which is $L_1(Q, \lambda)$ and $L_2(Q, \eta)$ which is the efficiency. The maximization of the efficiency requires $\partial L_2(Q, \eta)/\partial Q = 0$. Hence, we get

$$\frac{W_M [H_{avg}] \{Q\} - W_T [H_M] \{Q\}}{W_M^2} = 0$$

Combining maximization of L_1 and L_2 yields

$$\pi \rho U^2 b \langle [H_{avg}] + W_M(\{Q\}) [H_{avg}] - W_T(\{Q\}) [H_M] \rangle \{Q\} - \lambda \{Q\} = 0$$

as a new eigenvalue problem which is nonlinear due to presence of $W_M(\{Q\})$ and $W_T(\{Q\})$. These new set of equations converge to 4 decimal places in 4 iterations to give the last row of Table 9.3. For this case, efficiency is the highest but the average thrust is the lowest as given in Table 9.3.

Zero freestream: The thrust generated at zero freestream represents the case which corresponds to the motion starting from the rest. The reduced frequency at zero free stream becomes infinity, therefore, the Theodorsen function $C(\infty) = 0.5$. This gives us the Table 9.4 as follows

Figure 9.1. shows the various positions of the deforming thin airfoil while pitching and plunging with maximum thrust.

Comparison with experimental results: Optimum thrust values obtained above for a pitching plunging flat plate are compared with the experimental results obtained under similar conditions given in Bulut et al. (2016). The comparison is made for three different cases and the agreements are quite satisfactory as shown in Table 9.5.

In Table 9.5, all cases are the optimum cases where we, naturally, have theoretical values higher than the experimental values, the difference between the two increases as the frequency of the motion increases.

Non-sinusoidal path optimization: We can have an airfoil which may not necessarily flap simple harmonically to produce high aerodynamic thrust. For that

Fig. 9.1 Motion and deformation of thin airfoil for zero free stream at a = −0.5 and $\omega = 1$

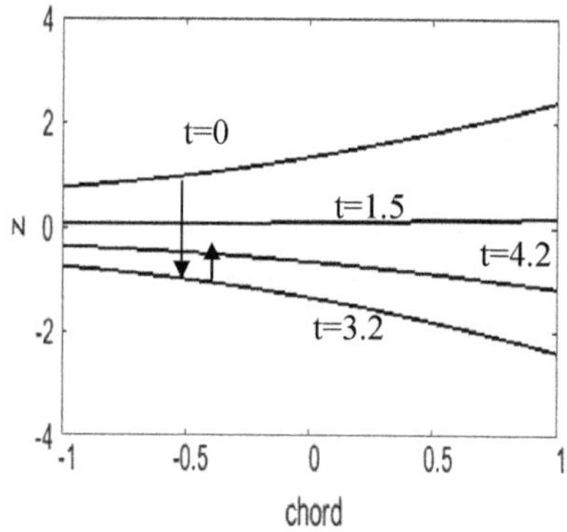

purpose we can employ non-uniform rational B splines (NURBS) like described and implemented in Kaya and Tuncer (2007). Let in two dimensions S(u) = [x(u), y (u)] be the smooth curve described with the following parametric representation

$$x(u) = \frac{2P_1u(1-u)^2 + 2P_2u^2(1-u)}{(1-u)^3 + u(1-u)^2 + u^2(1-u) + u^3}$$
$$y(u) = \frac{-(1-u)^3 - u(1-u)^2 + u^2(1-u) + u^3}{(1-u)^3 + u(1-u)^2 + u^2(1-u) + u^3}$$

The non-sinusoidal periodic function is then defined as

$$f[u(\omega t)] = y(u) \text{ where } \tan(\omega t) = -\frac{x(u)}{y(u) - P_0}$$

For a known ωt the above equation is solved for u to determine y(u) = f(ωt). Here, x and y determine a closed curve where P_0 defines the center and P_1 and P_2 indicates the flatness of the NURBS curve. Once the NURBS curve are found, the pitch-plunge motion is described as

$$h = -\bar{h}f_h(\omega t), \quad \alpha = -\bar{\alpha}f_\alpha(\omega t + \varphi) \tag{9.13}$$

where, ω is the angular frequency, $k = \omega b/U$ is the reduced frequency and φ is the phase difference with pitch and plunge.

The thrust optimized results are here, borrowed from Kaya and Tuncer (2007), for k = 0.5 and h = 0.5 reads as $\bar{\alpha} = 21.2^0$, $\varphi = 41.4^0$, $P_{0h} = -0.9$, $P_{1h} = P_{2h} = 3.5$, $P_{0\alpha} = -0.8$, $P_{1\alpha} = P_{2\alpha} = 0.2$.

Shown in Fig. 9.2 is the periodic path of the pitch and plunge obtained with above parameters.

Based on the Wagner function [BAH], now, we can evaluate the thrust induced by the nonsinusoidal motion as follows. The contribution of the lift to the thrust can be found directly by means of Wagner function. The leading edge suction's contribution, however, requires extra considerations. The leading edge suction velocity

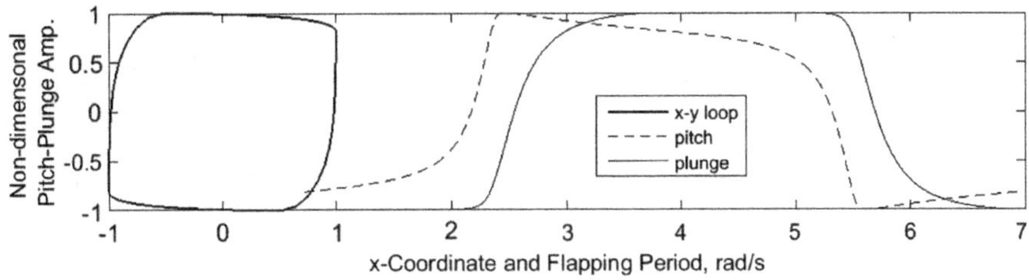

Fig. 9.2 Normalized x-y loop for plunge (left) and pitch-plunge amplitude variation for a period

for a pitching-plunging motion of an airfoil with the aid of Table 9.1 (for a = 0) is given as

$$P = \sqrt{2}C(k)\dot{h} + \sqrt{2}C(k)\,U\alpha + \sqrt{2}(C(k) - 1)\dot{\alpha}\,b/2 \tag{9.14}$$

If we collect the coefficients for the motion together we obtain

$$P = \sqrt{2}C(k)(\dot{h} + U\alpha + \dot{\alpha}\,b/2) - \dot{\alpha}\,b/2$$

The term in the parenthesis is nothing but the downwash of the arbitrary motion at the quarter chord of the airfoil. Hence,

$$P = \sqrt{2}C(k)\,w(t,\,b/2) - \dot{\alpha}b/2 \tag{9.15}$$

With the aid of Fourier transform the first term of P can be written for all the frequencies involved in the arbitrary non-sinusoidal unit change w_0 in downwash as follows

$$P(s) = w_0 \int_{-\infty}^{\infty} \frac{C(k)}{ik} e^{iks} dk - \dot{\alpha}\,b/2, \qquad k = \omega\,b/U \text{ and } s = Ut/b$$

Table 9.1 Sectional suction and lift coefficients

	$\bar{P}$	$\bar{\alpha}\bar{L}$
$\bar{h}$	$\sqrt{2}C(k)ik$	$2C(k)ik - k^2$
$\bar{\alpha}$	$\sqrt{2}(C(k) + (C(k) - 1)ik/2)$ $+ \sqrt{2}((1 - C(k))a - a\,ik)$	$2C(k)(ik/2 + 1) + ik$ $-2C(k)ika - ika + k^2a$
$\bar{q}$	$\sqrt{2}(C(k)ik/2 + C(k) - 1)$	$ik + (1 - C)(-1 - ik/2) + k^2$
$\bar{\beta}$	$\sqrt{2}(-2\sqrt{1-e^2} + T_4 ik)/2\pi$	$T_4 ik - T_1 k^2 + 2(T_{10} + T_{11} ik/2)/\pi$

Table 9.2 Maximum average thrust for various pitch points

a	T_{max}	$h*$	$\bar{\alpha}_r$	$\bar{\alpha}_i$	q_r	q_i
−1.0	1.9578	0.9368	0.1379	−0.2986	−0.0609	−0.1025
−0.5	1.5298	0.9577	0.1093	−0.2485	−0.0491	−0.0823
0.0	1.2505	0.9656	0.0804	−0.2380	−0.0325	−0.0590

Table 9.3 Optimum Values obtained with efficiency constraints η for k = 0.5

η	T_{max}	$h*$	$\bar{\alpha}_r$	$\bar{\alpha}_i$	λ_1	λ_2
0.30	0.2127	0.9752	0.2111	0.0663	0.2127	0.1795
0.40	0.2139	0.9760	0.2074	−0.0661	0.2139	−0.1332
0.50	0.2031	0.9677	0.1958	−0.1590	0.2031	−0.3174
0.60	0.1902	0.9592	0.1770	−0.2219	0.1901	−0.4014
0.70	0.1786	0.9520	0.1591	−0.2648	0.1783	−0.4263
0.11	0.2158	0.9776	0.2104	−0.0085	(no efficiency constraint)	
0.77	0.1397	0.9612	0.2174	−0.1699	(maximum efficiency)	

Table 9.4 Suction and lift coefficients for U = 0

	$\bar{F}$	$\bar{\alpha}\bar{L}$
$\bar{h}$	$\sqrt{2}Fi\omega b$	$-\omega^2 b^2$
$\bar{\alpha}$	$\sqrt{2}((F-1)/2-a)i\omega b$	$\omega^2 b^2 a$
$\bar{q}$	$\sqrt{2}Fi\omega b/2$	$\omega^2 b^2$
$\bar{\beta}$	$\sqrt{2}T_4 i\omega b/2\pi$	$-T_1\omega^2 b^2$

Table 9.5 Comparison with experimental results for $\alpha = 34^\circ$

Theo. T [N]	Test difference cases	Freq. [Hz]	$\omega = 2\pi f$ rad/s	h/c	H	Exp. T [N]	%
T1	0.1	0.63	0.4	0.8	0.0076	0.0089	17
T2	0.15	0.94	0.4	0.8	0.0162	0.0197	22
T3	0.2	1.26	0.4	0.8	0.0268	0.0351	30

The improper complex integral is related to the well known Wagner function $\phi(s)$ so that $P(s) = \sqrt{2}\, w_0\, \phi(s) - \dot{\alpha}\, b/2$ with $\phi(s) = \frac{1}{2\pi} \int\limits_{-\infty}^{\infty} \frac{C(k)}{ik} e^{iks} dk = 1 - 0.165e^{-0.0455s} - 0.335e^{-0.3s}$

Here, the Wagner function is the indicial admittance for unit excitation. For the arbitrary downwash w(t, b/2) the leading edge suction velocity becomes

$$P(s) = \sqrt{2}\left[w(b/2, s)\, \phi(0) + \int\limits_0^s w(b/2, \sigma)\, \phi'(s - \sigma)\, d\sigma \right] - \dot{\alpha}\, b/2 \qquad (9.16)$$

This is very similar to the lift and moment generated by the arbitrary motion which read as

$$L(s) = \pi\rho b\left[\ddot{h} + U\dot{\alpha}\right] - 2\rho U b\pi\left[w(b/2, s)\, \phi(0) + \int\limits_0^s w(b/2, \sigma)\, \phi'(s - \sigma)\, d\sigma \right] \qquad (9.17)$$

$$\begin{aligned} M(s) = & -\pi\rho b^2\left[b\ddot{\alpha}/4 + U\dot{\alpha}\right]/2 \\ & - \rho U b\pi\left[w(b/2, s)\, \phi(0) + \int\limits_0^s w(b/2, \sigma)\, \phi'(s - \sigma)\, d\sigma \right] \end{aligned} \qquad (9.18)$$

The total thrust due the leading edge suction and the lift is found from (9.1) and plotted on the Fig. 9.3.

The time averaged unsteady thrust coefficient, integrated over one period of time gives the following thrust average thrust coefficient $C_T = 0.72$ as opposed to the sinusoidally generated optimum thrust coefficient $C_t = 0.31$.

The aerodynamic efficiency is defined before for the sinusoidal motions. For the non-sinusoidal pitch-plunge the work done by the motion is the product of the force with the vertical velocity over a period of motion, hence

$$\begin{aligned} W_M &= \frac{1}{2}\rho U^2 \int\limits_0^T \int\limits_{-b}^{b} C_{pa}(x,t) \frac{\partial z_a(x,t)}{\partial t} dx\, dt \\ &= \frac{1}{2}\rho U^2 \int\limits_0^T \int\limits_{-b}^{b} C_{pa}(x,t)(-\dot{h} - U\alpha - U\dot{\alpha}x) dx\, dt \end{aligned}$$

This equation is simplified with performing the integral over the chord to get

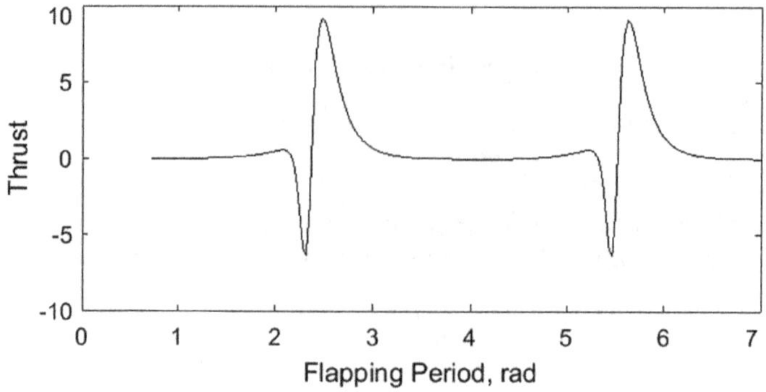

Fig. 9.3 Thrust variation for a period

$$W_M = \frac{1}{2}\rho U^2 \int_0^T C_L(-\dot{h} - U\alpha)\, dt + \frac{1}{2}\rho U^2 \int_0^T C_M(-U\dot{\alpha})\, dt \tag{9.19}$$

We can, now, calculate the value of the efficiency for the above example using the Wagner function to evaluate C_L and C_M from (9.17) and (9.18). Thus, the numerical value of the efficiency becomes

$$\eta = \frac{W_T}{W_M} = 0.28 \tag{9.20}$$

Thrust at Zero Free-Stream: Non-sinusoidal optimum solutions at zero free stream can be obtained with setting U = 0 for the leading edge suction velocity P and the sectional lift L from (9.16) and (9.17):

$$P(s) = \sqrt{2}\left[\dot{h}(s)\phi(0) + \int_0^s \dot{h}(\sigma)\,\phi'(s-\sigma)\,d\sigma\right] - \dot{\alpha}\,b/2 \;\text{ and }\; L(s) = \pi\,\rho\,b\,\ddot{h} \tag{9.21a, b}$$

Shown in Fig. 9.4 is the normalized thrust coefficient variation in one period of flapping.

The averaged thrust coefficient from Fig. 9.4 reads as 1.74 which is much higher than the thrust coefficient obtained with the sinusoidal flapping.

Above optimization is not based on any constraint. Next, we introduce the optimization with magnitude constraint.

Magnitude Constraint: The thrust function S can be optimized, using P and L expressions given above, for the non-sinusoidal flapping motion as an eigenvalue problem as follows.

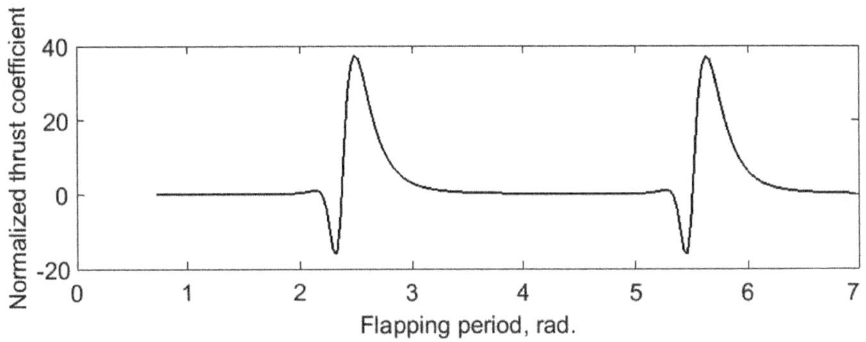

Fig. 9.4 Normalized thrust, $T/(\rho\omega^2 b^3)$, variation for one period of flapping

$$\begin{bmatrix} P_h^2 & -L_h \\ -L_h & P_\alpha^2 - L_\alpha \end{bmatrix} \begin{pmatrix} h \\ \alpha \end{pmatrix} = \lambda \begin{pmatrix} h \\ \alpha \end{pmatrix} \tag{9.22}$$

The eigenvalues of (9.22) read as $\lambda_{\min} = -0.7912$ and $\lambda_{\max} = 0.7869$, and the corresponding eigenvectors are $x_{\min} = (0.2646,\ 0.9644)$ and $x_{\max} = (-0.9644,\ 0.2646)$. Hence, example problem solved with the pitch-plunge motion constraint as described before gives the maximum thrust coefficient as $C_T = 0.78$ with

$$z_a(x,t) = 0.96 f_h(\omega t) - 0.26 x f_\alpha(\omega t + \varphi) \tag{9.23}$$

Shown in Fig. 9.5 is the thrust variation based on the magnitude constraint.

Power Extraction: By means of a periodic motion, sinusoidal or non-sinusoidal, it is possible to obtain negative thrust which implies power extraction from the airfoil. The optimum power extraction with non-sinusoidal oscillations is obtained with magnitude constraint as follows:

$$z_a(x,t) = -0.26 f_h(\omega t) - 0.96\, x f_\alpha(\omega t + \varphi) \tag{9.24}$$

The drag coefficient for this case reads as $C_D = -0.79$. The variation of the drag coefficient is shown in Fig. 9.6.

In this section, the aerodynamic thrust optimization of a thin airfoil is made for both simple harmonic and non-sinusoidal paths. Pitch, plunge and chord wise morphing of the airfoil are considered. The motion on-set from the rest is also studied under the thrust optimization at zero free stream. Finally, the maximum power extraction via flapping is studied. The following facts are observed:

(i) Pitch point location has an effect on the magnitude of the thrust; the thrust increases as the pitch point moves from mid-chord towards the leading edge,
(ii) the magnitude constraint renders the problem to an eigenvalue problem while maximizing the Lagrangian,

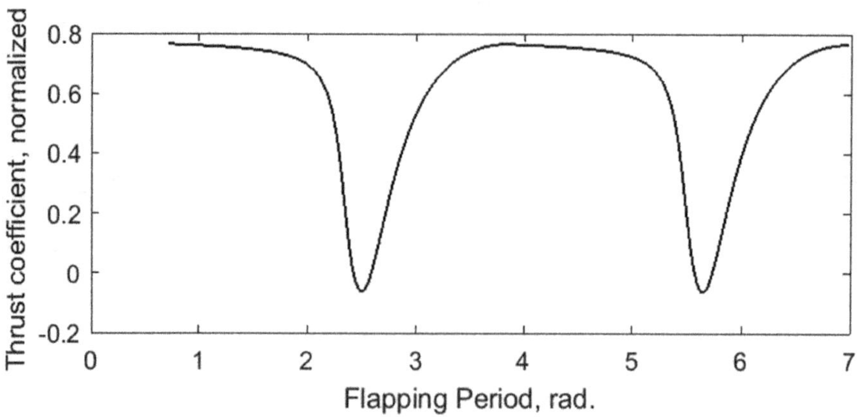

Fig. 9.5 Normalized thrust variation for the optimization with magnitude constraint

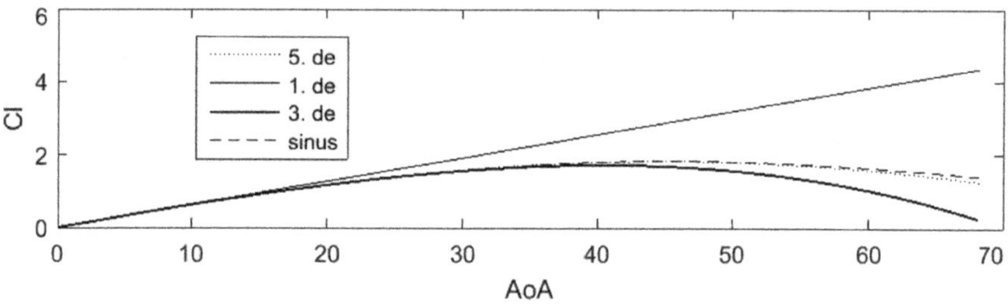

Fig. 9.6 **a** Normalized drag coefficient variation for the non-sinusoidal power extraction. **b** Linear, non-linear and the sine curves for the quasi-steady lift

(iii) comparison with the experimental study is satisfactory, especially at high reduced frequencies,
(iv) efficiency constraint is a good tool to increase the performance with small loss of thrust amplitude,
(v) imposing the efficiency constraint makes the problem solution in an iterative way which may yield to non-converging solutions if high efficiencies are sought,
(vi) maximization of the efficiency results in low average thrust value,
(vii) non-sinusoidal path optimization gives much more thrust compared to the sinusoidal path optimizations,
(viii) maximum thrust occurs at high plunge low pitch amplitudes, whereas maximum power extraction occurs at low plunge and high pitch amplitudes.

9.2 Thrust Optimization with Non-linear Modeling

The aim of this subsection is to calculate the optimum thrust with a nonlinear modeling of the leading edge vortex of a flapping wing in a periodic motion. The thrust optimization is reduced down to a solution of a nonlinear eigenvalue problem involving the amplitudes of the pitch and plunge where both amplitudes are subject to constrain. As an additional application no constraint on the pitch amplitude is also made. The establishment of the nonlinear eigenvalue problem, the solution technique and the results are presented here.

In order to express the expressions for the leading edge suction force, (9.16) and the lift (9.17) in terms of the quasi steady circulation first we have the equation for the pitching plunging airfoil as $z_a(x,t) = -h(t) - \alpha(t)(x-a)$, and the corresponding downwash as

$$w(b/2,t) = \frac{\partial z_a}{\partial t} + U\frac{\partial z_a}{\partial x} = -\dot{h}(t) - \dot{\alpha}(t)[b/2 - a] - U\alpha(t) \tag{9.25}$$

Wherein, a is the pitch point location. Now, we can write the expression for the quasi steady circulation as follows

$$\Gamma_{qs}(t) = b\,UC_{ls}(\alpha(t)) + \pi b(b/2 - a)\dot{\alpha}(t) + \pi b\dot{h}(t) \tag{9.26}$$

Here, $C_{ls} = A\sin 2\alpha$ (Taha et al. 2014a) is the sectional lift coefficient at high angles of attack including the effect of the leading edge vortex and it is the replacement for the convective term in (4). Comparing Eqs. (4) and (5) gives us the relation between the downwash and the circulation as follows

$$w(b/2,t) = -\Gamma_{qs}(t)/\pi b \tag{9.27}$$

If we use Eqs. (9.27) in (9.25) and (9.26), the leading edge suction velocity and the lift read as

$$P(s) = -\sqrt{2}/(\pi b)\left[\Gamma_{qs}(s)\phi(0) + \int_0^s \Gamma_{qs}(s)\phi'(s-\sigma)\,d\sigma\right] - \dot{\alpha}b/2 \tag{9.28}$$

and

$$L(s) = \pi\rho b\left[\vec{h} + U\dot{\alpha}\right] + 2\rho U\left[\Gamma_{qs}(s)\phi(0) + \int_0^s \Gamma_{qs}(s)\phi'(s-\sigma)\,d\sigma\right] \tag{9.29}$$

Expanding $C_{ls} = A\sin 2\alpha$ into the series in α we obtain the approximate expression for the steady lift coefficient as follows

$$C_{ls}(\alpha) \cong A\,(2\alpha - 4\alpha^3/3 + 4\alpha^5/15)$$

The relation between the steady lift coefficient and the quasi steady circulation (9.26) is now used in (9.28) and (9.29) to obtain the aerodynamic thrust, Eq. (9.1), as a nonlinear equation in terms of h *and* α which are functions of time. Then we can take the time average of this equation for over a period to have

$$\bar{S} = -(\pi\,\rho\,\bar{P}^2 - \bar{\alpha}\,\bar{L}) \tag{9.30}$$

For the optimization of (9.30) using $\vec{h}$ *and* $\bar{\alpha}$ only, equations similar to (9.7) and (9.8-a) yields a non-linear eigenvalue problem in the following matrix form

$$\begin{bmatrix} a_{11} - \lambda & a_{21} \\ a_{12} & a_{22} - \lambda \end{bmatrix} \begin{pmatrix} \bar{h} \\ \bar{\alpha} \end{pmatrix} = \begin{pmatrix} 0 \\ 0 \end{pmatrix} \qquad \bar{h}^2 + \bar{\alpha}^2 - 1 = 0 \tag{9.31a, b}$$

In Eq. (9.31a,b), except a_{11} all entries depend on $\bar{\alpha}$. The contribution of plunging to the leading edge velocity is denoted with P_h, to pitching is P_α, contribution of the lift to the plunging is L_h, and finally contribution to pitching is shown with L_α. Hence, the entries of the matrix read as follows: $a_{11} = \pi\,P_h^2$, $a_{12} = a_{21} = (1 - 2\bar{\alpha}^2/3)2\pi P_h P_\alpha - L_h/2$, $a_{22} = 16\pi(1 - 2\bar{\alpha}^2/3)^2 P_\alpha^2 - 2(1 - 2\bar{\alpha}^2/3)L_\alpha$. Now, with (9.31a,b) we have a non-linear system with 3 equations for 3 unknowns. The solution of this non-linear eigenvalue system gives the optimum thrust for the maximum eigenvalue and the corresponding eigenvector provides the amplitudes for the motion.

The method based on the non-linear modeling of the leading edge vortex is first implemented for a simple harmonically pitching plunging thin airfoil. The circulation created by the leading edge vortex is given as $C_l^{qs} = 1.833 \sin 2\alpha$ (Taha et al. 2014b), and the sine term is expanded into the series in terms of angle of attack, Fig. 9.1 shows the variation of quasi steady lift in terms of powers of angle of attack and shows with the expression given with sine.

As seen in (Fig. 9.6) the linear approach represents the sine curve in 0°–20° range, whereas the 3rd degree approach is good for 20°–50°. The iterative solution of the Eq. (9.31a,b) with the 3rd degree approach gives the value of maximum average thrust S = 3.63, in non dimensionalized form. The corresponding motion is determined from the corresponding eigenvector as

$$\begin{aligned} h(t) &= -0.48 \cos \omega\, t \\ \alpha(t) &= -49^\circ \cos(\omega\, t + \pi/2) \end{aligned} \tag{9.32a – b}$$

In (9.32a-b) the phase difference between the plunge and pitch is taken as 90° together with $\omega = 1$. The iterations begin with the values taken from the linear solution, and continues with these new values substituted in their proper places in the coefficient matrix. The linear solution gives the maximum value of the thrust as S = 4.26. The iterations, on the other hand, converges to 4 digit accuracy (Matlab 2015). If we increase the accuracy with the 5th degree approximation we obtain the value of the maximum thrust to read as S = 3.66. This means increasing the number of non-linear terms does not improve the accuracy of the thrust value that much. In (Fig. 9.7), the variation of the quasi steady circulation with respect to 3rd and 5th degree representations and also for the sine dependence. Accordingly, the 5th degree approximation and the sine representation gives almost the same circulation change by time.

The time variation of maximum thrust S, associated lift L and the effect of the leading edge suction force P*P, in non-dimensional forms, are shown in Fig. 9.8. As seen from Fig. 9.8, the propulsive force S is positive most of the period covered except for a very short duration at around the half and full period points on the curve.

Finite Wings: For the finite wings, the spanwise integration of the sectional value of the lift given with (9.29) provides the wing's lifting force. The time wise integration of the Duhamel integral is performed with Matlab as shown in the Appendix 15. Then, the optimization for the trust of the wings can be put into the following form

$$\begin{aligned} \pi\rho U^2 A/2[H]\{Q\} - \lambda\{Q\} &= 0 \\ \{Q\}^T\{Q\} - 1 &= 0 \end{aligned} \tag{9.32a, b}$$

Here, A is the wing area. For rectangular wings the Wagner function effect remains the same at every section to keep the Eq. (9.31a,b) the same. For the case of elliptical wings the Wagner function is altered to cause the changes in the coefficient matrix. There exist, however, expressions for the Wagner functions for the elliptical wings of the following aspect ratios (Bisblinghoff et al. 1996):

$$For \quad AR = 3 \quad \phi(s) = 0.60 - 0.17e^{-0.54s} \tag{9.33a, b}$$

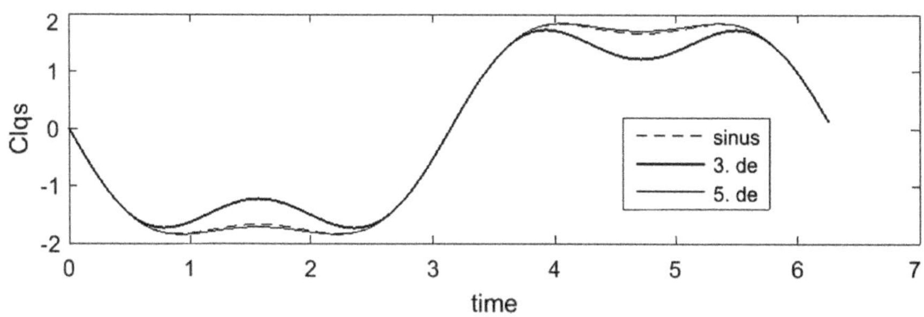

Fig. 9.7 Change in quasi steady circulation, Γ_{qs}, by time (sine —-, 3rd and 5th___ degree approximations)

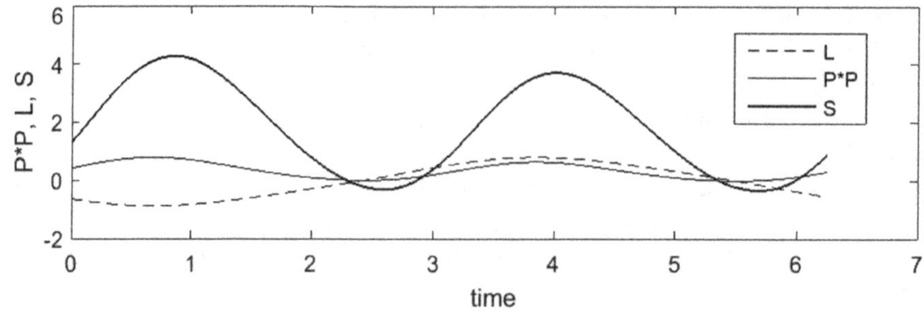

Fig. 9.8 Time variation of maximum aerodynamic thrust S, lift L, and P*P

$$For \quad AR = 6 \quad \phi(s) = 0.74 - 0.0.267e^{-0.381s}$$

For the elliptical wing the aerodynamic thrust S for a wing with aspect ratio of 3 is obtained from Eq. (9.32a,b) as shown in Fig. 9.4 wherein the shapes of the curves are similar to that of given in Fig. 9.9.

Here, the results obtained for Figs. 9.8, 9.9 are non-dimensional quantities. In Fig. 9.8 for non-dimensionalization the half chord is used for a charactestic length, whereas for Fig. 9.9 half of the wing area is taken as the characteristic area. In both figures the free stream speed is employed as the characteristic speed.

In short a non-linear model of the leading edge vortex is implemented to find the maximum aerodynamic thrust for a flapping wing. The implementation is based on a solution of a non-linear eigenvalue problem. The procedure is based on an iterative solution which converges to desired accuracy in 4–5 iterations. It is worth noting that the quasi steady formula gives the maximum lift at 45° angle of attack, whereas the nonlinear optimized unsteady solution gives the maximum lift at 49°.

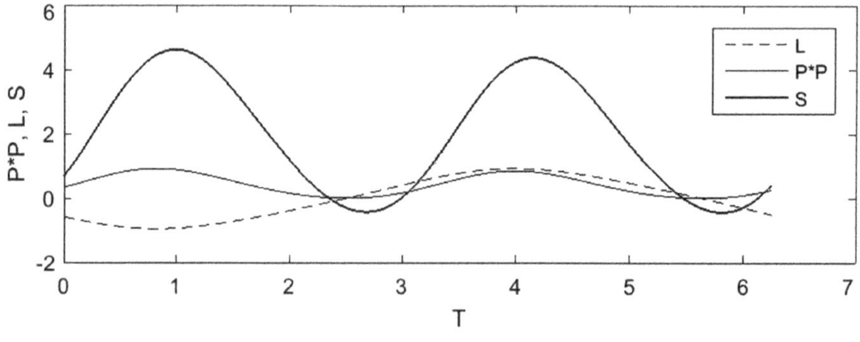

Fig. 9.9 Time variation of maximum aerodynamic thrust S, lift L, and P*P for an elliptical wing with AR = 3

The aerodynamic thrust optimization employs the Wagner function to consider the time lag between the flapping motion and the aerodynamic response of thin profiles or wings with a constraint on flapping.

9.3 Stability and Trim

The unsteady aerodynamic load predictions take considerably long time using numerical and experimental techniques. On the other hand fast methods are preferred for the flight dynamics and control applications. In unsteady aerodynamic applications the state-space representations of aerodynamic loadings result in very fast predictions compared to conventional methods. The state variable is either chosen formally or it is based on a change of a physical entity like the separation point location on an airfoil (Goman and Khrabrov 1994), (Gulcat 2011, 2016) and (Uhlig and Selig 2017). The timewise change of the state variable is given as the numerical solution of a first order ordinary differential equation. There are more complex applications, like spanwise morphing of a wing (Reich et al. 2011) and (Izraelevitz et al. 2017). The formal approach employs the Duhamel integral together with the Wagner function (Taha et al. 2014b) and (Gulcat 2017) or employs the Laplace transform (Leishman 2000). The Wagner function for a thin airfoil and for some elliptical wings with different aspect ratios are given in Bisblinghoff et al. (1996).

In this study the Wagner function is employed to predict the stability of a body whose wing is sweeping back and forth with varying angle of attack while hovering. The time lag between the wing motion and its aerodynamic response is given with the Wagner function. Here, because of sweeping motion the reduced speed changes periodically instead of being expressed with a constant free-stream speed. That is the reason for the exponents of the Wagner function to change periodically with time. The aerodynamic loads due to wing flapping can be determined in a fast manner to interact with the equation of motion for the body.

The system of coupled equations assuring the balance of aerodynamic loads with the body motion, involving the stability derivatives, are formed to represent the small perturbations from the equilibrium. The conditions to satisfy the vanishing of these small perturbations for the hover and trim are given in Taha et al. (2014b) and Mouy et al. (2017). The application of the method on the trim of a fruit fly in hover is made satisfactorily with the pertinent data obtained from Berman and Wang (2007) and Taha et al. (2014a) The resulting matrix system is 6×6 with time dependent quantities being the horizontal and vertical speeds, the pitch angle and its rate for the body and the pitch angle and its rate for the wing represented state-space. The resulting system for the flight dynamics coupled with the unsteady aerodynamic loads and its implementations will be provided in the following subsections.

For the variable free stream, $U(t)$, the derivative of the Wagner function Φ reads as

$$\frac{d\,\Phi(t-\tau)}{d\tau} = -a_i\frac{b_i}{b}U(\tau)e^{-\frac{b_i}{b}\int\limits_{\tau}^{t}U(\varsigma)d\tau}, \; i=1,2 \tag{9.34}$$

Then Eq. (9.34) gives the time dependent circulation with

$$\Gamma_c(t) = (1-a_1-a_2)\Gamma_{qs}(t) + x_i(t) \tag{9.35}$$

and with

$$x_i(t) = \int\limits_0^t \Gamma_{qs}(\tau)\; a_i\frac{b_i}{b}U(\tau)e^{-\frac{b_i}{b}\int\limits_{\varsigma}^{t}U(\varsigma)d\tau} d\tau$$

Using the Leibnitz's rule for the derivation with respect to time makes Eq. (8) to read

$$\dot{x}_i(t) = \frac{b_i}{b}U(t)\;(-x_i(t) + a_i\,\Gamma_{qs}(t)), \quad i=1,2 \tag{9.36}$$

Here, $x_i = x_i(t)$ time dependent state variable which is responsible for the unsteady lift, and it is obtained as the solution of a first order ODE, Eq. (9.36), with the initial condition of $x_i(0) = 0$. Now, the circulatory unsteady lift becomes

$$l_c(t) = \rho\,U(\tau)\left[(1-a_1-a_2)\,\Gamma_{qs}(t) + x_1(t) + x_2(t)\right] \tag{9.37}$$

The non-circulatory lift on the other hand reads as

$$l_{nc}(t) = -\pi\,\rho\,b^2 a_z(t)\;\cos\eta(t) \tag{9.38}$$

Wherein, $a_z(t)$ is the vertical acceleration and $\eta(t)$ is the pitch angle. The acceleration is created by: (i) free stream speed change $-r\ddot{\phi}\sin\eta$, and (ii) angle of attack change $-r\dot{\phi}\dot{\eta}\cos\eta$ make the total acceleration to become. From Eqs. (9.37) and (9.38) the total lift coefficient reads

$$C_l(t) = \frac{l_c(t) + l_{nc}(t)}{\rho U_m^2 b_m} \tag{9.39}$$

Here, U_m is the midspan maximum free stream speed, and b_m is half-chord at the midspan. The solution for a rectangular wing with certain aspect ratio is given in Gulcat (2017). Here, the unsteady aerodynamic loads are going to be calculated for a flapping elliptical wing of aspect ratio 3 as depicted in (Fig. 9.10).

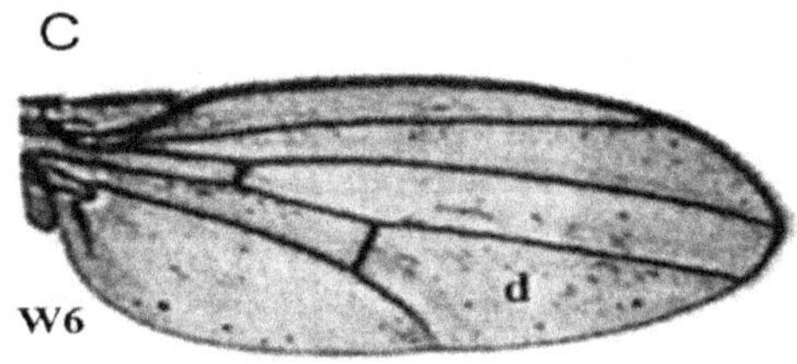

Fig. 9.10 Wing of a fruit fly. (Dimensions: body to root: 0.20 mm, span: 2.02 mm, mid-chord: 0.86 mm, S = 1.36 mm^2) (http://www.google.com.tr/search?q=shape+of+the+fruit+fly+wing)

Finite Wing: The Wagner function for the elliptical wing of aspect ratio 3 is given as $\Phi_W(s) = 0.6 - 0.17e^{-o.54s}$ (Bisplinghoff et al. 1996). The sectional lift coefficients given with (9.37) and (9.38) if integrated with respect to span give these values for the wing as follows

$$\int_{r_1}^{r_1+R} (l_c(t) + l_{nc}(t))dr = \rho \int_{r_1}^{r_1+R} \{U(\tau)[(0.6 - 0.17)\,\Gamma_{qs}(t) + x_1(t) + x_2(t)] - \pi\,\rho\, b^2 a_z(t)\ \cos\eta(t)\}dr \tag{9.40}$$

There are 3 resulting terms at the right hand side of (14) if we take $U(\tau) = \dot{\phi}\, r$. These are: (i) circulatory terms from the pitch and pitch rate, and (ii) non circulatory term. Hence,

Pitch: $l_{cp}(t) = \rho I_2\dot{\phi}[0.43C_L(\alpha,t) + x_1(t)], \quad I_2 = \int_{r_1}^{r_1+R} r^2 b\, dr$

Pitch rate: $l_{cpr}(t) = \rho I_1\dot{\phi}[0.43\pi\dot{\alpha}(t)/2 + x_2(t)], \quad I_1 = \int_{r_1}^{r_1+R} rb^2 dr$

Here,

$\dot{x}_1(t) = \frac{b_1\bar{r}\dot{\phi}(t)}{\bar{b}}\left[-x_1(t) + 0.17\dot{\phi}(t)C_L(\alpha,t)\right]$ *and* $\dot{x}_2(t) = \frac{b_1\bar{r}\dot{\phi}(t)}{\bar{b}}[-x_2(t) + 0.17\dot{\alpha}(t)]$

Non-circulatory: $l_{nc}(t) = \pi\rho I_1\left[\ddot{\phi}(t)\sin\eta(t) + \dot{\phi}(t)\dot{\eta}(t)\cos\eta(t)\right]\cos\eta(t)$

And $\bar{b}$ *and* $\bar{r}$ are the averaged half chord and half span and the static lift coefficient is $C_l = A\sin 2\alpha$, with A = 1.833 (Taha et al. 2014b). Figure 9.11 shows the angle of attack change while the profile makes the sweeping motion.

Shown in (Fig. 9.12) is the simplified wing-body configuration to represent the fruit fly. The coordinate axis shown are suitable for calculating the aerodynamic forces and moments.

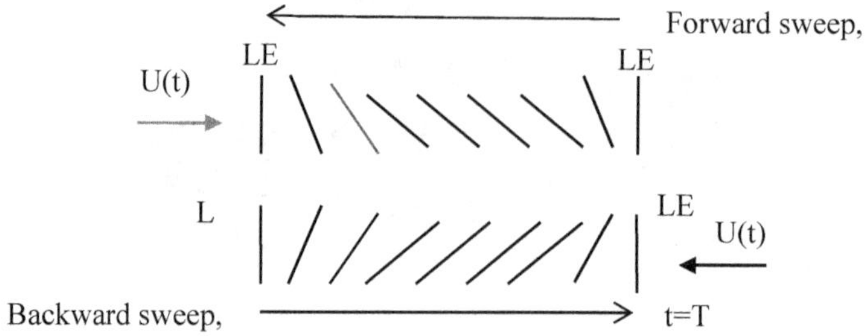

Fig. 9.11 Angle of attack change during forward and backward sweep, 90°–40°–90°

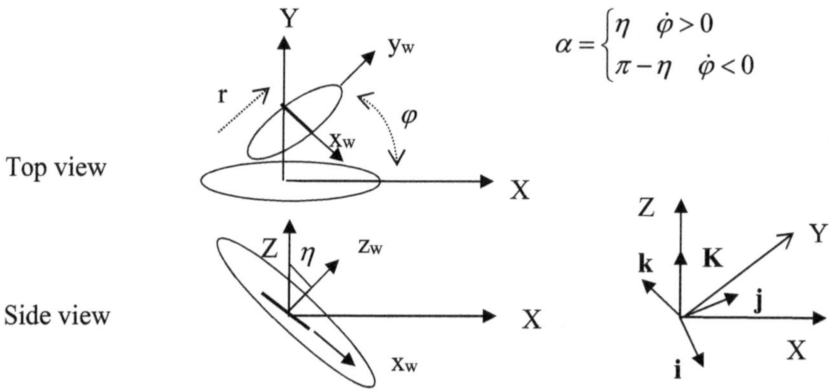

Fig. 9.12 Wing-body configuration, top and side views

Appendix

Appendix 16 provides the necessary information about the numerical values of I_1, I_2 and $\bar{b}$ to be used in the following equation

$$C_L(t) = \frac{l_{cr}(t) + l_{cp}(t) + l_{nc}(t)}{\rho U_m^2 S/2} \tag{9.40}$$

The variation of the total lift coefficient for one period is shown in (Fig. 9.13) for the thin airfoil and for the wing.

Using (Fig. 9.13) and Eq. (9.40) for a period of the motion we get the result of the following integration as

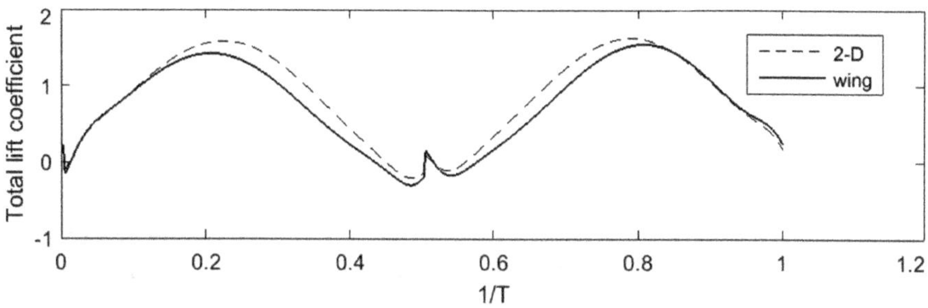

Fig. 9.13 Variation of the total lift coefficient for a period

$$\bar{C}_L = \frac{1}{T}\int_{T}^{2T} C_L(t)\,dt = 0.75 \tag{9.41}$$

Accordingly, the lifting force generated for both wings reads as $F = 2\bar{C}_L\rho U_m^2 S/2 = 7.11\ \mu\text{N}$,

Here, the flapping frequency for the wing is f = 240 Hz which is sufficient to lift the fruit fly which weighs about $W = 7.06\ \mu\text{N}$ (Berman and Wang 2007).

The wing sweep here given by $\phi = -75^o \cos\omega t$, and the change of the pitch angle is provided with the arctangent and sine functions as shown in (Fig. 9.14). The free stream velocity span -wise variation is given by $\dot{\phi}\, r = \bar{\phi}\omega\, r \sin\omega t$.

Flight Stability: Study of the longitudinal stability of a body with a flapping wing requires the coupled treatment of unsteady aerodynamics with the parameters of the flight mechanics. Shown in (Fig. 9.15) are the necessary parameters to prescribe the hovering body under the gravity where the longitudinal and vertical velocity perturbation velocities are u and v respectively with also are the pitch rate of q and the pitch angle θ. The dynamic equilibrium equation using the notation of Nelson (1998) reads as follows

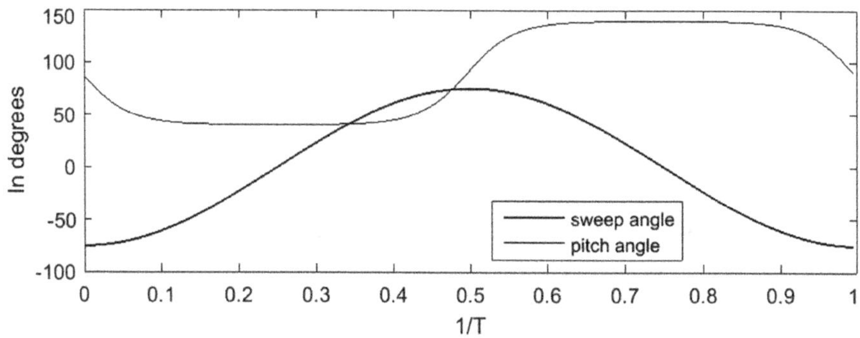

Fig. 9.14 Sweep angle: $-75^\circ < \phi < 75^\circ$, and the pitch angle, $40^\circ < \eta < 140^\circ$, variations

$$\begin{pmatrix} \dot{u} \\ \dot{w} \\ \dot{q} \\ \dot{\theta} \end{pmatrix} = \begin{pmatrix} -qw - g\sin\theta \\ qu + g\cos\theta \\ 0 \\ q \end{pmatrix} + \begin{pmatrix} X/m \\ Z/m \\ M/I_y \\ 0 \end{pmatrix} \tag{9.42}$$

Here, X, Z and M are the horizontal and the vertical forces and the pitching moment respectively, and m is the mass, I_y is the rotational moment of inertia and finally g is the gravitational acceleration.

Unsteady aerodynamics of a flapping wing gives the lift, moment in terms of the pitch and the pitch rate. Here, we implement the state variables concept, x_1 and x_2, with two ordinary differential equations in time involving the pitch and its rate:

$$\dot{x}_1(t) = \frac{b_1 U_{ref}}{\bar{b}}\left[-x_1(t) + a_1\dot{\phi}(t)C_L(\alpha, t)\right] \ \ and \ \ \dot{x}_2(t) = \frac{b_1 U_{ref}}{\bar{b}}\left[-x_2(t) + a_1\dot{\alpha}(t)\right]$$

If we let the sate variables interact with the equation of motion, (9.42), the perturbation equations for the system involves the rate of 6 variables, $\dot{\chi} = \begin{pmatrix} \dot{u} & \dot{w} & \dot{q} & \dot{\theta} & \dot{x}_1 & \dot{x}_2 \end{pmatrix}^T$, and their stability derivative matrix as follows

$$\begin{pmatrix} \dot{u} \\ \dot{w} \\ \dot{q} \\ \dot{\theta} \\ \dot{x}_1 \\ \dot{x}_2 \end{pmatrix} = \begin{pmatrix} -qw - g\sin\theta \\ qu + g\cos\theta \\ 0 \\ q \\ 0 \\ 0 \end{pmatrix} + \begin{pmatrix} X_0 \\ Z_0 \\ M_0 \\ 0 \\ X_{1_0} \\ 0 \end{pmatrix} + \begin{bmatrix} X_u & X_w & X_q & 0 & X_{x_1} & X_{x_2} \\ Z_u & Z_w & Z_q & 0 & Z_{x_1} & Z_{x_2} \\ M_u & M_w & M_q & 0 & M_{x_1} & M_{x_2} \\ 0 & 0 & 0 & 0 & 0 & 0 \\ X_{1_u} & X_{1_w} & X_{1_q} & 0 & X_{1_{x_1}} & 0 \\ 0 & 0 & X_{2_q} & 0 & 0 & X_{2_{x_2}} \end{bmatrix} \begin{pmatrix} u \\ w \\ q \\ 0 \\ x_1 \\ x_2 \end{pmatrix} \tag{9.43}$$

In Eq. (9.43), the column vector expressed with subscript o shows the effect of pitch rate and the subscript for the coefficient matrix indicates the derivatives (Appendix 17). The horizontal sectional force is shown by, X', and the vertical sectional force by, Z'. The sectional lift *l* and the drag *d* are employed to give these horizontal and vertical forces as follows

$$X' = -\mathrm{sgn}(\dot{\phi})(d - l\alpha_i) \ \ and \ \ Z' = -(l + d\alpha_i)\ , \ \ \alpha_i \ induced\ angle\ of\ attack \tag{9.44a, b}$$

Here, two differen contributions to the sectional forces are possible: (i) from pitching l_P,
(ii) from pitch rate l_{Pr}, which are determined with x_1 and x_2. On the other hand the induced angle of attack is small and can be approximated as

$$\alpha_i \cong \frac{w_{eff}}{|U|}, \quad and \quad w_{eff} = w - q(r\sin\phi + a), \quad U \cong r\dot{\phi} + u\cos\phi$$

The perturbation velocities u, w and the pitch rate q, all contribute to the sectional lift coefficient in terms of pitch and the pitch rate. Considering the effective free stream velocities, the sectional lift forces read as

$$\begin{aligned} l_P &= \rho(r\dot{\phi} + u\cos\phi)\mathrm{sgn}(\dot{\phi})\left[b(r\dot{\phi} + u\cos\phi)A(\sin 2\eta + 2\alpha_i\cos 2\eta)\Phi(0) + x_1(t)\right], \text{ and} \\ l_{\Pr} &= \rho(r\dot{\phi} + u\cos\phi)\mathrm{sgn}(\dot{\phi})\pi b^2[q\cos\phi + (1/2 - a)q\cos\phi\,\Phi(0) + x_2(t)] \end{aligned} \tag{9.45a, b}$$

Similarly, the sectional drag becomes

$$d_P = \rho(r\dot{\phi} + u\cos\phi)\mathrm{sgn}(\dot{\phi})\left[b(r\dot{\phi} + u\cos\phi)A(\sin^2\eta + \alpha_i\sin 2\eta)\Phi(0) + x_1(t)\right]$$

Since the induced angle of attack is small the second order terms are neglected, and this gives the induced lift and the drag as follows

$$l_P\alpha_i = \rho\, w_{eff}\left[brA\dot{\phi}\sin 2\eta\,\Phi(0)\right], \qquad d_P\alpha_i = \rho\, w_{eff}\left[2brA\dot{\phi}\sin^2\eta\,\Phi(0)\right] \tag{9.46a, b}$$

Averaging

Equation (9.43) is now utilized for the stability analysis of the time dependent periodic system. For this purpose, we take the time average of the quantities for a period of time T as

$$\dot{\bar{\chi}} = f(\bar{\chi}) + h(\bar{\chi}), \quad here \quad h(\bar{\chi}) = \frac{1}{T}\int_0^T h(\chi, t)dt$$

Now, the averaged system reads as

$$\begin{pmatrix} \dot{\bar{u}} \\ \dot{\bar{w}} \\ \dot{\bar{q}} \\ \dot{\bar{\theta}} \\ \Delta\dot{\bar{x}}_1 \\ \dot{\bar{x}}_2 \end{pmatrix} = \begin{pmatrix} -\bar{q}\bar{w} - g\sin\bar{\theta} \\ \bar{q}\bar{u} + g\cos\bar{\theta} \\ 0 \\ 0 \\ 0 \\ 0 \end{pmatrix} + \begin{pmatrix} \bar{X}_0 + \bar{X}_{x_1}\bar{x}_{1eq} \\ \bar{Z}_0 + \bar{Z}_{x_1}\bar{x}_{1eq} \\ \bar{M}_0 + \bar{M}_{x_1}\bar{x}_{1eq} \\ 0 \\ 0 \\ 0 \end{pmatrix} + \begin{bmatrix} \bar{X}_u & \bar{X}_w & \bar{X}_q & 0 & \bar{X}_{x_1} & \bar{X}_{x_2} \\ \bar{Z}_u & \bar{Z}_w & \bar{Z}_q & 0 & \bar{Z}_{x_1} & \bar{Z}_{x_2} \\ \bar{M}_u & \bar{M}_w & \bar{M}_q & 0 & \bar{M}_{x_1} & \bar{M}_{x_2} \\ 0 & 0 & 1 & 0 & 0 & 0 \\ \bar{X}_{1_u} & \bar{X}_{1_w} & \bar{X}_{1_q} & 0 & \bar{X}_{1_{x_1}} & 0 \\ 0 & 0 & \bar{X}_{2_q} & 0 & 0 & \bar{X}_{2_{x_2}} \end{bmatrix} \begin{pmatrix} \bar{u} \\ \bar{w} \\ \bar{q} \\ \bar{\theta} \\ \Delta\bar{x}_1 \\ \bar{x}_2 \end{pmatrix} \tag{9.47}$$

Here, $\Delta\bar{x}_1 = \bar{x}_1 - \bar{x}_{1eq}$ *and* $\bar{x}_{1eq} = -\bar{X}_{1_0}/\bar{X}_{1_{x1}}$. The algebraic Eq. (9.47) are solved for $(\bar{u} = 0, \ \bar{w} = 0, \ \bar{q} = 0, \ \bar{\theta} = 0)$ in hover.

Trim in hover: Time averaged equation of motion is now implemented for the trimming of the hovering body while flapping its wings. For the wing flapping, there are two different types; (i) symmetric flapping, and (ii) anti-symmetric flapping.

(i) Trim with symmetric flapping

For the sake of simplicity, we choose the time dependent sweeping motion as a saw-tooth shaped which is expressed as Mouy et al. (2017)

$$\phi(t) = \begin{cases} 4\bar{\phi}/T(t - T/4), & 0 \le t \le T/2 \\ -4\bar{\phi}/T(t - 3T/4), & T/2 \le t \le T \end{cases}$$

During the sweeping motion of the wing, a piecewise constant angle of attack is considered as follows

$$\eta(t) = \begin{cases} \bar{\alpha}, & 0 \le t \le T/2 \\ \pi - \bar{\alpha}, & T/2 \le t \le T \end{cases}$$

During the symmetric sweeping, the full unsteady treatment gives a non-zero value for the X force component, whereas the quasi-steady approach yields 0 result (Mouy et al. 2017). The full unsteady treatment results in Eq. (9.47) for the x direction, in terms of averaged values, as follows (Appendix 18):

$$\bar{X}_{x_1}\bar{x}_{1eq} = 8\frac{\rho I_{10}}{mT^2}\bar{r}2\bar{b}\bar{\phi}\sin\bar{\phi}Aa_1\sin 2\bar{\alpha} \tag{9.48}$$

The right hand side of Eq. (9.48) is 0 only for the average sweep or the angle of attack, $\bar{\phi} = 0 \ or \ \bar{\alpha} = 0$. This means trimming is possible only for the no sweep or no lift! For this reason we have to resort to antisymmetric sweep.

(ii) Trim with anti-symmetric flapping

In order to achieve trim during hover the sweeping motion is modified as follows:

$$\phi(t) = \begin{cases} \bar{\phi}_0 + 4\bar{\phi}/T(t - T/4), & 0 \le t \le T/2 \\ \bar{\phi}_0 - 4\bar{\phi}/T(t - 3T/4), & T/2 \le t \le T \end{cases}$$

The angle of attack:

$$\eta(t) = \begin{cases} \bar{\alpha}_d, & 0 \le t \le T/2 \\ \pi - \bar{\alpha}_u, & T/2 \le t \le T \end{cases}$$

The trim equations then read:

$$\begin{aligned}\bar{X}_0 + \bar{X}_{x1}\bar{X}_{1eq} &= 0\\ \bar{Z}_0 + \bar{Z}_{x1}\bar{X}_{1eq} &= -g\\ \bar{M}_0 + \bar{M}_{x1}\bar{X}_{1eq} &= 0\end{aligned} \qquad (9.49\text{a, b, c})$$

In Eq. (9.49a,b,c), there are 3 equations and 3 unknowns; $\bar{\alpha}_d,\ \bar{\alpha}_u\ and\ \bar{\phi}_0$. We use the time averaged values for the stability derivatives to obtain following non-linear expressions (Appendix 3).

Force balance in X:

$$\sin 2\bar{\alpha}_d + \sin 2\bar{\alpha}_u = -\frac{I_{21}}{I_{10}\bar{r}b}(\sin^2\bar{\alpha}_d - \sin^2\bar{\alpha}_u)(0.6 - a_1)/a_1 \qquad (9.50)$$

Force balance in Z:

$$\sin 2\bar{\alpha}_d + \sin 2\bar{\alpha}_u = \frac{mgT^2}{4\rho A(0.43I_{21} + 0.17I_{10}\bar{r}2\bar{b})\bar{\phi}^2} \qquad (9.51)$$

Moment balance in M:

$$\begin{aligned}&a\sin\bar{\phi}\cos\bar{\phi}[0.43I_{22}(\sin\bar{\alpha}_d - \sin\bar{\alpha}_u) - 0.17I_{11}\bar{r}2\bar{b}(\sin 2\bar{\alpha}_d + \sin 2\bar{\alpha}_u)]x\\ &(\cos\bar{\alpha}_d + \sin\bar{\alpha}_d - \cos\bar{\alpha}_u + \sin\bar{\alpha}_u) + 0.43x_{cg}\bar{\phi}(\sin 2\bar{\alpha}_d + \sin 2\bar{\alpha}_u)(I_{21} - 2I_{10}\bar{r}2\bar{b})\\ &+ \sin\bar{\phi}(0.43I_{31}\cos\bar{\phi}_0 - 0.17I_{20}4\bar{r}\bar{b}\sin\bar{\phi}_0)(\sin 2\bar{\alpha}_d + \sin 2\bar{\alpha}_u) = 0\end{aligned} \qquad (9.52)$$

Here, a is distance between the pitch point and the quarter chord.

We solve for X and Z force balances (9.50) and (9.51), to obtain $\bar{\alpha}_d\ and\ \bar{\alpha}_u$ in terms of the sweep angle and the period, $\bar{\phi}\ and\ T$, respectively. Hence, we can calculate the angle of attack for the forward and backward sweeps. Afterwards, by the aid of (9.52) the initial sweep angle $\bar{\phi}_0$ is determined to complete the solution for the trim in hover.

Trimming of fruit-fly in hover: The pertinent parameters for a fruit-fly is provided as follows (Berman and Wang 2007):

m = 0.72 mg, f = 254 Hz, $\bar{\phi} = 75°$, a = 0, $x_{cg} = 0.5$, $I_{10} = 4.89$, $I_{11} = 3.30$, $I_{21} = 4.69$, $I_{20} = 7.29$, $I_{22} = 3.32$, $I_{31} = 7.36$

From the simultaneous solution of (9.50), (9.51) and (9.52), the anti-symmetric trim results are obtained as follows $\bar{\alpha}_d = 34.3°$, $\bar{\alpha}_u = 55.6°$ and $\bar{\phi}_0 = 4°$

For this case the angle of attack differs for the forward and the backward sweeps. The angle of attack becomes 34.3° for the forward sweep and 55.6° in backward sweep. During sweeping the forward sweep angle changes $-71° < \phi(t) < 79°$ and in backward sweep $79° > \phi(t) > -71°$. The time change of sweep and angles of

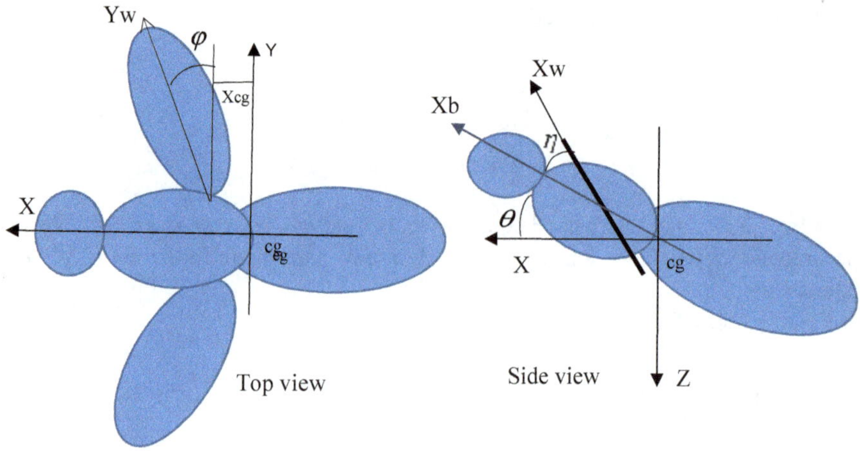

Fig. 9.15 Reference frame and the parameters used for the flight stability

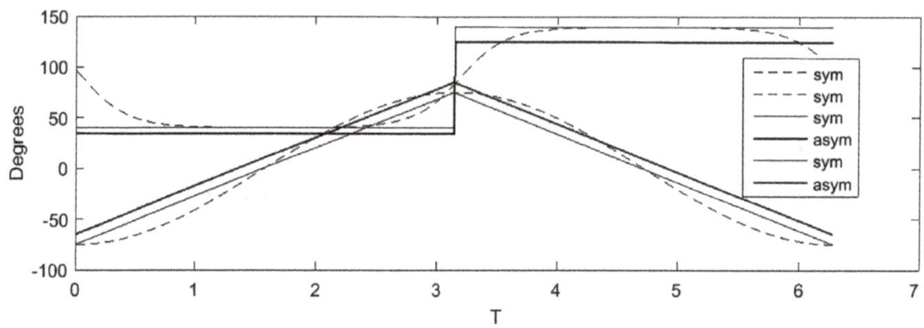

Fig. 9.16 Trimmed antisymmetric sweep and angle of attack____, un-trimmable symmetric sweep

attack are given for both symmetric and antisymmetric cases in (Fig. 9.16). The deviation from the symmetric case is about 5° for the angle of attack and 4° for the sweep, which can be applied easily for the control purposes.

References

Berman, G.J., Wang, Z.J.: Energy-minimazing kinematics in hovering flight. J. Fluid Mech. **582**, 153–168 (2007)

Bisplinghoff, R.L., Ashley, H., Halfman, R.L.: Aeroelasticity, pp. 393–394. Dover Publications Inc., New York (1996)

Bulut, J., Karakas, F., Fenercioglu, I., Gulcat, U.: A numerical and experimental study for aerodynamic thrust optimization. J. Aeronaut. Space Technol. **9**, 55–62 (2016). Turkish Air Force Academy

Garrick, I.E.: *Propulsion of a Flapping and Oscillating Airfoil*, NACA R-567 (1936)

Goman, M., Khrabrov, A.: State-space representation of aerodynamic characteristics of an aircraft in high angles of attack. J. Aircraft **31**(5), 1109–1115 (1994)

Gulcat, U.: Propulsive force of a flexible flapping thin airfoil. J. Aircraft **46**, 465–473 (2009)

Gulcat, U.: (2011) Shortcuts in unsteady and flapping wing aerodynamics, Invited Lecture, AIAC'2011–002, September 14-16 2011, METU, Ankara, TURKEY

Gulcat, U.: *Fundamentals of Modern Unsteady Aerodynamics*, Second Ed. Springer, Verlag, p. 298 (2016)

Gulcat, U.: Aerodynamic thrust optimization of a flapping thin airfoil. In: 9th Ankara International Aerospace Conference, 20–22 September 2017, METU Ankara TURKEY, AIAC-2017-27 (2017a)

Gulcat, U.: State-space representation of flapping wings in Hover. In: 9th Ankara International Aerospace Conference, 20–22 September 2017-METU, Ankara TURKEY, AIAC-2017-60 (2017b)

Gulcat, U.: Aerodynamic thrust optimization with nonlinear modeling for the leading edge vortex of a flapping wing. In: 10th Ankara International Aerospace Conference, 18–20 September 2019-METU, Ankara TURKEY, AIAC-2019-27 (2019a)

Gulcat, U.: Aerodynamic stability analysis of a flapping wing in hover using state-space representation. In: 10th Ankara International Aerospace Conference, 18–20 September 2019-METU, Ankara TURKEY, AIAC-2019-90 (2019b)

Izraelevitz, J.S., Quiang, Z., Triantafyllou, M.S.: State-space adaptation of unsteady lifting line theory twisting/flapping wings of finite span. AIAA J. **55**(4), 1279–1294 (2017)

Jones, K.D., Dohring, C.M., Platzer, M.F.: Experimental and computational investigation of Knoller-Betz effect. AIAA J. **36**, 1240–1246 (1998)

Kaya, M., Tuncer, I.H.: Nonsinusoidal path optimization of a flapping airfoil. AIAA J. **45**, 2075–2082 (2007)

Katzmayer, R.: *Effect on Periodic Changes of Angle of Attack on Behavior of Airfoils*, NACA TM-147 (1922)

Leishman, G.: *Principle of Helicopter Aerodynamics,* pp. 341–342. Cambridge University Press (2000)

Mouy, A., Rossi, A., Taha, H.E.: Coupled unsteady aero-flight dynamics of hovering insects/ flapping micro air vehicles. J. Aircraft **54**(5), 1738–1749 (2017)

Nelson, R.C.: Flight Stability and Automatic Control, vol. 2, p. 217. McGraw-Hill, New York (1998)

Platzer, M.F., Jones, K.D., Young, J., Lai, C.S.: Flapping-wing aerodynamics: progress and challenges. AIAA J. **46**, 2136–2149 (2008)

Reich, G.W., Eastep, F.E., Altman, A., Alberani, R.: Transient poststall aerodynamic modeling for extreme maneuvers in micro air vehicles. J. Aircraft **48**(2), 403–411 (2011)

Taha, H.E., Hajj, M.R., Beran, P.S.: State-space representation of the unsteady aerodynamics of flapping flight. Aerosp. Sci. Technol. **34**, 1–11 (2014a)

Taha, H.E., Hajj, M.R., Nayfeh, A.H.: Longitudinal flight dynamics hovering MAVs/Insects. J. Guid. Control Dyn. **37**(3), 970–978 (2014b)

Tuncer, I.H., Kaya, M.: Optimization of flapping airfoils for maximum thrust and propulsive efficiency. AIAA J. **43**, 2329–2336 (2005)

Uhlig, D.V., Selig, M.S.: Modeling micro air vehicle aerodynamics in unsteady high angle of attack flight. J. Aircraft **54**(3), 1064–1075 (2017)

Walker, W.P.: Optimization of harmonically deforming thin airfoils and membrane wings for optimum thrust and efficiency, Ph.D. Thesis, Virgina Polytechnic Institute and State University, May 2012

Walker, W.P., Patil, M.J.: Unsteady aerodynamics of deformable thin airfoils. J. Aircraft **51**, 1673–1680 (2014)

Wang, J.Z.: Vortex shedding and frequency selection in flapping flight. J. Fluid Mech. **410**, 323–341 (2000)

Chapter 10
Aerodynamics: The Outlook for the Future

In previous chapters, we have seen how the foundations of the aerodynamics were established and the developments were made in a little more than a century in this discipline in relation to the Aerospace Engineering applications. The progress is still continuing thanks to the advances made in wind tunnel and flight test measurements as well as the remarkable improvements achieved in computational means implemented in numerical simulations.

The knowledge provided by the classical aerodynamics is sufficient to determine the aerodynamic performances of the high aspect ratio wings at low subsonic speeds and the low aspect ratio wings at supersonic speeds. On the other hand, as the speed or angle of attack increases and/or the aspect ratio decreases, we need modern concepts for aerodynamic analysis. The increase in cruise speeds causes unsteady fluid–structure interaction because of unavoidable elastic behavior of high aspect ratio wings, and it also causes the wing to reach critical Mach numbers because of compressibility effects at high subsonic speeds. The three dimensional aeroelastic analyses of such wings can be done with reasonable computational effort because of advances made in modern aerodynamics. In addition, the design of supercritical airfoils, which has the geometry to delay the critical Mach number, has made the high subsonic cruise speed possible for the civilian and military aircrafts with wings having high aspect ratio, low sweep, low induced drag and high L/D for almost more than a quarter of a century.

During the last quarter of the twentieth century, the numerical and experimental studies performed for predicting the extra lift caused by the strong suction of a separated flow from the sharp leading edge made the design and construction of the planes with delta wings which are highly maneuverable at high angles of attack possible. At higher angles of attack the wing rock may occur depending on the sweep angle. The recent studies emphasize the effect of the leading edge sharpness or roundness on the wing rock phenomenon.

Ü. Gülçat, *Fundamentals of Modern Unsteady Aerodynamics*,
https://doi.org/10.1007/978-3-030-60777-7_10

One of the ultimate and ambitious aims of the aerospace industries is to design and construct very fast vehicles which are to take the vast distances between the major cities on earth in a few hours. The research and development branches of major aerospace companies have been conducting research to design a fast aerospace plane which can travel a distance equivalent of the half of the earth circumference in a couple of hours. All these designs are based on the sustainable hypersonic flight at upper levels of atmosphere. The concept of 'wave-rider' which was introduced more than half a century ago has become hot again because of its considerably high L/D values for sustainable hypersonic flight. The continuous hypersonic flight, on the other hand is possible only with powerful engines based on the supersonic combustion of fuels with very high heating capacities. The sustainable supersonic combustion, once thought to be out of question because of being unstable, first became possible under laboratory conditions since 1990′s, and then were tested on small unmanned hypersonic vehicles for short durations after the introduction of flame control devices which can provide controls over time intervals less than a millisecond. However, so far most of the attempts made in sustainable hypersonic flight tests have failed. Since the costs of these tests are too high, to reduce the risk of failure it is necessary to go through intense and time consuming studies. In order to have most risk free tests, it is necessary to start with an adequate data base for the relevant flight conditions. This, naturally, requires large data base exchange among the countries which allocate substantial budgets for their aerospace development programs.

The advances made, during last two decades, in research and development indicate that the interest in aerodynamics is in two opposite directions. The first direction is the steady or unsteady flow analysis for very small sized objects, which may even operate indoors at low Reynolds number and at moderate to high angles of attack. The second one is the aerothermodynamics of the large sized aerospace vehicles which can cruise at very high altitudes with very high speeds.

The design and construction of unmanned light small sized air vehicles fall under the first direction mentioned above. Shown in Fig. 10.1 are comparative positions of the flying objects, ranging from very small to large, on a graph represented as the relation between the flight Reynolds number and the mass as modified from Mueller and DeLaurier.

The small unmanned air vehicles are to fly and operate in Laminar flow regime as seen from Fig. 10.1. The flight of birds, however, occurs in laminar to turbulent transition. Both the small planes and the large jumbo jets flying in subsonic regimes function totally in turbulent flows. Shown in the left corner of Fig. 9.1, the flying insects, with their mass being less than a gram, generate lift and propulsion with flapping wings. In a hovering flight of insects, the free stream speed is zero; therefore, the maximum wing tip speed is taken as the characteristic speed for determining the Reynolds number. The flapping frequency of the wings is quite high for the considerably small wing span which makes the tip velocity still to yield a laminar flow. The flapping of wings for a hovering flight either occurs in a symmetrical forward and backward fashion with respect to a horizontal plane, or asymmetrical upstrokes and downstrokes with respect to an almost vertical plane (Wang). In the first type of

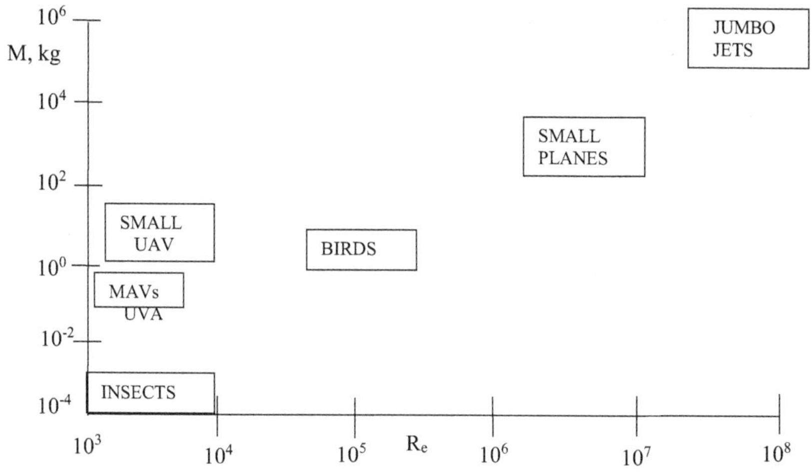

Fig. 10.1 Mass vs. Reynolds numbers for the flying objects varying from very small to very large

flapping the lifting force of the profile provides the hover, whereas in the second kind of flapping the hover is maintained with the drag generated by the profile. In addition, the experiments show that there is a sufficient lifting force generated by the wings flapping with amplitudes larger than their chords.

The sustainable forward flight with wing flapping is possible if the Reynolds number based on the free stream speed is larger than a critical value. Actually, for a thin airfoil at an effective angle of attack less than the dynamic stall angle, the product of the reduced frequency with the dimensionless plunge amplitude, kh, plays also an important role to get a propulsive force, Fig. 8.31, adapted from Gulcat. The empirical criteria, in a laminar flow regime, to obtain a propulsive force with flapping becomes: $log_{10}(R_e)*kh > 0.72$, where R_e is the Reynolds number based on the free stream speed. Below this value, negative propulsion is created. At higher angles of attack, where there is a strong leading edge vortex formation at very slow free stream velocities, the criteria to generate a propulsive force are based on the Reynolds number expressed, independently from the free stream speed, in terms of the frequency ω and the airfoil chord c reads as: $R_e = \omega c^2/\nu > 50$ (Wang). The first criterion is useful for cruising of the micro air vehicles, whereas the second criterion is helpful during the transition from hover to forward flight.

The purpose of defining a criterion for the sustainable flight conditions is to design, construct and operate small size air vehicles mainly capable of hover and/or fly forward with flapping wings as is done in nature. In this respect, the principal aerodynamic challenge in Micro Air Vehicle design is recently stated, in the conclusions and recommendation section of NATO TR-AVT-101 publication, as the search for *the greater robustness; namely, gust tolerance, maneuverability, and more predictable handling quantities such as capacity to hover or even perch rather than the pursuit of greater efficiency!* (TR-AVT-101).

The second direction in aerodynamic research is the design of very large and very fast aerospace planes which operate in high altitudes. Obviously, because of compressibility, heating and the chemical decomposition of the air at very high speeds, the multidisciplinary concepts from thermodynamics and the chemistry must also be considered. Shown in Fig. 10.2 is the historical and comparative development of the air vehicle range, speed and the cruising altitudes adapted from Küchemann and Noor.

The air vehicles shown in Fig. 10.2 travel their indicated ranges R, which are expressed in terms of the earth's diameter D, at about same time duration with cruising at given Mach numbers. At the upper right corner of Fig. 10.2, the 'wave rider' concept takes its position as the future aerospace plane to cruise at hypersonic speeds. The necessary steps to be taken with specific consideration to aerothermochemistry to develop such hypersonic planes are described in a paper by Tincher and Burnett. In their work, they further study the capabilities of such a plane to maneuver with assistance of the gravity in the atmosphere of a planet while making interplanetary travel in the future. The research related to the hypersonic aerodynamics made in Europe and USA during last two decades is published under the title of 'Sustainable Hypersonic Flight' in AGARD-CP-600. The national and/or multinational aerospace programs mentioned in this conference proceedings, however, are either continuing with delay or postponed or even canceled due to budgetary restrictions at the start of the new millennium. The more up to date version of Fig. 10.2 is given by Ahmed K. Noor and Venneri in their book '*Future Aeronautical and Space System*' published in AIAA series. In their work, the design and performance characteristics of single or multistage, faster than 12 Mach planes, which can orbit in the outside of our atmosphere, are provided. In this context, at Mach numbers less than 12, only the sub-orbital flights in the upper atmosphere seem to be possible. In this context, the most recent review of the

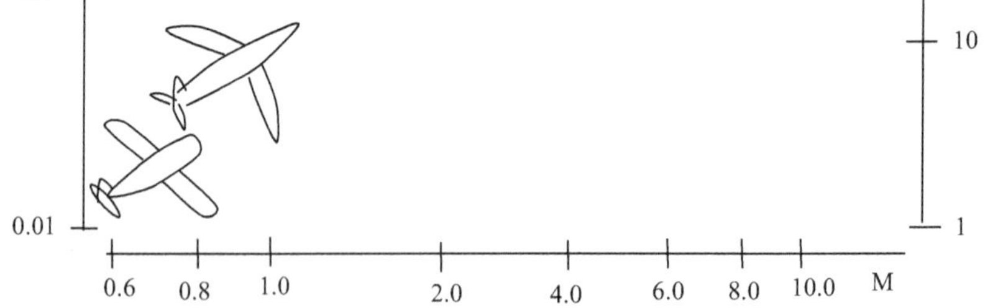

Fig. 10.2 Advances in Aerospace vehicles: Range and Altitude vs. Mach number

challenges and the critical issues concerning the reliability of a computational data and the limitations of the experimental data for hypersonic aerothermodynamics is provided in the extensive summary by Bertin and Cummings.

Speculative and overall predictions based on the various sources about the future aerospace projects as well as on the different scientific endeavors are provided by physicist Michio Kaku, who is a renowned Popular Science writer, in his recent book on visions (Kaku). As a futurologist, Kaku's predictions on the future extends to the end of the twenty-first century wherein he sees the realization of projects related to even interstellar travel, which will increase our level of civilization to type I civilization according to the classification of civilizations defined by Nicolai Kardeshev.

At the beginning of this millennium an abominable act of terror committed with four hijacked midsize jetliners shocked the whole world and changed the direction of research and development in the western world drastically. This change, mainly concerning national security, affected the research areas in many disciplines as well as the direction of research in aerodynamics. The necessity of developing MAVs functioning outdoors as well as indoors have become significant in operations related to the security of humankind for many years to come (TR-AVT-101). In this context, the unsteady aerodynamic tools are not only applied to analyze propulsive forces for aerial vehicles but also for the possible presence of explosive trace detection at the human aerodynamic wake (Settles) for aviation security applications in a nonintrusive and reliable manner. Similar particle detection studies are performed because of an outburst of a pandemic at the beginning of the 3rd decade of the 3rd millennium. A virus labeled COVID-19 (COrona VIrus Disease *20*19) spread to more than 180 countries in 3 months. This type of virus spreads very effectively by means of micro-droplets of saliva exhaled, sneezed or coughed out by a person. These droplets can be received by person with inhaling or with a direct contact with surfaces contaminated with the virus. In order to prevent the reception by inhaling a minimum safe distance is set as the 'social distance' between the two stationary persons..However, under non-stationary conditions and movements, like walking or running, the droplets' trajectories are predicted with wake aerodynamics via CFD (Blocken, 2020) and validated and calibrated at wind tunnels. It seems the quantity of the droplets received by the person is also important for this COVID-19 virus to be effective. This makes the further computations essential for determination of the safe social distances which restricts our life styles for years to come.

Related to the MAV technologies the recent developments has been in the field of unsteady aerodynamic optimization and sensitivity, especially, in flapping (Kaya et al. 2009) and deforming thin airfoils and membranes for optimum thrust and efficiency (Walker 2012) and [Walker and Patil, 2014]. Furthermore, Gulcat applied a similar optimization technique with both linear and nonlinear modeling of the leading edge vortex effect on the optimum thrust in (Gulcat 2017a) and (Gulcat 2019a) respectively. Also, for nonlinear unsteady theory of fish swimming and bird/insect flight an iterative approach to the study of flexible bodies and wings is given by Wu (2011), wherein meso Reynolds numbers of $1000 < Re < 10{,}000$ and incident angles of $20^0 - 40^0$ range are considered. In this context, the fluid-body interaction is studied, experimentally and theoretically, for flapping flags, fluttering objects and folding of leaves as self-streamlining bodies as wind-driven configurations by Shelley and Zhang (2011), wherein Reynolds number range lies in 2000–40,000.

Real time applications of unsteady aerodynamics on flying vehicles in extreme maneuvers require simple and efficient mathematical models which was first utilized as state-space representation of aerodynamics for airfoils and low aspect ratio wings by Goman and Khrabrov (1994). More recent application is given by Reich et.al. (2011) in an attempt to determine the flight control parameters of a wing concept for a perching MAV. For future applications, a detailed model study of state-space representation of aerodynamics is provided by Brunton (2012). In this context, aerodynamic stability and trim applications are presented in (Taha et al. 2014a, b) and (Gulcat 2017b, 2019b).

There have been and there will be many bio-inspired unsteady aerohydrodynamic flow studies related to the principles of the animal locomotion as in bird/insect fight or fish swim. There is, however, a relevant research area in biology related to the mammalian spermatozoa motility which is a subject of growing importance because of rising infertility and the possibility of improving animal breeding (Gafney et al. 2011). In their study, interestingly, human sperm migrations are found to be unaffected by the viscosities of the range $0.0007 - 0.14$ Pa-s, during the sperm's journey towards the egg with the speed of $25 - 45\ \mu m\ \mathrm{s}^{-1}$ and with its flagellum of length $55\ \mu m$ beating with the frequency of 60–120 rad s^{-1} corresponding to different mode shapes, all happening in unsteady microscopic scales in female reproductive tract!

The last but not the least of many applications of unsteady aerodynamics is the studies of power extraction from an aerohydrodynamically controlled oscillating-wing for the purpose of clean energy generation for prevention of climate change. The possibility of producing energy from sailing ships or from tethered power generators flying in the global jet streams may increase the available energy densities one order of magnitude higher than the current energy densities available with conventional windmilling techniques on the surface of earth or power from rivers and tides (Platzer and Klijn 2009).

Summary: *The outlook and future of the aerodynamics are discussed. The present advances imply that aerodynamics in the future is heading towards the analysis of unmanned very small and very slow vehicles, and manned or unmanned very fast and large vehicles. For the former Nature is closely observed. For the latter, however, several countries are collaborating, since the financial requirements are enormous for such projects. It seems some such aerospace projects have been canceled already because of financial burdens. Before the turn of the twentieth century, some scientific views were presented for the forecast of the development in aerospace projects during the twenty-first century. However, after 9/11 many military, and civilian, projects and studies were devoted to the national security. Inevitably, the discipline of aerodynamics will be influenced with this change and the research needed for increasing the wind and flowing water energy densities one order of magnitude.*

References

AGARD-CP-600, Future Aerospace Technology in the Service of the Alliance: Sustained Hypersonic Flight, V.3, December 1997

Bertin, J.J. and Cummings, R.M.: Critical Hypersonic Aerothermodynamic Phenomena, Annual Review of Fluid Mechanics, 2006

Blocken, B., Malizia, F., van Druenen, T., Marchal, T. 2020, Towards Aerodynamically Equivalent COVİD19 1.5m Social Distancing for Walking and Running, https://www.urbanphysics.net/Social%20Distancing%20v20_White_Paper.pdf

Brunton, S.L., 2012. Unsteady Aerodynamic Models for Agile Flight at Low Reynolds Numbers, PhD Dissertaion; Aerospace Engineering Department, Virgina Polytechnique Institute and State University

Gaffne, E.A., Gadelha, H., Smith, D.J., Blake, J.R., and Kirkman-Brown, J.C., 2011. Mammalian Sperm Motility: Observation and Theory, Annual Review of Fluid Mechanics

Goman, M. and Khrabrov, A., Sept.-Oct. 1994, State-Space Representation of Aerodynamic Characteristics of an Aircraft at High Angles of Attack, Journal of Aircraft, Vol. 31, No. 5

Gulcat, U.: Propulsive Force of a Flexible Flapping Thin Airfoil, Journal of Aircraft, Vol. 46, No. 2, March-April 2009

Gulcat, U. (2017a) Aerodynamic Thrust Optimization of a Flapping Thin Airfoil, 9thAnkara International Aerospace Conference, 20–22 September 2017, METU AnkaraTURKEY, AIAC-2017–27

Gulcat, U. (2017b) *State-Space Representation of Flapping Wings in Hover,* 9thAnkara International Aerospace Conference, 20–22 Sept. 2017-METU, Ankara TURKEY, AIAC-2017–60

Gulcat, U. (2019a) *Aerodynamic Thrust Optimization with Nonlinear Modeling for the leading edge vortex of a flapping wing,* 10th Ankara International Aerospace Conference, 18- 20 Sept. 2019-METU, Ankara TURKEY, AIAC-2019–27

Gulcat, U. (2019b) *Aerodynamic Stability Analysis of a Flapping Wing in Hover Using State-Space Representation,* 10th Ankara International Aerospace Conference, 18–20 Sept. 2019-METU, Ankara TURKEY, AIAC-2019–90

Kaku, Michio, How Science Will Revolutionize the 21st Century and Beyond: *Visions*, Oxford University Press, 1998

Kaya, M., Tuncer, I.H., Jones, K.D. and Patzer, M.F., March-April 2009, Optimization ofFlapping Motion Parameters for Two Airfoils in a Biplane Configurations,Journal of Aircraft, Vol. 46, No. 2

Küchemann, D.: Aerodynamic Design of Aircraft. Pergamon Press, Oxford (1978)

Mueller, Thomas J. and DeLaurier J.D.: Aerodynamics of Small Vehicles, Ann. Rev. Fluid Mech, 2003

NATO TR-AVT-101, 'Experimental and Computational Investigations in Low Reynolds Number Aerodynamics, with Application to Micro Air Vehicles (MAVs), June 2007

Noor, Ahmed K. and L.S. Venneri: *Future Aeronautical and Space Systems*, V.172, Progress in Astronautics and Aeronautics, AIAA, 1997

Platzer, M. and N. Sarigul-Klijn, 'A novel Approach to Extract from Free-Flowing Water and High Altitude Jet Streams, Proceedings of ES2009, Energy Sustainability2009, July 19–23, San Francisco, California USA

Taha, H.E., Hajj, M.R., Beran, P.S.: State-Space Representation of the Unsteady Aerodynamics of Flapping Flight. Aerosp. Sci. Technol. **34**, 1–11 (2014)

Taha, H.E., Hajj, M.R., Nayfeh, A.H.: Longitudinal Flight Dynamics Hovering MAVs/Insects. Journal of Guidence, Control, and Dynamics **37**(3), 970–978 (2014)

Settles, G.S.: Fluid Mechanics and Homeland Security, Annual Review of Fluid Mechanics, 2006

Reich, G.W., Eastep, F.E., Altman, A., Albertani, R., March-April. : Transient Poststall Aerodynamic Modeling for Extreme Maneuvers in Micro Air Vehicles. Journal of Aircraft **48** (2), 403–411 (2011)

Shelley, M.J. and Zhang, J., 2011. Flapping and Bending Bodies Interacting with Fluid Flows, Annual Review of Fluid Mechanics

Tincher, Douglas J. and D.W. Burnett: Hypersonic Waverider Test Vehicle: A LogicalNext Step, Journal of Spacecraft and Rockets, V.31, No3, May-June 1994

Optimization of Harmonically Deforming Thin Airfoils and Membrane Wings for Optimum Thrust and Efficiency, PhD Dissertaion; Aerospace Engineering Department, Virgina Polytechnique Institute and State University

Wang, Jane Z.: Dissecting Insect Flight, Annual Review of Fluid Mechanics, 2005

Wu, T.Y., 2011. Fish Swimming and Bird/Insect Flight, Annual Review of Fluid Mechanics

Appendices

A1: Generalized Curvilinear Coordinate Transform

Let the transformation from rectangular, xyz, to curvilinear, $\xi\eta\varsigma$, coordinates be

$$\begin{aligned} \xi &= \xi(x, y, z, t) \\ \eta &= \eta(x, y, z, t) \\ \zeta &= \zeta(x, y, z, t) \\ \tau &= t \end{aligned}$$

as shown in the following Fig. A1.1.

The differential lengths in curvilinear coordinates then become

$$\begin{aligned} d\xi &= \xi_x dx + \xi_y dy + \xi_z dz + \xi_t dt \\ d\eta &= \eta_x dx + \eta_y dy + \eta_z dz + \eta_t dt \\ d\varsigma &= \varsigma_x dx + \varsigma_y dy + \varsigma_z dz + \varsigma_t dt \\ d\tau &= dt \end{aligned} \rightarrow \begin{pmatrix} d\xi \\ d\eta \\ d\varsigma \\ d\tau \end{pmatrix} = \begin{bmatrix} \xi_x & \xi_y & \xi_z & \xi_t \\ \eta_x & \eta_y & \eta_z & \eta_t \\ \varsigma_x & \varsigma_y & \varsigma_z & \varsigma_t \\ 0 & 0 & 0 & 1 \end{bmatrix} \begin{pmatrix} dx \\ dy \\ dz \\ dt \end{pmatrix} \quad \text{(A1.1)}$$

In Eq. A1.1, the determinant of the coefficient matrix is named as Jacobian determinant, which in open form reads as

$$\begin{aligned} J &= \partial(\xi, \eta, \varsigma, \tau)/\partial(x, y, z, t) \\ &= \xi_x(\eta_y\zeta_x - \eta_z\varsigma_{yx}) - \xi_y(\eta_x\varsigma_x - \eta_z\varsigma_x) + \xi_z(\eta_x\zeta_y - \eta_y\zeta_x). \end{aligned}$$

In rectangular coordinates the flux vectors are defined to be

$$\vec{U}_t + \vec{F}_x + \vec{G}_y + \vec{H}_z = \vec{R}.$$

Ü. Gülçat, *Fundamentals of Modern Unsteady Aerodynamics*,
https://doi.org/10.1007/978-3-030-60777-7

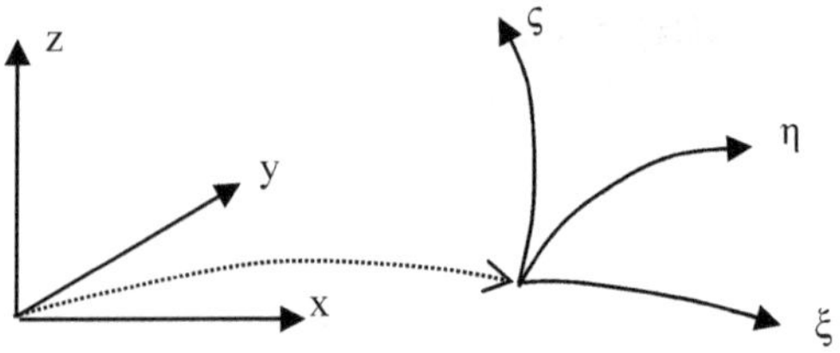

Fig. A1.1 Generalized curvilinear coordinate transformation

These flux vectors, using the chain rule, in curvilinear coordinates become

$$\frac{\partial}{\partial x}=\frac{\partial}{\partial \xi}\xi_x+\frac{\partial}{\partial \eta}\eta_x+\frac{\partial}{\partial \varsigma}\varsigma_x$$
$$\frac{\partial}{\partial y}=\frac{\partial}{\partial \xi}\xi_y+\frac{\partial}{\partial \eta}\eta_y+\frac{\partial}{\partial \varsigma}\varsigma_y$$
$$\frac{\partial}{\partial z}=\frac{\partial}{\partial \xi}\xi_z+\frac{\partial}{\partial \eta}\eta_z+\frac{\partial}{\partial \varsigma}\varsigma_z$$
$$\frac{\partial}{\partial t}=\frac{\partial}{\partial \tau}+\frac{\partial}{\partial \xi}\xi_t+\frac{\partial}{\partial \eta}\eta_t+\frac{\partial}{\partial \varsigma}\varsigma_t$$

The equation of motion in curvilinear coordinates then becomes

$$\vec{U}_t+\vec{F}_\xi\xi_x+\vec{F}_\eta\eta_x+\vec{F}_\varsigma\varsigma_x+\vec{G}_\xi\xi_x+\vec{G}_\eta\eta_x+\vec{G}_\varsigma\varsigma_x+\vec{H}_\xi\xi_x+\vec{H}_\eta\eta_x+\vec{H}_\varsigma\varsigma_x=\vec{R} \tag{A1.2}$$

The strong conservative form of Eq. A1.2 is obtained by dividing A1.2 with J and rearranging as follows

$$\left(\frac{\vec{U}}{J}\right)_\tau+\left(\frac{\vec{F}_\xi\xi_x+\vec{F}_\eta\eta_x+\vec{F}_\varsigma\varsigma_x+\vec{G}_\xi\xi_x}{J}\right)_\xi+\left(\frac{\vec{G}_\xi\xi_x+\vec{G}_\eta\eta_x+\vec{G}_\varsigma\varsigma_x}{J}\right)_\eta+\left(\frac{\vec{H}_\xi\xi_x+\vec{H}_\eta\eta_x+\vec{H}_\varsigma\varsigma_x}{J}\right)_\varsigma=\frac{\vec{R}}{J} \tag{A1.3}$$

If in Eq. A1.3:

$$\vec{U}_1=\frac{\vec{U}}{J},\quad \vec{F}_1=\left(\frac{\vec{F}_\xi\xi_x+\vec{F}_\eta\eta_x+\vec{F}_\varsigma\varsigma_x+\vec{G}_\xi\xi_x}{J}\right),\vec{G}_1=\left(\frac{\vec{G}_\xi\xi_x+\vec{G}_\eta\eta_x+\vec{G}_\varsigma\varsigma_x}{J}\right)$$
$$\text{and}\quad \vec{H}_1=\left(\frac{\vec{H}_\xi\xi_x+\vec{H}_\eta\eta_x+\vec{H}_\varsigma\varsigma_x}{J}\right)$$

then it becomes

$$\frac{\partial\vec{U}_1}{\partial\tau}+\frac{\partial\vec{F}_1}{\partial\xi}+\frac{\partial\vec{G}_1}{\partial\eta}+\frac{\partial\vec{H}_1}{\partial\varsigma}=\frac{\vec{R}}{J}$$

Let us now, rewrite the equation of continuity in the strong conservative form

$$\frac{\partial\rho}{\partial t}+\frac{\partial\rho u}{\partial x}+\frac{\partial\rho v}{\partial y}+\frac{\partial\rho w}{\partial z}=\frac{\partial\rho}{\partial\tau}+\frac{\partial\rho}{\partial\xi}\xi_t+\frac{\partial\rho}{\partial\eta}\eta_t+\frac{\partial\rho}{\partial\varsigma}\varsigma_t+\frac{\partial\rho u}{\partial\xi}\xi_x+\frac{\partial\rho u}{\partial\eta}\eta_x+\frac{\partial\rho u}{\partial\varsigma}\varsigma_x$$
$$+\frac{\partial\rho v}{\partial\xi}\xi_y+\frac{\partial\rho v}{\partial\eta}\eta_y+\frac{\partial\rho v}{\partial\varsigma}\varsigma_y+\frac{\partial\rho w}{\partial\xi}\xi_z+\frac{\partial\rho w}{\partial\eta}\eta_z+\frac{\partial\rho w}{\partial\varsigma}\varsigma_z=0. \tag{A1.4}$$

If we divide A1.4 by J, and note that $(\xi_i/j)_\xi+(\eta_i/J)_\eta+(\varsigma_i/J)_\varsigma=0,\ i=x,y,z$ we obtain

$$\frac{\partial\rho/J}{\partial\tau}+\frac{\partial\rho(\xi_t+\xi_x u+\xi_y v+\xi_x w)/J}{\partial\xi}+\frac{\partial\rho(\eta_t+\eta_x u+\eta_y v+\eta_x w)/J}{\partial\eta}+\frac{\partial\rho(\varsigma_t+\varsigma_x u+\varsigma_y v+\varsigma_x w)/J}{\partial\varsigma}=0$$

The derivation of the flux terms are performed similarly.

Summary: *The basics of generalized curvilinear coordinate transformation is provided.*

A2: Carleman Formula

The integral transform: If at x = ξ $g(x)=\frac{1}{2\pi}\int\limits_{-1}^{1}\frac{f(\xi)d\xi}{x-\xi}$ is singular then what is *f(x)* ?

Let us take the inverse of this integral. Let $g_1(\theta)$ be a regular function in the interval 0 < θ < π. The Hilbert integral form of this function reads as

$$g_1(\theta)=\frac{1}{\pi^2}\int_0^\pi\int_0^\pi K(\theta,\alpha)K(\phi,\alpha)g_1(\phi)d\alpha d\phi+\frac{1}{\pi}\int_0^\pi g_1(\phi)d\phi,\quad K(\theta,\phi)=\frac{\sin\phi}{\cos\phi-\cos\theta}$$

Which is singular at θ = ϕ. If we write

$f_1(\alpha)=\frac{1}{\pi}\int\limits_0^\pi K(\phi,\alpha)g_1(\phi)d\phi$, then

$$g_1(\theta)=\frac{1}{\pi}\int_0^\pi K(\theta,\alpha)f_1(\alpha)d\alpha+\frac{1}{\pi}\int_0^\pi g_1(\phi)d\phi$$

This gives us for the pair of functions f_1 and g_1 the following integral relations

$$f_1(\theta)=\frac{1}{\pi}\int_0^\pi g_1(\phi)\frac{\sin\theta}{\cos\theta-\cos\phi}d\phi \tag{A2.1a, b}$$

and $g_1(\theta)=\frac{1}{\pi}\int_0^\pi f_1(\phi)\frac{\sin\phi}{\cos\phi-\cos\theta}d\phi+\frac{1}{\pi}\int_0^\pi g_1(\phi)d\phi$

Here, the function f_1 is named as the Hilbert transform of g_1 in the interval $(0,\pi)$. Now, let $\cos\theta = x$ and $\cos\phi = \xi$ to yield

$$F(x) = \frac{f_1(\arccos x)}{\sqrt{1-x^2}} \quad \text{and} \quad G(x) = \frac{g_1(\arccos x)}{\sqrt{1-x^2}}.$$

The integral 1 – a in terms of x reads as $F(x) = \frac{1}{\pi}\int_{-1}^{1}\frac{G(\xi)d\xi}{\xi - x} \quad (-1<x<1),$

$$\tag{A2.2}$$

and its inverse reads as $\sqrt{1-x^2}G(x) = \frac{1}{\pi}\int_{-1}^{1}\sqrt{1-\xi^2}\frac{F(\xi)d\xi}{x-\xi} + \frac{1}{\pi}\int_{-1}^{1}G(\xi)d\xi.$

$$\tag{A2.3}$$

The last term of the right hand side of Eq. A2.3 is equal to an arbitrary constant and it makes the integral non-unique. The aerodynamically meaningful result can be reached with assigning proper value to this constant. In rectangular coordinates, if the free stream direction is in the direction of x axis G(1) value must be finite in order to satisfy the Kutta condition. Hence, the arbitrary constant can be chosen as follows.

$$\int_{-1}^{1} G(\xi)d\xi = \int_{-1}^{1}\sqrt{1-\xi^2}F(\xi)\frac{d\xi}{1-\xi}$$

Using the expression above in Eq. A2.3 gives

$$G(x) = \frac{1}{\pi}\sqrt{\frac{1-x}{1+x}}\int_{-1}^{1}\sqrt{\frac{1+\xi}{1-\xi}}F(\xi)\frac{d\xi}{x-\xi} \tag{A2.4}$$

As the last step if take $\boldsymbol{G(\xi) = -f(\xi)/2}$ in A1 and $\boldsymbol{g(x) = F(x)}$ in A2.3 we get

$$g(x) = \frac{1}{2\pi}\int_{-1}^{1}\frac{f(\xi)d\xi}{x-\xi} \tag{A2.5}$$

and

$$f(x) = -\frac{2}{\pi}\sqrt{\frac{1-x}{1+x}}\int_{-1}^{1}\sqrt{\frac{1+\xi}{1-\xi}}g(\xi)\frac{d\xi}{x-\xi} \tag{A2.6}$$

The pair of Eqs. A2.5 and A2.6 gives us the Schwarz's inverse integral transform for the thin airfoil theory.

Summary: *The Hilbert Integral representation of a suitably regular function is utilized to obtain the Schwarz solution of a thin airfoil problem. The original integral equation solution is not unique (1-a,b). The integral inversion which is suitable for the aerodynamics must satisfy the Kutta condition while providing a unique solution (3-a and b) which are expressed in terms of the Cauchy principle value of the integrals.*

A3: Cauchy Integral

The singular or non-singular definite integrals used to determine the aerodynamic coefficients are evaluated with the aid of complex integrals.

Example: The following integral I(x) for x either in or out of the interval [−1,1]:

$$\int_{-1}^{1} \sqrt{(1+\xi)/(1-\xi)}.d\xi/(\xi - x)$$

The above integral is singular for x being in –1 and 1. It can be evaluated using the Cauchy integral theorem. For this let us consider the complex plane $\zeta = \xi + i\eta$ and take the integral on a closed curve.

If we take $F(\zeta) = \sqrt{(1+\zeta)/(\zeta-1)}/(\zeta - x)$ the closed integral becomes $I = \oint F(\xi)d\xi$

Let us express the term under the radical in r-θ polar coordinates

$$\sqrt{(1+\zeta)/(\zeta-1)} = (R^2 + I^2)^{1/2} e^{i\theta} \quad ve \quad \theta = a\tan(I/R).$$

Here:
$R = ((\xi+1)(\xi-1)+\eta^2)/((\xi-1)^2+\eta^2)$ *and* $I = -2\xi/((\xi-1)^2+\eta^2)$.

Let the closed curve C1 be in −1 and 1 as shown in Fig. A3.1. At the top line of the curve $\eta = 0^+$ and $\xi - 1 \leq 0$ makes $R \leq 0, I = 0^-$ *and* $\theta = -\pi$. At the bottom line of the curve however $\eta = 0^-$ and $\xi - 1 \leq 0$ we have $R \leq 0, I = 0^+$ *and* $\theta = \pi$. Hence at the upper part we have

$$\sqrt{(1+\zeta)/(\zeta-1)} = \sqrt{(1+\xi)/(1-\xi)}e^{-i\pi/2} = -i\sqrt{(1+\xi)/(1-\xi)}$$

and at the lower part

$$\sqrt{(1+\zeta)/(\zeta-1)} = \sqrt{(1+\xi)/(1-\xi)}e^{i\pi/2} = i\sqrt{(1+\xi)/(1-\xi)}$$

Since in upper and lower lines $\zeta - x = \xi - x$ then $d\zeta = d\xi$. Around $\xi = x$ the arc radius ρ gives $\zeta - x = \rho e^{i\theta}$ and $d\zeta = \rho e^{i\theta}$ i dθ. The integral I then becomes

$$\mathrm{I} = \int_{-1}^{x-\rho} i\sqrt{(1+\xi)/(1-\xi)}.d\xi/(\xi - x) + \int_{x+\rho}^{1} i\sqrt{(1+\xi)/(1-\xi)}.d\xi/(\xi - x) + \int_{1}^{x+\rho} -i\sqrt{(1+\xi)/(1-\xi)}.d\xi/(\xi - x) + \int_{x-\rho}^{-1} -i\sqrt{(1+\xi)/(1-\xi)}.d\xi/(\xi - x) + \text{arc integrals.}$$

The counterclockwise line integral's first two terms come from the bottom line and the last two terms come from the upper line, and the upper and lower arc integrals cancel each other. If we let the arc radius go to zero and take the limit the singular integral becomes

$$\mathrm{I} = 2\,\mathrm{i} \int_{-1}^{1} \sqrt{(1+\xi)/(1-\xi)}.d\xi/(\xi - x) \tag{A3.1}$$

According to the Cauchy integral theorem, the integral I evaluated around the closed curve C will be the same as the integral evaluated around C1. Here, the curves C and C1 are the non-intersecting closed curves and the region enclosed between these two curves must be analytic.

$$I = \oint_{C_1} F(\zeta)d(\zeta) = \oint_{C} F(\zeta)d(\zeta)$$

Let us evaluate the integral about C as a circle whose radius is approaching infinity. Now, we observe that since $F(\zeta) \rightarrow 1/\zeta$ then the Cauchy theorem gives us

$$I = \oint_{C} (1/\zeta)d(\zeta) = 2\pi i \tag{A3.2}$$

If we equate integrals A3.1 and A3.2, we obtain

$$\int_{-1}^{1} \sqrt{(1+\xi)/(1-\xi)}.d\xi/(\xi - x) = \pi$$

Let us evaluate the same integral for the non-singular case where $x > 1$.

The closed curve C1, this time, can be taken without any arcs as a straight line as shown in Fig. A3.2. The value of integral I is found as A3.1. The closed curve C consist of C2, C3 which is around point x and the straight lines, right of point x, joining C2 and C3. The integral for the closed curve C can be written using residue theorems as follows

$$I = \oint_C F(\zeta)d(\zeta) = \oint_{C_2} F(\zeta)d(\zeta) - \oint_{C_3} F(\zeta)d(\zeta)$$

The value of the integral around C2 becomes 2πi if we let the radius go to infinity. The integral bout C3 becomes negative since the direction is clockwise. Using the residue theorem it reads as

$$\oint_{C_3} F(\zeta)d(\zeta) = 2\pi i\sqrt{(\xi+1)/(\xi-1)}$$

Equating the value of integrals gives

$$\int_{-1}^{1} \sqrt{(1+\xi)/(1-\xi)}.d\xi/(\xi-x) = \pi(1-\sqrt{(\xi+1)/(\xi-1)})$$

Summary: *Cauchy principle value of some real definite integrals is taken via line integrals of complex functions. Cauchy's integral formula and the residue theorems are utilized for this purpose.*

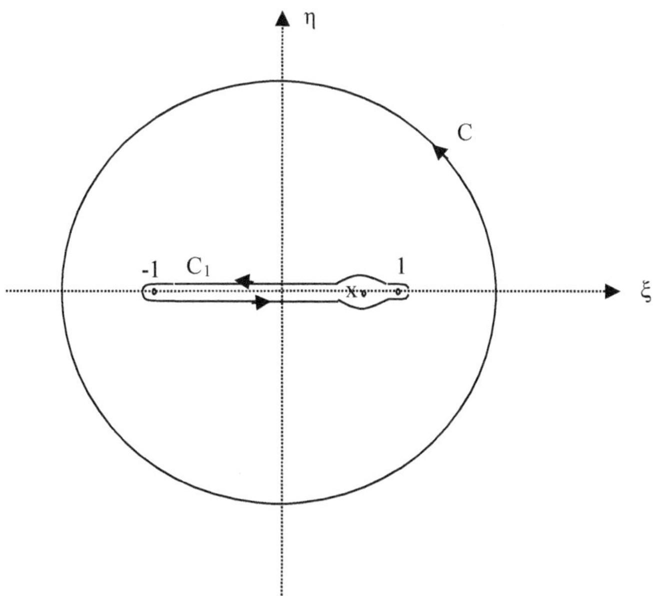

Fig. A3.1 The closed integrals for the interval $-1<x<1$

A4: Integral Tables

Singular Integrals $(-1 \leq x,\ \xi \leq 1)$

(1) $\int_{-1}^{1} \sqrt{\frac{1+\xi}{1-\xi}} \frac{d\xi}{x-\xi} = -\int_{-1}^{1} \sqrt{\frac{1-\xi}{1+\xi}} \frac{d\xi}{x-\xi} = -\pi$

(2) $\int_{-1}^{1} \sqrt{\frac{1+\xi}{1-\xi}} \frac{\xi\, d\xi}{x-\xi} = -\pi(1+x)$

(3) $\int_{-1}^{1} \sqrt{\frac{1-\xi}{1+\xi}} \frac{\xi\, d\xi}{x-\xi} = -\pi(1-x)$

(4) $\int_{-1}^{1} \sqrt{\frac{1-\xi}{1+\xi}} \frac{\xi^2\, d\xi}{x-\xi} = -\pi(1/2+x+x^2)$

(5) $\int_{-1}^{1} \sqrt{\frac{1-\xi}{1+\xi}} \frac{\xi^3\, d\xi}{x-\xi} = -\pi(1/2+x/2+3x^2/2+x^3)$

Non-Singular Integrals

(6) $\int_{-1}^{1} \sqrt{\frac{1+\xi}{1-\xi}} d\xi = \int_{-1}^{1} \sqrt{\frac{1-\xi}{1+\xi}} d\xi = \pi$

(7) $\int_{-1}^{1} \sqrt{\frac{1+\xi}{1-\xi}} \xi\, d\xi = -\int_{-1}^{1} \sqrt{\frac{1-\xi}{1+\xi}} \xi\, d\xi = \pi/2$

(8) $\int_{-1}^{1} \sqrt{\frac{1+\xi}{1-\xi}} \xi^2\, d\xi = \int_{-1}^{1} \sqrt{\frac{1-\xi}{1+\xi}} \xi^2\, d\xi = \pi/2$

(9) $\int_{-1}^{1} \sqrt{\frac{1+\xi}{1-\xi}} \xi^3\, d\xi = -\int_{-1}^{1} \sqrt{\frac{1-\xi}{1+\xi}} \xi^3\, d\xi = 3\pi/8$

(10) $\int_{-1}^{1} \sqrt{1-\xi^2}\, d\xi = \pi/2$

(11) $\int_{-1}^{1} \sqrt{1-\xi^2}\, \xi\, d\xi = 0$

(12) $\int_{-1}^{1} \sqrt{1-\xi^2}\, \xi^2\, d\xi = \pi/8$

Singular integrals with arbitrary integral limits for $\xi_1 \leq \xi \leq \xi_2$ and $-1 \leq x \leq 1$ can be evaluated as follows. Let

$$\mathrm{I} = \int_{\xi_1}^{\xi_2} \sqrt{\frac{1+\xi}{1-\xi}} \frac{\xi^n d\xi}{x-\xi}.$$

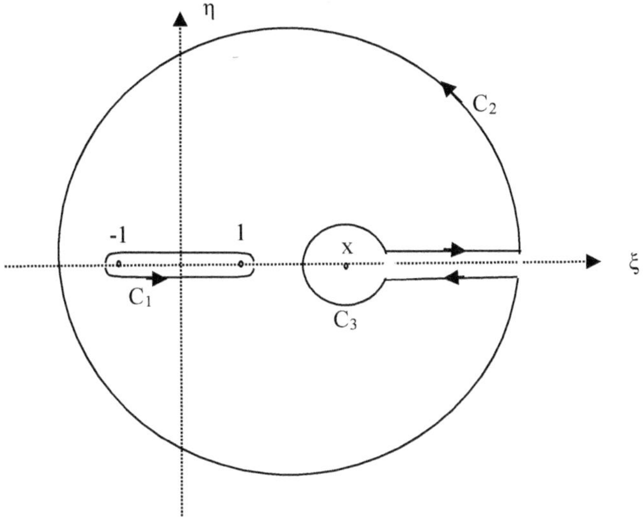

Fig. A3.2 The closed integrals for $x > 1$

For $n = 0$ multiplying the integrand by $\sqrt{(1+\xi)/(1+\xi)}$ yields

$$\mathrm{I} = \int_{\xi_1}^{\xi_2} \frac{1+\xi}{\sqrt{1-\xi^2}} \frac{d\xi}{x-\xi} = \int_{\xi_1}^{\xi_2} \frac{1}{\sqrt{1-\xi^2}} \frac{d\xi}{x-\xi} + \int_{\xi_1}^{\xi_2} \frac{\xi}{\sqrt{1-\xi^2}} \frac{d\xi}{x-\xi}.$$

Letting $t = \frac{1}{\xi - x}$ gives $\xi = 1/t + x$ and $d\xi = -dt/t^2$. The right hand side of the above integral, in terms of the new variable t, becomes

$\mathrm{I} = (1+x) \int_{t_1}^{t_2} \frac{dt}{\sqrt{R}} + \int_{t_1}^{t_2} \frac{dt}{t\sqrt{R}}$ where $R = -1 - 2xt + (1-x^2)t^2$. Above integrals are evaluated from the tables of Gradshteyn and Ryzhik (2000) as follows

$$\mathrm{I}_1 = \int \frac{dt}{\sqrt{R}} = \frac{1}{\sqrt{1-x^2}} \ln\left| 2\left[\sqrt{(1-x^2)(1-\xi^2)} + 2 - 2x\xi\right] / (\xi - x) \right| + c,$$

and

$$\mathrm{I}_2 = \int \frac{dt}{t\sqrt{R}} = -\arcsin(\xi) + c$$

Applying the integral limits ξ_1 *and* ξ_2 gives finally the value of integral I. In order to check the value of the integral let us substitute 1, that yields I = $-\pi$, which is equal to the value given by the Integral 1) of this Appendix.

For $n > 0$, the exponent ξ of the integrand, the singular integrals can be evaluated using a similar approach as described above.

A5: Hankel Functions

By definition, the first and the second kind of Hankel functions in terms of the Bessel functions read

$$H_n^{(1)}(x) = J_n(x) + iY_n(x)$$
$$H_n^{(2)}(x) = J_n(x) - iY_n(x)$$

Here, the Bessel functions J and Y, are the series solutions of the differential equations with variable coefficients given below

$$x^2\frac{d^2y}{dx^2} + x\frac{dy}{dx} + (x^2 - n^2)y = 0. \tag{A5.1}$$

The first solution of the Eq. A5.1 is called the Bessel function of the first kind nth order, and it is given as

$$J_n(x) = \sum_{k=0}^{\infty}\frac{(-1)^k (x/2)^{2k+n}}{k!(k+n)!}. \tag{A5.2}$$

The Bessel function of the second kind nth order is given as

$$Y_n(x) = \frac{2}{\pi}\left[(\log\frac{x}{2} + \gamma)J_n(x) - \frac{1}{2}\sum_{k=0}^{n-1}\frac{(n-k-1)!(x/2)^{2k-n}}{k!} + \frac{1}{2}\sum_{k=0}^{\infty}(-1)^{k+1}[\varphi(k) + \varphi(k+n)]\frac{(x/2)^{2k+n}}{k!(n+k)!}\right]$$

Here,: $\varphi(k) = \sum_{m=1}^{k}\frac{1}{m} = 1 + 1/2 + \ldots + 1/k$, $\varphi(0) = 0$, and $\gamma = \lim_{k\to 0}[\varphi(k) - \log k]$ $= 0.577215$, is given as Euler constant.

In addition, the modified Bessel functions from the normal Bessel functions can be written as: $I_n(x) = i^{-n} J_n(ix)$. Hence

$$I_n(x) = \sum_{k=0}^{\infty}\frac{(x/2)^{2k+n}}{k!(k+n)!}$$

Summary: *Hankel function of first and second kind of order n is defined in terms of the Bessel function of the first and second kind of order n. Modified Bessel function is also given.*

A6: The Response Function in a Linear System

The response of linear systems to the unit step function can be determined by means of the superpositioning technique. In this respect, let A(t) be the response of any linear system to the unit step function. Let us find the response of the same linear system to an arbitrary function f(t). First, let unit step function 1(t), which is discontinuous at t = 0, be defined as follows

$$I(t) = \begin{cases} 0, & t<0 \\ 1, & t \geq 0 \end{cases}$$

Since the response of the system to 1(t) is A(t), the response to Δf(t) which acts in time interval Δt can be found from the graph shown below:

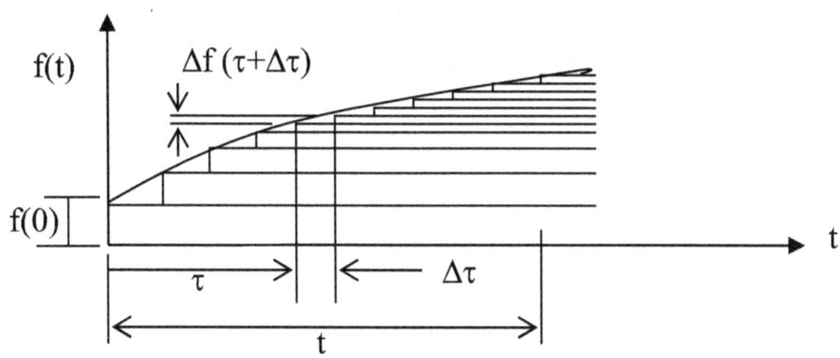

If for the system x is the dependent variable and t is the independent variable, and if at time level τ + Δτ Δf is acting on the system then at any time level t the response of the system will be

$$\Delta x(t, \tau + \Delta\tau) = \Delta f(\tau + \Delta\tau) A[t - (\tau + \Delta\tau)].$$

If we add up all responses which are due to the effect of f(t) before the time level t we obtain the following

$$x\,(t_2\tau+\Delta\tau) = f(0)\;A(t) + \sum_{r=0}^{r=t-\Delta\tau} \Delta f(\tau+\Delta\tau)A[t-(\tau+\Delta\tau)]$$

$$= f(0)\;A(t) + \sum_{z=0}^{r=t-\Delta\tau} \frac{\Delta f(\tau+\Delta\tau)}{\Delta\tau}A[t-(\tau+\Delta\tau)]\Delta\tau.$$

If we take the limit of the summation given above as $\Delta\tau$ goes to zero we obtain

$$x(\mathrm{t}) = \lim_{\Delta\tau\to 0}\{x(t,\tau+\Delta\tau)\} = f(0)\;A(t) + \int_0^t \frac{df(\tau)}{d\tau}A(t-\tau)\Delta\tau \qquad (A6.1)$$

The integral at the right hand side of Eq. A6.1 is called Duhemal's integral. Since f(t) is arbitrary a better version of it is given as follows

$$\mathrm{x(t)} = \mathrm{A(0)\;f(t)} + \int_0^t \mathrm{f(\tau)\;A'(t-\tau)\;\Delta\tau} \qquad (A6.2)$$

Integral at A6.2 is also referred as convolution integral.

Summary: *Indicial admittance function is given as the response of a linear system to unit excitation.*

A7 The Guderly Profile

When the free stream Mach number approaches unity, the transonic flow problem can be handled as a channel flow for the flows past symmetric airfoils. K.G. Guderly was one of the pioneering aerodynamicists who implemented that idea (Guderly 1962). The flow is subsonic at the leading edge of the Guderly profile and because of thickness effect the flow speeds up and reaches to supersonic regime afterwards. The pressure decrease during the flow speed up is linear for the Guderly profile. The geometry after the sonic region is determined such a way that it produces minimum wave drag in the supersonic region while causing no wave reflection from the profile surface with almost a constant pressure distribution. The symmetric surface equation in terms of the specific heat ratio of the air reads as

$$y = 3/2\,(\gamma+1)^{1/3}\left[1+(\gamma+1)^{-1/3}4/9x\right]^{3/2}1/5\left[3-2(\gamma+1)^{-1/3}4/9x\right]$$

The change of x until the maximum thickness is given by $-9/4\,(\gamma+1)^{1/3} \leq x \leq 1.7\,(\gamma+1)^{1/3}$. In the rest of the profile the necessary expansion is provided. Shown in Fig. A7.1 is the surface of the profile.

Summary: *Geometry of a special profile, named after K.G. Guderly, which has a unique surface pressure distribution at Mach numbers near unity.*

A8: Vibrational Energy

The calculation of the internal energy of polyatomic gases at high temperatures is rather complex because of the inadequacy of the classical mechanical concepts in handling the vibrational energy between the atoms of the molecules. Thus, we have to resort to the quantum mechanics for implementing the complex wave function ψ which gives probability distribution of quanta in terms of the potential V as Schrödinger's equation as follows

$$-\frac{h^2}{8\pi^2 m}\nabla^2\Psi + V\Psi = -\frac{h}{2\pi i}\frac{\partial\Psi}{\partial t} \tag{A8.1}$$

Here, h is the Planck's constant. The separation of variables for the complex function, i.e. $\Psi(x,y,z,t) = \psi(x,y,z)\,\phi(t)$, to solve the Schrödinger's equation gives

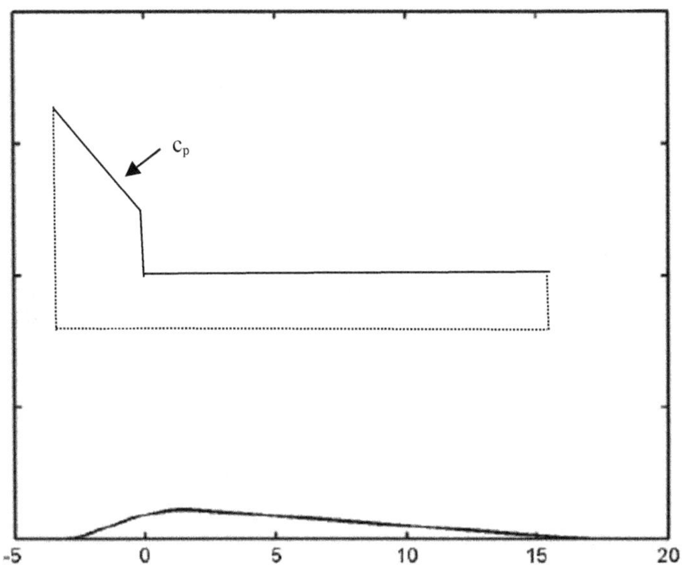

Fig. A7.1 The Guderly profile and the surface pressure distribution

$$\frac{1}{\psi}\left[-\frac{h^2}{8\pi^2 m}\,\nabla^2\psi + V\psi\right] = -\frac{h}{2\pi i\phi}\frac{d\phi}{dt} \tag{A8.2}$$

In order to satisfy Eq. A8.2 with a physically meaningful solution we have to equate both sides of the equation to a real constant positive ε. This makes the time dependent part of the wave function to satisfy a first order ordinary differential equation whose solution is

$$\phi(t) = C\exp(-2\pi i\varepsilon t/h) \tag{A8.3}$$

So far, we have seen the general solution for the wave equation. Now, let us represent the vibrating atoms of a diatomic gas as one dimensional harmonic oscillator. The potential function for the one dimensional oscillator can be written in terms of the vibration frequency ν as $V(x) = 2\pi^2 m\nu^2 x^2$ (Lee et. al. 1973). Here, we can consider the function V as the potential of an oscillating pendulum with mass m whose minimum value is at $x = 0$. Thus, Eq. A8.2 reads as

$$\frac{d^2\psi}{dx^2} + \frac{8\pi^2 m}{h^2}(\varepsilon - 2\pi^2 m\nu^2 x^2)\psi = 0 \tag{A8.4}$$

The complex wave function which satisfies Eq. A8.4 must also satisfy the following: $\int_{-\infty}^{+\infty} \psi\overline{\psi}\, dx = 1$, and $\lim_{x\to\pm\infty} \psi(x) = 0$. At the limits of these conditions, as x is very large, in the second term of Eq. A8.4 ε can be neglected compared to x to give the solution behaving as $\alpha = (m\nu/h)^{1/2}\ \theta(\alpha x)\exp(-\alpha^2 x^2/2)$ with $\alpha = (m\nu/h)^{1/2}$. Thus, substituting this solution into A4, in terms of the series solution the eigenvalues of ε showing the various energy levels read as

$$\varepsilon_n = (n + 1/2)h\nu, \quad n = 0, 1, 2, \ldots \tag{A8.5}$$

At each energy level we get the complex wave function, depending on n, as $\psi_n(x,t) = \psi_n(x)\phi_n(t)$. Here, $\psi_n(x)$ is expressed in terms of Hermit polynomials.

The effect of the energy levels expressed in terms of n are used to find the total internal energy of the molecules. If N_i be the number of molecules whose internal energy is ε_i then the number of total molecules will be $N = \Sigma N_i$ and the total internal energy will be $E = \Sigma N_i\varepsilon_i$. Let us examine, with the quantum statistics, the physics behind the thermodynamic equilibrium for which the total internal energy and the total number of molecules remain unchanged. First, we recall the Heisenberg's principle of uncertainty. In one dimensional space, the position of the molecule is given with x and its momentum is given with p = mu. According to the Heisenberg's principle the product of the uncertainties Δx and Δp, in terms of the Planck's constant h, reads as

$$\Delta x\,\Delta p \approx h = 6.6237x10^{-34} J.s \tag{A8.6}$$

In Eq. A8.6, the product $\Delta x\,\Delta p$ describes a very small area given with h, in 2-D phase space, which we call compartment. On the other hand, the product A = dx.dp indicates the cell area determined by the small increments in position and momentum. This product A is much larger than h even at molecular levels. Therefore, at any given time the number of compartments g in cell A satisfies $g = A/h >> 1$. A complete specification of the coordinates of the phase space in a compartment defines the 'microstate' of the system in a detail which is unnecessary in determining the observable properties of the gas. Knowing the number of molecules N_i in each cell enables us to know the 'macrostate' of the gas. Let us define the number of microstate in a macrostate as the 'thermodynamic property' and indicate with W. In order to derive an expression for W in terms of N_i, let us denote the compartments 1, 2, 3, …, g_i in cell i, and number of molecules in each cell with I, II, III,…N_i. In cell i some of the compartments may be empty. Starting with the compartment number we can identify each compartment with {..}. For example, if we have molecules I and II in compartment 1 we identify it with {1 I II}, in second compartment only III molecule then with {2 III}, and empty 3rd compartment with {3}, and so on. In these representations, if the numbers and the Roman numerals are arranged in all possible sequences, each sequence starting with a numeral will represent a microstate. Therefore, there are g_i ways for a sequence to begin for each one of g_i compartments, and in each of these compartments the remaining ($g_i + N_i - 1$) numbers and Roman numerals can be arranged in any order. On the other hand, n objects can be arranged in sequences as many as n!. Therefore, in g_i compartments, the number of different compartments which begin with a number is

$$g_i(g_i + N_i - 1)! \tag{A8.7}$$

Some of these sequences represent the same microstate. These representations are repeated g_i! times for g_i number of compartments. Therefore, we need to divide A8.7 with g_i!. In addition, the indistinguishable molecules of specie are considered here. There can be any two molecules in compartment 1, any single molecule in 2, and no molecule in 3, etc., to yield N_i! repeatings for a microstate in the cell *i*. Therefore, the number of microstate for a cell *i* is obtained by dividing A7 with N_i! as follows

$$W_i = \frac{g_i(g_i + N_i - 1)!}{g_i!N_i!} = \frac{(g_i + N_i - 1)!}{(g_i - 1)!N_i!} \tag{A8.8}$$

In this case, if we consider the same number of microstate for each cell then the number of total microstate will be given with the product of all microstates. That means the thermodynamic probability is determined as $W = \Pi\ W_i$. Using A8.8 gives us the thermodynamic probability W as follows

$$W = \Pi \frac{(g_i + N_i - 1)!}{(g_i - 1)!N_i!} \tag{A8.9}$$

The Stirling formula for very large x gives us $\ln(x!) \cong x \ln x - x$ (Lee et al. 1973). Since N_i and g_i are very large numbers, 1 can be neglected compared to them. If we take the logarithm of both sides of A8.9 we obtain

$$\ln W = \sum [(g_i + N_i)\ln(g_i + N_i) - g_i \ln g_i - N_i \ln N_i] \tag{A8.10}$$

The number of molecules in a cell varies with time. Therefore, N_i changes with time. The thermodynamic probability of the system is a maximum when the variation of A8.10 vanishes. If we take the variation of A8.10 and equate to zero after some manipulations we get

$$\sum \ln(\frac{g_i + N_i^o}{N_i^o})\ \delta N_i = 0 \tag{A8.11}$$

Here, N_i^o, is the number of molecules in a cell when the thermodynamic probability is maximum. The variation δN_i in Eq. A8.11 shows the changes in N_i. Since the total number of molecules N in the system is constant, the variation of N, $\delta N = 0$. This tells us that the variations of N_i must satisfy

$$\sum \delta N_i = \delta N_1 + \delta N_2 + \ldots = 0 \tag{A12}$$

The meaning of A8.12 is that N_i are dependent. If ε_i is the internal energy of the molecules in each cell then the total internal energy of the system reads as $E = \Sigma\ \varepsilon_i\ N_i$. In thermodynamic equilibrium at macro level the total internal energy is constant. This gives us $\delta E = 0$. Which means

$$\sum \varepsilon_i \delta N_i = \varepsilon_1\, \delta N_1 + \varepsilon_2\, \delta N_2 + \ldots = 0 \tag{A8.13}$$

As Lagrange multipliers if multiply A8.12 with -lnB and A8.13 with $-\beta$, and add the both into A8.11 we get

$$\sum (\ln \frac{g_i + N_i^o}{N_i^o} - \ln B - \beta\varepsilon_i)\ \delta N_i = 0 \tag{A8.14}$$

Equation A8.14 makes δN_i independent of each other. Therefore, in order to satisfy the Eq. A8.14 we have to set the expression in parenthesis equal to zero. After some manipulations we obtain

$$\frac{N_i^o}{g_i} = \frac{1}{B\ \exp(\beta\varepsilon_i) - 1} \tag{A8.15}$$

Equation A8.15 gives us the Bose-Einstein distribution function. On the other hand, if the number of molecules in a cell is much smaller than the number of compartments then the value given with Eq. A8.15 becomes very small which enables us to neglect 1 at the denominator of the term at the right hand side. Hence, we get the Maxwell-Boltzmann distribution for N_i^o at the thermodynamic equilibrium as follows

$$\frac{N_i^o}{g_i} = \frac{1}{B\,\exp(\beta\varepsilon_i)} \tag{A8.16}$$

In Eq. A8.16 the Lagrange multipliers B and β appear as unknowns. Let $N = \Sigma N_i$ be the number of molecules in the system. Using A16 we get

$$\sum N_i = N = \frac{1}{B}\sum g_i \exp(-\beta\varepsilon_i) \tag{A8.17}$$

In Eq. A8.17 the quantity given with $Z = \Sigma g_i \exp(-\beta\varepsilon_i)$, is known as the partition function. The unknown B in terms of the partition function is determined as B = Z/N. The number of molecules in each cell for the maximum thermodynamic probability reads as

$$N_i = \frac{Ng_i}{Z}\exp(-\beta\varepsilon_i) \tag{A8.18}$$

Here in Eq. A8.18, value of β remains as unknown.

In statistical mechanics, the entropy S of a system with maximum thermodynamic probability W is defined with Boltzmann constant, $k = 1.3803 \times 10^{-23}$ J/mole K, as

$$S = k\ln W \tag{A8.19}$$

Here, $k = 1.3803 \times 10^{-23}$ J/mole K is the Boltzmann constant (Lee et al. 1973).

The partition function Z obtained from the Maxwell-Boltzmann distribution is used in Eq. A8.19 gives us the expression for the entropy of the system in terms of the internal energy and the number of molecules in the system. This gives

$$S = kN\,\ln\frac{Z}{N} + k\beta E + kN \tag{A8.20}$$

The relation between the entropy, internal energy and the temperature under constant volume reads as dE/dS = T. The reciprocal, according to the classical thermodynamics give this relation as $(\partial S/\partial E)_V = 1/T$. In Eq. A8.20, for number a constant number of molecules we have $(\partial S/\partial E)_V = k\beta$. Hence, the classical and the statistical thermodynamics are tied together with

$$\beta = \frac{1}{kT} \tag{A8.21}$$

Then, for a harmonic oscillator the partition function reads as

$$Z = \sum g_i \exp(-\frac{\varepsilon_i}{kT}) = \sum g_i \exp(-\frac{(i+1/2)hv}{kT}) \tag{A8.22}$$

The relation between the Partition function and the internal energy can now be written using Eq. A8.18 as follows

$$E = \sum \varepsilon_i N_i = \frac{N}{Z} \sum g_i \varepsilon_i \exp(-\frac{\varepsilon_i}{kT}) \tag{A8.23}$$

If we take the derivative of Z with respect to T in Eq. A8.22 then we get

$$\frac{dZ}{dT} = \frac{1}{kT^2} \sum g_i \varepsilon_i \exp(-\frac{\varepsilon_i}{kT}) \tag{A8.24}$$

The internal energy E, from A8.23 and A8.24 reads as

$$E = \frac{kNT^2}{Z} \frac{dZ}{dT} \tag{A8.25}$$

Defining g specific energy as e = E/M, since M = Nm then

$$e = \frac{RT^2}{Z} \frac{dZ}{dT} \tag{A8.26}$$

Here, R = k/m, is the gas constant.

Let us finalize the partition function expression for diatomic gases using the expression A8.22 given for the harmonic oscillator. In a molecular level the statistical weight of a given level or the degeneracy g = A/h goes to 1 to give a final form to Eq. A8.22involving infinitely many cells of which has equally distributed internal energies. This gives

$$Z_{vib} = \sum_{i=0}^{\infty} \exp(-\frac{(i+1/2)hv}{kT}) \tag{A8.27}$$

Here, $1/(1-x) = 1 + x + x^2 + \ldots$ expansion, Eq. A8.27 is simplified to

$$Z_{vib} = \frac{\exp(-hv/2kT)}{1 - \exp(-hv/kT)} \tag{A8.28}$$

The specific internal energy e can be found in terms of the temperature T from Eq. A8.28 with the aid of A8.26.

Summary: *Vibrational energy formula for diatomic gases like N_2 and O_2 is provided. The formula is applicable when the air temperature is higher than 2000 K. The quantum mechanical approach as opposed to the classical mechanics is given in utilization of the energy of harmonic oscillators. The Maxwell–Boltzmann statistics is used together with the statistical definition of entropy to obtain the partition function for vibrational energy.*

A9: The Leading Edge Suction

The Blasius theorem of the potential theory gives us the force acting on a 2-D body enclosed by a closed surface in a complex velocity field W = u + iv as follows (Milne-Thomson 1973),

$$X + iY = i\rho/2 \oint W^2 dz \tag{A9.1}$$

Here, u and v are the x and y components of the velocity, X and Y are the x and y components of the forces acting on the body, and z = x + iy indicates the complex variable in the x–y plane. S simple proof of the theorem in the pressure field p is as follows (Fig. A9.1.

The differential force at any point on the differential surface dz = dx + idy because of the pressure is written as d(X + iY) = −pdy−ipdx. The pressure acting on the surface is given with Bernoulli's equation, in terms of the velocity square q^2 is $p = p_o - 1/2\ \rho q^2$. Since the stagnation pressure p_o, has a constant effect on the closed surface its total effect becomes zero. Therefore, the differential force reads as

$$d(X + iY) = 1/2\rho\, q^2(dy + idx) = 1/2\rho\, iq^2(dx - idy) = 1/2\rho\, iq^2 d\bar{z} \tag{A9.2}$$

If we define the complex velocity potential as $F = \phi + i\psi$, then the square of the velocity reads as

$$q^2 = \frac{dF}{dz} \cdot \frac{d\bar{F}}{d\bar{z}} \tag{A9.3}$$

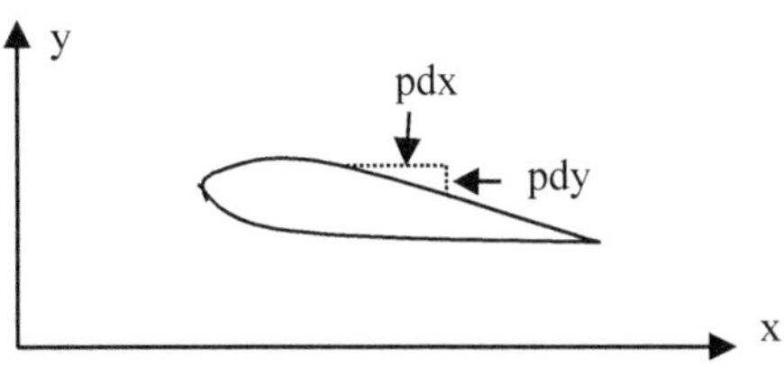

Fig. A9.1 Differential force acting on the surface: d (X + iY) = −pdy−ipdx

If we substitute Eq. A9.3 in A2 we obtain

$$d(X+iY) = 1/2\rho\, i\frac{dF}{dz}.d\overline{F} \tag{A9.4}$$

On the profile surface the stream function is constant. Therefore dψ = 0. Hence,

$$d\overline{F} = dF = \frac{dF}{dz}dz \tag{A9.5}$$

Now, using Eq. A9.5 in A9.4 makes the differential complex force to read in terms of W

$$d(X+iY) = 1/2\rho\, iW^2 dz \tag{A9.6}$$

The total force becomes the closed integral of Eq. A9.6 over the airfoil surface

$$X+iY = 1/2\rho\, i\oint W^2 dz \tag{A9.7}$$

On the other hand, for an airfoil simple harmonically pitching with α about a point ab and plunging with h as shown in Fig. A9.2, the complex velocity field reads as

$$W(z) = [U\alpha + \dot{h} + b(1/2 - a)\dot{\alpha}]\, C(k)\frac{\sqrt{2}}{\sqrt{z+b}} \tag{A9.8}$$

Here, C(k) is the Theodorsen function.

Substituting Eq. A9.8 into A9.7 gives the formula for the leading edge suction as follows

$$X = -\pi\rho\, 2[[U\alpha + \dot{h} + b(1/2 - a)\dot{\alpha}]\, C(k)]^2 \tag{A9.9}$$

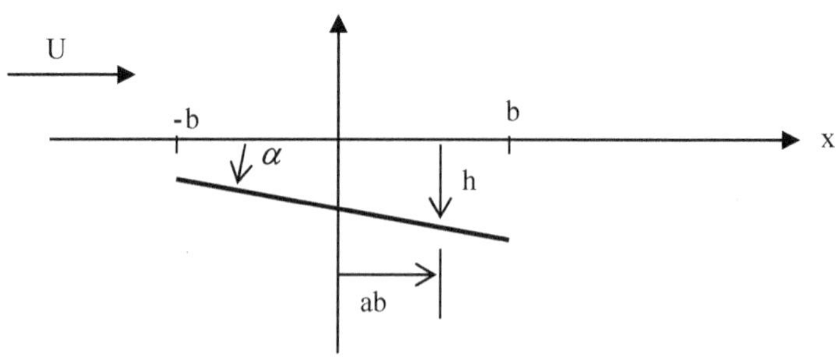

Fig. A9.2 Pitching-plunging airfoil

Here, the complex integral reads as $\oint \frac{dz}{z+b} = 2\pi i$.

Summary: *The leading edge suction force based on the potential theory is derived.*

A10: The Finite Difference Solution of the Boundary Layer Equations

The unsteady potential flow solution gives us the time dependent value of the surface vortex sheet strength. The velocity component tangent to the airfoil surface can be obtained from the surface vortex sheet strength. This tangent velocity is nothing but the boundary layer edge velocity which is to be used as a boundary condition for the vorticity transport equation. At the edge of the boundary layer the vorticity value becomes zero. The boundary layer equations, Eqs. 8.4–8.7, can be solved with marching in the main flow direction as follows.

If we discretize time with Δt, space with Δx and Δy, then Eq. 8.5 becomes an algebraic equation with superscript n showing the time step, and i, j indicating the discrete locations in x, y directions, as follows

$$\begin{aligned}\frac{\omega^n_{i,j} - \omega^{n-1}_{i,j}}{\Delta t} &= -u^{n-1}_{i,j}\frac{\omega^n_{i,j} - \omega^n_{i-1,j}}{\Delta x} \\ &\quad - v^{n-1}_{i,j}\frac{\omega^n_{i,j+1} - \omega^n_{i,j-1}}{2\Delta y} + \frac{1}{\mathrm{Re}}\frac{\omega^n_{i,j+} - 2\omega^n_{i,j} + \omega^n_{i,j-1}}{(\Delta y)^2}\end{aligned} \tag{A10.1}$$

Organizing Eq. A10.1 for the unknown values of $\omega^n_{i,j}$ in j at a station i gives

$$A_j\omega^n_{i,j-1} + B_j\omega^n_{i,j} + C_j\omega^n_{i,j+1} = D_j\,,\ \ j = 2, J \tag{A10.2}$$

Here, ω_1 is the unknown wall vorticity value and $\omega_{J+1} = 0$ is the vorticity at the edge of the boundary layer. This makes the number of unknowns, J, one more than the number of equations given by A10.2. If we find one more equations we can close the problem, i.e., have equal number of equations with unknowns.

If we show the free stream velocity with U the velocity at the upper surface of the profile becomes

$$\mathrm{V_u(x, y, t) = U + u' = U + \gamma_a(x, y, t)/2} \tag{A10.3}$$

and at the lower surface

$$\mathrm{V_4(x, y, t) = U - u' = U - \gamma_a(x, y, t)/2} \tag{A10.4}$$

Integrating the vorticity values normal to the surface as shown in Fig. A10.1 gives

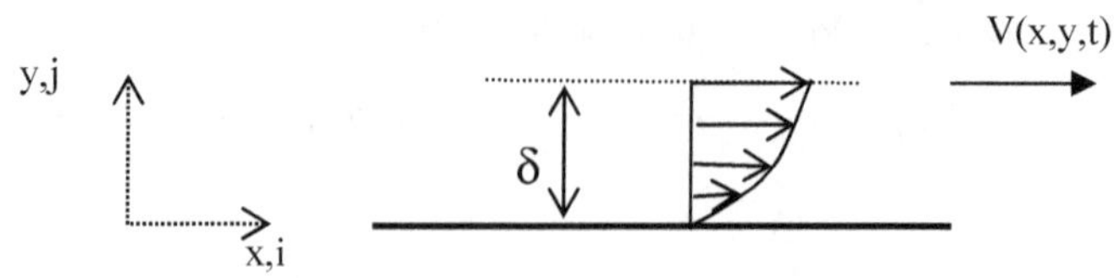

Fig. A10.1. The boundary layer velocity profile

$$-\omega_1/2 - \omega_2 - \omega_3 - \ldots - \omega_I = V/\Delta n \tag{A10.5}$$

Hence, from the simultaneous solution of Eqs. A10.2 and A10.5 we obtain the vorticity values.

Once we know the vorticity profile at a station we can obtain the tangential velocity components at a point *i*, *j* by numerical integration as follows

$$u_{i,j} = -\left[u_{i,j-1} + (\omega_0/2 + \sum_{k=1}^{j-1} \omega_{i,k})\,\Delta n\right] \tag{A10.6}$$

The vertical velocity components, on the other hand, are obtained with the proper discretezation of the continuity equation as follows.

$$v_{i,j} = v_{i,j-1} - \frac{\Delta y}{\Delta x}\left(u_{i,j} + u_{i,j-1} - u_{i-1,j} - u_{i-1,j-1}\right) \tag{A10.7}$$

The continuity equation is discretized involving the points shown in the molecule below.

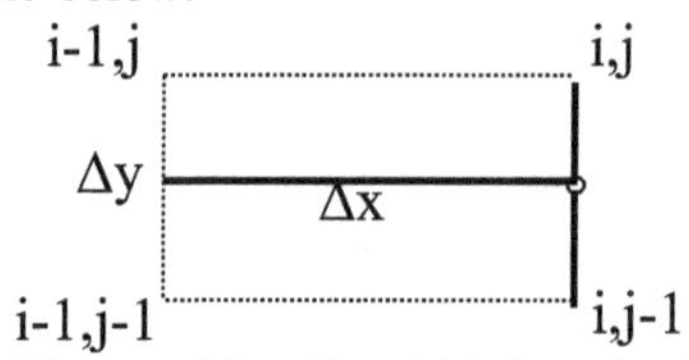

Now, writing Eq. A10.5 as the first line and the open form of Eq. A10.2 as the rest of the lines, the matrix form of those become

$$\begin{bmatrix} 1/2 & 1 & 1 & 1 & . & 1 \\ A_2 & B_2 & C_3 & . & . & . \\ . & A_3 & B_3 & C_3 & . & . \\ . & . & . & . & . & . \\ . & . & . & . & . & . \\ . & . & . & . & . & . \\ . & . & . & . & A_J & B_J \end{bmatrix}^n \begin{pmatrix} \omega_0 \\ \omega_2 \\ \omega_3 \\ . \\ . \\ \omega_{J-1} \\ \omega_J \end{pmatrix}^n = \begin{pmatrix} -V/\Delta n \\ D_2 \\ D_3 \\ . \\ . \\ D_{J-1} \\ D_J \end{pmatrix}^n \tag{A10.8}$$

Wherein the entries of the coefficient matrix are:

$$A_j = -v_{i,j}^{n-1}\frac{\Delta t}{2\Delta y} - \frac{\Delta t}{R_e \Delta y^2}, \quad C_j = v_{i,j}^{n-1}\frac{\Delta t}{2\Delta y} - \frac{\Delta t}{R_e \Delta y^2}$$

$$B_j = \omega_{i,j}^{n-1} + u_{i,j}^{n-1}\omega_{i-1,j}^{n}\frac{\Delta t}{\Delta x}, \quad D_j = 1 + u_{i,j}^{n-1}\frac{\Delta t}{\Delta x} + \frac{2\Delta t}{R_e \Delta y^2}$$

Equation A10.8 is almost tri-diagonal except at the first line which is a full line. It has a special way of solution with direct inversion based on the elimination of unknowns starting from the last line, or it can be solved with Sherman-Morison formula (Press et.al. 1992). As the test case, steady state mid-chord velocity profile of an impulsively started flat plate Re = 1000 is shown in Fig. A10.2. In discretization, a 10 × 10 coarse mesh with $\Delta x = 0.1$L, $\Delta y = 0.04$L, and $\Delta t = 0.04$ is used for marching 50 steps. As seen from Fig. A10.2the numerical solution is closer to the Blasius solution (Schlichting 1968) than the solution obtained with a Navier-Stokes solver (Sankar 1977).

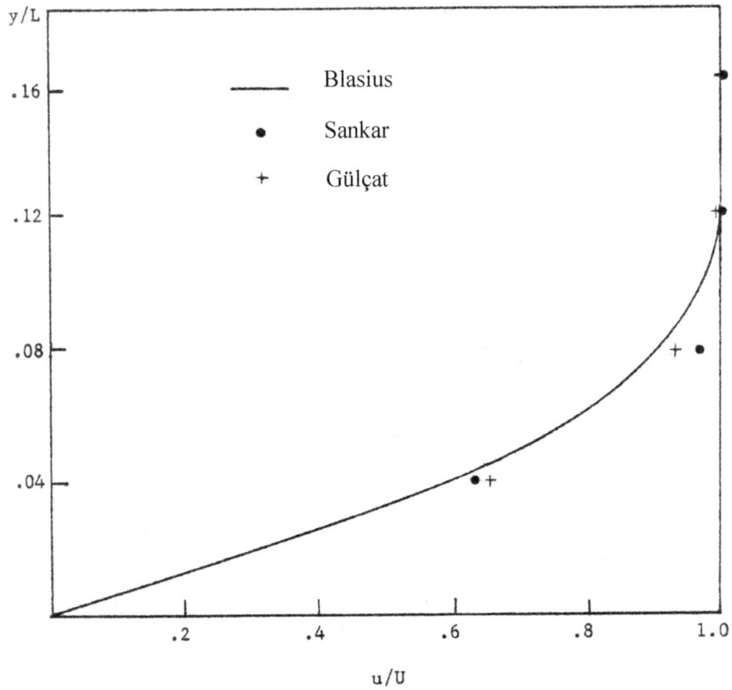

Fig. A10.2 Velocity profile at the mid-chord of a flat plate at Re = 1000

Summary: *A numerical technique for the solution of the unsteady boundary layers is provided. The technique is based on the finite difference method which marches step by step along the boundary layer. The procedure utilizes the solution of a special tri-diagonal system which involves a coefficient matrix whose first row is full.*

A11: 3-D Boundary Layer Solution

The finite difference solution of time dependent boundary layer Eqs. (8.59) are performed for point *i,j,k* at a time level n as follows

$$d_{jk}(.)_{i,j-1,k}+a_{jk}(.)_{i,j,k-1}+b_{jk}(.)_{ijk}+c_{jk}(.)_{i,j,k+1}+e_{jk}(.)_{i,j+1,k}=r_{jk} \qquad \text{(A11.1)}$$

The algebraic equation above is a block pseudo penta diagonal one with following constants together with the right hand side which is expressed in terms of the flow variables of previous time level $n-1$ and the previous chord wise station $i-1$ in the rectangular *x-y-z* coordinates.

$$d_{jk}=-\frac{\Delta t}{2\Delta y}v^n_{ijk} \qquad a_{jk}=-\frac{\Delta t}{2\Delta z}w^n_{ijk}-\frac{\Delta t}{\mathrm{Re}\,\Delta z^2} \qquad b_{jk}=1+\frac{\Delta t}{\Delta x}u^n_{ijk}+\frac{2\Delta t}{\mathrm{Re}\,\Delta z^2}$$
$$c_{jk}=\frac{\Delta t}{2\Delta z}w^n_{ijk}-\frac{\Delta t}{\mathrm{Re}\,\Delta z^2}\,;\quad e_{jk}=\frac{\Delta t}{2\,\Delta y}v^n_{ijk}\,;\quad r_{jk}=\left(1+\frac{\Delta t}{\Delta x}\right)u^n_{ijk}(.)^{n+1}_{i-1,j,k}+\omega^n_{x,ijk}\frac{(.)_{ijk}-(.)_{i-1,jk}}{\Delta x}$$
$$+\,\omega^n_{y,ijk}\frac{(.)_{i,j+1,k}-(.)_{,j-1,ki}}{2\Delta y} \qquad \text{(A11.2)}$$

The *u* and *v* values at any point *i,j,k* are found from the integral relation wich is valid for the boundary layer

$$u=\int_0^z \omega_y d\varsigma \quad \text{and} \quad v=-\int_0^z \omega_x d\varsigma \quad \text{as}$$

$$u_{i,j,k}=-\left[u_{i,j,k-1}+\left(\omega_{y,ij,1}/2+\sum_{n=2}^{k}\omega_{y,ij,n}\right)\Delta z\right] \quad \text{and} \qquad \text{(A11.3a, b)}$$

$$v_{i,j,k}=-\left[v_{i,j,k-1}+\left(\omega_{x,ij,1}/2+\sum_{n=2}^{k}\omega_{x,ij,n}\right)\Delta z\right]$$

Then, the continuity Eq. (8.58) solved for the vertical velocity component in terms of *u* and *v* yields

$$w_{i,j,k+1} = w_{i,j,k} - \left(u_{i,j,k} + u_{i,j,k-1} - u_{i-1,j,k} - u_{i-1,j,k-1}\right)\tfrac{\Delta z}{2\Delta x}$$
$$-\left(v_{i,j,k} + v_{i,j,k-1} - v_{i,j,k-1} - u_{i,j-1,k-1}\right)\tfrac{\Delta z}{2\Delta y}$$

For given edge velocities U_e and V_e, the integral relations (8.60a,b) with the vorticity components at any station i for bs number of normal points read as

$$0.5\omega_{x,i,j,1} + \omega_{x,i,j,2} + \omega_{x,i,j,3} + \ldots + \omega_{x,i,j,bs} = -V_e/\Delta z \qquad (8.61a)$$

and

$$0.5\omega_{y,i,j,1} + \omega_{y,i,j,2} + \omega_{y,i,j,3} + \ldots + \omega_{y,i,j,bs} = U_e/\Delta z \qquad (8.61b)$$

For j = 1,…,m span wise locations. The matrix equation for *m* x *n* unknowns then with Eqs. (8.59-, 8.61a,b) becomes

$$\begin{bmatrix} B_1 & C_1 & & \\ A_2 & B_2 & C_2 & \\ \cdots & \cdots & \cdots & \\ A_{n-1} & B_{n-1} & C_{n-1} & \\ & A_n & B_n & \end{bmatrix} \begin{pmatrix} \Omega_1 \\ \Omega_2 \\ \cdots \\ \Omega_{n-1} \\ \Omega_n \end{pmatrix} = \begin{pmatrix} R_1 \\ R_2 \\ \cdots \\ R_{n-1} \\ R_n \end{pmatrix} \qquad (A11.4)$$

Where Ω_i and R_i are the unknown and known subvectors resepectively, and for any k:

$$A_k = \begin{bmatrix} 0 & & & & \\ & d_{k2} & & & \\ & & d_{k3} & & \\ & & & . & \\ & & & & d_{kn} \end{bmatrix} B_k = \begin{bmatrix} 0.5 & 1 & 1 & 1 & . & 1 \\ a_{k2} & b_{k2} & c_{k2} & 0 & 0 & . \\ 0 & a_{k3} & b_{k3} & c_{k3} & 0 & . \\ . & . & . & . & . & . \\ . & . & . & . & a_{kn} & b_{kn} \end{bmatrix}, C_k$$
$$= \begin{bmatrix} 0 \\ e_{k2} \\ e_{k3} \\ . \\ e_{kn} \end{bmatrix}$$

The relation between the lifting pressure value p and the surface velocity components as the backwash $u_{u,l}$ and the spanwash $v_{u,l}$ for the unsteady flow are given as follows:

$$u_{u,l}(x,y) = \pm\frac{1}{4}\left[p(x) - i(\omega/U)e^{-i\omega x/U}\int_{x_{le}}^{x} p(\xi)e^{i\omega\xi/U}d\xi\right] \tag{A11.5}$$

and,

$$v_{u,l}(x,y) = \pm\frac{1}{4}\frac{\partial}{\partial s}\left[i(\omega/U)e^{-i\omega x/U}\int_{x_{le}}^{x} p(\xi)e^{i\omega\xi/U}d\xi\right] \tag{A11.6}$$

where, s denotes the span wise direction. Here, we have to make a note that the lifting pressure of the planform must be provided at each node for numerical evaluation of the integrals given above. The edge velocities U_e and V_e are found numerically by integrating the lifting pressure using the trapezoidal rule as follows.

At a spanwise station j the chordwise variation of the edge velocity for a fixed spanwise location j reads as

$$U_{eij} = 1 \pm \left[p_{ij} - \sqrt{-1}\,\frac{\omega}{U}\exp(-\sqrt{-1}\omega x_i/U)\right]\sum_{k=1}^{i}\left[p_{k+1,j}\exp(\sqrt{-1}\omega x_{k+1}/U) + p_{kj}\exp(\sqrt{-1}\omega x_k/U)\right]\Delta x/8 \tag{A11.7}$$

and

$$V_{eij} = 1 \pm \exp(-\sqrt{-1}\omega x_i/U)\sum_{k=1}^{i}\left[\frac{dp_{k+1,j}}{dy}\Big|_j \exp(\sqrt{-1}\omega x_{k+1}/U) + \frac{dp_{k,j}}{dy}\Big|_j \exp(\sqrt{-1}\omega x_k/U)\right]\Delta x/8 \tag{A11.8}$$

Where, $\left.\frac{dp}{dy}\right|_j = \frac{p_{i,j+1} - p_{i,j-1}}{2\Delta y}$

The lifting pressure values used in (A11.7–8) at point on the wing is found either with Kernel Function method or Doublet Lattice method described in Chap. 5.

The spanwise vorticity obtained on the upper and lower surfaces gives the surface friction coefficient via solution of (A11.4) as follows:

$$C_{Dv} = \int_A C_f\,dA \tag{A11.8}$$

Integration of the surface friction coefficient over the upper and lower surfaces gives the time variation of the viscous drag, i.e.

$$C_{Dv} = \int_A C_f\,dA \tag{A11.9}$$

A12: Calculation for the lift

I. Unsteady contributions by the image vortex sheets

The kernel of the integral (8.62) is expanded into the Taylor series given below

$$H(x-\xi) = \frac{-2h_g\alpha}{(x-\xi)^2+4h_g^2} = \alpha h^{-1}\sum_{n=0}^{\infty} H_n\left(\frac{x-\xi}{h}\right)^n, \quad H_n = (-1)^{1+n/2}2^{-(n+1)}, \; n = 0,\, 2,\, 4\ldots.$$

$$K(x-\xi) = \frac{1}{x-\xi} - \frac{x-\xi}{(x-\xi)^2+4h^2} = \frac{1}{x-\xi} + h^{-1}\sum_{n=0}^{\infty} K_n\left(\frac{x-\xi}{h}\right)^n, \quad K_n = (-1)^{(n+1)/2}2^{-(n+1)}, \; n = 1,\, 3,\, 5\ldots.$$

Note that, the same function also acts as the coefficient of the exponent in the second tem of left hand side of (8.62).

The vortex sheet strength with two terms becomes

$$\overline{\gamma}_a(x) = \sum_{n=1}^{\infty} h_g^{-n}\overline{\gamma}_n \cong \overline{\gamma}_o + \overline{\gamma}_2/h_g^2$$

Then, for $\overline{\gamma}_2$, in terms of $\overline{\gamma}_o$ which is the out of ground effect, we have for the first term in the series as follows

$$\int_{-1}^{1} \frac{\overline{\gamma}_2(\xi)}{x-\xi}\,d\xi = -\int_{-1}^{1} [2H_o + K_1(x-\xi)]\gamma_o(\xi)\,d\xi = f_2(x), \quad H_o = -1/2, \quad K_1 = -1/4 \tag{A12.1}$$

Inverting (A12.1) gives

$$\overline{\gamma}_2(x) = -\frac{1}{\pi^2}\sqrt{\frac{1-x}{1+x}}\int_{-1}^{1}\sqrt{\frac{1+\xi}{1-\xi}}\frac{f_2(\xi)}{x-\xi}\,d\xi$$

The series in the kernel function is utilized to determine the ground effect. The vortex sheet strength which is to be used in (A12.1), caused by the first term of the kernel without ground effect, reads as (Bisplinghoff et al. 1996)

$$\bar{\gamma}_o(x) = \frac{2}{\pi}\sqrt{\frac{1-x}{1+x}}\left\{\int\limits_{-1}^{1}\sqrt{\frac{1+\xi}{1-\xi}}\frac{\overline{w}_a(\xi)}{x-\xi}d\xi + \frac{ik\overline{\Omega}}{4}\int\limits_{1}^{\infty}\sqrt{\frac{\lambda+1}{\lambda-1}}\frac{e^{-ik\lambda}}{x-\lambda}d\lambda\right\} \quad \text{(A12.2)}$$

In addition, the effect of the image of the wake vortex has to be taken into account.

The effect of the image of the wake vorticity using (A12.2) gives

$$\bar{\gamma}_2^{(1)} = \frac{ik\overline{\Omega}e^{-ik}}{16\pi}\left(-\frac{1}{2ik}+\frac{1}{k^2}\right)(2+x)\sqrt{\frac{1-x}{1+x}} \quad \text{(A12.3)}$$

and, similarly from the unsteady wake vorticity itself we get

$$\bar{\gamma}_2^{(2)} = \frac{ik\overline{\Omega}}{4\pi}\left\{\left[\frac{x+2}{ik}-\frac{1}{k^2}\right]e^{-ik} + (x+2)C_1(k) - C_2(k)\right\}\sqrt{\frac{1-x}{1+x}} \quad \text{(A12.4)}$$

where

$C_1(k) = \frac{\pi}{2}\left[H_1^{(2)}(k) + iH_o^{(2)}(k)\right]$ *and* $C_2(k) = \frac{\pi}{4}\left[2H_1^{(2)}(k) + iH_o^{(2)}(k) + iH_2^{(2)}(k)\right]$

The effect of the image of the bound vortex for the pitching plunging airfoil is found similarly as in terms of the amplitudes of pitch and plunge as follows

$$\bar{\gamma}_2^h(x) = U\frac{ik\bar{h}}{2}\sqrt{\frac{1-x}{1+x}}\left(x+\frac{3}{2}\right) \quad \text{and} \quad \bar{\gamma}_2^\alpha(x) = U\bar{\alpha}\sqrt{\frac{1-x}{1+x}}\left[\frac{1}{2}\left(x+\frac{3}{2}\right)-\frac{ik}{4}(x+2)\right]$$

The contributions from αh terms of the bound and wake vorticity respectively reads as

$$\bar{\gamma}_2^{\alpha h}(x,t) = -2U\bar{\alpha}\alpha(x,t)h_g\sqrt{\frac{1-x}{1+x}} \quad \text{and} \quad \bar{\gamma}_2^w(x,t) = \frac{\overline{\Omega}\alpha(x,t)h_g}{4\pi}e^{-ik}\sqrt{\frac{1-x}{1+x}}$$

The lifting pressure coefficient in terms of the surface vortex sheet strength reads as (Bisplinghoff et al. 1996)

$$\bar{C}_{p_a}(x) = \frac{2}{U}\left[\bar{\gamma}_a(x) + ik\int\limits_{-1}^{x}\bar{\gamma}_a(\xi)d\xi\right], \quad \bar{\gamma}_a = \bar{\gamma}_o + \left(\bar{\gamma}_2^h + \bar{\gamma}_2^\alpha + \bar{\gamma}_2^{(1)} + \bar{\gamma}_2^{(2)} + \bar{\gamma}_2^{\alpha h} + \bar{\gamma}_2^w\right)/h_g^2 \quad \text{(A12.5a, b)}$$

After the integration of the second term, having improper inegrals to be taken in (A12.5b), we have the following

$$\overline{C}_{p_a}(x) = \overline{C}_{po}(x) + \tfrac{2}{U}(\overline{\gamma}_2^h + \overline{\gamma}_2^\alpha + \overline{\gamma}_2^{(1)} + \overline{\gamma}_2^{(2)} + \overline{\gamma}_2^{ah} + \overline{\gamma}_2^w)/h_g^2 + \tfrac{1}{2h_g^2}\left[-k^2\overline{h}(3a_1(x)+a_2(x)) + k^2\overline{\alpha}(2a_1(x)+a_2(x)/2) + ik\overline{\alpha}(3a_1(x)+a_2(x))\right]$$
$$-2a_1(x)\tfrac{\overline{\alpha}\alpha}{h_g} - \tfrac{k^2\overline{\Omega}}{\pi h_g^2 U}\left\langle\left\{\tfrac{1}{8}[2a_1(x)+a_2(x)/2](1/k^2 - 1/2ik)e^{-ik} + \tfrac{1}{2}[2a_1(x)+a_2(x)/2](\tfrac{e^{-ik}}{ik} + C_1(k)) - a_1(x)(\tfrac{e^{-ik}}{k^2} + C_2(k))\right\} - \tfrac{i}{2k}(1+x)\alpha h_g\right\rangle \quad \text{(A12.6)}$$

where,

$$a_1(x) = \sin^{-}(x) + \sqrt{1-x^2} + \pi/2 \quad and \quad a_2(x) = (x-2)\sqrt{1-x^2} - \sin^{-}(x) - \pi/2$$

Integrating (A12.5b) along the full chord gives us the circulation as follows

$$\overline{\Gamma}_a/b = \int_{-1}^{1}\overline{\gamma}_a(x)dx = \overline{\Omega}e^{-ik} = -2\int_{-1}^{1}\sqrt{\tfrac{1+\xi}{1-\xi}}\overline{w}_a(\xi)d\xi + U\left[\tfrac{\pi ik}{2}\overline{h} + \tfrac{\pi}{2}(1-\tfrac{3ik}{4})\overline{\alpha}\right]/h_g^2 - \tfrac{2U\overline{\alpha}\alpha\pi}{h_g}$$
$$+ ik\overline{\Omega}\left[C_1(k) + (1+\tfrac{\alpha}{2\pi h_g})\tfrac{e^{-ik}}{ik}\right] + \left\{\tfrac{ik\overline{\Omega}}{16}(\tfrac{3}{2k^2} - \tfrac{3}{4ik})e^{-ik} + \tfrac{ik\overline{\Omega}}{4}\left[(-\tfrac{1}{k^2} + \tfrac{3}{2ik})e^{-ik} + \tfrac{3}{2}C_1(k) - C_2(k)\right]\right\}/h_g^2$$

After cancellations, the dimensionless reduced circulation for pitching plunging airfoil becomes

$$\frac{\overline{\Omega}}{U} = \frac{-2\pi\left[\overline{\alpha} + ik(\overline{h} + \overline{\alpha}/2)\right] - \frac{\pi}{2}\left[ik\overline{h} + (1-\frac{3ik}{4})\overline{\alpha} - 4\overline{\alpha}\alpha h_g\right]/h_g^2}{ik\left[C_1(k) + \frac{\alpha}{2\pi h_g}\frac{e^{-ik}}{ik}\right] + ik\left\{\frac{3}{16}(\frac{1}{2k^2} - \frac{1}{4ik})e^{-ik} + \frac{1}{4}\left[(-\frac{1}{k^2} + \frac{3}{2ik})e^{-ik} + \frac{3}{2}C_1(k) - C_2(k)\right]\right\}/h_g^2}$$

The sectional lift coefficient by integration of the total lifting pressure (A12..6) then reads

$$\overline{C}_l = \overline{L}'/(\rho U^2 b) = \frac{1}{2}\rho U^2 \int_{-b}^{b}\frac{\overline{C}_{p_a}(x)dx}{\rho U^2 b} = \frac{1}{2}\int_{-1}^{1}\overline{C}_{p_a}(\xi)d(\xi) \quad \text{(A12.7)}$$

to yield (8.63), wherein

$$C_1(k) = \frac{\pi}{2}\left[H_1^{(2)}(k) + iH_o^{(2)}(k)\right] \quad and \quad C_2(k) = \frac{\pi}{4}\left[2H_1^{(2)}(k) + iH_o^{(2)}(k) + iH_2^{(2)}(k)\right]$$

II. Improper Integrals

Integrals in (A12.5b-6) are evaluated using the Hankel functions of the second kind, utilized in (Theodorsen 1949) as follows

$$\int_1^\infty \frac{e^{ikx}dx}{\sqrt{x^2-1}} = -\frac{\pi}{2}H_o^{(2)}$$

$$\int_1^\infty \frac{xe^{ikx}dx}{\sqrt{x^2-1}} = -\frac{\pi}{2}H_1^{(2)}$$

$$\int_1^\infty \frac{x^2e^{ikx}dx}{\sqrt{x^2-1}} = -\frac{\pi}{4}H_o^{(2)} - \frac{\pi}{4}iH_2^{(2)}$$

III. Wake Integral

The integral effect of the wake on the vortex sheet strength, which gives the x component of the perturbation velocity, is calculated using a converging series approach as follows.

$$\begin{aligned} I(k,x) &= \int_1^\infty \sqrt{\tfrac{\lambda+1}{\lambda-1}}\tfrac{e^{-ik\lambda}}{x-\lambda}d\lambda = -\int_1^\infty \tfrac{\lambda+1}{\sqrt{\lambda^2-1}}\tfrac{1}{\lambda}(1+\tfrac{x}{\lambda}+\tfrac{x^2}{\lambda^2}+\tfrac{x^3}{\lambda^3}+\ldots)e^{-ik\lambda}d\lambda \\ &= -\int_1^\infty \tfrac{1}{\sqrt{\lambda^2-1}}\Big[1+\tfrac{1}{\lambda}(x+1)+\tfrac{1}{\lambda^2}(x^2+x)+\tfrac{1}{\lambda^3}(x^3+x^2)+\ldots\Big]e^{-ik\lambda}d\lambda, \quad -1<x<1 \\ &= i\tfrac{\pi}{2}H_0^{(2)}(k) - \int_1^\infty \tfrac{1}{\sqrt{\lambda^2-1}}\Big[\tfrac{1}{\lambda}(x+1)+\tfrac{1}{\lambda^2}(x^2+x)+\tfrac{1}{\lambda^3}(x^3+x^2)+\ldots\Big]e^{-ik\lambda}d\lambda \end{aligned} \tag{A12.8}$$

Where, $H_0^{(2)}$ is the zeroth order Hankel function of the second kind. In short, (A12.9) becomes

$$I(k,x) = i\frac{\pi}{2}H_0^{(2)}(k) - \sum_{j=1}^{\infty} J_j(k)B_j(x), \;\; J_j(k) = \int_1^\infty \frac{1}{\sqrt{\lambda^2-1}}\frac{e^{-ik\lambda}d\lambda}{\lambda^j} \;\; and \;\; B_j(x) = (x+1)x^{j-1}, \; j = 1,2,3,\ldots$$

The improper integrals $J_j(k)$ in (A1) can be obtained as hypergeometric functions (Matlab 2015) of the reduced frequency k. Term by term integration of (A1) for $k = 0.5$ yields the converging values as given in the following Table A12.1.

Table A12.1 Values of the Integral $J_i(k = 0.5)$

$-i\pi H_o^{(2)}/2$	J1	J2	J3	J4	J5	J6	J7	J8	J9
0.6982–1.4741i	0.8016–0.8824i	0.7144–0.5911i	0.6215–0.4453i	0.5491–0.3640i	0.4945–0.3133i	0.4525–0.2785i	0.4192–0.2530i	0.3922–0.2373i	0.3505–0.2046i

A13. Evaluation of double integrals

Simplification of integrals involving (8.73) is done as follows

$$w_g(x,y) = \frac{1}{4\pi}\int_{-b}^{b}\int_{-l}^{l}\frac{\partial\gamma_a}{\partial\eta}\frac{(x-\xi)(y-\eta)}{\left[(x-\xi)^2+4h^2\right]R}d\xi\, d\eta + \frac{1}{4\pi}\int_{-b}^{b}\int_{-l}^{l}\frac{\partial\delta_a}{\partial\xi}\frac{(x-\xi)(y-\eta)}{\left[(y-\eta)^2+4h^2\right]R}d\xi\, d\eta$$
$$+\frac{1}{4\pi}\int_{-l}^{l}\delta_a(b,\eta)\frac{-(y-\eta)d\eta}{\left[(y-\eta)^2+4h^2\right]}$$

Noting that in (8.73): $\delta_a(b,\eta) = d\Gamma/d\eta$ and $\frac{\partial\gamma_a}{\partial\eta} = \frac{\partial\delta_a}{\partial\xi}$ combining first and second terms of the downwash, it reads as

$$w_g(x,y) = \frac{1}{4\pi}\int_{-b}^{b}\int_{-l}^{l}\frac{\partial\gamma_a}{\partial\eta}\frac{(x-\xi)(y-\eta)\left[(x-\xi)^2+4h^2+(y-\eta)^2+4h^2\right]}{\left[(x-\xi)^2+4h^2\right]\left[(y-\eta)^2+4h^2\right]R}d\xi\, d\eta$$
$$+\frac{1}{4\pi}\int_{-l}^{l}\frac{d\Gamma}{d\eta}\frac{-(y-\eta)d\eta}{\left[(y-\eta)^2+4h^2\right]} \tag{A13.1}$$

Since, for small aspect ratio wings we have $(x-\xi)^2 >> (y-\eta)^2+4h^2$ the first term of the downwash simplifies to

$$\frac{1}{4\pi}\int_{-b}^{b}\int_{-l}^{l}\frac{\partial\gamma_a}{\partial\eta}\frac{(x-\xi)(y-\eta)}{\left[(y-\eta)^2+4h^2\right]|x-\xi|}d\xi\, d\eta$$
$$=\frac{1}{4\pi}\int_{-l}^{l}\frac{y-\eta}{(y-\eta)^2+4h^2}\frac{\partial}{\partial\eta}\left(\int_{x_l(\eta)}^{b_0}\gamma_a(\xi,\eta)\frac{x-\xi}{|x-\xi|}d\xi\right)d\eta \tag{A13.2}$$

Evaluation of the inner integral with paying attention to the absolute valued term the upper limit of the integral changes to x from b_0 to give

$$=\frac{1}{2\pi}\int_{-l}^{l}\frac{y-\eta}{(y-\eta)^2+4h^2}\frac{\partial}{\partial\eta}\left(\int_{x_l(\eta)}^{x}\gamma_a(\xi,\eta)d\xi\right)d\eta \cong \frac{1}{2\pi}\int_{-\beta(x)}^{\beta(x)}\frac{y-\eta}{4h^2}\frac{\partial\Delta\phi'(x,\eta)}{\partial\eta}d\eta$$

Here, $\Delta\phi' = \phi'_u - \phi'_l$ with $\int\limits_{x_l(\eta)}^{x} \gamma_a(\xi,\eta)d\xi = \int\limits_{x_l(\eta)}^{x} (u'_u - u'_l)d\xi = \int\limits_{x_l(\eta)}^{x} \frac{\partial\Delta\phi'}{\partial\xi} d\xi =$ $\Delta\phi'$ and the kernel is approximated with $\frac{y-\eta}{4h^2}$. This gives us the effect of the ground on the downwash as follows:

$$w_g(x,y) = \frac{1}{2\pi} \int\limits_{-\beta(x)}^{\beta(x)} \frac{y-\eta}{4h^2} \frac{\partial\Delta\phi'(x,\eta)}{\partial\eta} d\eta$$

with the transformation of the integral limits similar to that given in (Bisplinghoff et al. 1996) i.e.$y^* = \frac{y}{\beta(x)}$, $\eta^* = \frac{\eta}{\beta(x)}$, $h^* = \frac{h}{\beta(x)}$

they become from −1 to 1 to yield

$w_g(x,y) = \frac{1}{2\pi} \int\limits_{-1}^{1} \frac{y^*-\eta^*}{4h^{*2}} \frac{\partial\Delta\phi'}{\partial\eta^*} d\eta^*$ with first approximation for oge $\frac{\partial\Delta\phi'}{\partial y} = \frac{2U}{\pi\sqrt{1-y^{*2}}} \int\limits_{-1}^{1} \frac{\partial z_a}{\partial x} \frac{\sqrt{1-\eta^{*2}}}{y^*-\eta^*} d\eta^*$.

We get $w_g(x,y) = \dfrac{-2\,U\alpha}{4\pi\,h^{*2}} \displaystyle\int\limits_{-1}^{1} \frac{y^*-\eta^*}{\sqrt{1-\eta^{*2}}} \eta^* d\eta^* = \frac{U\alpha}{4h^{*2}}$ with $\dfrac{\partial z_a}{\partial x} = -\alpha$

(A13.3)

The vorticity induced x component of the perturbation velocity reads as

$$u'(x,y) = \frac{1}{4\pi} \int\limits_{-b}^{b} \int\limits_{-l}^{l} \frac{2h_g\gamma_a}{R^3} d\xi\, d\eta \tag{A13.4}$$

Integrating by parts with respect to ξ we obtain

$$u'(x_l,y) = -\frac{2h_g}{4\pi} \int\limits_{-l}^{l} \gamma_a \frac{(x_l-\xi)}{\left[(y_l-\eta)^2+4h^2\right]R}\Bigg|_{\xi=x_l}^{\xi=b_0} d\eta + \frac{2h_g}{4\pi} \int\limits_{x_l}^{b_0} \int\limits_{-l}^{l} \frac{\partial\gamma_a}{\partial\xi} \frac{(x_l-\xi)}{\left[(y_l-\eta)^2+4h^2\right]R} d\xi\, d\eta$$

The first integral on the right hand side vanishes at the leading and the trailing edges. The second integral simplifies for low aspect ratio wings with $R \cong |x_l - \xi|$ to give

$$u'(x_l,y)=\frac{2h_g}{4\pi}\int_{x_l}^{b_0}\int_{-l}^{l}\frac{\partial\gamma_a}{\partial\xi}\frac{(x_l-\xi)}{\left[(y_l-\eta)^2+4h^2\right]|x_l-\xi|}d\xi\,d\eta$$
$$=\frac{2h_g}{4\pi}\int_{x_l}^{b_0}\int_{-l}^{l}\frac{\partial\gamma_a}{\partial\xi}\frac{(x_l-\xi)}{\left[(y_l-\eta)^2+4h^2\right]|x_l-\xi|}d\xi\,d\eta$$

Further simplification and integration with respect to ξ gives

$$u'(x_l,y)=\frac{h_g}{\pi}\int_{-\beta(x)}^{\beta(x)}\frac{\gamma_a(x,\eta)}{\left[(y_l-\eta)^2+4h^2\right]}\,d\eta\cong\frac{1}{4\pi h_g}\int_{-\beta(x)}^{\beta(x)}\gamma_a(x,\eta)\,d\eta$$
$$=\frac{U}{4\pi h_g}\int_{-\beta(x)}^{\beta(x)}\frac{C_{pa}(x,\eta)}{2}\,d\eta$$

Using the value of pressure coefficient in oge, (Bisplinghoff et al. 1996) we have

$$u'\frac{\partial z_a}{\partial x}=\frac{U}{2\pi h_g}\frac{\partial z_a}{\partial x}\int_{-\beta(x)}^{\beta(x)}\frac{\partial}{\partial x}(\sqrt{\beta^2(x)-\eta^2}\,\frac{dz_a}{dx})d\eta=\frac{U}{2\pi h_g}\frac{\partial z_a}{\partial x}\frac{\partial}{\partial x}(\frac{\pi\beta^2(x)}{2}\frac{dz_a}{dx})$$

$$u'\frac{\partial z_a}{\partial x}=\frac{U}{2\pi h_g}\frac{\partial z_a}{\partial x}\int_{-\beta(x)}^{\beta(x)}\frac{\partial}{\partial x}(\sqrt{\beta^2(x)-\eta^2}\,\frac{dz_a}{dx})d\eta=\frac{U}{2\pi h_g}\frac{\partial z_a}{\partial x}\frac{\partial}{\partial x}(\frac{\pi\beta^2(x)}{2}\frac{dz_a}{dx})$$

The original equation now reads as

$$u'\frac{\partial z_a}{\partial x}=\frac{U}{4h_g}\frac{\partial z_a}{\partial x}\frac{\partial}{\partial x}(\beta^2(x)\frac{dz_a}{dx})=-\frac{1}{2\pi}\int_{-\beta(x)}^{\beta(x)}\frac{1}{y-\eta}\frac{\partial\Delta\phi'(x,\eta)}{\partial\eta}d\eta \qquad \text{(A13.5)}$$

The inversion of (A13.5) gives the corrected pressure coefficient as, (Bisplinghoff et al. 1996)

$$C_{pa}=\frac{2}{U}\frac{\partial\Delta\phi'(x,y)}{\partial x}=\frac{2}{U}\frac{\partial}{\partial x}\int_{-\beta(x)}^{y}\frac{\partial\Delta\phi'(x,\bar{y})}{\partial\bar{y}}d\bar{y}$$
$$=\frac{1}{2h_g}\frac{\partial}{\partial x}\left\{\frac{\partial z_a}{\partial x}\frac{\partial}{\partial x}\left[\sqrt{\beta^2(x)-y^2}(-\beta^2(x)\frac{\partial z_a}{\partial x})\right]\right\} \qquad \text{(A13.6)}$$

A14. Efficiency Constraint

The elements of $H_{avg}$and H_m matrices are

H_{avg}(1,1)=2*(F*F+G*G)*k*k;
H_{avg} (1,2)=2*k*F*F-k*F-G*k*k+2*G*G*k;

H_{avg}(1,3)=(F*F+G*G)*k*k-F*k*k+k*k/2+G*k;

H_{avg}(2,2)=0.5*(F*F+G*G)*k*k+2*F*F-k*k*F-2*F+2*G*G+k*k/2-G*k;
H_{avg}(3,3)= H_{avg}(2,2);

H_{avg}(2,1)= H_{avg}(1,2); H_{avg}(3,1)= H_{avg}(1,3);

and,

H_m(1,1)=4*F*k*k; H_m(1,2)=2*G*k+k*k; H_m(1,3)=2*F*k-2*G*k*k;

H_m(2,1)= H_m(1,2); H_m(3,1)= H_m(1,3);
H_m(2,2)=-2*G*k+k*k-F*k*k; H_m(3,3)= H_m(2,2);
Now, the set of equations were

$$\pi\rho U^2 b\left[H_{avg}\right]\{Q\} - \lambda\{Q\} - \lambda_\eta\left\langle\left[H_{avg}\right] - \eta_{ref}[H_M]\right\rangle\{Q\} = 0$$

$$\{Q\}^T\{Q\} - 1 = 0$$

$$\{Q\}^T\left\langle\left[H_{avg}\right] - \eta_{ref}[H_M]\right\rangle\{Q\} = 0$$

Let a_{ij} *be the elements of* $\left[H_{avg}\right]$ and b_{ij} *be the elements of* $\left\langle\left[H_{avg}\right] - \eta_{ref}[H_M]\right\rangle$

Setting the unknown vector $\boldsymbol{x} = \boldsymbol{Q}$, then the equations in indicial notain (repeated index implies summation) read as

$$(a_{ij} - \lambda\delta_{ij} - \lambda_\eta b_{ij})x_j = 0$$
$$x_j x_j - 1 = 0$$
$$x_i b_{ij} x_j = 0$$

Wherein, λ *and* λ_η are additional unknowns which make the system totally quadratic.

For 3-degrees of freedom we have 3 parameters of motion, one for plunging and two for pitching, and 2 Lagrange multipliers as unknowns. Hence, the iterative equations with variable coefficient matrix and the unknown vector read as

$$\begin{bmatrix} a_{11} & a_{12} & a_{13} & -x_1 & -b_{1j}x_j \\ a_{21} & a_{22} & a_{23} & -x_2 & -b_{2j}x_j \\ a_{31} & a_{32} & a_{23} & -x_3 & -b_{3j}x_j \\ 2x_1 & 2x_2 & 2x_2 & 0 & 0 \\ b_{1j}x_j & b_{2j}x_j & b_{3j}x_j & 0 & 0 \end{bmatrix} \begin{pmatrix} \delta x_1 \\ \delta x_2 \\ \delta x_3 \\ \delta\lambda \\ \delta\lambda_\eta \end{pmatrix} = - \begin{pmatrix} f_1 \\ f_2 \\ f_3 \\ f_4 \\ f_5 \end{pmatrix}$$

Where,

$$\begin{pmatrix} f_1 \\ f_2 \\ f_3 \\ f_4 \\ f_5 \end{pmatrix} = \begin{pmatrix} (a_{1j} - \lambda\delta_{1j} - \lambda_\eta b_{1j})x_j \\ (a_{2j} - \lambda\delta_{2j} - \lambda_\eta b_{2j})x_j \\ (a_{3j} - \lambda\delta_{3j} - \lambda_\eta b_{3j})x_j \\ x_j x_j - 1 \\ x_i b_{ij} x_j \end{pmatrix}$$

Here, known values of x_i, λ *and* λ_4 from the previous iteration level are used in the coefficient matrix and in the load vector.

A.15: Duhamel integral with the sine term calculated for the elliptical wing with AR = 3.

$$\int_0^t \sin(\sigma)\varphi'(t-\sigma)\, d\sigma$$

faa = int((sin(x))^1*(0.17*0.54*exp(0.54*(x-t))),x,0,t)

fs(t) = (201*exp(-(3*t)/10))/2180 + (5282933469138125*exp(-(91*t)/200))/849368334410448896 - (45559957544805335149*cos(t))/462905742253694648320 + (564570522729881746163*sin(t))/18516229690147785932800;

A16: The properties of the eliptical wing shape of the fruit fly is given in Fig. A16.1. Accordingly, the first and the second moment of inertia for the wing read as

$I_1/2 = \int_{r_1}^{r_1+R} rb^2(r)dr$ = eval(int((−(x−1.01−0.20)^2/1.01^2 + 1)*0.43^2*x, 0.20, 2.22)) = 0.3013 mm^4

$I_2/2 = \int_{r_1}^{r_1+R} r^2 b(r)dr$ = eval(int(sqrt(−(x−1.01−0.20)^2/1.01^2 + 1) *0.43^2*x^2,0.20,2.22) = 0.5043 mm^4

The equation of the ellipse is $\frac{(r-1.01-0.20)^2}{(1.01)^2} + \frac{b^2}{(0.43)^2} = 1.$

A17: Aerodynamic forces and moments at a constant AoA

$$K_{mn} = \rho A I_{mn}/2, \quad I_{mn} = 2\int_{r_1}^{R} r^m c^n(r)dr, \quad I_y = Moment\ of\ inertia\ of\ the\ body$$

$$X_{P0} = -2\frac{K_{21}}{m}\dot{\varphi}|\dot{\varphi}|\cos\varphi\sin^2\eta\,\Phi(0) \quad Z_{P0} = -\frac{K_{21}}{m}\dot{\varphi}|\dot{\varphi}|\sin 2\eta\,\Phi(0)$$

$$M_{P0} = 2|\dot{\varphi}|\dot{\varphi}\sin\eta\,(\frac{K_{21}}{I_y}x_{cg}\cos\eta + \frac{K_{22}}{I_y}a\cos\varphi + \frac{K_{31}}{I_y}\sin\varphi\cos\eta)\Phi(0).$$

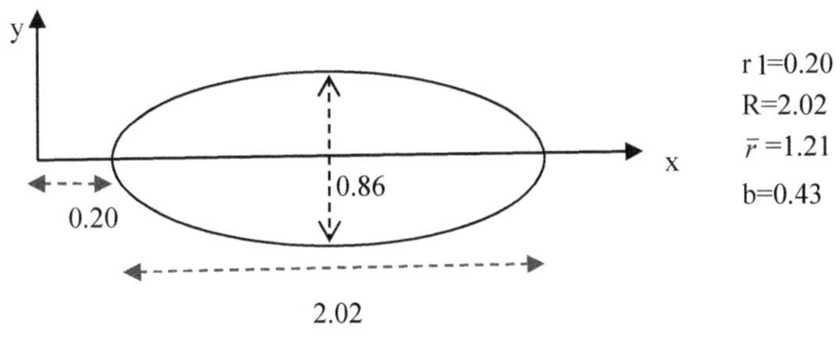

Fig. A16.1 Pertinent dimensions, in mm, of the elliptical wing of a fruit fly

Partial derivatives of forces and moments:
x-dir:

$$X_{Pu} = -4\frac{K_{11}}{m}|\dot{\varphi}|\cos^2\varphi\sin^2\eta\,\Phi(0), \quad X_{Pw} = -\frac{K_{11}}{m}|\dot{\varphi}|\cos\varphi\sin 2\eta\,\Phi(0),$$

$$X_{Pq} = (\frac{K_{21}}{m}|\dot{\varphi}|\cos\varphi\sin\varphi\sin 2\eta + x_{cg}\frac{K_{11}}{m}|\dot{\varphi}|\cos\varphi\sin 2\eta)\Phi(0), X_{Px_1} = -\rho\frac{I_{10}}{m}|\dot{\varphi}|\cos\varphi, X_{Px_2} = 0.$$

z-dir;

$$Z_{Pu} = -4\frac{K_{11}}{m}|\dot{\varphi}|\cos^2\varphi\,\Phi(0), Z_{Pw} = Z_{Pu}/2, Z_{Pq} = \frac{K_{21}}{m}|\dot{\varphi}|\sin\varphi\sin^2\eta\,\Phi(0) - x_{cg}Z_{Pw}, Z_{Px_1} = -\rho\frac{I_{10}}{m}|\dot{\varphi}|, Z_{Px_2} = 0.$$

Moments:

$$M_{Pu} = 4\frac{K_{12}}{I_y}a\,|\dot{\varphi}|\cos^2\varphi\sin\eta\,\Phi(0) + 2\frac{m}{I_y}X_{Pq},$$

$$M_{Pw} = 2(\frac{K_{12}}{I_y}a\,|\dot{\varphi}|\cos\varphi\cos\eta + \frac{K_{21}}{I_y}a\,|\dot{\varphi}|\cos\varphi\cos^2\eta\,)\Phi(0) - \frac{m}{I_y}x_{cg}Z_{Pw}$$

$$M_{Pq} = 2|\dot{\varphi}|(a\cos\varphi\cos\eta\,(\frac{K_{12}}{I_y}x_{cg} + \frac{K_{22}}{I_y}\sin\varphi) - \sin\varphi\cos^2\eta\,)(\frac{K_{21}}{I_y}x_{cg} + \frac{K_{31}}{I_y}\sin\varphi))\Phi(0) - \frac{m}{I_y}x_{cg}Z_{Pq}$$

$$M_{Px_1} = \rho\frac{I_{11}}{I_y}a\,|\dot{\varphi}|\cos\varphi(\cos\eta + \sin\eta) + \rho\frac{I_{20}}{I_y}|\dot{\varphi}|\sin\varphi\cos^2\eta\, - \frac{m}{I_y}x_{cg}Z_{Px_1}, M_{P_2} = 0$$

Partial derivatives of aerodynamic forces and moments wrt pitch rate i:

x-dir: 0 contribution to the derivatives of X.

z-dir:

$$Z_{\mathrm{Pr}\,q} = -\frac{K_{\mathrm{Pr}\,12}}{m}\dot{\varphi}\cos\varphi\,\Phi(0) - \rho\frac{I_{12}}{m}\pi\dot{\varphi}\cos\varphi$$

$$Z_{\mathrm{Pr}_{x2}} = -\rho\frac{I_{21}}{m}\dot{\varphi}, K_{\mathrm{Pr}_{mn}} = \pi\rho(1/2 - a)I_{mn}$$

Rest of the derivatives of Z is 0.

Moment: $M_{\mathrm{Pr}\,q} = \dot{\varphi}\cos\varphi(\frac{K_{\mathrm{Pr}\,13}}{I_y} a\cos\varphi\cos\eta + \frac{K_{\mathrm{Pr}\,22}}{I_y}\sin\varphi)\Phi(0) - \rho\pi\frac{I_{13}}{I_y}\dot{\varphi}\cos\varphi - \frac{K_v}{I_y}\mu f\cos^2\varphi,\ K_v = \frac{\pi}{16}\rho I_{04}$

$$M_{\mathrm{Pr}\,x_2} = \rho\frac{I_{11}}{I_y} a\,|\dot{\varphi}|\cos\varphi(\cos\eta + \sin\eta) + \rho\frac{I_{20}}{I_y}\,|\dot{\varphi}|\sin\varphi - \frac{m}{I_y} x_{cg} Z_{\mathrm{Pr}\,x_2}$$

Pitch derivatives:

$$X_{1_u} = 2b_1\overline{r}|\dot{\varphi}|Aa_1\sin 2\eta\,\cos\varphi, \quad X_{1_w} = 2b_1\overline{r}|\dot{\varphi}|\dot{\varphi}Aa_1\cos 2\eta,$$

$$X_{1q} = -2b_1\overline{r}|\dot{\varphi}|Aa_1(x_{cg} + \overline{r}\sin\varphi), \quad X_{1_{x1}} = -b_1\overline{r}|\dot{\varphi}|/b$$

Last 2 derivatives realted to pitch:

$$X_{2_q} = 2b_1\overline{r}|\dot{\varphi}|\pi(b/2 - a)a_1\cos\varphi + b_1\overline{r}|\dot{\varphi}|\pi\, a_1\cos\varphi, \quad X_{2_{x2}} = -b_1\overline{r}|\dot{\varphi}|/b$$

A18: Trim Parameters:

(i) Symmetric:

$$\overline{X}_{x1} = -\frac{1}{T}\int_0^T \rho\frac{I_{10}}{m}|\dot{\varphi}|\cos\varphi\,dt = \rho\frac{I_{10}}{m}\frac{1}{T}\frac{4\overline{\varphi}}{T}\left[\int_0^{T/2}\cos\frac{4\overline{\varphi}}{T}(t - T/4)\,dt + \int_{T/2}^{T}\cos\frac{4\overline{\varphi}}{T}(t - 3T/4)\,dt\right]$$

$$= -\rho\frac{I_{10}}{m}\frac{1}{T}\left[\sin\frac{4\overline{\varphi}}{T}(t - T/4)\Big|_0^{T/2} + \sin\frac{4\overline{\varphi}}{T}(t - 3T/4)\Big|_{T/2}^{T}\right] = \rho\frac{I_{10}}{m}\frac{1}{T}(\sin\overline{\varphi} + \sin\overline{\varphi} + \sin\overline{\varphi} + \sin\overline{\varphi})$$

$$= 4\rho\frac{I_{10}}{mT}\sin\overline{\varphi}$$

$$\overline{X}_{1eq} = \frac{\overline{X}_{10}}{\overline{X}_{1x_1}} = \frac{\frac{1}{T}\int_0^T b_1\overline{r}^2|\dot{\varphi}|\dot{\varphi}\,Aa_i\sin 2\eta\,dt}{-2b_i\overline{r}|\dot{\varphi}|/\overline{c}} = \frac{Aa_1}{2T}\left(\int_0^{T/2}\dot{\varphi}\sin 2\overline{\alpha}dt + \int_{T/2}^{T}\dot{\varphi}\sin 2(\pi - \overline{\alpha})dt\right)$$

$$= \frac{Aa_1}{2T}\left(\frac{4\overline{\varphi}}{T}\sin 2\overline{\alpha}\,t\Big|_0^{T/2} - \frac{4\overline{\varphi}}{T}\sin(-2\overline{\alpha}\,t)\Big|_{T/2}^{T}\right) = \frac{2}{T}\overline{\varphi}Aa_1\sin 2\overline{\alpha}$$

After summing up we obtain (9.48).

$$\overline{X}_{x1}\overline{X}_{1eq} = 8\rho\,\overline{r}2\overline{b}\frac{I_{10}}{mT^2}\overline{\varphi}\sin\overline{\varphi}Aa_1\sin 2\overline{\alpha}$$

(ii) Antisymmetric: Integrating for a period:
Equation (24a)

$$\overline{X}_{P0} = -16\frac{\rho A I_{21}}{mT^2}\overline{\varphi}\cos\overline{\varphi}\sin\overline{\varphi}(\sin^2\overline{\alpha}_d - \sin^2\overline{\alpha}_u)\Phi(0)/2$$

$$\overline{X}_{x1}\overline{X}_{1eq} = 8\rho\,\overline{r}2\overline{b}\frac{I_{10}}{mT^2}\overline{\varphi}\sin\overline{\varphi}\cos\overline{\varphi}Aa_1(\sin 2\overline{\alpha}_d - \sin 2\overline{\alpha}_u)$$

$\overline{X}_{P0} + \overline{X}_{x1}\overline{X}_{1eq} = 0$ gives (9.49a), and (9.49b,c) which are obtained similarly.

References

Anderson, D.A., Tannehill, J.C., Pletcher, R.H.: Computational Fluid Mechanics and Heat Transfer, Hemisphere, New York (1984)
Bisplinghoff, H.A., Halfman, R.L.: Aeroelasticity. Dover Publications Inc., New York (1996)
Dowell, E.H. (ed.): A Modern Course in Aeroelasticity. Kluwer Academic
Gradshteyn, L.S., Ryzhik, I.M.: Tables of Integrals, Series and Products, 6th Ed. Academic Press (2000)
Guderly, K.G.: The Theory of Transonic Flow. Pergamon Press, Oxford (1962)
Gülçat, Ü.: Separate Numerical Treatment of Attached and Detached Flow Regions in General Viscous Flows, Ph.D. Dissertation, Georgia Institute of Technology, School of Aerospace Engineering, Atlanta (1981)
Hildebrand, F.B.: Methods of Applied Mathematics, Prentice-Hall Inc., Englewood Cliffs, N. J. (1965)
Hildebrand, F.B.: Advanced Calculus for Applications, Prentice-Hall Inc. Engelwood Cliffs, New Jersey (1976)
Lee, J.F., Sears, F.W., Turcotte, D.L.: Statistical Thermodynamics. Addison-Wesley, Reading Mass. (1973)
Matlab, R.L.: The MathWorks, Inc. (2015)
Milne-Thomson, L.M.: Theoretical Aerodynamics. DoverPublications, New York (1973)
Press, W.H., Flannery, B. P., Teukolsky, S.A., Vetterling, W.T.: Numerical Recipes, Cambridge University Press, London, Chap. 2 (1992)
Sankar, L.: Numerical Study of Laminar Unsteady Flow Over Airfoils, Ph.D. Dissertation, Georgia Institute of Technology, School of Aerospace Engineering, Atlanta (1977)
Schlichting, H.: Boundary Layer Theory, Mc-Graw Hill, New York (1968)
Şuhubi, E.S.: Functional Analysis, Kluwer Academic Publishers (2003)

Index

Ü. Gülçat, *Fundamentals of Modern Unsteady Aerodynamics*,
https://doi.org/10.1007/978-3-030-60777-7

CPSIA information can be obtained
at www.ICGtesting.com
Printed in the USA
LVHW082347020722
722645LV00004B/57